ALGORITHMIC CRYPTANALYSIS

CHAPMAN & HALL/CRC
CRYPTOGRAPHY AND NETWORK SECURITY

Series Editor
Douglas R. Stinson

Published Titles

Jonathan Katz and Yehuda Lindell, Introduction to Modern Cryptography

Antoine Joux, Algorithmic Cryptanalysis

Forthcoming Titles

Burton Rosenberg, Handbook of Financial Cryptography

Maria Isabel Vasco, Spyros Magliveras, and Rainer Steinwandt, Group Theoretic Cryptography

Shiu-Kai Chin and Susan Beth Older, Access Control, Security and Trust: A Logical Approach

CHAPMAN & HALL/CRC
CRYPTOGRAPHY AND NETWORK SECURITY

ALGORITHMIC CRYPTANALYSIS

Antoine Joux

 CRC Press
Taylor & Francis Group
Boca Raton   London   New York

CRC Press is an imprint of the
Taylor & Francis Group an **informa** business
A CHAPMAN & HALL BOOK

Chapman & Hall/CRC
Taylor & Francis Group
6000 Broken Sound Parkway NW, Suite 300
Boca Raton, FL 33487-2742

International Standard Book Number: 978-1-4200-7002-6 (Hardback)

Library of Congress Cataloging-in-Publication Data

Joux, Antoine.
  Algorithmic cryptanalysis / Antoine Joux.
    p. cm. -- (Chapman & Hall/CRC cryptography and network security)
  Includes bibliographical references and index.
  ISBN 978-1-4200-7002-6 (hardcover : alk. paper)
  1. Computer algorithms. 2. Cryptography. I. Title. III. Series.

  QA76.9.A43J693 2009
  005.8'2--dc22
                                                              2009016989

Visit the Taylor & Francis Web site at
http://www.taylorandfrancis.com

and the CRC Press Web site at
http://www.crcpress.com

À Katia, Anne et Louis

Contents

II Algorithms

III Applications

Preface

The idea of this book stemmed from a master's degree course given at the University of Versailles. Since most students in this course come from a mathematical background, its goal is both to prime them on algorithmic methods and to motivate these algorithmic methods by cryptographically relevant examples. Discussing this course with colleagues, I realized that its content could be of interest to a much larger audience. Then, at Eurocrypt 2007 in Barcelona, I had the opportunity to speak to Sunil Nair from Taylor & Francis. This discussion encouraged me to turn my course into a book, which you are now holding.

This book is intended to serve several purposes. First, it can be a basis for courses, both at the undergraduate and at the graduate levels. I also hope that it can serve as a handbook of algorithmic methods for cryptographers. It is structured in three parts: background, algorithms and applications. The background part contains two chapters, a short introduction to cryptography mostly from a cryptanalytic perspective and a background chapter on elementary number theory and algebra. The algorithms part has nine chapters, each chapter regroups algorithms dedicated to a single topic, often illustrated by simple cryptographic applications. Its topics cover linear algebra, sieving, brute force, algorithms based on the birthday paradox, Hadamard-Fourier-Walsh transforms, lattice reduction and Gröbner bases. The applications part takes a different point-of-view and uses recipes from several chapters in the algorithms part to address more advanced cryptographic applications. This final part contains four chapters dealing with linear feedback shift register based stream ciphers, lattice methods for cryptanalysis, elliptic curves and index calculus methods.

All chapters in the algorithms and applications parts have an exercise section. For all exercises whose number is marked with an "h" exponent, e.g., exercise 1^h, hints and solutions are given on the book's website whose address is http://www.joux.biz/algcrypt. To allow the book to serve as a textbook, about half of the exercises have neither hints nor solutions.

The content of this book should not necessarily be read or taught in linear order. For a first reading or an introductory course, the content of Chapters 2, 3 and 6 covering basic number theory, linear algebra and birthday paradox algorithms should suffice. For a longer course, the choice of chapters depends on the background of the reader or students. With a mathematical background, I would recommend choosing among Chapters 4, 7, 10 and 11. Indeed, these chapters are based on mathematical premises and develop algorithms on this basis. With a computer science background, Chapters 5, 8 and 9 are more suited. Finally, the applications presented in the last part can be used for dedicated graduate courses. Alternatively, they can serve as a basis for course end projects.

Throughout this book, we discuss many algorithms. Depending on the specific aspect that needs to be emphasized, this is done using either a textual description, an algorithm in pseudo-code or a C code program. The idea is to use pseudo-code to emphasize high-level description of algorithms and C code to focus on lower-level implementation details. Despite some drawbacks, the C programming language is well suited for programming cryptanalytic applications. One essential advantage is that it is a relatively low-level programming language that allows to tightly control the behavior of the code that is executed by the target processor. Of course, assembly language would give an even tighter control. However, it would be much harder to read and would only be usable on a single microprocessor or family of microprocessors.

Note that for lack of space, it was not possible to present here C programs for all algorithms that are discussed in this book. Several additional codes are available for downloading on the book's website. All these codes were developed and tested using the widely available Gnu GCC compiler. Note that these codes are not optimally tuned, indeed, fine tuning C code is usually specific to a single compiler version and often hurt the code's legibility. Where timings are given, they were measured on an Intel Core 2 Duo at 2.4 Ghz.

Writing this book was a long and challenging undertaking. It would not have been possible without the help of many people. First, I would like to thank my Ph.D. advisor, Jacques Stern, without his guidance, I would not have taken the path of research and cryptography. I also wish to thank all my colleagues and co-authors, for discussing fascinating research problems. It was a great source of inspiration while writing this book. All my students and former students deserve special thanks, especially for forcing me to reconsider previous knowledge again and again. Through sheer coincidence, I happened to be the program chair of Eurocrypt 2009 while writing this book, it was a very nice experience and I am extremely grateful to the wonderful people who accepted to serve on my committee. During the finalization of the manuscript, I attended a seminar on "Symmetric Cryptography" at the "Leibniz-Zentrum für Informatik" in Schloss Dagstuhl, Germany. Attending this seminar and discussing with all the participants was extremely helpful at that time, I would like to give due credit to the organizers and to the wonderful staff at Schloss Dagstuhl. A few of my colleagues helped me during proofreading, thanks to Johannes Buchmann, Pierre-Alain Fouque, Steven Galbraith, Louis Goubin, Reynald Lercier, Michael Quisquater, Michael Schneider and Nicolas Sendrier, this book contains much fewer typos than it would have. Thanks to Michel Abdalla for putting together a large bibliography of cryptography-related articles and for letting me use it. Last but not least, I would like to express all my gratitude to my family for supporting me all these years and for coping with my occasional absentmindedness.

Finally, I wish to acknowledge institutional support from the Délégation Générale pour l'Armement and the University of Versailles and Saint-Quentin-en-Yvelines.

Existing programs or libraries

Many of the algorithms presented here have been programmed, in very efficient ways, into existing computer packages. In many cases, reprogramming the methods might not be needed or might even be counter-productive when the available programs are very efficient.

We give here a short discussion of available programs and libraries which contain algorithmic methods discussed in this book. This discussion does not pretend to exhaustivity. We regroup the stand-alone tools on one side and libraries that need to be used in conjunction with a user written program on the other. Note that stand-alone tools usually incorporate a programming language to allow the development of user's applications. Some of the programs offer both options, a stand-alone tool and a library; we list them in the stand-alone category. The various programs are listed in alphabetical order. We recommend using them for benchmarking and before considering to write user's specific code.

Stand-alone tools

- **GAP** This computer algebra system is developed by the GAP group, its home page is `http://www.gap-system.org/`. It includes many features and offers very useful group theoretic algorithms. In particular, it is able to manipulate group characters and group representation.

- **MAGMA** Magma is a computer algebra system that can be bought online at `http://magma.maths.usyd.edu.au/`. An online calculator, with limited computing power, is also available. The Magma language is mathematically oriented and every object belongs to a rigourously defined structure. Magma includes a large number of features. In particular, it offers algebraic geometry tools and knows how to compute with elliptic curves and divisors. Magma also contains a fast implementation of F_4 Gröbner basis algorithm and lattice reduction tools.

- **Maple** Maple computer algebra is a very well-known and versatile system, used in a large variety of applications. The current version contains a very efficient implementation of the F_5 Gröbner basis algorithm.

- **PARI/GP** This computer algebra system was initiated by Henri Cohen and is currently maintained by Karim Belabas under the GPL license. It offers both a stand-alone tool and a C library. In addition to classical features such as modular computation, linear algebra, polynomials, it offers some specific functionalities to compute information about general number fields and elliptic curves over the complex field. For more information, look up the webpage at `http://pari.math.u-bordeaux.fr/`.

- **SAGE** Sage is an open-source mathematics software system `http://www.sagemath.org/` based on the Python language. It incorporates many efficient implementations of algorithms for algebra. One specificity of Sage is that it offers the option of interfacing with other computer algebra systems and of incorporating functionalities from existing libraries.

Libraries

- **FFTW** This library developed at MIT by Matteo Frigo and Steven G. Johnson is dedicated to high-performance computation of Fourier transforms. The home page of the library is located at `http://www.fftw.org/`.

- **NTL** This library written by Victor Shoup and available at `http://www.shoup.net/ntl/` is based on the C++ language. It implements finite fields, routines for univariate polynomials, linear algebra and several lattice reduction algorithms.

Part I

Background

Chapter 1

A bird's-eye view of modern cryptography

Since cryptanalysis cannot exist without cryptography, this background chapter aims at making a brief, necessarily incomplete survey of modern cryptography, recalling some essential definitions and facts for the perusal of this book and laying down the notational ground. In particular, it presents various security notions, corresponding to several classes of adversaries. Modern cryptanalysis is the counterpart to these security notions. The fundamental goal of a cryptanalyst is to violate one or several of these security notions for algorithms that claim, implicitly or explicitly, to satisfy these security notions. This can be achieved in two main ways, either by overcoming an underlying security hypothesis or by exhibiting a specific flaw in the considered algorithm or protocol.

This chapter only intends to serve as an introduction to the topic and certainly to give a complete description of modern cryptography. The reader may wish to consult a reference book on cryptography. There are many such books, a few examples are [Buc04, MvOV97, Sch96, Sti02].

1.1 Preliminaries

Cryptography is a ubiquitous tool in the world of information security. It is required when trying to keep the secrecy of communications over open channels or to prove the authenticity of an incoming message. It can be used to create many multiparty protocols in a way that makes cheating difficult and expensive. In fact, its range of applicability is very wide and it would not be possible to give a complete list of functionalities that can be achieved through the use of cryptography. Instead, we are going to focus on a small set of fundamental goals and see how they can be formalized into precise security notions. From an historical perspective, the oldest and foremost cryptographic goal is confidentiality.

Confidentiality appeared quite early in human history. At that time, messengers were regularly sent between troops or traders to carry important messages. They were also regularly captured by enemies and they sometimes

3

turned out to be spies or traitors. In this context, the basic idea was to be able to write messages in a way that would preserve the secrecy of the message meaning against these events. Later, with the invention of postal services, telegraphs, radio communications and computer networks, it became easier to send messages and at the same time easier to intercept or copy these messages. Thus, the basic question remains: how can we make sure that messages will not be read by the wrong person? One option is to hide the very existence of the message through various means, this is called steganography. We will not consider this option any further. Another option does not try to hide the message but simply to make sure that it cannot be understood except by the intended recipient, using something akin to a scrambling process, called encryption.

This notion of confidentiality is trickier than it may first appear. What precisely can we hide about a message? Is it possible to be sure that nothing can be learned about it? A first limit is that it is not possible to hide everything about a given message, looking at the encrypted message, an attacker can always learn or at least estimate the length of the message. The only way to avoid this would be to output ciphertexts of the maximum accepted input length for all messages. This would, of course, yield utterly impractical cryptosystems. Moreover, the attacker may have some prior information and seeing the message is not going to make him forget it. As a consequence, it is convenient to assume that the length of the message is not hidden by the encryption and to measure the amount of new knowledge that can be extracted by the attacker from the message. Similarly, the attacker may obtain prior information about the encryption system. As a consequence, to make cryptography useful in a wide variety of contexts, it is necessary to assume that the specifications of the cryptosystem are public, or could be leaked to the adversary. The security of the system should only rely on a short secret: the key of the system. This essential principle was proposed by Auguste Kerckhoffs in 1883 and published in [Ker83].

This approach and its limits were further studied by Shannon in 1945 in a confidential report titled *A Mathematical Theory of Cryptography*. This report was declassified after World War II and the results published in [Sha49]. In order to study the security of cryptographic systems, this paper introduced a new mathematical theory: information theory. In a nutshell, information theory contained good news and bad news about cryptography. The good news is that perfect confidentiality is possible and can be achieved using a simple encryption algorithm called the One Time Pad. The bad news is that the One Time Pad is impractical for most applications and that according to information theory nothing more practical can be secure. Indeed, the One Time Pad views messages as sequences of symbols (bits or characters) and encrypts them by a simple mixing of each symbol with a corresponding symbol extracted from the key. However, it is crucial for the security of this scheme to use a random key of the same length as the message to encrypt. With any shorter key, the One Time Pad degenerates into a variation of the

Vigenere cipher and becomes very weak. Of course, transmitting very long keys securely is rarely easier than directly transmitting messages securely. Moreover, this system is error prone and any key reuse dooms the security of the corresponding messages. In practice, a user would expect to use a relatively short key for the transmission of long messages. Using information theory, Shannon showed that this not possible. Indeed, a powerful enough cryptanalyst can always try to decrypt the transmitted message using all possible keys. The only key that yields a meaningful message is the correct one.

In order to bypass this impossibility result, modern cryptography takes into account the amount of work required from the cryptanalyst and assumes that, even for relatively short key lengths, trying all keys costs too much and is not an option. This idea is at the core of computationally based cryptography. An asymptotically oriented approach to this idea can be obtained by using complexity theory. In this approach, easy tasks such as encryption or decryption are modeled by polynomial time computations and hard tasks are assumed to be in harder classes of complexity[1]. This approach has an essential drawback, complexity classes are too coarse and they do not always finely reflect the hardness of real computation. For example, a polynomial time algorithm of complexity n^{100} is usually totally impractical, while an exponential time algorithm of complexity $2^{n/100}$ is often useful. A more concrete approach was proposed by Bellare, Kilian and Rogaway in [BKR00] and aims at giving a more precise information about the cost of attacks for real life parameters of cryptographic schemes. However, even this concrete approach is not complete and comparing the practicality and the full cost [Wie04] of attacks is a difficult art.

Pushing the idea of computationally based cryptography a bit further, in 1976, Diffie and Hellman invented public key cryptography [DH76]. The basic idea is to use trapdoor one-way functions, i.e., functions which are easy to compute, hard to invert and which become easy to invert once a secret value, the trapdoor, is known.

Note that, in spite of achieving perfect confidentiality, the One Time Pad is not perfectly secure. Indeed security is more than simply confidentiality, it also covers the concept that an attacker should not be able to tamper with messages without being detected. Clearly, this is not true with the One Time Pad, since changing any bit of the ciphertext has a simple effect: changing the same bit in the corresponding plaintext. This property allows an attacker to perform any change of his choice on the transmitted message. To prevent this, it is necessary to invoke another cryptographic functionality: integrity.

---

[1]At most, one can hope for NP-complete cryptanalysis, since guessing the correct key suffices to break any cryptographic scheme.

1.1.1 Typical cryptographic needs

These two basic functionalities, confidentiality and integrity, give a first criteria to classify cryptographic algorithms. Another essential criterion is the distinction between secret key and public key algorithms. Secret key algorithms use the same key, or sometimes distinct but equivalent keys, to encrypt and decrypt, to authenticate or verify authentication. Public key algorithms use different keys, the public key to encrypt or verify signatures, the private key to decrypt or sign.

Using these two criteria, we obtain four classes of cryptographic systems.

1.1.1.1 Secret key encryption

Typical secret key algorithms encrypt messages using a short secret key common to the sender and the recipient of the secret message. Typically, secret keys of recent algorithm are often between 128 and 256 bits. Secret key encryption algorithms are further divided into two main categories: stream ciphers based and block ciphers based.

Stream ciphers combine a pseudo-random generator of cryptographic quality, also called a keystream generator, together with One Time Pad encryption.

Block ciphers are keyed permutations which act on blocks of bits; blocks of 128 bits are a frequent choice. In order to encrypt messages, they are combined with a mode of operation which describes how to parse the messages into blocks and decompose the encryption of a message into encryption of blocks.

Some of the basic mode of operations have been known for a long time and were already standardized for use with the DES algorithm. More recently, the NIST[2] encouraged research for new modes of operation in order to propose them as standards for use together with the AES block cipher. To illustrate modes of operation and their importance in secret key encryption, let us describe three well-known modes (see Figure 1.1): Electronic Code Book (ECB), Cipher Block Chaining (CBC) and Counter mode (CTR).

The ECB mode works as follows: first it pads the plaintext message P to ensure that its length becomes a multiple of the block length, some care should be taken to make sure that the padding can be reversed after decryption to recover the original message. A standard solution is to add a single 1 after the original message, followed by the number of zeros needed to fill the last message block. Note that with this padding, messages whose original length is already an entire number of blocks are enlarged by one full block. After padding, the ECB mode parses the padded message in n-bit blocks, where n is the length of the cipher's blocks. Let the i-th block be denoted by $P^{(i)}$. To encrypt P, each block $P^{(i)}$ is encrypted separately.

Another very common encryption mode is the Cipher Block Chaining (CBC) mode. To add security, this encryption mode is randomized. The randomiza-

---

[2]National Institute of Standards and Technology

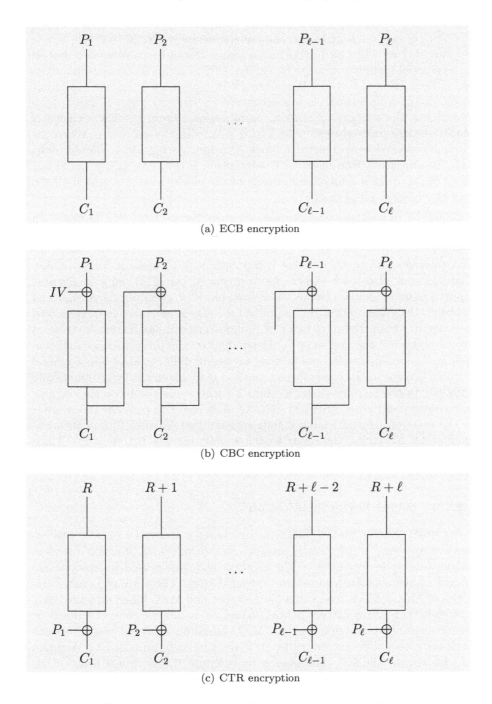

(a) ECB encryption

(b) CBC encryption

(c) CTR encryption

Figure 1.1: Some classical encryption modes

tion is added at the very beginning of the encryption process by simply adding one block of random initial value (IV) at the beginning of the message. There are two options when using this initial value, it can be considered either as an additional plaintext message block, say $P^{(0)}$ or as an additional ciphertext block, then denoted by $C^{(0)}$. When the IV is considered as an extra plaintext block, the first ciphertext block is set to $C^{(0)} = \Pi(P^{(0)})$ where Π denotes the underlying block cipher or random permutation. From the first ciphertext block, we then proceed iteratively, letting $C^{(i)} = \Pi(P^{(i)} \oplus C^{(i-1)})$. When the IV is considered as a ciphertext block, the first encryption is simply omitted. An important fact about CBC encryption is that the encryption of any block of plaintext is a function not only of the block value, but also of all the previous blocks and of the IV.

As modes of encryption go, we also consider the Counter (CTR) mode. In this mode, the block cipher is used to generate a pseudo-random sequence which is then used similarly to a one-time pad in order to encrypt the plaintext message. Thus, CTR mode is a simple way to make a stream cipher algorithm out of a block cipher. More precisely, the CTR mode is given as input a starting counter value. The first block of pseudo-random material is obtained by encrypting this input value. Then the value is incremented in order to obtain the next block of pseudo-randomness, incremented again for the following one and so on... Depending on the precise implementation choice, the incrementation can be done in several different ways. On a general purpose processor, the most efficient method is to increment by arithmetically adding 1 to the counter value, modulo 2^b, where b is the block size in bits. In hardware, either on ASICs or FPGAs, it is faster to consider the counter as the state of a linear feedback shift register (see Chapter 2) and to increment it by advancing the linear feedback shift register by one step. Thus, the exact specifications of the CTR mode may vary depending on the target architecture.

1.1.1.2   Secret key authentication

In [Sim82, Sim85, Sim86], Simmons developed a theory for perfect authentication systems, which can be seen as an equivalent of Shannon's perfect encryption. The secret key authentication algorithms used in practice are known as Message Authentication Codes (MACs). There are two main categories of MACs, MAC based on a block cipher and MAC based on a universal hash function. To construct a MAC based on a block cipher, it suffices to devise a specific mode of operation. MAC based on universal hash functions work on a very different principle; they were initially proposed by Wegman and Carter in [WC81]. The idea is to compute the universal hash of the message to authenticate and then to encrypt this value. This method yields very fast MAC algorithms. Indeed, there exist some very fast universal hashing algorithms that only cost a few processor operations per message block, see [NP99].

To illustrate MACs based on a block cipher, let us consider the CBC encryption mode once more. Another interesting feature of this mode is that a very simlar variation can be used as a Message Authentication Code (MAC). In this alternative mode called CBC-MAC, we very closely follow the CBC encryption process with a couple of simple changes. The first change is that CBC-MAC does not need an IV. Moreover, adding an IV would make CBC-MAC insecure if the IV is processed as a ciphertext block. The second change is that in CBC-MAC, we do not output any intermediate block encryption but only the value of the last block. The third and final change concerns the output of the final block. If this block is directly given as MAC value, then the resulting authentication mode is only secure for messages of fixed length. In practice, it is usually required to have the ability to process messages of arbitrary length. In that case, the last encrypted block should be post-processed before being used as a MAC. The most common post-processing simply reencrypts this value with the block cipher keyed with another independent key.

1.1.1.3 Public key encryption

Public key encryption algorithms mostly rely on number theoretic hard problems. One approach to public key encryption, first proposed in [DH76], is to directly rely on a trapdoor one-way permutation. In that case, the one-way permutation is made public and used for encryption. The trapdoor is kept private and used for decryption. The typical example is the famous cryptosystem of Rivest, Shamir and Adleman (RSA). Another approach is the key exchange algorithm of Diffie and Hellman, also introduced in [DH76], which does not encrypt messages but lets two users agree on a common secret key. Once a common secret key has been agreed upon, the users can encrypt messages using a secret key algorithm. As a consequence, key exchange algorithms suffice to offer the public key encryption functionality.

Moreover, note that for performance reasons, even trapdoor one-way permutations are rarely used to directly encrypt messages or message blocks. It is more practical to build a hybrid cryptosystem that encrypts a random key with the trapdoor one-way permutation and encrypts the message using a secret key encryption scheme.

In addition, when using the RSA public key cryptosystem, special care should be taken not to simply encrypt small keys. Indeed, such a direct approach opens the way to multiplicative attacks. This is further developed in Chapter 8.

1.1.1.4 Public key signature

The most frequently encountered public key signatures algorithms are counterparts of the public key encryption algorithms stated above. The RSA signature algorithm follows the approach proposed in [DH76] and inverses the one-way permutation, thanks to the trapdoor in order to sign. Verification is achieved by computing the one-way permutation in the forward direction.

Note that in the case of RSA, this approach needs to be applied with care in order to avoid multiplicative attacks. Before going through the inverse one-way permutation, the information to be signed needs to be carefully prepared using a padding algorithm. Typical approaches are the full domain hash (FDH) and the probabilistic signature scheme (PSS) described in [BR96].

The Diffie-Hellman key exchange algorithm also has corresponding signature algorithms. These algorithms are based on a modified zero-knowledge proof of knowledge of a discrete logarithm. The algorithm of Schnorr [Sch91] and the NIST standard Digital Signature Algorithm are two examples. Zero-knowledge proofs of knowledge are not further discussed in this book.

This idea of using modified zero-knowledge proofs to build a signature scheme can be applied with a very large variety of hard computational problems. It was introduced by Fiat and Shamir in [FS87]. Using this approach signature algorithms have been based on many hard computational problems.

For the same reason that public encryption is rarely used to directly encrypt messages, public key signature schemes are rarely[3] applied directly to messages. Instead, the message to be signed is first transformed using a cryptographic hash function. Here, the goal of the hash function is to produce a short unique identifier for the message. In order to yield such an identifier, the hash function should be constructed in a way that prevents a cryptanalyst to efficiently build two messages hashing to the same value. In other words, the hash function should be collision resistant.

1.2   Defining security in cryptography

In the framework of computationally based cryptography, an important task is to define what kinds of actions can be considered as attacks. Clearly, recovering the key from one or several encrypted messages is an attack. However, some tasks may be easier and remain useful for an adversary. Along the years, a complex classification of attacks appeared. This classification describes attacks by the type of information they require: there are ciphertext only attacks, known plaintext attacks, chosen plaintext attacks and even chosen ciphertext attacks. Also, by the amount of effort the adversary uses to intercept messages or temper with the cryptosystem: this yields the notions of passive, lunchtime and active attacks. Finally, by the type of information that the attack outputs: there are key recovery attacks, message recovery attacks and distinguishers. A key recovery allows the adversary to compute the key or some equivalent information which can afterwards be used to decrypt any

---

[3]One notable exception to this general rule is signature with message recovery, which embeds a (short) message within the signature, thus avoiding separate transmission.

message. A message recovery attack aims at deciphering a single message. The goal of a distinguisher is to learn a small amount of information about the encryption process.

Modern cryptographers have learned that, as illustrated by many historical examples [Kah67], where cryptography is concerned it is preferable to err on the side of caution. Indeed, the state-of-the-art of attacks against a given cryptographic scheme can only move forward yielding better and better attacks. Often, when faced with an incomplete attack which could easily be dismissed as practically irrelevant, cryptographers prefer to consider it as an advanced warning signal that indicates that further attacks may be forthcoming. As a consequence of this cautiousness, a very strong definition of confidentiality is used in cryptography. When a cryptographic scheme fails to achieve this definition, it calls for a reaction. In the early stages, the best reaction is to patch or dump the system, depending on the exact nature of the attack. After the system has been widely deployed, unless it is utterly broken and calls for immediate replacement, the best reaction is to start preparing a replacement and a phase-out strategy.

Another reason for choosing a strong definition of confidentiality is that it facilitates the work of cryptanalysts. Indeed, it takes much less work to simply point out an intrinsic weakness of a cryptographic scheme with a so-called certification attack than to turn this weakness into a full-fledged key recovery attack. As a consequence, when several algorithms need to be compared, it is very useful to use certification attacks as criteria to prune out the least plausible candidates. For example, this approach was followed by NIST for the selection of the AES encryption standard.

1.2.1   Distinguishers

The strongest definitions of confidentiality which are currently available rely on the notion of distinguishers. Depending on the exact characteristics of the system being considered, the precise definition of distinguishers possibly needs to be adapted. However, the basic style of the definitions is always preserved. All distinguishers share some basic properties:

- A distinguisher, also called a distinguishing adversary, is a computational process, often modeled by a Turing machine.

- A distinguisher $\mathcal{A}$ interacts in a black box manner with an environment $\mathcal{E}$ that encapsulates the cryptographic scheme under attack and in particular chooses random keys for this cryptographic scheme.

- The behavior of the environment depends on the value of a control bit c, chosen uniformly at random upon the first call to the environment.

- The adversary outputs a single bit, 0 or 1, and the goal of the adversary is to determine the value of c with a probability of success greater than $1/2$, i.e., to achieve a better success rate than by blindly guessing c.

- The advantage of the adversary $\mathbf{adv}(\mathcal{A})$ is defined as:

$$\mathbf{adv}(\mathcal{A}) = |2\Pr(\mathcal{A} \text{ outputs } c) - 1|. \qquad (1.1)$$

These basic properties already call for some comments. A first remark concerns the presence of an absolute value in the definition of the advantage. This is useful because it ensures that the advantage is always non-negative. Moreover, it makes sense because when $2\Pr(\mathcal{A} \text{ outputs } c) - 1 < 0$, we can construct a new adversary $\mathcal{A}'$ by reversing the output of $\mathcal{A}$. This adversary succeeds when $\mathcal{A}$ fails and vice versa. As a consequence:

$$2\Pr(\mathcal{A}' \text{ outputs } c) - 1 = 1 - 2\Pr(\mathcal{A} \text{ outputs } c) > 0. \qquad (1.2)$$

Another important remark is that:

$$\mathbf{adv}(\mathcal{A}) = |\Pr(\mathcal{A} \text{ outputs } 0 \mid c = 0) - \Pr(\mathcal{A} \text{ outputs } 0 \mid c = 1)|. \qquad (1.3)$$

In this equation, the notation $\Pr(|)$ denotes a conditional probability, conditioned by the event written at the right of $|$. It is a simple consequence of the two following facts:

$$\Pr(\mathcal{A} \text{ outputs } c) = \Pr(\mathcal{A} \text{ outputs } 0 \mid c = 0)/2 + \Pr(\mathcal{A} \text{ outputs } 1 \mid c = 1)/2,$$
$$1 = \Pr(\mathcal{A} \text{ outputs } 0 \mid c = 1) + \Pr(\mathcal{A} \text{ outputs } 1 \mid c = 1). \qquad (1.4)$$

Also, when using distinguishers, we should remember that in addition to the trivial adversary that simply guesses c, we can devise a generic adversary that models exhaustive key search. This adversary simply guesses the key material that has been chosen by the environment for the underlying cryptographic scheme. Using this key, it tries to determine whether c equal 0 or 1. If the key is correct, this is usually easy. Note, however, that the details depend on the exact flavor of distinguisher we are considering. Moreover, it is also easy to determine that the guessed key is incorrect. In that case, the adversary reverses to the trivial strategy of guessing c. This key guessing adversary obtains an advantage of the order of 2^{-k}, where k is the bit length of the key. This shows that in the definition of confidentiality we should not consider adversaries with an exponentially small advantage. Two different kinds of advantages are usually considered: advantages above a constant larger than $1/2$, such as $2/3$ for example, and advantages exponentially close to one, such as $1 - 2^{-k}$. In fact, these two kinds of advantages yield the same security notion and an adversary with a constant advantage can be converted into an adversary with advantage exponentially close to one by repeating it enough times using different random coins.

Distinguishing attacks against ECB encryption

To illustrate distinguishing attacks, let us consider distinguishers against the ECB. These attacks rely on the fact that encryption with a block cipher cannot hide equalities between blocks. As a consequence, an adversary can often gain some information about the encrypted messages. A very classical example of this weakness consists in encrypting a bitmap picture in ECB mode and remarking that the general shape of the picture remains visible. In particular, large zones of uniform color remain uniform. To formalize this weakness into a distinguishing attack, let us consider an adversary that does not query the encryption mode and directly proposes two messages M_0 and M_1 consisting of 2 blocks each after padding. M_0 is chosen to ensure that its two blocks are equal, and M_1 to ensure that they are different. When the adversary is given back the encryption of one message, he simply checks whether the two ciphertext blocks are equal. In case of equality, he announces that M_0 was encrypted and otherwise that M_1 was. The adversary succeeds with probability 1 and, thus, has advantage 1. Since the total number of blocks involved in the attack is very small, this shows that ECB encryption is generically insecure.

ECB encryption can also be shown insecure by using a different chosen message attack. In this attack, the adversary first queries the encryption mode for the encryption of any message of his choice M. Then, he sends two messages M_0 and M_1, where M_0 is equal to M and M_1 is any other message of the same length. When he receives the encryption of one among M_0 and M_1, he compares this encryption to the encryption of M he already had. If both are equal, he announces that M_0 was encrypted and otherwise that it was M_1. This attack also succeeds with probability one. The main interest of this second attack is that it can be generalized to any deterministic mode. To thwart this attack, it is important to make sure that encrypting twice the same message does not usually output twice the same ciphertext. This can be achieved by adding randomness to the message during the encryption process. A typical way of adding randomness is the use of an IV as in CBC encryption. This simple randomization prevents the above attacks against the ECB mode to work against CBC encryption.

1.2.1.1 Allowed queries for distinguishers

In cryptography, two different types of distinguishers are alternatively encountered, chosen plaintext adversaries (CPA) and chosen ciphertext adversaries (CCA). These distinguishers differ by the type of queries they are allowed to perform. Chosen plaintext adversary can query an encryption oracle and obtain encryptions of arbitrary messages they construct. In addition, chosen ciphertext adversaries can also ask for decryption of arbitrary strings they construct. After considering chosen ciphertext adversaries, designers of cryptographic systems have introduced the idea of authenticating correctly constructed ciphertexts, this allows their systems to answer `invalid` when asked to decrypt arbitrary strings. This is a key idea to design CCA-secure cryptographic schemes.

1.2.1.2   Three flavors of distinguishers

We now informally describe three frequent flavors of distinguishers.

1.2.1.2.1   Find then guess distinguishers   The simplest flavor of distinguishers is called "find-then-guess" or FTG distinguishers. After initialisation of the environment, the distinguishing adversary interacts with the environment in three consecutive phases.

1. The adversary sends messages of his choice to the environment and receives the corresponding ciphertexts, encrypted by the cryptographic scheme using the key chosen during initialization. This phase behaves independently of the bit c chosen by the environment. It is also possible to allow the adversary to ask for decryption of arbitrary ciphertexts of his choice when considering chosen ciphertext attacks. Each message can be chosen interactively after receiving the encryption for the previous message.

2. The adversary produces two test messages M_0 and M_1 of the same length. It sends the messages to the environment and receives a ciphertext C corresponding to an encryption of M_c.

3. The adversary may once again ask for encryption and/or decryption of messages of his choice, with a single, essential, exception: it is not allowed to ask for the decryption of the message C itself. Note that for chosen ciphertext attacks, requesting the decryption of messages derived from C is acceptable, as long as they differ from C. Typically, truncated, padded or slightly different copies of C are allowed in that case.

After the three phases, the adversary outputs his guess for c.

1.2.1.2.2   Left or right distinguishers   A (polynomially) more powerful flavor of distinguishers than FTG distinguishers are "left-or-right" or LOR distinguishers. It consists of a single phase, where the adversary sends pairs of messages (M_0, M_1) of the same length and receives the encryption of M_c. Pairs of messages are chosen interactively after receiving previous encryption. In the case of chosen ciphertext attacks, the adversary may also send pairs of ciphertexts (C_0, C_1) and learn the decryption of C_c. To avoid trivial attacks, redundant queries are forbidden, i.e., an adversary is not allowed to request the decryption of a ciphertext returned by a previous query as part of a pair of ciphertexts.

At the end of the interaction the adversary produces a guess for c, i.e., tries to determine whether the left-hand side or the right-hand side of queries was processed by the environment. This explains the name of "left-or-right" distinguishers.

To show that LOR adversaries are more powerful than FTG adversaries, it suffices to prove that any FTG adversary can be transformed into an LOR adversary which is as powerful. The proof is very simple, it suffices to embed the FTG adversary in a LOR-wrapper which runs it in a black box way. In the first and final phase, when the FTG adversary requests an encryption of M, the wrapper forwards the pair (M, M) to the environment and returns the answer. In the middle phase, the FTG adversary produces a pair of messages (M_0, M_1). The wrapper simply forwards this pair and the environment's answer. At the end, the wrapper copies the output of the FTG adversary. Clearly, the wrapper in the LOR context is as successful as the original adversary in the FTG context. Moreover, the number and length of queries and the running times are essentially identical.

1.2.1.2.3 Real or random distinguishers The FTG and LOR distinguishers both test the ability of an adversary to extract information from ciphertexts when a very small amount of information remains unknown. "Real-or-Random" or ROR distinguishers are based on a different paradigm and try to distinguish between real encrypted messages and purely random encrypted messages. As usual, during initialization, the environment chooses a random bit c and random keys for its embedded cryptographic scheme. During interaction, the adversary sends messages of his choice to the environment. If $c = 0$, the environment is in real mode and returns the encryption of each message it receives. If $c = 1$, the environment is in random mode, in that case, it returns the encryption of a uniformly distributed random string of the same length.

In fact, it was shown in [BDJR97] that ROR security is equivalent to LOR security. In [RBBK01], a variation of the ROR security is proposed, it is called indistinguishability from random strings and often denoted by IND$. In this variation, depending on the value of its inner bit, the environment either returns the encryption of the message it received or a purely random string of the same length as the encrypted message.

This style of distinguisher is very useful for some security proofs, because there are more tools for showing that a string is indistinguishable from a random string than for addressing environment with two sides, where each side has its own specific description. However, IND$ security is stronger than LOR security or, equivalently, than ROR security.

Indeed, assuming that LOR secure cryptosystems exist, it is possible to construct examples of schemes which are LOR secure but not IND$ secure. The basic idea is very simple. Starting from any LOR secure encryption scheme S, we construct a new scheme S', which encrypts a message M under key k as $0\|S_k(M)$, i.e., it simply prepends a 0 to the encryption of M using S. It is clear that the LOR security of S' is the same as the LOR security of S. However, S' is not IND$ secure because any output of the ROR environment that starts with 1 is necessarily coming from the random mode. This example shows that requiring IND$ security is in some sense too much.

1.2.2   Integrity and signatures

In modern times, cryptography deals with more than confidentiality. It is also used to protect messages or files against tempering. This protection can be based either on secret key or on public key algorithms. In secret key cryptography, we saw that this protection is offered by message authentication codes. With public key cryptography, the protection is based on a stronger mechanism called signature. The essential difference between MACs and signatures is that message authentication codes protect their users against attacks by third parties but offer no protection against dishonest insiders, while signatures offer this additional protection. This difference is reflected when defining the security notions for integrity. Integrity mechanisms of both types rely on two algorithms. The first algorithm takes as input a message and outputs an authentication tag. It also uses some key material, either the common key in the secret key framework or the private key of the signer in the public key framework. The second algorithm takes as input a message and an authentication tag and returns either valid or invalid. It uses either the common secret key or the public key of the signer. In both frameworks, the goal of an attacker is to construct a forgery, i.e., a valid authentication on a message, without knowledge of the secret or private keys. As with confidentiality, the attacker is also allowed to first make queries, more precisely, he can obtain authentication tags for any message of his choice. For the security notion to make sense, the produced forgery should not simply be a copy of one of these tags but should be new. This can be made precise in two different ways. One option is to ask the attacker to output a valid authentication tag for a new message, which has not been given during the queries. The alternative is to also allow additional tags for messages which have already been authenticated, as long as the forged tag has never been produced as answer to a query on this message. For example, in this alternative, if a tag σ has been produced for M and a tag σ' for M' (with $\sigma \neq \sigma'$ and $M \neq M'$), assuming that σ' is also a valid authentication tag for M, it counts as a valid forgery, despite the fact that M was already authenticated and that σ' was already produced, because the pair (M, σ') is new.

To measure the efficiency of an attacker in the case of forgeries, we define its advantage as the probability that its output (M, σ) is a valid forgery. Note that, here, there is no need to subtract $1/2$ because the output no longer consists of a single bit and is thus much harder to guess. For example, guessing a valid authentication tag on t-bits at random succeeds with low probability $1/2^t$. A forgery attack is considered successful when its complexity is low enough and when its probability of success is non-negligible, for example larger than a fixed constant $\epsilon > 0$.

1.2.3 Authenticated encryption

After seeing the definitions of confidentiality and integrity/signatures, a natural question is to consider authenticated encryption. Is it possible to construct cryptographic systems that meet both the requirements of confidentiality and integrity/signature? In particular, is there a generic approach to compose secure cryptographic methods that individually ensure confidentiality and integrity/signature and construct a new cryptosystem which ensures both?

In the context of authenticated encryption, it is interesting to consider some natural methods to compose an encryption scheme and an authentication scheme and see why these methods are not generically secure. We start in the context of secret key cryptography, i.e., with secret key encryption and MACs. We discuss the case of public key primitives afterwards.

1.2.3.1 Authenticated encryption in the secret key setting

The goal of authenticated encryption is to perform encryption and authentication of messages, while guaranteeing the security of both primitives simultaneously. This can be done by composing two preexisting cryptographic primitives or by devising a new specific algorithm (for some examples, see [Jut01, KVW04, Luc05, RBBK01]). The generic composition approach, i.e., for arbitrary preexisting primitives, was studied in detail by Bellare and Namprempre in [BN00] and raises some deep questions about the relations between confidentiality and integrity.

1.2.3.1.1 Encrypt and MAC Given a secret key encryption scheme and a MAC, the first idea that comes to mind in order to encrypt and protect the integrity of a message M at the same time is simply to concatenate an encryption of M and a MAC of M. The reason that makes this simple idea insecure is that a MAC algorithm does not necessarily hide the complete content of the message. For example, if we are given a secure MAC algorithm, we can easily construct another secure MAC based on it in a way that completely destroys confidentiality. It suffices to form an extended MAC by concatenating the original one with the first few bits of the message. The reader may check that this yields another secure MAC and that it cannot preserve confidentiality. Moreover, MAC algorithms are usually deterministic algorithms that compute a short tag from the input message and verify the correctness of the received tag by recomputing it and comparing values. With deterministic MAC algorithms, the simple concatenation construction always fails to be secure. Indeed, it is clear that the following adversary is always a successful find-the-guess distinguisher:

- The adversary asks for authenticated encryption of random messages of the same length until two messages with a different MAC are found. Let M_0 and M_1 be these two messages and (C_0, m_0) and (C_1, m_1) be the corresponding authenticated encryptions. In these encryptions, C_i is the regular ciphertext and m_i the MAC tag. We have $m_1 \neq m_2$ with high probability.

- The adversary sends (M_0, M_1) to the environment and receives an encrypted message (C_c, m_c). Since the encryption algorithm is secure, C_c does not permit to distinguish which message is encrypted. However, since the MAC algorithm is deterministic, the MAC tag m_c is either m_0 or m_1. If $m_c = m_0$, the adversary announces that M_0 is the encrypted message. If $m_c = m_1$, it announces M_1. Clearly, this guess is always correct.

1.2.3.1.2 MAC then Encrypt The reason why the previous approach fails is that MACs are not intended to protect the confidentiality of messages. To avoid this issue, one possible approach is the MAC then Encrypt paradigm where we concatenate the MAC tag m to the message M and encrypt $(M\|m)$ into a ciphertext C. This clearly prevents the MAC tag from leaking information about the encrypted message. However, this composition is not secure either. To understand why, given a secure encryption scheme Enc, we can construct a new encryption scheme Enc$'$ that encrypts M into $(\mathsf{Enc}(M)\|1)$, simply adding an additional bit after the message encrypted by Enc. The corresponding decryption Dec$'$ strips the last bit, without checking its value, and applies the regular decryption Dec.

When Enc$'$ is used together with any MAC scheme in a MAC then encrypt construction, the resulting scheme does not ensure authenticity. Indeed, an adversary can forge a valid encryption message in the following way:

- Send an arbitrary message M and obtain a ciphertext $C = \mathsf{Enc}'(M\|m)$.

- Replace the final bit of C by a zero, thus forming a message C'.

- Give C' as forgery.

Clearly, Dec$'$ decrypts C' into $(M\|m)$ since the last bit is discarded anyway. As a consequence, the MAC tag is accepted as valid. Thus, C' is a legitimate forgery.

It is important to remark that the above attack is quite artificial. However, other reasons why this order of composition is not generically secure are discussed in [Kra01]. Another interesting property shown in this paper is that in the context of secure channels, the MAC then Encrypt composition is secure for some specific encryption algorithms, including CBC encryption.

1.2.3.1.3 Encrypt then MAC After MAC then Encrypt, we can try the other direction of composition, first encrypt the message M into a ciphertext C, then compute a MAC m of the ciphertext. Bellare and Namprempre showed in [BN00] that the Encrypt then MAC approach allows to construct a secure authenticated encryption given any secure encryption and any secure MAC, under the condition that independent keys are used for the two schemes. To sketch the proof, let us start with integrity. We claim that an adversary cannot form a new valid ciphertext by himself, unless he forges a valid MAC for some string (the corresponding unauthenticated ciphertext). Concerning confidentiality, it is clear that the MAC cannot help. Otherwise, it would be possible to attack the confidentiality of the encryption scheme simply by adding a MAC tag to it. Since this operation could easily be performed by an adversary, we see that the Encrypt then MAC composition is also secure from the confidentiality point-of-view.

1.2.3.2 Authenticated encryption in the public key setting

In the public key setting, the adversary is granted more power, since he has access to the public keys and can encrypt and verify signatures by himself. Thus, any generic composition insecure in the secret key setting is also insecure in the public key setting. However, additional attacks exist in the public key setting. We now explain why neither "Encrypt then Sign" nor "Sign then Encrypt" are secure and discuss secure methods.

1.2.3.2.1 Sign then Encrypt Of course, the Sign then Encrypt composition inherits the weakness of MAC then Encrypt. However, other weaknesses appear in the public key setting. In particular, the Sign then Encrypt composition suffers from a forwarding attack. Indeed, the legitimate recipient of a message can after decryption decide to reencrypt the same message for a third party, whose public key is known to him. For the third party, since the signature is valid, the message seems to come from the initial sender and the forwarding leaves no tracks. It is easy to come with contexts where this forwarding attack can be considered an attack. Anyway, it is clearly an undesirable property for a secure cryptographic scheme.

1.2.3.2.2 Encrypt then Sign The Encrypt then Sign composition fails to be secure for another reason. Indeed, this composition is subject to a ciphertext stealing attack. The ciphertext stealing works as follows: the attacker intercepts a message from a sender to a receiver and prevents this message from reaching the receiver. After interception, the attacker strips the signature from the original encrypted message and replaces it by his own signature. After that, he resends the modified message to its intended recipient. Since the signature is valid and since the message can be correctly decrypted, the recipient logically assumes that this is a legitimate message from the attacker.

Depending on the application, this ciphertext stealing attack can be used to break confidentiality or for other malicious purposes. A breach of confidentiality may occur when the recipient answers the message, especially if he quotes it. In a different context, if the recipient is a timestamping or registering authority, the attacker could falsely claim ownership of the encrypted information that the original sender wanted to register.

Since Encrypt then Sign is a straightforward adaptation of Encrypt then MAC to the public key context, it is interesting to precisely identify the reason that prevents this attack from applying in the secret key context. Trying to apply the attack in the secret key setting, we see that nothing prevents the attacker from removing the original MAC tag or from adding a new one. This new MAC tag passes the verification on the recipient tag. Moreover, the encrypted message could be correctly decrypted with the secret key shared by the original sender and the recipient. In truth, the attack fails because of the natural hypothesis that, in the secret key setting, each secret key belongs to a pair of users and is not shared by anyone else. Under this hypothesis, it is highly unlikely that the recipient accepts to verify the MAC tag using a key shared with a user and then to decrypt using a key shared with another user.

1.2.3.2.3 Signcryption In the public key setting, in order to avoid the above attacks, it is essential to precisely define the expected security properties and to carefully check that they are satisfied. The name signcryption for such cryptographic schemes was proposed in [Zhe97]. A formal treatment of signcryption was first given in [ADR02].

To avoid the above weaknesses of the encrypt then sign and sign then encrypt composition, other methods have often been proposed for applications. A first idea is to bind the signature and encryption together by adding fields, for example at the beginning of the message, explicitly identifying the two participants of the exchange, sender and recipient. With this additional precaution, both sign-then-encrypt and encrypt-then-sign resist the above attacks. A slight variation of this idea which adds the identities of the sender and recipient in various places is proven secure in [ADR02]. The drawback of this solution is that it needs to mix up routing information together with the message itself. This is often judged to be unsatisfactory by application developers who prefer to manage the message at the application layer and the routing information at a lower layer. It is also inconvenient if the users desire to archive a signed copy of the message after stripping it from the routing information.

Another option relies on triple wrapping. Two flavors are possible: sign-encrypt-sign and encrypt-sign-encrypt. They are resistant to ciphertext stealing and forwarding. Note that sign-encrypt-sign is usually preferred, since it allows the recipient to archive a signed copy of the original message. With the triple wrapping method, the sender performs three cryptographic operations in sequence on the message, encrypting with the recipient public key and signing with his own private key. The recipient performs the complementary operations on the received message. In the sign-encrypt-sign, the recipient also needs to check that both signatures were issued by the same person.

1.2.4 Abstracting cryptographic primitives

In order to construct secure cryptosystems, cryptographers often start from small building blocks and put them together to assemble these cryptosystems. Of course, it is essential for these building blocks to satisfy relevant security properties. We now briefly describe how the security of two essential building blocks, block ciphers and hash functions is often modelled.

1.2.4.1 Blockciphers

As said in Section 1.1.1.1, a block cipher is a keyed family of permutations that operate on blocks of n bits. To select a permutation in the family, one uses a k-bit key. To model the security of a block cipher, two models are often used. The first approach considers pseudo-random permutation families. It is based on distinguishers. In this approach, the adversary knows the algorithmic description of a family of pseudo-random permutations and its goal is to determine whether a permutation chosen by the environment is a truly random permutation or a permutation selected from the family by choosing a random key. A good block cipher aims at being a pseudo-random permutation family. Another, much stronger, approach is the ideal cipher model. In this model, mostly used in proofs, a block cipher is idealized as a family of purely random permutations. Note that, while very convenient for proofs, this cannot be achieved by any concrete block cipher. Indeed, every block cipher has a short algorithmic description, which is not the case for a family of purely random permutations.

In addition, some other properties of block ciphers are sometimes considered in cryptanalysis. A typical example considers related key attacks. Here, the adversary is no longer limited to querying the blockcipher with a fixed key. Instead, he is allowed to make queries using several related keys, obtained, for example, by xoring or adding fixed constants to an initial key. A formal treatment of this notion is given in [BK03]. One difficulty with this notion of related key attacks is that, unless the allowed operations on keys are very limited, these attacks are too strong. For example, if the attacker is allowed both adding and xoring constants, any block cipher can easily be attacked. Indeed, adding '1' and xoring '1' to the initial key yields the same new key, if and only if the low order bit of the key is a '0'. Adding and xoring other powers of two permit the adversary to learn each bit of the key. Of course, once the key is known, the adversary wins.

1.2.4.2 Hash functions

Cryptographic hash functions have two different flavors. For theoretical purposes, one usually encounters keyed family of hash functions. This flavor is very close to block ciphers, except for two noticeable differences: hash functions are modelled by random functions instead of permutations and their inputs have variable length instead of fixed length. Where needed, hash functions are idealized by random oracles. It was recently proven that the random oracle model and the ideal cipher model are equivalent [CPS08].

The other flavor of hash functions is used for practical purposes. In that context, it is very useful to have access to an unkeyed hash function. This is, for example, the case of signature schemes, as discussed in Section 1.1.1.4. With unkeyed hash functions, specific security properties need to be introduced. Three very useful properties are collision resistance, preimage resistance and second preimage resistance. Preimage and second preimage resistance can easily be defined. For preimage resistance, we simply say that H is preimage resistant if there exists no efficient adversary that given a value h can output a message M such that $H(M) = h$. Similarly, H is second preimage resistant if no efficient adversary can, given a message M, find a different message M' such that $M \neq M'$ and $H(M) = H(M')$. These two definitions are straightforward to formalize, even with unkeyed hash functions.

However, collision resistance is a trickier property. For any unkeyed hash function H, there exists an efficient adversary which simply prints out two messages M and M' contained in its code, such that $H(M) = H(M')$. For keyed family, the problem vanishes, which explains why they are preferred for theoretical purposes. Of course, the existence of the above efficient adversary does not help to find collision in practice. Thus, the common answer is to overlook the above problem and to simply keep the definition informal: a hash function is then said to be collision resistant when no practical method can efficiently yield collisions.

Chapter 2

Elementary number theory and algebra background

Number theory is at the core of many cryptographic systems, as a consequence, it is routinely used in many cryptanalysis. In this chapter, we discuss some elementary but crucial aspects of number theory for cryptographers, including a basic description of the RSA and Diffie-Hellman public key cryptosystems.

2.1 Integers and rational numbers

The construction of the ring of integers $\mathbb{Z}$ is out of the scope of this book and we simply take it for granted. We recall a few elementary facts:

1. $\mathbb{Z}$ possesses two commutative laws called addition and multiplication, respectively, denoted by "+" and "×" (the symbol × is often removed from equations or replaced by "·" or even by nothing as in xy). Commutativity means that for any x and y, $x + y = y + x$ and $xy = yx$. In addition the operations are associative, i.e., $(x + y) + z = x + (y + z)$ and $(xy)z = x(yz)$.

2. The neutral element of addition is 0.

3. For all x in $\mathbb{Z} : 0 \cdot x = 0$.

4. The neutral element of multiplication is 1.

5. For any element x in $\mathbb{Z}$, we can construct an element denoted by $-x$ and called the opposite of x, such that $x + (-x) = 0$. The subtraction of two elements $x - y$ is defined as the sum $x + (-y)$.

6. The notation $\mathbb{Z}^*$ denotes the set of non-zero elements of $\mathbb{Z}$.

7. For any element x in $\mathbb{Z}^*$ and any pair (y, z) of elements of $\mathbb{Z}$, $xy = xz$ if and only if $y = z$.

8. The multiplication distributes with respect to addition, i.e., for all x, y and z:

$$(x + y)z = xz + yz.$$

9. $\mathbb{Z}$ is totally ordered by an order $\geq$ compatible with the ring operations, i.e.:

   (a) For all $x : x \geq x$.

   (b) For all x and y, if $x \geq y$ and $y \geq x$ then $x = y$.

   (c) For all x, y and z, if $x \geq y$ and $y \geq z$ then $x \geq z$.

   (d) For all x and y, either $x \geq y$ or $y \geq x$ hold.

   (e) For all x, y and z, $x \geq y$ if and only if $x + z \geq y + z$.

   (f) The notation $x > y$ indicates that $x \geq y$ and $x \neq y$.

   (g) For all x, y and for all $z > 0$, $x \geq y$ if and only if $xz \geq yz$.

   (h) For all x, y and for all $z < 0$, $x \geq y$ if and only if $xz \leq yz$.

10. The absolute value of x, denoted by $|x|$ is defined as x when $x \geq 0$ and as $-x$ otherwise.

11. For all $x \neq 0$ and y, there exist two integers q and r, called the quotient and remainder of the (Euclidean) division of y by x such that $0 \leq r < |x|$ and:

$$y = qx + r.$$

12. When the remainder of the division of y by x is 0, i.e., when $y = qx$, we say that x divides y, that x is a divisor of y or that y is a multiple of x. Note that when x divides y, $-x$ also divides y. Thus, it is convenient to consider positive divisors only.

13. 1 (and -1) divides all integers.

14. For all $x \neq 0$, x divides itself, since $x = 1 \cdot x$.

15. A prime is an integer $x > 1$ with no non-trivial divisor, i.e., with no positive divisor except 1 and x.

16. A positive integer $x > 1$ which is not a prime is said to be composite.

17. Any composite number $N > 1$ can be written as

$$N = \prod_{i=1}^{t} p_i^{e_i}, \tag{2.1}$$

where each p_i is a prime and $e_i > 0$ is called the multiplicity of p_i in N and where no two p_is are equal. Moreover, up to the order of factors, this decomposition is unique. This statement is called the fundamental theorem of arithmetic.

Among the above notions, primes and the Euclidean division both play an essential role in cryptography.

From the integers in $\mathbb{Z}$, constructing the set $\mathbb{Q}$ of rational numbers (or simply rationals) is quite easy and uses a classical construction of quotienting a set by an equivalence relation. We describe this process in detail here since we use it again in a more complicated case in Chapter 14. Let us consider the set $\mathbb{Z} \times \mathbb{Z}^*$ containing pairs of integers, with a non-zero second member. We first define an equivalence relation $\equiv$ on this set. For all pairs (x_1, x_2) and (y_1, y_2), we say that $(x_1, x_2) \equiv (y_1, y_2)$ if and only if $x_1 y_2 = y_1 x_2$. This clearly is an equivalence relation, i.e., a relation which is reflexive, symmetric and transitive:

- **Reflexivity** For all pairs (x_1, x_2), we have $(x_1, x_2) \equiv (x_1, x_2)$ since $x_1 x_2 = x_1 x_2$.

- **Symmetry** For all pairs (x_1, x_2) and (y_1, y_2), the equivalence $(x_1, x_2) \equiv (y_1, y_2)$ implies $(y_1, y_2) \equiv (x_1, x_2)$.

- **Transitivity** For all pairs (x_1, x_2), (y_1, y_2) and (z_1, z_2), if $(x_1, x_2) \equiv (y_1, y_2)$ and $(y_1, y_2) \equiv (z_1, z_2)$ then $(x_1, x_2) \equiv (z_1, z_2)$. Indeed, $x_1 y_2 = y_1 x_2$ implies $x_1 z_2 y_2 = y_1 x_2 z_2$ and $y_1 z_2 = z_1 y_2$ implies $x_2 z_2 y_1 = x_2 z_1 y_2$. Thus, $x_1 z_2 y_2 = x_2 z_1 y_2$ and since $y_2 \neq 0$, we find $x_1 z_2 = x_2 z_1$.

The set $\mathbb{Q}$ is defined as the set of equivalence classes of $\mathbb{Z} \times \mathbb{Z}^*$ under the equivalence relation $\equiv$. The equivalence class of (x_1, x_2) is denoted by x_1/x_2 or $\frac{x_1}{x_2}$. Since $(x_1, x_2) \equiv (-x_1, -x_2)$, it is possible to assume that $x_2 > 0$. Elements of $\mathbb{Q}$ written as x_1/x_2 with $x_2 > 0$ are called fractions.

It is clear that for any integer $\lambda > 0$, $(x_1, x_2) \equiv (\lambda x_1, \lambda x_2)$ or equivalently:

$$\frac{x_1}{x_2} = \frac{\lambda x_1}{\lambda x_2}.$$

When a fraction x_1/x_2 cannot be written as y_1/y_2 with $0 < y_2 < x_2$, we say that the fraction is in irreducible form. In that case, there exists no integer $\lambda > 1$ that divides both x_1 and x_2. Every fraction has a unique representation in irreducible form.

The set of integers $\mathbb{Z}$ is naturally embedded into $\mathbb{Q}$ by sending the integer x to the fraction $x/1$. In the sequel, we simply use the symbol x when referring to the fraction $x/1$. The set $\mathbb{Q}$ inherits, the addition, the multiplication and the order from $\mathbb{Z}$ as follows:

- Define addition by saying that the sum of the equivalence classes of (x_1, x_2) and (y_1, y_2) is the equivalence class of $(x_1 y_2 + y_1 x_2, x_2 y_2)$. This definition is compatible with the equivalence relation and identical to the integer addition on the image of $\mathbb{Z}$ by the natural embedding.

- Define multiplication by saying that the product of the equivalence classes of (x_1, x_2) and (y_1, y_2) is the equivalence class of $(x_1 y_1, x_2 y_2)$.

This definition is compatible with the equivalence relation and identical to the integer multiplication on the image of $\mathbb{Z}$ by the natural embedding.

- Define the order for equivalence classes (x_1, x_2) and (y_1, y_2), with $x_2 > 0$ and $y_2 > 0$ as follows: $(x_1, x_2) \geq (y_1, y_2)$ if and only if $x_1 y_2 \geq y_1 x_2$. Once again, this is compatible with the equivalence relation and gives the same order as in $\mathbb{Z}$ after the natural embedding.

We now recall the basic properties of these operations in $\mathbb{Q}$:

1. Addition and multiplication in $\mathbb{Q}$ are commutative and associative.

2. The neutral element of addition is $0 = 0/1$.

3. For all x in $\mathbb{Q} : 0 \cdot x = 0$.

4. The neutral element of multiplication is $1 = 1/1$.

5. For any element x/y in $\mathbb{Q}$, its opposite is $(-x)/y$, denoted by $-x/y$.

6. The notation $\mathbb{Q}^*$ denotes the set of non-zero elements of $\mathbb{Q}$.

7. Any element x/y of $\mathbb{Q}^*$ has a multiplicative inverse y/x that satisfies $(x/y) \times (y/x) = 1$.

8. The multiplication distributes with respect to addition in $\mathbb{Q}$.

9. $\mathbb{Q}$ is totally ordered by the order $\geq$, and $\geq$ is compatible with the ring operations.

10. The absolute value of x/y, denoted by $|x/y|$ is defined as $|x|/|y|$.

Since every non-zero element has an inverse, $\mathbb{Q}$ is not only a ring but also a field. Note that the above construction can be applied not only to $\mathbb{Z}$ but to any entire ring. In general, the resulting field is called the field of fractions of the entire ring.

2.2   Greatest common divisors in $\mathbb{Z}$

Given two integers x and y, we say that an integer z is a common divisor of x and y if z divides both x and y. The set of common divisors of x and y is a finite set and thus contains a largest element, called the greatest common divisor or GCD of x and y. The computation of GCDs is one of the most frequent algorithmic tasks encountered in cryptography. One of its most basic application is the rewriting of fractions in irreducible form. Indeed, for any fraction x/y, if λ is a common divisor of x and y, we can write $x = \lambda x'$ and $y = \lambda y'$ and remark that $x/y = x'/y'$. If λ is the greatest common divisor of x and y, then x'/y' is in irreducible form.

To compute the greatest common divisor of x and y, it would, of course, be possible to consider each number between 1 and the minimum of $|x|$ and $|y|$ and to test whether it divides x and y. The largest of these numbers is then the required GCD. However, when x and y become large in absolute value, this algorithm requires a very large number of division and becomes extremely inefficient. A much more efficient method, called Euclid's algorithm, is based on Euclidean division. This algorithm is based on the following fact. If λ is a common divisor of x and y; if, q and r are the quotient and remainder in the Euclidean division of y by x, then λ is a common divisor of x and r. Conversely, any common divisor of x and r is a common divisor of x and y. Indeed, writing $x = \lambda x'$, we have the equality:

$$y = (qx')\lambda + r. \tag{2.2}$$

Thus, y is a multiple of λ if and only if r is a multiple of λ. Since this is true for all divisors, it is true in particular for the greatest. Thus, the GCD of x and y is equal to the GCD of x and r. Applying this idea repeatedly and assuming, without loss of generality, that $y \geq x \geq 0$ we can define a sequence of integers z, starting with $z_0 = y$, $z_1 = x$ and letting z_{i+1} be the remainder of the Euclidean division of z_{i-1} by z_i. This sequence of integers is decreasing, thus at some point we find $z_k = 0$ and stop the sequence. Thanks to the above remark, we know that the greatest common divisor of z_{i-1} and z_i is identical to the greatest common divisor of z_i and z_{i+1}. Thus, by induction, the GCD of x and y is equal to the GCD of z_{k-1} and z_k. Since $z_k = 0$, this GCD is z_{k-1}. The definition of this sequence can be rewritten in algorithmic form as Algorithm 2.1. In this algorithm, we express the quotient in the Euclidean division of y by x as the fraction y/x rounding down to the nearest integer.

The fact that the sequence z is decreasing shows that algorithm terminates and that it stops after at most $|x|$ steps. However, it is much more efficient than that. Indeed, the number of steps can be bounded by $O(\log |x|)$. A simple way to prove this fact is to show that for all values of i we have $z_{i+2} \leq z_i/2$, when both values of the sequence are defined. Indeed, either $z_{i+1} \leq z_i/2$ and we are done because the sequence is decreasing, or $z_{i+1} > z_i/2$ in which case $z_{i+2} = z_i - z_{i+1} < z_i/2$. Since z is a sequence of integers, we cannot have more than $\log_2 |x|$ divisions by two. Studying the complexity of GCD algorithms is a research topic in its own right.

In addition to the GCD of x and y, Euclid's algorithm can also be used to recover additional information by keeping track of the transformation used from an element of the sequence z to the next. From a mathematical point-of-view, let us define two additional sequences α and β in parallel to z. First, we explicitly write:

$$z_{i+2} = z_i - q_i z_{i+1}, \tag{2.3}$$

Algorithm 2.1 Euclid's greatest common divisor algorithm

Require: Input two integers X and Y

    Let $x \longleftarrow |X|$

    Let $y \longleftarrow |Y|$

    if $x > y$ **then**

        Exchange x and y

    end if

    while $x > 0$ **do**

        Let $q \longleftarrow \lfloor y/x \rfloor$ (Quotient of Euclidean division)

        Let $r \longleftarrow y - qx$ (Remainder of Euclidean division)

        Let $y \longleftarrow x$

        Let $x \longleftarrow r$

    end while

    Output y (GCD of X and Y)

---

where q_i is the quotient in the Euclidean division of z_i by z_{i+1}. Using this quotient, we now define α by the equations:

$$\alpha_0 = 1$$
$$\alpha_1 = 0 \qquad\qquad\qquad (2.4)$$
$$\alpha_{i+2} = \alpha_i - q_i\alpha_{i+1},$$

and β by the equations:

$$\beta_0 = 0$$
$$\beta_1 = 1 \qquad\qquad\qquad (2.5)$$
$$\beta_{i+2} = \beta_i - q_i\beta_{i+1}.$$

These two sequences have the property that for all $0 \le i \le k$:

$$z_i = \alpha_i z_0 + \beta_i z_1. \qquad\qquad\qquad (2.6)$$

This is true for $i = 0$ and $i = 1$ and readily follows by recurrence for greater values of i. We can also show that for all values $0 \le i \le k - 1$:

$$\det\begin{pmatrix} \alpha_i & \beta_i \\ \alpha_{i+1} & \beta_{i+1} \end{pmatrix} = \alpha_i\beta_{i+1} - \alpha_{i+1}\beta_i = (-1)^i.$$

Indeed, this is clear when $i = 0$ and follow by induction when remarking that:

$$\begin{pmatrix} \alpha_{i+1} & \beta_{i+1} \\ \alpha_{i+2} & \beta_{i+2} \end{pmatrix} = \begin{pmatrix} 0 & 1 \\ 1 & -q_i \end{pmatrix} \cdot \begin{pmatrix} \alpha_i & \beta_i \\ \alpha_{i+1} & \beta_{i+1} \end{pmatrix}.$$

Indeed, the transition from one step to the next is a multiplication by a matrix of determinant -1.

In particular, the GCD of α_i and β_i is 1. Looking at Equation (2.6) for $i = k$ and using the fact that $z_k = 0$ we see $-\beta_k/\alpha_k$ is an irreducible expression for z_0/z_1. Note that it might be necessary to change the signs of both α_k and β_k in order to enforce a positive denominator. The values α_{k-1} and β_{k-1} are also very useful, they are called Bézout's coefficients and they allow us to express the GCD z_{k-1} of z_0 and z_1 as $\alpha_{k-1}z_0 + \beta_{k-1}z_1$. Euclid's extended algorithm is a variation of Algorithm 2.1 that in addition to the GCD computes the coefficients α_{k-1} and β_{k-1}. We give a description as Algorithm 2.2.

Algorithm 2.2 Euclid's extended algorithm

Require: Input two integers X and Y
    Let $\alpha_y \longleftarrow 0$ and $\beta_x \longleftarrow 0$.
    if $X \geq 0$ then
        Let $x \longleftarrow X$ and $\alpha_x \longleftarrow 1$.
    else
        Let $x \longleftarrow -X$ and $\alpha_x \longleftarrow -1$.
    end if
    if $Y \geq 0$ then
        Let $y \longleftarrow Y$ and $\beta_y \longleftarrow 1$.
    else
        Let $y \longleftarrow -Y$ and $\beta_y \longleftarrow -1$.
    end if
    if $x > y$ then
        Exchange x and y
        Exchange α_x and α_y
        Exchange β_x and β_y
    end if
    while $x > 0$ do
        Let $q \longleftarrow \lfloor y/x \rfloor$ (Quotient of Euclidean division)
        Let $r \longleftarrow y - qx$ (Remainder of Euclidean division)
        Let $\alpha_r \longleftarrow \alpha_y - q\alpha_x$
        Let $\beta_r \longleftarrow \beta_y - q\beta_x$
        Let $y \longleftarrow x$, $\alpha_y \longleftarrow \alpha_x$ and $\beta_y \longleftarrow \beta_x$
        Let $x \longleftarrow r$, $\alpha_x \longleftarrow \alpha_r$ and $\beta_x \longleftarrow \beta_r$
    end while
    Output y and (α_y, β_y)
    Optionally output (α_x, β_x)

The notion of greatest common divisor can be generalized to sets of integers. It is the largest positive integer that divides all the elements of the state. It is easy to compute it iteratively once we remark that given a set S containing more than 2 elements, the GCD of S can be obtained as follows: Let a be any element of S, let b be the GCD of the set $S - \{a\}$, then the GCD of S is equal to the GCD of a and b. As a consequence, we can write Algorithm 2.3 to compute the GCD of a set or list of numbers.

Algorithm 2.3 GCD of a list of numbers

Require: Input a list of integers $X_1, \ldots, X_t$

    Let $g \longleftarrow X_1$

    for i from 2 to t **do**

        Let $g \longleftarrow GCD(g, X_i)$ (using Euclid's GCD algorithm)

    end for

    Output g

2.2.1　Binary GCD algorithm

One drawback of Euclid's algorithm to compute GCDs is the need to perform Euclidean divisions. When faced with large integers, this can be a worry, especially if one wants to avoid using a preexisting library for manipulating large integers. In that case, Stein's binary GCD algorithm is a nice alternative solution. This algorithm computes GCDs without any Euclidean division, using a few simple properties, that holds for all integers $a \geq 0$ and $b \geq 0$:

- If $a = 0$ and $b \neq 0$ then $GCD(a, b) = b$;

- If $a \neq 0$ and $b = 0$ then $GCD(a, b) = a$;

- If a and b are both even then $GCD(a, b) = 2\, GCD(a/2, b/2)$;

- If a is odd and b is even then $GCD(a, b) = GCD(a, b/2)$;

- If a is even and b is odd then $GCD(a, b) = GCD(a/2, b)$;

- If a and b are both odd, with $a \geq b$ then $GCD(a, b) = GCD((a-b)/2, b)$;

- If a and b are both odd, with $a \leq b$ then $GCD(a, b) = GCD(a, (b-a)/2)$.

Using all of these properties in a proper sequence yields an algorithm whose efficiency is similar to Euclid GCD. Depending on the exact architecture of the computer running the algorithms, it might be somewhat faster or slightly slower. An algorithmic description is given in Algorithm 2.4.

It is also useful to know how an extended version of Stein's algorithm can also compute Bézout's coefficients. Since Stein's algorithm uses subtractions and divisions by two, by following the same approach as in the extended Euclidean Algorithm 2.2, it is easy, given two integers A and B to obtain three integers α, β and e such that:

$$\alpha A - \beta B = 2^e GCD(A, B). \tag{2.7}$$

In order to obtain Bézout's coefficients, it suffices to modify the above coefficients in order to get rid of the unwanted factor 2e. After possibly exchanging the role of A and B, we may assume that $\alpha \geq 0$ and $\beta \geq 0$. Without loss of generality, we may also assume that A and B are not both even. Indeed, the Bézout's coefficients of $2A$ and $2B$ are the same as the Bézout's coefficients of A and B. Let us start with the simple case where A and B are both odd. In that case, let us define:

$$\alpha' = \alpha \left(\frac{B+1}{2} \right)^e \bmod B \text{ and} \tag{2.8}$$

$$\beta' = \beta \left(\frac{A+1}{2} \right)^e \bmod A.$$

It is clear that $\alpha' A - \beta' B = GCD(A,B) \pmod{AB}$. Moreover, since α' is reduced modulo B and β' modulo A, $\alpha' A - \beta' B$ lies in $] - AB, AB[$. As a consequence, $\alpha' A - \beta' B = GCD(A,B)$, not only modulo AB but also in the ring of integers, and (α', β') are Bézout's coefficients.

To treat the general case, it suffices to show how to obtain the coefficients for the pair $(A, 2B)$ from the coefficients (α', β') for (A, B), when A is odd. Remember that in that case, $GCD(A, 2B) = GCD(A, B)$. If β' is even, the coefficients are simply $(\alpha', \beta'/2)$. If β' is odd, they are $(\alpha' + B, (\beta' + A)/2)$. Note that there exist algorithms for computing GCDs with a better asymptotic complexity than Euclid's or Stein's algorithm.

2.2.2 Approximations using partial GCD computations

The extended Euclidean Algorithm 2.1 yields more information about the relation between its inputs X and Y than simply its outputs $GCD(X, Y)$, (α_y, β_y) and (α_x, β_x). Going back to the mathematical representation by sequences, we see that all intermediate values of the sequence (α, β) are interesting; more precisely, at each step $i \geq 2$ at the sequence, the fraction $-\beta_i/\alpha_i$ is a good approximation of z_0/z_1, i.e. X/Y in Algorithm 2.1. More precisely, we have:

$$\left| \frac{z_0}{z_1} + \frac{\beta_i}{\alpha_i} \right| = \left| \frac{z_i}{\alpha_i z_1} \right| \leq \frac{1}{|\alpha_i \alpha_{i+1}|}. \tag{2.9}$$

As a consequence, it can be very useful to use intermediate values of α_i, β_i and z_i to obtain a small linear combination of z_0 and z_1 whose coefficients are also small. Roughly, one can expect to achieve values of all coefficients around the square root of the initial numbers.

2.2.2.1 Application to real numbers

It is also possible to run the Euclidean algorithm on inputs which are rational of even real numbers instead of integers. When the inputs are two rational numbers, say a and b, the output is the smallest positive rational

---

Algorithm 2.4 Stein's binary greatest common divisor algorithm

---

Require: Input two integers X and Y

  Let $x \longleftarrow |X|$

  Let $y \longleftarrow |Y|$

  if $x = 0$ **then**

    Output y

  end if

  if $y = 0$ **then**

    Output x

  end if

  Let $p \longleftarrow 1$

  while x is even and y is even **do**

    Let $x \longleftarrow x/2$

    Let $y \longleftarrow y/2$

    Let $p \longleftarrow 2p$

  end while

  while x is even **do**

    $x \longleftarrow x/2$

  end while

  while y is even **do**

    $y \longleftarrow y/2$

  end while

  if $x > y$ **then**

    Exchange x and y

  end if

  while $x > 0$ **do**

    Let $r \longleftarrow (y - x)/2$

    while r is even **do**

      $r \longleftarrow r/2$

    end while

    if $r \geq x$ **then**

      Let $y \longleftarrow r$

    else

      Let $y \longleftarrow x$

      Let $x \longleftarrow r$

    end if

  end while

  Output py (GCD of X and Y)

---

c such that $a\mathbb{Z} + b\mathbb{Z} = c\mathbb{Z}$. If $a = n_a/d_a$ and $b = n_b/d_b$, we find that $c = GCD(n_a d_b, n_b d_a)/d_a d_b$. Thus, using the GCD algorithm with rationals is essentially the same as using it with integers.

When the inputs are real numbers x and y, matters are more complicated. Assuming temporarily that computations are performed with infinite precision, two cases arise, either x/y is a rational, in which case there exists a smallest positive real z such that $x\mathbb{Z} + y\mathbb{Z} = z\mathbb{Z}$, or x/y is irrational, in which case $x\mathbb{Z} + y\mathbb{Z}$ contains arbitrarily small positive numbers. In this second case, the extended Euclidean algorithm never terminates and can be used to produce good rational approximation of x/y. Note that running the algorithm on the pair of inputs $(x/y, 1)$ yields the same result as on the pair (x, y).

2.2.2.2 Alternative approaches for approximations

Approximations of fractions or of irrationals can also be obtained using different approaches. We briefly present these alternatives here.

A first possibility is the continued fraction algorithm, which can be seen as a rewriting of the extended Euclidean algorithm for inputs of the form $(x, 1)$. Given an arbitrary real x, we define two sequences, one formed of reals r and one formed of integers c. The two sequences are obtained from the following rules:

- The sequence r is initialized by setting $r_0 = x$.

- The two sequences are computed recursively, for all i:

$$c_i = \lfloor r_i \rfloor \quad \text{and} \tag{2.10}$$

$$r_{i+1} = \frac{1}{r_i - c_i}. \tag{2.11}$$

Note that, if $r_i = c_i$ for some value of i, we cannot define r_{i+1} and the two sequences have a finite number of terms.

The sequence c is called the continued fraction expansion of x. This sequence is finite if and only if x is rational.

The second possibility to obtain an approximation of a real by a rational fraction with small coefficients is to use lattice reduction in dimension two (see Chapter 10, Exercise 1).

2.3 Modular arithmetic

Modular arithmetic is a very important tool in cryptography. In particular, it is the keystone of RSA, Diffie-Hellman and elliptic curve cryptosystem. We first look at it from a mathematical perspective, before addressing the computational aspects. Choose an integer $N > 1$ and consider the set $N\mathbb{Z}$ obtained by multiplying all integers by N. This set is an ideal of $\mathbb{Z}$, i.e., it satisfies the following properties:

- For any x and y in $N\mathbb{Z}$, $x + y$ is in $N\mathbb{Z}$.

- For any x in $N\mathbb{Z}$ and any α in $\mathbb{Z}$, αx is in $N\mathbb{Z}$.

Thanks to these properties, it is possible to construct the quotient of $\mathbb{Z}$ by $N\mathbb{Z}$, denoted by $\mathbb{Z}/N\mathbb{Z}$. More precisely, this quotient is obtained by considering equivalence classes for the equivalence relation $\equiv$, defined by $x \equiv y$ if and only if $x - y \in N\mathbb{Z}$. As usual, the quotient $\mathbb{Z}/N\mathbb{Z}$ inherits the addition and multiplication from $\mathbb{Z}$. In order to represent $\mathbb{Z}/N\mathbb{Z}$, it is often convenient to choose a representative element in the interval $[0, N - 1]$. In some cases, an interval with positive and negative numbers such as $]-N/2, N/2]$ might be used instead of $[0, N - 1]$. Most of the time, equivalence classes of $\mathbb{Z}/N\mathbb{Z}$ are identified with their representative. For example, we speak of the element 2 in $\mathbb{Z}/7\mathbb{Z}$ instead of using the precise but cumbersome expression "the class of the integer 2 in $\mathbb{Z}/7\mathbb{Z}$."

In $\mathbb{Z}/N\mathbb{Z}$, any element has an opposite. For example, the opposite of the class represented by x can be represented by $N-x$. When considering inverses, the situation is more complicated; of course, 0 does not have an inverse, but neither does an element x of $\mathbb{Z}/N\mathbb{Z}$ when x and N are not coprime. Because of this, we need to make a clear distinction between $\mathbb{Z}/N\mathbb{Z}$ with N composite and $\mathbb{Z}/p\mathbb{Z}$ with p prime. Indeed, when p is prime, all non-zero elements have inverses and $\mathbb{Z}/p$ is a field. On the contrary, when N is composite $\mathbb{Z}/N\mathbb{Z}$ is a ring with non-trivial divisors of 0. For some basic operations, this difference is not too significant and the same algorithms are used in both cases. However, for some more advanced operations, efficient algorithms are only available for the $\mathbb{Z}/p\mathbb{Z}$ case. Conversely, the problem of finding alternate representations only arises for the composite case $\mathbb{Z}/N\mathbb{Z}$.

2.3.1   Basic algorithms for modular arithmetic

Most basic algorithms for modular arithmetic are extremely simple, this is in particular the case of modular addition, modular subtraction and modular multiplication. The only real difficulty is that these algorithms need to perform Euclidean divisions, which are not easy to implement for large integers. However, this difficulty is usually hidden. Indeed, for small integers, Euclidean division is directly available at the hardware level in most processors and can be accessed in C using the operators / and %. For large integers, the usual approach is to use one of the numerous existing libraries for manipulation of large integers, which include Euclidean division. The computation of modular inverses is more complicated than the above elementary operations; however, it entirely relies on the extended Euclidean algorithm which

---

Algorithm 2.5 Addition modulo N

Require: Input x and y in the range $[0, N-1]$

 Let $s \longleftarrow x + y$
 if $s \geq N$ **then**
  Let $s \longleftarrow s - N$
 end if
 Output s

Algorithm 2.6 Subtraction modulo N

Require: Input x and y in the range $[0, N-1]$

 if $y > x$ **then**
  Let $x \longleftarrow x + N$
 end if
 Let $s \longleftarrow x - y$
 Output s

---

we already addressed. For reference, we describe the modular operations as Algorithms 2.5, 2.6, 2.7 and 2.8.

Another computation is frequently encountered in modular arithmetic: exponentiation. Given an element x of $\mathbb{Z}/N\mathbb{Z}$ and an integer n we need to define and compute x^n. The mathematical definition is easily obtained by considering three cases:

- for $n = 0$, we define $x^0 = 1$;

- for $n > 0$, we define $x^n = x \cdot x \cdots x$, with n occurrences of x;

- for $n < 0$, we define $x^n = y \cdot y \cdots y$, with $-n$ occurrences of the inverse y of x.

To compute x^n efficiently, some care needs to be taken. In particular, the elementary approach that consists in computing x^n over the integer ring $\mathbb{Z}$ followed by a reduction modulo N does not work in the general case. Indeed, when n is even moderately large, x^n is a huge number which cannot even be stored, let alone computed on any existing computer. Since x^n in $\mathbb{Z}/N\mathbb{Z}$ is routinely computed in all implementations of the RSA cryptosystem, another approach is required.

This approach needs to reduce the number of multiplications that are performed ($|n|$ in the definition) and at the same time to prevent the growth of the intermediate values that appear in the computation. Since, $x^{-n} = (1/x)^n$, it suffices to deal with positive values of n. The most frequently encountered algorithm is the "square and multiply" method, it is based on the following equation:

$$x^n = ((((x^{n_\ell})^2 x^{n_{\ell-1}})^2 \cdots)^2 x^{n_1})^2 x^{n_0}, \tag{2.12}$$

Algorithm 2.7 Multiplication modulo N

Require: Input x and y
    Let $p \longleftarrow xy$
    Using Euclidean division, write $p = qN + r$
    Output r

Algorithm 2.8 Multiplicative inverse modulo N

Require: Input x
    Compute extended GCD g of x and N ($g = ax + bN$)
    if $g \neq 1$ **then**
        Output "x is non-invertible modulo N"
    else
        Output a
    end if

---

where $n = \sum_{i=0}^{\ell} n_i 2^i$ is the binary expansion of n. Taking care to perform modular reduction after each operation to avoid an uncontrolled growth of representative, this equation can be used as a basis for an efficient exponentiation algorithm in two different ways, reading it either from left to right or from right to left. This yields Algorithms 2.9 and 2.10.

Algorithm 2.9 Exponentiation in $\mathbb{Z}/N\mathbb{Z}$, left-to-right version

---

Require: Input N, $x \in \mathbb{Z}/N\mathbb{Z}$ and integer n on ℓ bits
    if $n < 0$ **then**
        Let $n \longleftarrow -n$
        Let $x \longleftarrow (1/x) \pmod{N}$
    end if
    Write n in binary $n = \sum_{i=0}^{\ell-1} n_i 2^i$
    Let $y \longleftarrow 1$
    for i from $\ell - 1$ downto 0 **do**
        Let $y \longleftarrow y^2 \pmod{N}$
        if $n_i = 1$ **then**
            Let $y \longleftarrow xy \pmod{N}$
        end if
    end for
    Output y

Algorithm 2.10 Exponentiation in $\mathbb{Z}/N\mathbb{Z}$, right-to-left version

---

Require: Input N, $x \in \mathbb{Z}/N\mathbb{Z}$ and integer n on ℓ bits
  if $n < 0$ **then**
    Let $n \longleftarrow -n$
    Let $x \longleftarrow (1/x) \pmod{N}$
  end if
  Write n in binary $n = \sum_{i=0}^{\ell-1} n_i 2^i$
  Let $y \longleftarrow 1$
  for i from 0 to $\ell - 1$ **do**
    if $n_i = 1$ **then**
      Let $y \longleftarrow xy \pmod{N}$
    end if
    Let $x \longleftarrow x^2 \pmod{N}$
  end for
  Output y

2.3.1.1 Invertible elements in $\mathbb{Z}/N\mathbb{Z}$

Another aspect of modular arithmetic is the set of elements in $\mathbb{Z}/N\mathbb{Z}$ that are invertible, i.e., coprime to N. This set forms a multiplicative group denoted by $(\mathbb{Z}/N\mathbb{Z})^*$. The cardinality of this multiplicative group is called the Euler's totient function or Euler's phi function and usually denoted by $\phi(N)$. If N is decomposed into primes as:

$$N = \prod_{i=1}^{n} p_i^{e_i}, \tag{2.13}$$

then

$$\phi(N) = \prod_{i=1}^{n} (p_i - 1) p_i^{e_i - 1}. \tag{2.14}$$

A well-known theorem from group theory is the following:

THEOREM 2.1
Let G be a multiplicative group, e the neutral element in G and $|G|$ the cardinality of G, then for any element x in G we have $x^{|G|} = e$.

PROOF See [Lan05, Chapter I, Proposition 4.1].   ☐

Applying this theorem to $(\mathbb{Z}/N\mathbb{Z})^*$, we find that for any x such that $0 < x < N$ and $GCD(x, N) = 1$, we have:

$$x^{\phi(N)} = 1 \pmod{N}. \tag{2.15}$$

2.3.2   Primality testing

Since primes are used in many cryptosystems, the ability to efficiently detect that numbers are primes is an essential task in cryptography. Note that, since the density of primes is quite high[1], knowing how to test primality is enough to generate prime numbers: it suffices to pick numbers at random and test their primality. After testing about $\ln(N)$ candidates in the range $[N/2, N]$, we obtain a prime number of the prescribed size.

Known primality tests are of two kinds: pseudo-primality tests and true primality tests. Pseudo-primality tests rely on a basic randomized subroutine which always returns `true` when its input is a prime and which returns `false` with noticeable probability when its input is a composite. After calling the basic subroutine a large enough number of times and obtaining `true` each time, we know that either the input is a prime or we have been unlucky to an unbelievable extreme. Note that we never know for sure when the input is prime. However, a single `false` output guarantees that the input is composite. The most frequently used pseudo-primality test is the Miller-Rabin primality test, it is based on the fact that for any prime p, if we write $p - 1 = 2^e q$, with q odd, and define for any integer $0 < z < p$, the sequence (z_i) for $0 \leq i \leq e$ by:

$$z_0 = z^q \pmod{p} \text{ and} \tag{2.16}$$
$$z_{i+1} = z_i^2 \pmod{p}.$$

Then, we either have $z_0 = 1$ or there exists an integer $i < e$ such that $z_i = -1$. Moreover, if N is an odd composite, defining $N - 1 = 2^e q$ and proceeding as above yields a negative result for at least $3/4$ of the integers $0 < z < N$.

From a practical point-of-view and especially for cryptography, pseudo-primality tests are enough. However, a true polynomial time algorithm for deciding primality was discovered in 2002 by Agrawal, Kayal and Saxena [AKS02]. It is known as the AKS primality test.

2.3.2.1   Computing square roots modulo primes

Computing modular square roots or, more generally, finding roots of modular polynomials (see Section 2.5.3) is one specific task which is easy to perform modulo primes and, in general, hard to perform modulo composites, unless their factorization is known. Depending on the value of the prime p, the available algorithms for computing square roots vary. The simplest case occurs with primes p such that $p = 3 \pmod 4$. Indeed, for such primes, if z is a quadratic residue (i.e., a square), then $z^{(p+1)/4}$ is a square root for z. This is easily seen by writing $z = u^2$ and remarking that:

$$z^{(p+1)/4} = u^{(p+1)/2} = u \cdot u^{(p-1)/2} = \pm u. \tag{2.17}$$

---

[1]For numbers around N, the fraction of primes is essentially $1/\ln(N)$.

Indeed, $u^{(p-1)} = 1$ and consequently $u^{(p-1)/2} = \pm 1$. Note that, when $p = 3$ (mod 4), then -1 is a quadratic non-residue. Thus, exactly one of u and $-u$ is a square. In addition, $z^{(p+1)/4}$ is the only square root of z which is itself a square.

When $p = 1$ (mod 4), matters become more complicated, especially if a large power of 2 divides $p - 1$. In that case, we need to write $p - 1 = 2^e q$ with q odd. The method of choice to compute square roots in this case is Shanks-Tonelli algorithm. This method is based on the remark that for any quadratic residue $z = u^2$ (mod p), the value $z^{(q+1)/2}$ is "almost" a square root for z. More precisely, if we let $\theta = z^{(q+1)/2}/u$, then $\theta^{2^e} = 1$. Indeed:

$$\theta^{2^e} = \left(\frac{u^{q+1}}{u}\right)^{2^e} = u^{2^e q} = u^{p-1} = 1 \quad (\text{mod } p). \tag{2.18}$$

As a consequence, to obtain a correct square root, it suffices to multiply $z^{(q+1)/2}$ by an adequate 2^e-th root of unity in $\mathbb{F}_p$. If we are given a primitive 2^e-th root of unity, this can be done efficiently using a special case of Pohlig-Hellman method for discrete logarithm that is described in Chapter 6. This is precisely described as Algorithm 2.11.

2.3.2.2 Jacobi symbols

In Shanks-Tonelli square root algorithm, we simply assumed that the input is a quadratic residue. However, it is useful to have an algorithm for testing this property. Modulo a prime p, testing whether an element z is a quadratic residue is easy, it suffices to check that $z^{(p-1)/2}$ is equal to 1 modulo p. However, there exists a more efficient way to compute this information. Given a prime p and an integer z, we define the Legendre symbol $\left(\frac{z}{p}\right)$ to be 0 if $z = 0$ (mod p), 1 if z is a quadratic residue and -1 if z is a quadratic non-residue. Given two odd primes p and q, the values $\left(\frac{q}{p}\right)$ and $\left(\frac{p}{q}\right)$ are related by the law of quadratic reciprocity which asserts that:

$$\left(\frac{q}{p}\right) \cdot \left(\frac{p}{q}\right) = (-1)^{(p-1)(q-1)/4}. \tag{2.19}$$

Another noteworthy property of the Legendre symbol is its multiplicativity:

$$\left(\frac{ab}{p}\right) = \left(\frac{a}{p}\right) \cdot \left(\frac{b}{p}\right). \tag{2.20}$$

In order to compute the Legendre symbol, it is first generalized into the Jacobi symbol $\left(\frac{a}{b}\right)$, defined for arbitrary non-negative integers a and b as:

$$\left(\frac{a}{b}\right) = \prod_i \left(\frac{a}{p_i}\right)^{e_i}, \tag{2.21}$$

Algorithm 2.11 Shanks-Tonelli algorithm for square roots in $\mathbb{F}_p$

Require: Input $p = 2^e q + 1$, p prime, q odd, z a quadratic residue in $\mathbb{F}_p$
  repeat
    Pick a random g in $\mathbb{F}_p^*$
    Let $g \longleftarrow g^q$
    Let $h \longleftarrow g$, $i \longleftarrow 0$
    while $h \neq 1$ **do**
      Let $h \longleftarrow h^2$, $i \longleftarrow i + 1$
    end while
  until $i = e$ {Here g is a primitive 2^e-th root of unity}
  Let $h \longleftarrow z^{(q+1)/2}$
  Let $\theta \longleftarrow h^2/z$
  while $\theta \neq 1$ **do**
    Let $k \longleftarrow \theta^2$, $i \longleftarrow 1$
    while $k \neq 1$ **do**
      Let $k \longleftarrow k^2$, $i \longleftarrow i + 1$
    end while
    Let $k \longleftarrow g$
    for j from 1 to $e - i - 1$ **do**
      Let $k \longleftarrow k^2$
    end for
    Let $h \longleftarrow hk$
    Let $\theta \longleftarrow \theta k^2$
  end while
  Output h

---

where $\prod_i p_i{}^{e_i}$ is the decomposition of b into primes. With this generalized definition, the law of quadratic reciprocity still holds for arbitrary odd integers a and b. Using this generalization, it is possible to write down Algorithm 2.12 in order to compute Jacobi symbols.

2.3.3 Specific aspects of the composite case

In the composite case, using knowledge of the factorization of N, it is possible to represent $\mathbb{Z}/N\mathbb{Z}$ in several different ways. The first available representation simply uses a representative in the interval $[0, N-1]$ (or $]-N/2, N/2]$). The other representations correspond to factorizations of N in a product of pairwise coprime factors. Mathematically, they come from the Chinese remainder theorem:

THEOREM 2.2
Let N be a composite integer and let $N = N_1 N_2 \cdots N_\ell$, be a factorization of N into factors such that all pairs (N_i, N_j) are coprime. Then the two rings $\mathbb{Z}/N\mathbb{Z}$ and $\mathbb{Z}/N_1\mathbb{Z} \times \mathbb{Z}/N_2\mathbb{Z} \times \cdots \mathbb{Z}/N_\ell\mathbb{Z}$ are isomorphic. Moreover, this isomorphism can be explicitly and efficiently computed.

PROOF Working by induction on the number of factors, it suffices to address the basic case of a two-factor decomposition $N = N_1 N_2$ with N_1 and N_2 coprime.

In one direction, going from $\mathbb{Z}/N\mathbb{Z}$ to $\mathbb{Z}/N_1\mathbb{Z} \times \mathbb{Z}/N_2\mathbb{Z}$ is easy, it suffices to send x to $(x \bmod N_1, x \bmod N_2)$. Clearly, two different representatives for the same equivalence class in $\mathbb{Z}/N\mathbb{Z}$ are sent to the same element in $\mathbb{Z}/N_1\mathbb{Z} \times \mathbb{Z}/N_2\mathbb{Z}$, since $N = 0 \pmod{N_1}$ and $N = 0 \pmod{N_2}$. Moreover, it is clear that mapping elements in this way is compatible with the ring operations. In the reverse direction, since N_1 and N_2 are coprime, we first use the extended Euclidean algorithm and write:

$$\alpha_1 N_1 + \alpha_2 N_2 = 1. \qquad (2.22)$$

We can remark that $\alpha_1 N_1$ is congruent to 0 modulo N_1 and congruent to 1 modulo N_2. Conversely, $\alpha_2 N_2$ is congruent to 1 modulo N_1 and to 0 modulo N_2. As a consequence, to any pair (x_1, x_2) in $\mathbb{Z}/N_1\mathbb{Z} \times \mathbb{Z}/N_2\mathbb{Z}$, we can associate the element $x = x_1 \alpha_2 N_2 + x_2 \alpha_1 N_1$ in $\mathbb{Z}/N\mathbb{Z}$; this element x satisfies:

$$x \equiv x_1 \pmod{N_1} \text{ and} \qquad (2.23)$$
$$x \equiv x_2 \pmod{N_2}.$$

Instead of using induction, it is also possible to derive direct coefficients to make the isomorphism between $\mathbb{Z}/N\mathbb{Z}$ and $\mathbb{Z}/N_1\mathbb{Z} \times \mathbb{Z}/N_2\mathbb{Z} \times \cdots \mathbb{Z}/N_\ell\mathbb{Z}$ explicit. For any i in $[1, \ell]$, let $M_i = \prod_{j \neq i} N_j$ and remarking that N_i and M_i

---

Algorithm 2.12 Computation of Jacobi symbols

Require: Input $a \geq 0$ and $b \geq 0$

  if $b = 0$ **then**
    if $a = 1$ **then**
      Output 1
    else
      Output 0
    end if
  end if
  if a and b are even **then**
    Output 0
  end if
  Let $i \longleftarrow 0$
  while b is even **do**
    Let $b \longleftarrow b/2$ and $i \longleftarrow i + 1$
  end while
  if i is even **then**
    Let $S \longleftarrow 1$
  else
    Let $S \longleftarrow 1$ if a is either 1 or 7 modulo 8
    Let $S \longleftarrow -1$ if a is either 3 or 5 modulo 8
  end if
  while $a \neq 0$ **do**
    Let $i \longleftarrow 0$
    while a is even **do**
      Let $a \longleftarrow a/2$ and $i \longleftarrow i + 1$
    end while
    if i is odd **then**
      Let $S \longleftarrow -S$ if b is either 3 or 5 modulo 8
    end if
    if $(a-1)/2$ and $(b-1)/2$ are both odd **then**
      Let $S \longleftarrow -S$
    end if
    Let $c \longleftarrow a$
    Let $a \longleftarrow b \bmod c$
    Let $b \longleftarrow c$
  end while
  if $b = 1$ **then**
    Output S
  else
    Output 0
  end if

---

are coprime write:

$$\beta_i M_i + \gamma_i N_i = 1. \qquad (2.24)$$

Since $\beta_i M_i$ is congruent to 1 modulo N_i and to 0 modulo each N_j for $j \neq i$, the isomorphism between x and $(x_1, \cdot, x_\ell)$ is given by reduction modulo each N_i in the forward direction and by:

$$x = \sum_{i=1}^{\ell} x_i \beta_i M_i \quad (\text{mod } N) \qquad (2.25)$$

in the reverse direction.   ⬜

We leave it to the reader to express the Chinese remainder theorem in effective algorithmic form.

2.3.3.1 Square roots and factoring from $\phi(N)$

Let N be a composite number and, for simplicity, assume that $N = pq$ is a product of two primes[2]. One key observation is that, modulo N, 1 has four square roots, 1, -1 and the two numbers obtained using the Chinese remainder theorem on 1 mod p and -1 mod q or on -1 mod p and 1 mod q. We call 1 and -1 the trivial square roots of 1. Let z be a non-trivial square root of 1, then z can be used to factor N. Indeed, $z - 1$ is a multiple of p or q, but not both. Thus, the GCD of N and $z - 1$ is a proper divisor of N.

This observation can be used in several ways. A first application is to show that any algorithm for computing square roots modulo N can be used to factor N. Note that it does not suffice to call this algorithm on the value 1 because it may then return 1. Instead, we use a self-randomizing property of modular square roots, choose a random value r and ask for a square root s of r^2 (mod N), while keeping r secret. After this operation, r/s is a random square root of 1. Thus, it is non-trivial with probability 1/2 (or more, if N has more factors) and leaks the factorization of N.

Another application is to show that the knowledge of $\phi(N)$ is also enough to factor N. Indeed, writing $\phi(N) = 2^e I$, we may see that for any number x in $\mathbb{Z}/N\mathbb{Z}$, letting $y = x^I$ and squaring y repeatedly, we obtain 1. Thus, somewhere in the path between y and 1, we encounter a square root of 1. If this square root is trivial or in the rare case where $y = 1$, we simply choose another value for x. In truth, the knowledge of $\phi(N)$ is not really needed to use this argument: it is also possible to factor N using the same method when a multiple of $\phi(N)$ is known. Also note that in the two-factor case, there is an easy deterministic method to factor N when $\phi(N)$ is known. Indeed, in that case, we know that $pq = N$ and $p + q = N + 1 - \phi(N)$. As a consequence, p and q are the roots of the quadratic equation $x^2 - (N + 1 - \phi(N))x + N = 0$.

---

[2]If N has more factors, the same kind of argument applies. The only difference is that there are more non-trivial square roots of 1.

2.4   Univariate polynomials and rational fractions

Polynomials in a single variable are also very frequently encountered in cryptography. Given a ring $\mathcal{R}$, we define the ring $\mathcal{R}[X]$ of polynomials in the variable X as follows:

- A polynomial P is a formal sum $P(X) = \sum_{i=0}^{N_p} p_i X^i$, where p_i is a sequence of elements of $\mathcal{R}$. The value p_i is called the coefficient of X^i in P. For convenience, it is useful to define $p_i = 0$ for all $i > N_p$, to omit N_p and write $P(X) = \sum_{i=0}^{\infty} p_i X^i$. However, the reader should keep in mind that P only contains finitely many non-zero coefficients.

- The degree of a polynomial P is the largest integer $d(P)$ such that the coefficient of $X^{d(P)}$ is non-zero. For the zero polynomial 0 corresponding to the null sequence, we define $d(P) = -\infty$.

- The sum of two polynomials $P(X) = \sum_{i=0}^{\infty} p_i X^i$ and $Q(X) = \sum_{i=0}^{\infty} q_i X^i$ is defined as $(P + Q)(X) = \sum_{i=0}^{\infty}(p_i + q_i)X^i$. Each polynomial has an opposite, $(-P)(X) = \sum_{i=0}^{\infty}(-p_i)X^i$.

- The degree of a sum is at most the largest among the degrees of the summands P and Q, i.e., $d(P+Q) \leq \max((d(P), d(Q))$. When, the degrees of P and Q are different, equality holds: $d(P+Q) = \max((d(P), d(Q))$.

- The product of two polynomials $P(X) = \sum_{i=0}^{\infty} p_i X^i$ and $Q(X) = \sum_{i=0}^{\infty} q_i X^i$ is defined as:

$$(PQ)(X) = \sum_{i=0}^{\infty} \left(\sum_{j=0}^{i} p_j q_{i-j} \right) X^i. \tag{2.26}$$

We leave it to the reader to check that the sum in the above equation is indeed finite and that $\mathcal{R}[X]$ is a ring. Note that no non-constant polynomial has an inverse.

- The degree of a product is the sum of the degrees $d(PQ) = d(P) + d(Q)$ under the convention that addition with $-\infty$ yields $-\infty$.

- A polynomial is called unitary when its coefficient of highest degree is 1. In other words, P is unitary when $P(X) = X^d + \sum_{i=0}^{d-1} p_i X^i$. When $\mathcal{R}$ is a field, any non-zero polynomial can be transformed into a unitary polynomial by multiplication by the inverse of its highest degree coefficient.

- Given a polynomial P and a unitary polynomial Q, it is possible to write $P = AQ + R$, with R a polynomial of smaller degree than Q:

$d(R) < d(Q)$. As in the integer case, A and R are called the quotient and remainder of the (Euclidean) division of P by Q. When $R = 0$, we say that Q divides P.

- When $\mathcal{R}$ is an entire ring, a unitary polynomial P is called reducible when there exists two non-constant polynomials Q and R, i.e., $d(Q) \neq 0$ and $d(R) \neq 0$ such that $P = QR$. Otherwise, P is called irreducible.

  When P is reducible, it is possible to choose Q and R unitary in the expression $P = QR$.

- Over an entire ring, any unitary polynomial P can be written as a product of irreducible polynomials, i.e.:

$$P = \prod_{i=1}^{t} P_i^{e_i}. \tag{2.27}$$

  Moreover, up to the order of factors, this decomposition is unique.

Note that when non-entire rings are involved, polynomials may factor in very strange ways. To take a simple example, over $\mathbb{Z}/15\mathbb{Z}$, one may see that $(3x^2 - 1) \cdot (5x^2 + 2) = x^2 - 2 \bmod 15$. Furthermore, $x^2 - 2$ is irreducible both modulo 3 and modulo 5.

2.4.1 Greatest common divisors and modular arithmetic

Thanks to the existence of a Euclidean division process for polynomials, it is possible to use Euclid's algorithm to compute the greatest common divisor of two polynomials over a field (or an entire ring). It is possible to directly use Algorithm 2.1 without any change, simply replacing numbers by polynomials where appropriate. Similarly, the extended Algorithm 2.2 can also be used directly for polynomials. However, it is interesting to consider the analog of the results of Section 2.2.2 and look at the degrees of intermediate polynomials that appear during the extended GCD algorithm. Instead of having a multiplicative relation relating the product of the current coefficients and of the current reduced value to the initial values of the algorithm as in the integer case, we obtain a similar additive relation involving the degrees of the corresponding polynomials. In particular, we can expect along the GCD computation a partial relation where all the polynomials have half degree, when compared to the algorithm's input.

A variation of Stein's binary GCD algorithm can also be used for polynomials. Instead of giving a special role to the prime 2, this variation gives this role to the irreducible polynomial x. It is described as Algorithm 2.13. An extended version can also be obtained.

Algorithm 2.13 Stein's greatest common divisor algorithm for polynomials

---

Require: Input two polynomials $A(x)$ and $B(x)$

  Let $a(x) \longleftarrow A(x)$

  Let $b(x) \longleftarrow B(x)$

  if $a(x) = 0$ **then**

    Output $b(x)$

  end if

  if $b(x) = 0$ **then**

    Output $a(x)$

  end if

  Let $p \longleftarrow 0$

  while $a(0) = 0$ and $b(0) = 0$ **do**

    Let $a(x) \longleftarrow a(x)/x$

    Let $b(x) \longleftarrow b(x)/2$

    Let $p \longleftarrow p + 1$

  end while

  while $a(0) = 0$ **do**

    $a(x) \longleftarrow a(x)/x$

  end while

  while $b(0) = 0$ **do**

    $b(x) \longleftarrow b(x)/x$

  end while

  if $\deg a > \deg b$ **then**

    Exchange $a(x)$ and $b(x)$

  end if

  while $a(x) \neq 0$ **do**

    Let $c(x) \longleftarrow (b(0)a(x) - a(0)b(x))/x$

    while $c(0) = 0$ **do**

      $c(x) \longleftarrow c(x)/x$

    end while

    if $\deg c \geq \deg a$ **then**

      Let $b(x) \longleftarrow c(x)$

    else

      Let $b(x) \longleftarrow a(x)$

      Let $a(x) \longleftarrow c(x)$

    end if

  end while

  Output $x^p b(x)$ (GCD of $A(x)$ and $B(x)$)

2.4.2   Derivative of polynomials

A very useful operation when working with polynomials is the computation of derivatives. We quickly recall the basic definition and properties of derivatives. To any polynomial f in a polynomial ring $\mathcal{R}[X]$ we associate the polynomial $D(f)$ or $D(f)$, also denoted f', defined as follows: If $f(x) = \sum_{i=0}^{\infty} f_i x^i$, we let:

$$D(f)(x) = \sum_{i-1}^{\infty} i f_i x^{i-1}. \tag{2.28}$$

Note that since f only has finitely many non-zero coefficients, so does $D(f)$. Thus $D(f)$ is a polynomial. Moreover, for any pair of polynomials f and g, we have the following properties:

$$D(f+g) = D(f) + D(g) \quad \text{and} \tag{2.29}$$
$$D(fg) = D(f)g + fD(g). \tag{2.30}$$

The Property (2.29) is clear from the definition. Moreover, thanks to this property, it suffices to prove Property (2.30) in the special case where f is a monomial, say $f(x) = ax^k$. In this special case, we see that:

$$D(fg) = D(ax^k g(x)) = \sum_{i=k}^{\deg(g)+k} aig_{i-k}x^{i-1} \tag{2.31}$$

$$= \sum_{i=0}^{\deg(g)} a(i+k)g_i x^{i+k-1}$$

$$= ax^k \sum_{i=1}^{\deg(g)} ig_i x^{i-1} + kax^{k-1} \sum_{i=0}^{\deg(g)} g_i x^i$$

$$= ax^k D(g)(x) + kax^{k-1}g$$

$$= fD(g) + D(f)g.$$

By definition, the derivative of any constant polynomial is zero. The converse is almost true. More precisely, see [Lan05, Chapter IV, Proposition 1.12], for any polynomial f of degree at least 1 over a finite field $\mathbb{F}_q$ of characteristic p, i.e., $q = p^n$ (see below), either $Df \neq 0$ or there exists another polynomial g such that $f(x) = g(x^p) = g(x)^p$ over $\mathbb{F}_q[x]$.

2.5   Finite fields

Finite fields are often considered in cryptography and it is useful to see how they can be constructed. We have already seen, in the context of modular arithmetic, that $\mathbb{Z}/p\mathbb{Z}$ is a field, if and only if, p is prime. Using a construction similar to modular arithmetic and based on polynomials, we are now going to construct some additional examples of finite fields.

2.5.1 The general case

In this section, we let p be a prime and consider polynomials over $\mathbb{Z}/p\mathbb{Z}$. We also choose an irreducible polynomial $I(x)$ of degree n in this polynomial ring. Using $I(x)$, we can define an equivalence relation between polynomials that says that $f \equiv g$ if and only if $f(x) - g(x)$ is a multiple of $I(x)$. Equivalently, this means that f and g have the same remainder in the Euclidean division by $I(x)$. Taking the quotient of the ring of polynomials by this equivalence relation, we obtain a ring containing p^n elements. Indeed, in the equivalence class of an arbitrary polynomial f, there is a polynomial of degree $< n$: the remainder in the Euclidean division of f by I. Moreover, no two polynomials of degree $< n$ may belong to the same class. Since a polynomial of degree $< n$ is completely described by n coefficients in $\mathbb{Z}/p\mathbb{Z}$, the quotient ring contains p^n classes. To prove that this ring is a field, we need to show that any polynomial f in $\mathbb{Z}/p\mathbb{Z}[x]$ is either a multiple of I or prime to I. Since I is irreducible, this holds by definition.

An important theorem stated in [Lan05, Chapter V, Theorem 5.1] says that all finite fields are either of the form $\mathbb{Z}/p\mathbb{Z}$ or obtained by the above construction. Moreover, for each possible value of p^n there is a unique field with p^n elements. A field with p^n elements, where p is a prime, is said to be of characteristic p. Alternatively, the characteristic can be defined as the order of the unit element 1 in the field $\mathbb{F}_{p^n}$ viewed as an additive group. Note that, two different irreducible polynomials of degree n over $\mathbb{Z}/p\mathbb{Z}$ give two different representations of the same finite field. Thanks to this remark, when using this construction directly to represent elements in finite fields and compute with them, it is possible to make a convenient choice for I in order to speed up the computations. One frequent option is to use $I(x) = x^n + i(x)$, where $i(x)$ is a polynomial of low degree containing a small number of coefficients. This specific choice allows to speed-up reduction modulo I. It also permits to multiply finite field elements faster.

The representation of a finite field $\mathbb{F}_q$ by a quotient of $\mathbb{F}_p[x]$ by an irreducible $I(x)$, is usually called a polynomial or a power basis for $\mathbb{F}_q$.

The Frobenius map In a finite field $\mathbb{F}_{p^n}$, the Frobenius map is the function ϕ which sends any element x to x^p. This map satisfies some essential properties:

- For all x and y in $\mathbb{F}_{p^n}$, $\phi(xy) = \phi(x)\phi(y)$.

- For all x and y in $\mathbb{F}_{p^n}$, $\phi(x + y) = \phi(x) + \phi(y)$.

The first property is a simple consequence of the commutativity of multiplication. The second property comes from the fact that p divides all coefficients in the expansion of $(x + y)^p$, except the coefficients of x^p and y^p.

The Frobenius map is a very useful tool when working with finite fields. We discuss it a bit further in Chapter 14, showing that it is also very important for elliptic curves. For now, let us give a simple example of its power. For any polynomial $f(X) = \sum_{i=0}^{d_f} f_i X^i$, we define the Frobenius of f as the polynomial $\phi(f)(X) = \sum_{i=0}^{d_f} \phi(f_i) X^i$. The reader can easily check that for any element x in $\mathbb{F}_{p^n}$ we have $\phi(f(x)) = \phi(f)(\phi(x))$.

The Frobenius map can also be used to characterize the subfields of $\mathbb{F}_{p^n}$. For any divisor m of n, let us denote by ϕ^m the composition of the Frobenius map with itself m times. Then, the subset of elements x in $\mathbb{F}_{p^n}$ such that $\phi^m(x) = x$ is the finite field $\mathbb{F}_{p^m}$. When n is prime, the only proper subfield of $\mathbb{F}_{p^n}$ is $\mathbb{F}_p$ itself.

Multiple extensions. In some cases, it can be useful to represent $\mathbb{F}_{p^{n_1 n_2}}$ as an extension of degree n_2 of $\mathbb{F}_{p^{n_1}}$. This can be done using an irreducible polynomial I of degree n_2 with coefficients in $\mathbb{F}_{p^{n_1}}$. In the special case where n_1 and n_2 are coprime, this can even be done with an irreducible polynomial I with coefficients in $\mathbb{F}_p$.

Normal bases. We saw that by choosing sparse irreducible polynomials, it is possible to obtain efficient representations of finite fields by a polynomial basis that allows faster multiplication. Another frequent representation is the use of normal bases which represent elements of $\mathbb{F}_{2^n}$ as sums of α, α^2, α^4, ..., $\alpha^{2^{n-1}}$, for a well-chosen value of α. With normal bases, the Frobenius map can be applied very easily, by simply rotating the coefficients. However, multiplications become more costly. For special finite fields, it is even possible to use optimal normal bases that allow faster computation of multiplication and Frobenius.

Recently, a new representation was proposed in [CL09] by Couveignes and Lercier that allows these fast operations for all finite fields. This representation is more complex that normal bases and makes use of elliptic curves over the base field to represent extension fields.

2.5.2 The special case of $\mathbb{F}_{2^n}$

Finite fields of characteristic 2 are frequently encountered in cryptographic algorithms. As a consequence, fast implementations of arithmetic operations in $\mathbb{F}_{2^n}$ have been developed for this special case, both in software and in hardware. One frequently encountered method makes use of linear feedback shift registers (LFSR). Since LFSR are also used as a basis for many stream ciphers, we now describe them in more details.

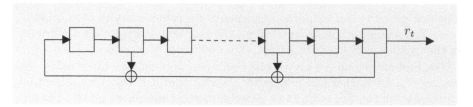

Figure 2.1: Ordinary LFSR

2.5.2.1   Representation by LFSRs

A LFSR on n bits is formed of n individual registers each able to store a single bit, say $r_0, r_1, \ldots, r_{n-1}$. The value contained in each register changes with time. For LFSRs, time is discretized and follows a clock signal. We assume that the clock is initially set to 0 and goes up by increments of 1. We denote the value contained in r_i at time t by $r_i^{(t)}$. There exist two different kinds of LFSRs, Galois LFSR and ordinary LFSR, which differ by the evolution of the register from time t to time $t+1$. They are represented in Figures 2.1 and 2.2.

From time t to time $t+1$, a Galois LSFR evolves according to the following rules:

- For all $0 \leq i < n - 1 : r_{i+1}^{(t+1)} = r_i^{(t)} \oplus c_{i+1}r_{n-1}^{(t)}$, where each c_i is a constant 0 or 1.

- Moreover, $r_0^{(t+1)} = r_{n-1}^{(t)}$.

The time advance of a Galois LFSR can be interpreted as multiplication as a polynomial by X modulo a polynomial of $\mathbb{F}_2[X]$. Indeed, if we let $I(X) = X^n \oplus \bigoplus_{i=1}^{n-1} c_iX^i \oplus 1$ and let $f^{(t)}$ for every value of t denote the polynomial $f^{(t)}(X) = \bigoplus_{i=0}^{n-1} r_i^{(t)}X^i$. We easily see that when $r_{n-1}^{(t)} = 0$, then $f^{(t+1)}(X) = Xf^{(t)}(X)$. Similarly, when $r_{n-1}^{(t)} = 1$ then $f^{(t+1)}(X) = Xf^{(t)}(X) \oplus I(X)$. As a consequence, assuming that I is irreducible, the simple rule for advancing the corresponding Galois LFSR corresponds to multiplication by α, a root of I in $\mathbb{F}_{2^n}$. This correspondence holds under the convention that the registers (r_i) of a Galois LFSR encode the finite field element $\bigoplus_{i=0}^{n-1} r_i\alpha^i$. Clearly, the sum of two finite field elements can be obtained by a simple XOR of the registers. Finally, multiplication by an arbitrary element can be obtained by a sequence of multiplications by α and XOR operations, similar to the exponentiation Algorithm 2.9 or 2.10. A simple implementation of $\mathbb{F}_{2^{32}}$ using a Galois LFSR in C code is given as Program 2.1.

To understand the other kind of LFSR called ordinary LFSR when opposed to Galois LFSR or simply LFSR in many contexts, it is useful to focus on the sequence of bits $s_t = r_{n-1}^{(t)}$ obtained by considering only the high order bit

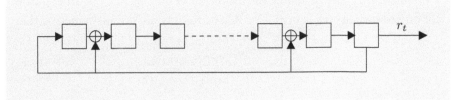

Figure 2.2: Galois LFSR

Program 2.1 Representation of $\mathbb{F}_{2^{32}}$ with a Galois LFSR

```
#include <stdio.h>
#include <stdlib.h>

/* GF(2^32) using x^32+x^7+x^3+x^2+1 -> 2^7+2^3+2^2+1=141*/
#define IRRED 141

typedef unsigned int state;

state Mul_by_alpha(state val)
{
  state res;

  res=(val<<1); if (val&0x80000000) res^=IRRED;
  return(res);
}

state Add(state val1, state val2)
{
  return(val1^val2);
}

state Mul(state val1, state val2)
{
  state res=0;

  while(val2!=0) {
    if (val2&1) res=Add(res,val1);
    val1=Mul_by_alpha(val1);
    val2>>=1;
  }
}
```

---

of a Galois LFSR. Let us follow the evolution of one bit $r_{n-1}^{(t)}$ as time passes. After one round, the bit is in $r_0^{(t+1)}$. After i rounds, we see by induction that:

$$r_{i-1}^{(t+i)} = r_{n-1}^{(t)} \oplus \bigoplus_{j=1}^{i-1} c_j r_{n-1}^{t+j}. \tag{2.32}$$

Finally, with the additional convention that $c_0 = 1$, we find that:

$$s_{t+n} = \bigoplus_{j=0}^{n-1} c_j s_{t+j}. \tag{2.33}$$

Ordinary LSFRs directly compute the sequence s from this recurrence relation. Once again we need a register able to store n bits. For simplicity, we reuse the same notation $r_i^{(t)}$. With ordinary LFSRs, with well-chosen initial values of the registers, we can achieve the relation $r_i^{(t)} = s_{t-i}$, for all acceptable values of i and t, i.e., $0 \leq i < n$ and $t - i \geq 0$. The evolution rules for ordinary LFSRs from time t to time $t + 1$ are the following:

- For all $0 \leq i < n - 1 : r_{i+1}^{(t+1)} = r_i^{(t)}$.

- Moreover, $r_0^{(t+1)}$ is obtained as the XOR of a subset of the values $r_i^{(t)}$. More precisely, we have:

$$r_0^{(t+1)} = \bigoplus_{i=0}^{n-1} c_i r_i^{(t)}, \tag{2.34}$$

where each c_i is the same constant as before.

The above discussion shows that extracting a single bit from a Galois LFSR or for the corresponding ordinary LFSR yields the same sequence. Both LFSRs are thus said to be equivalent. One very important property about LFSRs is that they produce periodic sequences whose lengths are equal to the multiplicative order of the root α of I in the finite field $\mathbb{F}_{2^n}$. Note that the all-zeros sequence can also be produced by any LFSR and is an exception to this rule since its periodicity is 1. If α is a generator of the multiplicative group $\mathbb{F}_{2^n}^*$, its order is $2^n - 1$ and the length of the sequence is maximal. When this condition is satisfied, the irreducible polynomial I is said to be primitive.

Letting $N = 2^n - 1$, to check whether α has order N, we need to know the factorization of N. Assuming that $N = \prod_{i=1}^{t} p_i^{e_i}$, we can check that no proper divisor of N is the order of α simply by checking that for all $1 \leq i \leq t$:

$$\alpha^{N/p_i} \neq 1. \tag{2.35}$$

2.5.2.1.1 Berlekamp-Massey algorithm Knowing that LFSR can generate long periodic sequences by using simple linear recurrences, it is also interesting to consider the dual question. Given a sequence of bits, can we determine the short linear recurrence expression that can generate this sequence?

The Berlekamp-Massey algorithm is a very efficient method to solve this question. It can be applied to find an optimal linear recurrence that generates a given sequence over any field $\mathbb{K}$. In order to describe this algorithm, it is useful to introduce a few notations. Given a sequence T of elements of the field $\mathbb{K}$, a nice way to describe the shortest linear recurrence is to introduce the notion of minimal polynomial of a sequence.

First, let us define the evaluation of a polynomial f of $\mathbb{K}[X]$ of degree d_f on a sequence T of elements of $\mathbb{K}$. Assuming that $f(X) = \sum_{i=0}^{d_f} \alpha_i X^i$, we define the sequence $S = f(T)$ by the following formula:

$$S_i = \sum_{j=0}^{d_f} \alpha_j T_{i+j}. \tag{2.36}$$

If T is an infinite sequence, S is also infinite. Otherwise, S contains d_f fewer elements than T. We say that f annihilates T if and only if $f(T)$ is the all zero sequence. We define I_T to be the set of all polynomials that annihilate T. This set is an ideal of $\mathbb{K}[X]$. When I_T is different from the zero ideal, we can find a generator f_T of I_T (unique up to multiplication by a constant of $\mathbb{K}^*$) is called the minimal polynomial of S. Note that f_T is the polynomial of smallest degree that annihilates T.

In this section, we assume for simplicity that T is an infinite sequence of numbers over a field $\mathbb{K}$, with minimal polynomial f_T of degree d. By definition, f_T annihilates T, thus $f_T(T) = 0$. It is useful to express the sequence T by forming series $T(x) = \sum_{i=0}^{\infty} T_i x^i$ in $\mathbb{K}[[x]]$. Matching the definition of multiplication for series and polynomials and the definition of evaluation of f_T at T, we see that i-th term of $f_T(T)$ is the coefficient of x^{d+i} in $\widetilde{f_T}(x)T(x)$, where $\widetilde{f_T}$ is the reciprocal polynomial of f_T, i.e., when $f_T = \sum_{i=0}^{d} f_t^{(i)} x^i$ we write $\widetilde{f_T} = \sum_{i=0}^{d} f_t^{(i)} x^{d-i}$. Since, $f_T(T)$ is the zero sequence, we see that $\widetilde{f_T}(x)T(x)$ is a polynomial $g_T(x)$ of degree at most $d-1$. In other words:

$$T(x) = \frac{g_T(x)}{\widetilde{f_T}(x)}, \tag{2.37}$$

using division according to increasing powers of x, i.e., considering each polynomial as a Taylor series. Note that we need to take some care with Equation (2.37). To see why, let us construct a sequence T, starting with d arbitrary values from T_0 to T_{d-1} and constant after that point. The minimal polynomial of such a sequence is $x^d - x^{d-1}$. Now the reciprocal polynomial is simply $x - 1$. The problem is that this polynomial no longer contains information about the degree d. Thankfully, in this case, the degree of g_T is exactly $d-1$ and this allows us to recover d. In order to compute $\widetilde{f_T}$ and g_T, it suffices to consider the first D terms of T and to decompose $T(x)$ as a sum $T(x) = T_D(x) + x^D \mathcal{T}_D(x)$. Thus:

$$T_D(x) = \frac{g_T(x)}{\widetilde{f_T}(x)} - x^D \mathcal{T}_D(x). \tag{2.38}$$

Multiplying by $\widetilde{f_T}(x)$ and matching terms of the same degrees, we see that:

$$\widetilde{f_T}(x)T_D(x) + x^D h_T(x) = g_T(x), \tag{2.39}$$

where h_T is a polynomial of degree at most $d-1$. When $D \geq 2d$, this expression can be obtained through a partial run of the extended Euclidean algorithm on the polynomials T_D and x^D and this is at the core of the Berlekamp-Massey algorithm.

A slightly different formulation allows to remove the cumbersome issue of having to alternate between the f_T and $\widetilde{f_T}$ representations. The idea is to take the reciprocal of all polynomials in Equation (2.39), with a small caveat on our use of the word reciprocal. As defined above, the reciprocal is defined as the sum up to the degree of the polynomial. Here, instead of using the effective value of the degree, we use the expected value and treat T_D as a polynomial of degree D even when its coefficient of x^D is zero; similarly, h_T is viewed as a polynomial of degree $d-1$. The case of g_T is slightly more subtle, it should be considered as a polynomial of degree $D + d - 1$, with at least its D upper coefficients equal to zero. To emphasize this fact, we write the reciprocal as $x^D \widetilde{g_T}(x)$, where $\widetilde{g_T}$ is a reciprocal of g_T viewed as a degree $d-1$ polynomial. After this, Equation (2.39) becomes:

$$f_T(x)\widetilde{T_D}(x) + \widetilde{h_T}(x) = x^D \widetilde{g_T}(x) \text{ or} \tag{2.40}$$
$$f_T(x)\widetilde{T_D}(x) - x^D \widetilde{g_T}(x) = \widetilde{h_T}(x).$$

Once again, we can use an extended Euclidean algorithm, this time on the polynomials $\widetilde{T_D}$ and x^D.

However, this is not the complete story. Indeed, both of the above approaches have a notable disadvantage; they require as input an upper bound d on the degree of f_T, in order to choose D and this value controls the whole computation. It is much preferable to rewrite the operation in a way that lets us increase D as the computation proceeds. The idea is to create a series of polynomials that each annihilate increasingly long subsequences of the original sequence. A typical example is a polynomial f such that:

$$f(x)T_N(x) = g(x) + x^N h(x), \tag{2.41}$$

with $\deg g < \deg f \leq N/2$ and $h(0) \neq 0$. Assume that we know an equation as above and have another equation:

$$F(x)T_M(x) = G(x) + x^M H(x), \tag{2.42}$$

with $M > N$. If $N - \deg f \geq M - \deg F$, we can improve the precision of Equation (2.42) by letting:

$$F'(x) = F(x) - \frac{H(0)}{h(0)} x^{M-N} f(x) \text{ and} \qquad (2.43)$$

$$G'(x) = G(x) - \frac{H(0)}{h(0)} x^{M-N} g(x).$$

Then, there exists H' such that

$$F'(x) T_{M+1}(x) = G'(x) + x^{M+1} H'(x).$$

Moreover, we can check that the degree of F' is equal to the degree of F and that $\deg G' < \deg F'$.

By repeating this improvement to Equation (2.42), we obtain an incremental way to compute f_T, the resulting process is given as Algorithm 2.14. In this algorithm, polynomials need to be encoded as arrays. However, for ease of reading, we write them as polynomials and denote by $[x^l]F$ the l-th coefficient of the polynomial F, with numbering starting at 0.

2.5.3 Solving univariate polynomial equations

Over finite fields, there are efficient algorithms, Berlekamp's and Cantor-Zassenhaus, for finding roots of polynomials equations $f(x) = 0$ and also to factor $f(x)$ into irreducible factors. Of course, we may, without loss of generality assume that f is unitary, after multiplying it by the inverse of its highest degree coefficient. Moreover, it is clear that factoring f into irreducible factors directly yields the roots of $f(x) = 0$, simply by considering the irreducible factors of degree 1. Indeed, r is a root of f, if and only if, $x - r$ divides $f(x)$.

In order to factor f, there are two essential intermediate steps, squarefree factorization and distinct degree factorization.

Squarefree factorization A polynomial is said to be squarefree when no irreducible factor appears with exponent greater than 1 in its factorization. The goal of squarefree factorization is to decompose f into a product:

$$f(x) = \prod_{i=1}^{m} f^{(i)}(x), \qquad (2.44)$$

where each factor is squarefree. Clearly, unless we give more precision, this does not define a unique decomposition for f. To make the decomposition unique, we require the squarefree decomposition to be minimal, i.e., to contain as few factors as possible.

In order to compute such a minimal squarefree factorization, we rely on the basic property that whenever the square of an irreducible polynomial divides

---

Algorithm 2.14 Berlekamp-Massey algorithm

Require: Input sequence S of N elements over $\mathbb{F}_q$
   Let $f(x) \longleftarrow 1$
   Let $L_0 \longleftarrow 0$
   while $S[L_0] = 0$ **do**
     Let $L_0 \longleftarrow L_0 + 1$
   end while
   Let $\alpha \longleftarrow S[L_0]$
   Let $F(x) \longleftarrow 1$
   Let $\delta \longleftarrow 1$
   for l from 0 to $N - 1 - L_0$ **do**
     Let $\beta \longleftarrow \sum_{i=0}^{l} S[l + L_0]([x^l]F)$ in $\mathbb{F}_q$
     if $\beta \neq 0$ **then**
       if $2L > l$ **then**
         Let $F(x) \longleftarrow F(x) - (\beta/\alpha)x^\delta f(x)$
         Let $\delta \longleftarrow \delta + 1$
       else
         Let $L \longleftarrow l + 1 - L$
         Let $g(x) \longleftarrow F(x) - (\beta/\alpha)x^\delta f(x)$
         Let $f(x) \longleftarrow F(x)$
         Let $F(x) \longleftarrow g(x)$
         Let $\delta \longleftarrow 1$
         Let $\alpha \longleftarrow \beta$
       end if
     else
       Let $\delta \longleftarrow \delta + 1$
     end if
   end for
   Output $F(x)x^{L_0}$ (lowest degree annihilator of S up to N elements)

---

f, then the irreducible polynomial also divides Df. Indeed, if $f = g^2 h$, we have:

$$D(f) = D(g(gh)) = gD(gh) + D(g)gh = g(D(gh) + D(g)h). \qquad (2.45)$$

Thus, g divides $D(f)$.

Conversely, if g is an irreducible polynomial that divides f, such that g^2 does not divide f, then g does not divide Df. Indeed, write $f = gh$ with h not a multiple of g. If g divides Df, the equality $Df = gD(h) + D(g)h$ implies that g divides $D(g)h$. Since g is irreducible and h not a multiple of g, we find that $D(g)$ is a multiple of g. Since the degree of $D(g)$ is lower than the degree of g, this is possible only if $D(g) = 0$. We saw previously that $D(g)$ can be zero, when g is either a constant polynomial or a p-th power. Since g is irreducible, none of these options is possible. We conclude that g cannot divide Df.

With this remark in mind, let us consider the GCD of f and Df. If $Df = 0$ then f is a p-th power. Otherwise, if g is an irreducible polynomial that divides f and g^t the greatest power of g that divides f, we find that g^{t-1} divides Df and thus the above GCD. Letting $f^{(1)} = f/GCD(f, Df)$ yields the first step of the squarefree factorization. $f^{(1)}$ is the product of all irreducible polynomials that divide f, without their exponents. The other factor $GCD(f, Df)$ contains the irreducible polynomials that appear at least twice in f, with an exponent equal to the corresponding exponent in f minus 1. Repeating the same process with $GCD(f, Df)$, we obtain $f^{(2)}$ that contains all the irreducible appearing at least twice, each with exponent 1 and so on. Finally, we obtain a decomposition as in Equation (2.44), where $f^{(i)}$ is the product of all irreducible polynomials that appear at least i times in f. This squarefree decomposition is described as Algorithm 2.15.

Distinct degree factorization The next step toward complete factorization of polynomials is to take as input a squarefree polynomial f, usually a polynomial coming out of Algorithm 2.15 and to write it as a product of several factors. The i-th factor in the decomposition is the product of all irreducible polynomials of degree i.

The key ingredient of distinct degree factorization is the fact that every element α in a finite field $\mathbb{F}_{q^n}$ of characteristic p, i.e., where q is either p or a proper power of p, satisfies:

$$\alpha^{q^n} - \alpha = 0. \qquad (2.46)$$

If $I(x)$ is an irreducible polynomial of degree n over $\mathbb{F}_q$, we know that I can be used to represent $\mathbb{F}_{q^n}$. Applying Equation (2.46) to a root α of I implies that $I(x)$ divides the polynomial $P_{q,n}(x) = x^{q^n} - x$. Since the argument applies to all irreducible of degree n, all these polynomials divide $P_{q,n}$. Moreover, no irreducible polynomial of degree higher than n may divide $P_{q,n}$.

As a consequence, taking the GCD of f with $P_{q,1}$ yields the product $f^{(1)}$ of all irreducible of degree 1 dividing f. Replacing f by $f/f^{(1)}$ and taking

---

Algorithm 2.15 Squarefree factorization of polynomials

Require: Input polynomial f, finite field $\mathbb{F}_{p^n}$

Let $m \longleftarrow 0$

Let a be the highest degree coefficient of f

Let $f \longleftarrow f/a$

repeat

    Let $g \longleftarrow D(f)$

    if $g = 0$ **then**

        Here f is a p-th power

        Recurse on $f^{1/p}$, multiply all multiplicities by p and return

    end if

    Let $h \longleftarrow GCD(f, g)$

    Let $m \longleftarrow m + 1$ and $f^{(m)} \longleftarrow f/h$

    Let $f \longleftarrow h$

until $f = 1$

Output a and $f^{(1)}, \ldots, f^{(m)}$

---

the GCD with $P_{q,2}$ yields the product $f^{(2)}$ of all irreducible of degree 2 and so on. For practicality of the resulting algorithm, it is essential to split the GCD computations encountered here into two parts. Indeed, the polynomials $P_{q,n}$ have a high degree q^n and it would not be practical to directly use Algorithm 2.1. Instead, we first compute the polynomial x^{q^n} modulo the other polynomial $f(x)$ using Algorithm 2.9 or 2.10. After subtracting x from the result, we may use Algorithm 2.1 to conclude the GCD computation. Distinct degree factorization is given as Algorithm 2.16.

Algorithm 2.16 Distinct degree factorization of a squarefree polynomial

---

Require: Input unitary squarefree polynomial f

Require: Input prime power q

  Let $m \longleftarrow 0$

  repeat

      Let $m \longleftarrow m + 1$

      Let $g \longleftarrow x^{q^m} \bmod f(x)$

      Let $g \longleftarrow g - x$

      Let $f^{(m)} \longleftarrow GCD(f, g)$

      Let $f \longleftarrow f/f^{(m)}$

  until $f = 1$

  Output $f^{(1)}, \ldots, f^{(m)}$

Concluding the factorization At this final stage, we receive as input a polynomial f, knowing that it has no repeated factors and that all its factors have the same degree k. The strategy is to split f into a product of polynomials of smaller degree until all the factors have degree k. At this point, we know that the factorization is complete and we stop. Thanks to distinct degree factorization, we know that the GCD of f and $x^{q^k} - x$ is equal to f. When q is odd, we see that q^k is also odd and we may write:

$$x^{q^k} - x = x(x^{\frac{q^k-1}{2}} - 1)(x^{\frac{q^k-1}{2}} + 1). \qquad (2.47)$$

This remark can be used to split f into polynomials of smaller degree. First, if $k = 1$ we can easily test whether x divides f or not, simply by looking at the constant coefficient of f. Second, we can compute the GCD of f with $x^{(q^k-1)/2} - 1$. We expect that this yields a factor of f, roughly of half degree.

This is a nice start, but does not suffice to conclude the factorization of f. In order to go further, we need to repeat the same idea with other polynomials. To illustrate this, let us consider the case $k = 1$. In this case, we know that f divides $x^q - x$ but also any polynomial of the form $(x - a)^q - (x - a)$. To prove this, it suffices to consider the polynomial g defined by $g(x) = f(x+a)$. Since f splits into factors of degree 1, g also splits into factors of degree 1, thus $x^q - x$ divides g. Replacing x by $x - a$ in the division yields the desired result. Since,

$$(x - a)^q - (x - a) = (x - a)((x - a)^{\frac{q-1}{2}} - 1)((x - a)^{\frac{q-1}{2}} + 1) \qquad (2.48)$$

we get another chance of decomposing f into smaller factors. We obtain an efficient probabilistic algorithm simply by trying a small number of random values for a.

When k is larger than 1, we can use the same argument and prove that f divides $(x - a)^{q^k} - (x - a)$. However, in general, this is not sufficient to fully factor f. Instead, we need to generalize the idea and remark that for any polynomial h, f divides $h(x)^{q^k} - h(x)$. Using random polynomials of degree k suffices to conclude the factorization using the identity:

$$h(x)^{q^k} - h(x) = h(x)(h(x)^{\frac{q^k-1}{2}} - 1)(h(x)^{\frac{q^k-1}{2}} + 1). \qquad (2.49)$$

The resulting probabilistic method is described as Algorithm 2.17.

In characteristic 2, with $q = 2^e$ the same idea is used with a different way of splitting $x^{2^{ke}} - x$. When ke is 1 or 2, it suffices to use exhaustive search. Indeed, the only possible roots when $e = 1$ and $k = 1$ are 0 and 1. When $e = 1$ and $k = 2$, we are looking for the only irreducible factor of degree 2 over $\mathbb{F}_2$: i.e., $x^2 + x + 1$. When $e = 2$ and $k = 1$, we search for roots over $\mathbb{F}_4$: i.e 0, 1, ρ or $\rho + 1$. Here ρ is a non-trivial cube root of unity in $\mathbb{F}_4$, satisfying $\rho^2 + \rho + 1 = 0$.

When ke is even and larger than 2, we use the same approach as in the odd characteristic case, but we use different relations. Two cases are distinguished.

Algorithm 2.17 Final splitting of polynomials

Require: Input unitary squarefree polynomial f with factors of degree k only
Require: Input prime power q
  Let $g(x)$ be a random unitary polynomial of degree k
  Create an empty list of factors L
  Let $f_0 \longleftarrow GCD(f(x), g(x))$
  if $\deg f_0 = k$ **then**
    Add f_0 to the list of factors L
  end if
  Let $G(x) \longleftarrow g(x)^{(q^k-1)/2}$ in $\mathbb{F}_q[X]/(f(x))$
  Let $f_1 \longleftarrow GCD(f(x), G(x) - 1)$
  if $\deg f_1 > k$ **then**
    Recursively call the final splitting on f_1 and obtain a list L_1
    Concatenate L_1 after L
  else
    if $\deg f_1 = k$ **then**
      Append f_1 to L
    end if
  end if
  Let $f_{-1} \longleftarrow GCD(f(x), G(x) + 1)$
  if $\deg f_{-1} > k$ **then**
    Recursively call the final splitting on f_{-1} and obtain a list L_{-1}
    Concatenate L_{-1} after L
  else
    if $\deg f_{-1} = k$ **then**
      Append f_{-1} to L
    end if
  end if
  Output the list of factors L

When e is odd or equivalently when $\mathbb{F}_{2^e}$ does not contains a cube root of unity, we use the identity:

$$x^{2^k} \oplus x = x(x^{(2^k-1)/3} \oplus 1)(x^{2(2^k-1)/3} \oplus x^{(2^k-1)/3} \oplus 1). \qquad (2.50)$$

When e is even, we let ρ denote a cube root of unity in $\mathbb{F}_{2^e}$ and we use the identity:

$$x^{2^k} \oplus x = x(x^{(2^k-1)/3} \oplus 1)(x^{(2^k-1)/3} \oplus \rho)(x^{(2^k-1)/3} \oplus \rho \oplus 1). \qquad (2.51)$$

Finally, when e and k are both odd, we remark that $x^{2^{ek}} + x$ divides $x^{2^{2ek}} + x$ modulo two and use the analog of Equation (2.50) for this higher degree polynomial.

2.6   Vector spaces and linear maps

A vector space over a field $\mathbb{K}$ is a set V, whose elements are called vectors, together with two operations, addition of two vectors and multiplication of a vector of V by an element of $\mathbb{K}$, satisfying the following properties:

- Addition has a neutral element called the zero vector and denoted by $\vec{0}$.

- Every vector $\vec{v}$ has an opposite $-\vec{v}$, which can be obtained by multiplying $\vec{v}$ by the constant -1 in $\mathbb{K}$. In particular, if $\mathbb{K}$ is a field of characteristic 2, every vector $\vec{v}$ is its own opposite.

- Multiplication by a constant is distributive, i.e., for any pair of vectors $(\vec{v}, \vec{w})$ and any scalar α in $\mathbb{K}$, we have:

$$\alpha(\vec{v} + \vec{w}) = \alpha\vec{v} + \alpha\vec{w}. \qquad (2.52)$$

- Similarly, for any vector $\vec{v}$ and any pair of scalars (α, β), we have:

$$(\alpha + \beta)\vec{v} = \alpha\vec{v} + \beta\vec{v}. \qquad (2.53)$$

A finite family of n vectors $(\vec{v_1}, \cdots, \vec{v_n})$ is a generating family for V, if and only if, for all vector $\vec{v}$ in V there exists (at least) one n-uple of scalars $(\alpha_1, \cdots, \alpha_n)$ such that:

$$\vec{v} = \sum_{i=1}^{n} \alpha_i \vec{v_i}. \qquad (2.54)$$

A family $(\vec{v_1}, \cdots, \vec{v_n})$ is a free family, if and only if, any n-uple of scalars $(\alpha_1, \cdots, \alpha_n)$ such that $\sum_{i=1}^{n} \alpha_i \vec{v_i} = \vec{0}$ is the all zero n-uple.

A family of vectors $(\vec{v_1}, \cdots, \vec{v_n})$ which is both free and a generating family for V is called a basis of V. For any vector $\vec{v}$ of V, there exists a unique n-uple $(\alpha_1, \cdots, \alpha_n)$ such that Equation (2.54) holds. All bases of V have the same cardinality, called the dimension of V and denoted $\dim V$. Note that, in this book, we only consider finite dimensional vector spaces.

For example, $\mathbb{K}$ is a vector space of dimension 1 over itself. More generally, $\mathbb{F}_{p^n}$ can be seen as a vector space of dimension n over $\mathbb{F}_p$.

A subset of V stable by addition and by multiplication by $\mathbb{K}$ is called a vector subspace of V.

Given two vector spaces V and W over $\mathbb{K}$, a map L from V to W is said to be linear, if and only if, for all pairs of scalars (α, β) and all pairs of vectors $(\vec{v}, \vec{w})$, we have:

$$L(\alpha\vec{v} + \beta\vec{w}) = \alpha L(\vec{v}) + \beta L(\vec{w}). \qquad (2.55)$$

The subset of vectors $\vec{v}$ of V such that $L(\vec{v}) = \vec{0}$ is a subspace of V called the kernel of L and denoted by $\text{Ker}(L)$. Similarly, the set of all vectors $L(\vec{v})$ is a subspace of W called the image of L and denoted by $\text{Im}(L)$. The dimensions of the kernel and image of L satisfy the relation:

$$\dim V = \dim \text{Im}(L) + \dim \text{Ker}(L). \qquad (2.56)$$

For any vector $\vec{y}$ in $\text{Im}(L)$, there exists (at least) one vector $\vec{x}$ such that $L(\vec{x}) = \vec{y}$. Moreover, the set of vectors that are mapped to $\vec{y}$ by L is $\vec{x} + \text{Ker}(L)$. If the dimension of $\text{Ker}(L)$ is zero, then $\text{Ker}(L) = \{\vec{0}\}$ and any vector in $\text{Im}(L)$ has a unique inverse image. In that case, we say that L is invertible or that it has full rank. We can then define the map that sends $\vec{y}$ to this inverse image, it is a linear map, called the inverse of L and denoted by L^{-1}. Note that L^{-1} is only defined on $\text{Im}(L)$. The composition $L^{-1} \circ L$ is the identity on V. Moreover, if $W = \text{Im}(L)$, then $L \circ L^{-1}$ is the identity on W.

If $(\vec{v_1}, \cdots, \vec{v_n})$ is a basis for V and $(\vec{w_1}, \cdots, \vec{w_m})$ is a basis for W, then a linear map from V to W can be described by a matrix M with m rows and n columns. Letting $M_{i,j}$ denote the entry on row i and column j of M, we define the entries of M by considering the unique decomposition of each $L(\vec{v_i})$ as a sum:

$$L(\vec{v_i}) = \sum_{j=1}^{m} M_{i,j} \vec{w_j}. \qquad (2.57)$$

With this representation, composition of linear maps can be computed using matrix multiplication.

Linear maps from V to $\mathbb{K}$ are called linear forms. The set of all linear forms on V is a vector space over $\mathbb{K}$, called the dual space of V. The dual of V has the same dimension as V.

Both for matrices and for linear maps from a vector space to itself, it is possible to define a deep mathematical invariant called the determinant, e.g., see [Lan05, Chapter XIII]. The determinant is non-zero, if and only if, the matrix or linear map is invertible. Moreover, the determinant of the product of two matrices or of the composition of two maps is equal to the product of the determinants. The computation of determinants is described in Chapter 3.

2.7 The RSA and Diffie-Hellman cryptosystems

To conclude this chapter, we use the number theory to briefly recall the description of the RSA and Diffie-Hellman cryptosystems.

2.7.1 RSA

The system of Rivest, Shamir and Adleman (RSA) is based on the structure of the multiplicative group $\mathbb{Z}/N\mathbb{Z}$, when N is the product of two large primes p and q. After choosing p and q, we also choose an encryption exponent e coprime to $\phi(N)$ and let d denote the inverse of e modulo $\phi(N)$. Thus, we have $ed = 1 + \lambda\phi(N)$ for some integer λ.

Since $\phi(N) = (p-1)(q-1)$, we know that for all integers $0 < x < N$, coprime with N, we have $x^{\phi(N)} = 1 \pmod{N}$. As a consequence:

$$x^{ed} = x^{1+\lambda\phi(N)} = x \cdot (x^{\phi(N)})^\lambda = x \pmod{N}. \qquad (2.58)$$

This shows that the two maps:

$$x \rightarrow x^e \pmod{N} \text{ and}$$
$$x \rightarrow x^d \pmod{N}$$

are inverses of each other in the group $(\mathbb{Z}/N\mathbb{Z})^*$. We leave it as an exercise to the reader to show that the two maps are inverses of each other everywhere in $\mathbb{Z}/N\mathbb{Z}$. Note that this is not really needed for the RSA cryptosystem. Indeed, we may assume that no one ever encrypts an integer $0 < x < N$ not coprime to N, since that would essentially reveal the factorization of N. We may also assume that no one ever encrypts 0, since it is a weak message for RSA.

RSA as a whole is an efficient system for the following reasons:

- RSA modulus can be efficiently constructed. Indeed, to construct RSA modulus, we need the ability to pick large prime at random. This can be done easily, thanks to the efficient primality testing algorithm described in Section 2.3.2. Note that many practical implementations of RSA key do not choose large primes uniformly at random in a specified interval, instead, the start from a random number in the interval and find the next prime after this number. Since primes are non-uniformly distributed, this procedure favors primes which are far apart from their immediate predecessor. However, RSA numbers of this type are not easier to factor than ordinary RSA numbers and no attack is known against this non-uniform generation.

- Computing decryption exponents is easy and results from a simple application of Euclid's algorithm.

- The RSA encryption and decryption permutations are, given N and e (or d), efficiently computed. If suffices to use a fast modular exponentiation algorithm such as Algorithms 2.9 or 2.10.

From the security point-of-view, it is clear that RSA can be completely broken using a modular e-th root algorithm. However, at the present time, the only known approach to compute arbitrary e-th roots without external help, i.e., given only N and e, is to factor N and then to proceed as the private key owner. Note that, the knowledge of both e and d can be used to compute a multiple of $\phi(N)$ and, as shown in Section 2.3.3.1, this multiple can be used to recover the factorization of N. For these reasons, it is usually said that the security of RSA relies on the hardness of factoring N.

In the above paragraph, the words "arbitrary" and "external help" are essential to the link between factoring and the security of RSA. Concerning "arbitrary," we see that since RSA encryption is deterministic, it is easy to decrypt a value x chosen from a small subset, simply by enumerating possible encryptions. A more advanced attack, described in Chapter 8 allows to decrypt x efficiently, with a non-negligible probability of success, when it belongs to a small interval. Concerning "external help," let us remark that plain RSA does not resist chosen ciphertext attacks, i.e., given access to a RSA decryption box, it is also possible to decrypt messages not submitted to the box. This is based on an essential property of RSA, which is not shared by trapdoor one-way permutations in general: its multiplicativity. By multiplicativity of RSA, we simply mean that given x and y in $\mathbb{Z}/N\mathbb{Z}$ the following relation is satisfied:

$$(xy)^e = x^e \cdot y^e \pmod{N}. \tag{2.59}$$

To realize a chosen ciphertext attack, the attacker receives a value z and tries to compute $x = z^d \pmod{N}$ without submitting this value to the decryption box. He chooses a random element r in $\mathbb{Z}/N\mathbb{Z}^*$ and computes $z_1 = zr^e \pmod{N}$. Sending z_1 to the decryption box, he receives a value x_1 such that $x_1 = xr \pmod{N}$. Thus, a modular division by r suffices to recover x.

Despite these attacks, the multiplicativity of RSA is not necessarily a bad thing. Indeed, it can also be used constructively. In particular, homomorphic encryption based on RSA or on similar computations, such as the Paillier's cryptosystem [Pai99], explicitly requires multiplicativity. Moreover, general methods have been devised to use RSA securely, both for signature, such as the full domain hash and probabilistic signature schemes of [BR96], and for encryption, such as the optimal asymmetric encryption scheme of [BR94].

2.7.2   Diffie-Hellman key exchange

Let $\mathbb{K}$ be any finite field and let g be a generator of a subgroup of $\mathbb{K}^*$ of order q, preferably prime. In order to create a common secret, two users A and B proceed as follows:

- Each user chooses a random integer in the range $[0, q - 1]$, a for user A and b for user B, and raises g to this power.

- Then A sends g^a to B and B sends g^b to A.

- Upon reception A computes $(g^b)^a$ and B computes $(g^a)^b$. Clearly, these two elements of $\mathbb{K}$ are equal and their common value is g^{ab}, where ab is computed modulo q.

- Finally, A and B extract their common secret from g^{ab}. Two frequently encountered approaches are to extract a few bits from the binary representation of g^{ab} or, alternatively, to use a hash value of g^{ab}. This last option is often used in proofs that rely on the random oracle model.

From a security point-of-view, an algorithm that could solve computational Diffie-Hellman, i.e., compute g^{ab} given g^a an g^b, but not a or b, would completely break Diffie-Hellman key exchange. At the present time, this only known general algorithm, involves computing discrete logarithms, i.e. recovering a or b. Computations of discrete logarithms are discussed in Chapters 6, 7 and 15.

Note that the Diffie-Hellman key exchange algorithm is not secure against active attackers. The simplest attack is probably to substitute g^a and g^b by 1 during transmission. After this tampering, A computes $1^a = 1$ as its common key and B computes $1^b = 1$ as its common key. Since the two users end up with the same key, the communication may proceed without problem. After that initial tampering, the attacker knows the common secret 1. As a consequence, he just needs to eavesdrop on the remainder of the communication. Of course, this can be avoided by excluding 0 from the allowed range of a and b and by aborting if the incoming value is a 1.

Another well-known active attack against Diffie-Hellman key exchange is the man-in-the-middle attack. Here, the attacker breaks the communication link between A and B. Afterwards, he communicates with A pretending to be B and with B pretending to be A. On each side, he participates in a copy of the Diffie-Hellman key exchange. After that, the attacker plays the role of a proxy between A and B, he decrypts with one common key any incoming message, reencrypts it with the other key and, finally, forwards the message. As a consequence, he listens to the whole communication. Neither A and B can detect the man-in-the-middle attack, unless some external mean of authentication is used.

Non interactive Diffie-Hellman

The Diffie-Hellman key exchange algorithm can be used as a hybrid cryptosystem in a very simple way. Each user simply needs to choose a pair (x, g^x) as in the Diffie-Hellman key exchange algorithm. However, instead of choosing a new pair for each run of the protocol, he keeps it fixed. As a consequence, x becomes the user's private key and g^x his public key. With this change, it becomes possible to exchange a secret in a non-interactive fashion. When a user wants to initiate a communication, he chooses a random value r and compute g^r. Then, considering that the public key of the intended recipient as the other party contribution to this run of the protocol, he can derive a common secret g^{rx} and, finally, extracting a key from this common secret, he can encrypt the message he wants to transmit using a secret key cryptosystem. Attaching g^r to the encrypted message allows the recipient to recover g^{rx} and decrypt.

ElGamal encryption [ElG85] is a variant that encrypts by multiplying g^{rx} with a message m consisting of a number in $\mathbb{F}_p$. As RSA, it suffers from several weaknesses related to multiplicativity, for an example see Chapter 8.

Zero-knowledge proofs of discrete logarithms

Once a user has a public key g^x, he can use it to prove his identity. Of course, this should be done without revealing the secret key x. An ideal tool for this purpose is the notion of zero-knowledge proof [GMR89], which formally defines the meaning of proving that x is known, without giving away information on x. For discrete logarithm, there is a simple and efficient protocol that achieves this. Assuming that the system parameters, i.e., the prime p and an element g of prime order q, are known and that the prover wants to prove knowledge of x such that $Y = g^x \pmod{p}$, the protocol proceeds as follows:

- The prover chooses a random number r in the range $[0, q - 1]$ and announces $R = g^r \bmod p$.

- The verifier chooses a random number c in the range $[0, q - 1]$, called the challenge and sends it to the prover.

- The prover computes $u = (r + cx) \bmod q$ and sends it to the verifier.

- The verify checks that $g^u = R \cdot Y^c \pmod{p}$ and, if so, accepts the proof.

It is easy to check that an honest prover, who knowns x and follows the protocol correctly always successfully convinces the verifier. Moreover, if, for a fixed value of R, a prover can give convincing answers u and u' for two different challenges c and c', he can recover x by remarking that $(u' - u) = (c' - c)x \pmod{q}$. As a consequence, a cheating prover, i.e., someone who does not know x and tries to run the protocol as prover, has at most 1 chance out of q to succeed. This probability is so low that it can be ignored for all practical purposes. This proof is zero-knowledge because a verifier can produce a convincing copy of a legitimate proof, by first choosing the challenge c, the answer u and then by computing R as $g^u \cdot Y^{-c}$ (mod p). Using this method, the verifier is able to simulate the prover on the single challenge c (for the value R) and this fake proof cannot be distinguished from a real one by a third party.

In cryptography, such knowledge proofs are often used as authentication protocols.

Signature based on discrete logarithm

The above zero-knowledge proof is easy to transform into a signature by replacing the interaction with the verifier by the output of a hash function H. Assuming that H is a random oracle with output in $[0, q-1]$, the simplest way is to use Schnorr's signature [Sch90] and let $c = H(R, M)$, where $R = g^r$ is produced by the signer and M is the message to be signed. After computing u, the pair (R, u) becomes a signature for M. There are several possible variations of this idea, including ElGamal's signature [ElG85] and NIST digital signature algorithm.

Part II

Algorithms

Chapter 3

Linear algebra

Linear algebra is a widely used tool in computer science and cryptography is no exception to this rule. One notable difference is that in cryptography and cryptanalysis, we mostly consider linear algebra over finite fields (or sometimes rings). Compared to linear algebra over real or complex numbers, there are two essential changes. One for the best: no stability problems can occur; one for the worst: the notion of convergence is no longer available. This makes linear algebra in cryptanalysis quite different from linear algebra for scientific computing. The reader interested in the scientific computing aspect of linear algebra may refer to [GL96].

3.1 Introductory example: Multiplication of small matrices over $\mathbb{F}_2$

In order to illustrate the specificities of the linear algebra problem encountered in cryptography, we first consider the multiplication of small matrices over the finite field $\mathbb{F}_2$. More precisely, we show how to optimize the implementation of the basic matrix multiplication Algorithm 3.1, directly derived from the mathematical definition of matrix multiplication, when multiplying Boolean matrices. We especially consider matrices of size 32×32, 64×64 or 128×128. The natural idea that first comes to mind is to directly follow the algorithm description and write the simple Program 3.1. As written, it works on 32×32 matrices, but this can be easily changed by defining DIM differently.

One important drawback of this elementary implementation is that it wastes memory. Indeed, each matrix entry is represented by a full integer. This is clearly suboptimal, since a single bit would suffice. This waste can be reduced by encoding the entries using shorter integers. However, even using 8-bit integers (the C type char) is already costly in terms of memory. As a consequence, before trying to improve the code running time, we replace the matrix representation by an optimal representation where several bits are packed into a single integer. With 32-bit integers, this is extremely well-suited to represent square matrices whose dimension is a multiple of 32. On comput-

---

Algorithm 3.1 Elementary square matrix multiplication

Require: Input $n \times n$ matrices M and N
  Create $n \times n$ matrix R initialized to 0
  for l from 1 to n **do**
    for c from 1 to n **do**
      for k from 1 to n **do**
        $R[l,c] \longleftarrow (R[l,c] + M[l,k] \cdot N[k,c]) \bmod 2$
      end for
    end for
  end for
  Output R

---

ers that offer larger 64- or 128-bit integers, matrices' dimension divisible by 64 or 128 are a natural choice. In particular, many recent processors have special instructions to work with such extended integers, interested readers may refer to Figure 3.1. Assuming 32-bit words, we are going to focus on 32×32 matrices. Changing the memory representation and using logical operations to compute arithmetic modulo 2 (logical AND for multiplication and XOR for addition) yields the code of Program 3.2. In order to access the individual elements of the matrices in each line, the program uses two preprocessors macros. The first macro bit(M,l,c) extracts the bit in line l and column c of the matrix M under the convention that each line of the matrix is stored in an unsigned integer of the correct size. The second macro flipbit(M,l,c) flips the value of the bit from 0 to 1 or from 1 to 0. We do not give a macro for writing a value into a specific bit, since flipping bits is sufficient for our purpose, easier to write and faster to execute.

In order to improve the matrix multiplication algorithm and compare the performance of various implementations, we need a reliable way of measuring the running times. Since the time for performing a single multiplication is negligible compared to the time of starting the program and reading the input matrices, it is useful to add a loop in order to repeat the multiplication a large number of times, say 100,000 times.

Our first optimization comes from an improvement of the binary scalar product of two 32-bit words. Instead of using a loop to extract one bit position from each word, multiply these bits together and add up the products, we remark that all the multiplications can be performed in parallel using a single wordwise logical AND of integers. Moreover, adding up the products requires a smaller number of XOR than initially expected. The key idea is to XOR the upper and lower halves of a word. This performs 16 additions in parallel using a single XOR and can be viewed as folding the word in two. Folding again, we perform 8 additions and obtain an intermediate result on a single byte. After three more folds, we end up with the expected result on a single bit. Clearly, since a matrix multiplication consists of computing

Program 3.1 Basic C code for matrix multiplication over $\mathbb{F}_2$

```c
#include <stdio.h>
#include <stdlib.h>
#define DIM 32

void input_mat(int mat[DIM][DIM])
{
  int l,c;
  for (l=0;l<DIM;l++) {
    for (c=0;c<DIM;c++) {
      scanf("%d",&mat[l][c]); } } }

void print_mat(int mat[DIM][DIM])
{
  int l,c;
  for (l=0;l<DIM;l++) {
    for (c=0;c<DIM;c++) {
      printf("%d ",mat[l][c]); }
    printf("\n"); } }

void Mul(int res[DIM][DIM],
 int mat1[DIM][DIM], int mat2[DIM][DIM])
{
  int l,c,k;
  for (l=0;l<DIM;l++) {
    for (c=0;c<DIM;c++) { res[l][c]=0;
      for (k=0;k<DIM;k++) {
        res[l][c]+=mat1[l][k]*mat2[k][c];
      }
      res[l][c]%=2; } } }

main()
{
  int mat1[DIM][DIM]; int mat2[DIM][DIM];
  int mat3[DIM][DIM]; int count;
  printf("Input Mat1\n");  input_mat(mat1);
  printf("Input Mat2\n");  input_mat(mat2);
  for (count=0;count<100000;count++) Mul(mat3,mat1,mat2);
  printf("Product :\n");  print_mat(mat3);
}
```

Program 3.2 Matrix multiplication over $\mathbb{F}_2$ with compact encoding

```
#include <stdio.h>
#include <stdlib.h>
#define DIM 32
#define WORD unsigned int
#define bit(M,l,c) ((M[l]>>c)&1)
#define flipbit(M,l,c) if (1) {M[l]^=(1UL<<c);} else

void input_mat(WORD mat[DIM])
{
  int l,c,val;
  for (l=0;l<DIM;l++) {
    mat[l]=0;
    for (c=0;c<DIM;c++) {
      scanf("%d",&val);
      if (val) flipbit(mat,l,c); } } }

void print_mat(WORD mat[DIM])
{
  int l,c;
  for (l=0;l<DIM;l++) {
    for (c=0;c<DIM;c++) {
      printf("%d ",bit(mat,l,c)); }
    printf("\n"); } }

void Mul(WORD res[DIM],WORD mat1[DIM],WORD mat2[DIM])
{
  int l,c,k,val;
  for (l=0;l<DIM;l++) {
    res[l]=0;
    for (c=0;c<DIM;c++) {
      val=0;
      for (k=0;k<DIM;k++) {
        val^=bit(mat1,l,k)&bit(mat2,k,c); }
      if (val) flipbit(res,l,c); } } }

main()
{
  WORD mat1[DIM]; WORD mat2[DIM]; WORD mat3[DIM]; int count;
  printf("Input Mat1\n");  input_mat(mat1);
  printf("Input Mat2\n");  input_mat(mat2);
  for(count=0; count<100000; count++) Mul(mat3,mat1,mat2);
  printf("Product :\n");  print_mat(mat3);
}
```

Recent microprocessors are becoming more and more powerful and manufac-
turers use this additional power to enrich the functionalities of processors.
MMX and SSE instructions have been introduced as part of this process.
They allow to compute on 64-bit and 128-bit specific registers. The advantage of these instructions, when optimizing programs, is twofold. First,
these instructions compute faster than ordinary microprocessor instructions.
Second, since these instructions operate on specific registers rather than on
the main registers, they are extremely useful for codes which require many
registers. For example, the bitslice technique, described in Chapter 5, requires a large number of registers and MMX or SSE instructions are very
useful in this case.

One important caveat is that MMX or SSE registers share their storage area
with floating point registers. As a consequence, both types of instruction
cannot be interleaved within a program and when starting a sequence of
MMX or SSE instructions or a sequence of floating point instructions, the
storage area needs to be cleared by using a specific instruction called EMMS.
Otherwise, we get either unpredictable results or unwanted errors.

MMX and SSE operations can view the specific registers in several ways.
For example, a 64-bit MMX register can be considered either as eight bytes,
four 16-bit words or two 32-bit words, optionally using saturated arithmetic
of signed or unsigned kind. There are instructions to add in parallel eight
bytes, four 16-bit words or two 32-bit words. SSE registers can also contain
these different types of contents; however, they contain twice as many of
each. One notable exception, which is very useful for cryptographic purposes,
concerns logical operations. Indeed, performing a logical AND, OR or XOR
on eight bytes or on two 32-bit words yields the same result. Thus, for logical
operations, it is not necessary to have three different types.

At the present time, with the GCC compiler, MMX and SSE operations can
only be used in C programs by using a specific mean of inlining assembly
code. The easiest way probably is to use the definitions from the include files
`mmintrin.h`, `emmintrin.h`, `pmmintrin.h`, `tmmintrin.h` and `xmmintrin.h`.
This files defines two new data types `__m64` for 64-bits MMX registers and
`__m128` for 128-bits SSE registers. These special data types can be operated
on using specific operations. A few of these operations are listed below.

Instruction name	Functionality	Data type
`__mm_empty`	Reset storage area	None
`__mm_add_pi16`	Addition	Four 16-bit words
`__mm_sub_pi8`	Subtraction	Eight bytes
`__mm_xor_pd`	XOR	128-bit word

Figure 3.1: MMX and SSE instructions

the scalar products of lines with columns, this can be very useful. Since in our representation words are used to encode lines, we first need to extract columns from the representation of the second matrix and encode them back into another integer. The multiplication subroutine thus obtained is given as Program 3.3.

Program 3.3 Matrix multiplication using fast scalar product

```
void Mul(WORD res[DIM], WORD mat1[DIM], WORD mat2[DIM])
{
  int l,c,k;
  WORD val,mask;
  for (l=0;l<DIM;l++) res[l]=0;
  for (c=0;c<DIM;c++) { mask=0;
    for (k=0;k<DIM;k++) {
      mask^=(bit(mat2,k,c)<<k); }
    for (l=0;l<DIM;l++) {
      val=mat1[l]&mask;
      val^=(val>>16); val^=(val>>8);
      val^=(val>>4);  val^=(val>>2);
      val^=(val>>1);
      if (val&1) flipbit(res,l,c);
    }
  }
}
```

In order to further improve this code, we are going to modify two parts of the program. The first modification replaces the extraction of columns from the second matrix by a faster approach. Since we need all the columns of this matrix, we are, in truth, computing the transpose of this matrix. Thus, our modification consists of writing a fast transpose procedure and applying it before calling the matrix multiplication. The second modification is to speed up the slow part of the scalar product. Clearly, when performing the scalar products, the multiplication part is optimal since we compute 32 parallel bit multiplications using the logical AND on 32-bit words. However, the addition part is suboptimal, we need five folds to compute a single bit. This can be improved by remarking that the second fold only uses 16 bits out of 32, the third only 8 bits and so on... Thus, it is possible to share the logical AND operations between several folds thus computing several bits of the resulting matrix at once.

Let us start by improving matrix transposition. In order to do this, we view the original matrix as a 2×2 block matrix. Transposing this matrix, we find that:

$$^\top\!\begin{pmatrix} A\ B \\ C\ D \end{pmatrix} = \begin{pmatrix} {}^\top\!A\ {}^\top\!C \\ {}^\top\!B\ {}^\top\!D \end{pmatrix}.$$

Using the above remark, transposing a 32×32 matrix can be done in five steps. The first step exchanges the upper right and lower left 16×16 matrices. The second step performs a similar exchange on four pairs of 8×8 matrices. Then 16 pairs of 4×4 matrices, followed by 64 pairs of 2×2 matrices and finally 256 pairs of 1×1 matrices are exchanged. By using logical operations and shifts, each of these steps costs a small number of operations per line. From an asymptotic point-of-view $2^t \times 2^t$ matrices can be transposed in $O(t\,2^t)$ operations on 2^t-bit words, instead of 2^{2t} operations for the basic algorithm. The transposition C code for 32 matrices is given as Program 3.4. Note the constants defined at the beginning of the program which are defined in order to extract the bits contained in the submatrices that need to be exchanged at each step.

Improving the additions during the scalar product multiplication is done in a similar way. The first fold is performed as before, since it uses up full words. The second fold regroups the computation of 2 second stage folds in a single operation, the third fold regroups 4 third stage folds, and so on. This allows us to compute 32 bits using a total of 63 elementary fold operations, thus the amortized cost is less than two operations per bit, instead of five in the previous approach. The corresponding code for matrix multiplication makes use of the fast transposition Program 3.4 and is included as Program 3.5. To illustrate the technique, the inner loops have been partially unrolled.

In order to compare the practical performance of these algorithms, we give in Table 3.1 a sample of running times on a laptop computer, with each program compiled using gcc's -O3 option. These timings are given using 3 different versions of gcc. They clearly illustrate that low-level optimization are not only machine dependent, but also deeply rely on the specific compiler being used. For example, the loop unrolling in Program 3.5 improves the running time with gcc 4.0.1 but makes it worse for the other two versions. With gcc 4.3.2, Programs 3.1 and 3.5 illustrate the fact that this compiler version can vectorize some programs using MMX and SSE instructions (see Figure 3.1).

3.2 Dense matrix multiplication

The complexity of matrix multiplication is a very important and difficult problem in computer science. Indeed, the complexity of many important linear algebra problems can be reduced to the complexity of matrix multiplication. As a consequence, improving matrix multiplication is a key problem.

Program 3.4 Fast transposition of 32×32 matrices over $\mathbb{F}_2$

```
#define DIM 32
#define Cst16 0xffff
#define Cst8 0xff00ff
#define Cst4 0xf0f0f0f
#define Cst2 0x33333333
#define Cst1 0x55555555

void Transpose(WORD transp[DIM], WORD mat[DIM])
{
  int l,c,l0,l1;
  WORD val1,val2;
  for (l=0;l<DIM/2;l++) {
    transp[l]=(mat[l]&Cst16)|((mat[l+(DIM/2)]&Cst16)<<16);
    transp[l+(DIM/2)]=((mat[l]>>16)&Cst16)|
                      (mat[l+(DIM/2)]&(Cst16<<16));
  }
  for(l0=0;l0<2;l0++)
    for (l1=0;l1<DIM/4;l1++) {
      l=l0*(DIM/2)+l1;
      val1=(transp[l]&Cst8)|((transp[l+(DIM/4)]&Cst8)<<8);
      val2=((transp[l]>>8)&Cst8)|(transp[l+(DIM/4)]&(Cst8<<8));
      transp[l]=val1; transp[l+(DIM/4)]=val2;
    }
  for(l0=0;l0<4;l0++)
    for (l1=0;l1<DIM/8;l1++) {
      l=l0*(DIM/4)+l1;
      val1=(transp[l]&Cst4)|((transp[l+(DIM/8)]&Cst4)<<4);
      val2=((transp[l]>>4)&Cst4)|(transp[l+(DIM/8)]&(Cst4<<4));
      transp[l]=val1; transp[l+(DIM/8)]=val2;
    }
  for(l0=0;l0<8;l0++)
    for (l1=0;l1<DIM/16;l1++) {
      l=l0*(DIM/8)+l1;
      val1=(transp[l]&Cst2)|((transp[l+(DIM/16)]&Cst2)<<2);
      val2=((transp[l]>>2)&Cst2)|(transp[l+(DIM/16)]&(Cst2<<2));
      transp[l]=val1; transp[l+(DIM/16)]=val2;
    }
  for (l=0;l<DIM;l+=2) {
    val1=(transp[l]&Cst1)|((transp[l+1]&Cst1)<<1);
    val2=((transp[l]>>1)&Cst1)|(transp[l+1]&(Cst1<<1));
    transp[l]=val1; transp[l+1]=val2;
  }
}
```

Program 3.5 Faster scalar product for multiplying of 32 × 32 matrices

```
void Mul(WORD res[DIM], WORD mat1[DIM], WORD mat2[DIM]) {
  int l,c,k; WORD transp[DIM]; WORD tmp[DIM]; WORD val;
  Transpose(transp,mat2);
  for (l=0;l<DIM;l++) {
    for (c=0;c<DIM;c+=4) {
      val=mat1[l]&transp[c];
      val^=(val>>16); val&=Cst16;
      tmp[c]=val;
      val=mat1[l]&transp[c+1];
      val^=(val>>16); val&=Cst16;
      tmp[c+1]=val;
      val=mat1[l]&transp[c+2];
      val^=(val>>16); val&=Cst16;
      tmp[c+2]=val;
      val=mat1[l]&transp[c+3];
      val^=(val>>16); val&=Cst16;
      tmp[c+3]=val; }
    for (c=0;c<DIM/2;c+=4) {
      val=tmp[c]|(tmp[c+(DIM/2)]<<(DIM/2));
      tmp[c]=(val&Cst8)^((val>>(DIM/4))&Cst8);
      val=tmp[c+1]|(tmp[c+1+(DIM/2)]<<(DIM/2));
      tmp[c+1]=(val&Cst8)^((val>>(DIM/4))&Cst8);
      val=tmp[c+2]|(tmp[c+2+(DIM/2)]<<(DIM/2));
      tmp[c+2]=(val&Cst8)^((val>>(DIM/4))&Cst8);
      val=tmp[c+3]|(tmp[c+3+(DIM/2)]<<(DIM/2));
      tmp[c+3]=(val&Cst8)^((val>>(DIM/4))&Cst8); }
    for (c=0;c<DIM/4;c+=2) {
      val=tmp[c]|(tmp[c+(DIM/4)]<<(DIM/4));
      tmp[c]=(val&Cst4)^((val>>(DIM/8))&Cst4);
      val=tmp[c+1]|(tmp[c+1+(DIM/4)]<<(DIM/4));
      tmp[c+1]=(val&Cst4)^((val>>(DIM/8))&Cst4); }
    for (c=0;c<DIM/8;c+=2) {
      val=tmp[c]|(tmp[c+(DIM/8)]<<(DIM/8));
      tmp[c]=(val&Cst2)^((val>>(DIM/16))&Cst2);
      val=tmp[c+1]|(tmp[c+1+(DIM/8)]<<(DIM/8));
      tmp[c+1]=(val&Cst2)^((val>>(DIM/16))&Cst2); }
    val=tmp[0]|(tmp[2]<<2);
    tmp[0]=(val&Cst1)^((val>>1)&Cst1);
    val=tmp[1]|(tmp[3]<<2);
    tmp[1]=(val&Cst1)^((val>>1)&Cst1);
    val=tmp[0]|(tmp[1]<<1);
    res[l]=val; } }
```

Program	Runtime (100,000 mult.)		
	gcc 4.0.1	gcc 4.2.1	gcc 4.3.2
3.1	9.01 s	8.16 s	3.68 s
3.2	8.09 s	12.51 s	12.39 s
3.3	1.38 s	1.38 s	1.30 s
3.5	0.32 s	0.38 s	0.27 s
3.5 without loop unrolling	0.38 s	0.24 s	0.11 s

Table 3.1: 32×32 **Boolean matmul. on Intel Core 2 Duo at** 2.4 **GHz**

Asymptotically, the basic Algorithm 3.1 multiplies $n \times n$ matrices in time $O(n^3)$. In the other direction, we have a trivial lower bound of the running time of matrix multiplication: $O(n^2)$ which is the time to read the input matrices and/or write the result. The first matrix multiplication that beats n^3 complexity was proposed in 1969 by Volker Strassen. After that initial step, several further asymptotic improvements were proposed; they are discussed in Section 3.2.2.

Throughout the current section, we focus on matrices represented by a dense array of coefficients. Matrices with sparse encoding are discussed in Section 3.4. We consider the complexity of matrix multiplication, detail many practical aspects of matrix multiplication algorithms, overview some asymptotic complexity results and show the relation between matrix multiplication and other linear algebra problems.

3.2.1 Strassen's algorithm

The first matrix multiplication with asymptotic complexity better than n^3 of Strassen [Str69] is a divide-and-conquer algorithm built on a "magic" recipe for multiplying 2×2 matrices. The ordinary formula for 2×2 matrices requires 8 multiplications and 4 additions. Strassen's formula requires 7 multiplications and 18 additions. For 2×2 matrices, this costs more than the usual algorithm; however, for large $2n \times 2n$ matrices, we need 7 multiplications of $n \times n$ matrices and 18 additions. Since matrices can be added together in time quadratic in n, the contribution of the additions to the asymptotic complexity is negligible and the analysis focuses on the number of multiplications. We see that the running time $T(n)$ of Strassen algorithm as a function of n satisfies a recurrence formula:

$$T(2n) = 7 \cdot T(n) + O(n^2)$$

and conclude that $T(n) = n^{\log 7 / \log 2} \approx n^{2.807}$.

Strassen's formula for multiplying $M = \begin{pmatrix} a & b \\ c & d \end{pmatrix}$ by $M' = \begin{pmatrix} a' & b' \\ c' & d' \end{pmatrix}$ are:

$$
\begin{aligned}
P_1 &= (a + c) \cdot (a' + b'), && \text{(3.1)} \\
P_2 &= (b + d) \cdot (c' + d'), \\
P_3 &= (b + c) \cdot (c' - b'), \\
P_4 &= c \cdot (a' + c'), \\
P_5 &= b \cdot (b' + d'), \\
P_6 &= (c - d) \cdot c', \\
P_7 &= (a - b) \cdot b' && \text{and}
\end{aligned}
$$

$$
M \cdot M' = \begin{pmatrix} P_1 + P_3 - P_4 - P_7 & P_5 + P_7 \\ P_4 - P_6 & P_2 - P_3 - P_5 + P_6 \end{pmatrix}.
$$

Note that another set of formulas with as many multiplications but fewer additions was later proposed as an alternative by Winograd and described further on in this section. To transform the formulas of Strassen into a matrix multiplication algorithm, the basic idea is to use a recursive algorithm that multiplies matrices using seven recursive calls to itself on matrices of half size. The problem is that while this approach works very well for matrices whose dimensions are powers of two, it needs to be modified to deal with matrices which do not obey this restriction. From a theoretical point-of-view, the easiest way to multiply $n \times n$ matrices is to embed them within matrices of dimension $2^t \times 2^t$ with the smallest possible value of t such that $2^t \geq n$. In the worst case, this embedding doubles the dimension of the matrix and thus this only affects the constant factor in the runtime complexity $O(n^{\log_2 7})$, not the exponent. However, from a practical point-of-view, it is better to deal with the difficulty progressively throughout the algorithm rather than once and for all at the beginning. To multiply matrices of even dimensions, we use a direct recursion. To multiply matrices of odd dimensions, we need to deal with the imperfect split. Two methods are possible, we can round the size of the half-size matrices either up or down. To round up, it suffices to add bordering zeros to increase the size of the original matrix. To round down, we use the ordinary matrix multiplication formulas to deal with the extra line and column. The two methods are respectively presented as Algorithms 3.2 and 3.3.

Practical aspects of Strassen's multiplication

When implementing Strassen's algorithm, it is clearly better to have a cutoff point and to turn back to ordinary matrix multiplication for small matrices, thus aborting the recursion earlier. Indeed, for small matrices, the overhead of dividing and reassembling the matrices dominates the running time. In this section, we give some explicit data comparing ordinary matrix multiplication with Strassen's multiplication for some typical matrices. This data is far from

---

Algorithm 3.2 Strassen matrix multiplication (rounding up)

Require: Input $n \times n$ matrices M and N
  if $n = 1$ **then**
    Return a 1×1 matrix with entry $M[1,1] \cdot N[1,1]$
  end if
  Let $h_1 \longleftarrow \lceil n/2 \rceil$
  Let $n_1 \longleftarrow 2h_1$
  If needed, add zeros to expand M and N into $n_1 \times n_1$ matrices
  Create R a $n_1 \times n_1$ matrix with zero entries
  Create a $h_1 \times h_1$ matrix M_1, with $M_1[i,j] = M[i,j] + M[h_1 + i, j]$
  Create a $h_1 \times h_1$ matrix N_1, with $N_1[i,j] = N[i,j] + N[i, h_1 + j]$
  Recursively compute $R_1 = M_1 \cdot N_1$
  Add R_1 to upper left quadrant of R, i.e., $R[i,j]+ = R_1[i,j]$
  Create a $h_1 \times h_1$ matrix M_2, with $M_2[i,j] = M[i, h_1 + j] + M[h_1 + i, h_1 + j]$
  Create a $h_1 \times h_1$ matrix N_2, with $N_2[i,j] = N[h_1 + i, j] + N[h_1 + i, h_1 + j]$
  Recursively compute $R_2 = M_2 \cdot N_2$
  Add R_2 to lower right quadrant of R, i.e., $R[h_1 + i, h_1 + j]+ = R_2[i,j]$
  Create a $h_1 \times h_1$ matrix M_3, with $M_3[i,j] = M[i, h_1 + j] + M[h_1 + i, j]$
  Create a $h_1 \times h_1$ matrix N_3, with $N_3[i,j] = N[h_1 + i, j] - N[i, h_1 + j]$
  Recursively compute $R_3 = M_3 \cdot N_3$
  Add R_3 to upper left quadrant of R, i.e., $R[i,j]+ = R_3[i,j]$
  Subtract R_3 from lower right quadrant of R, i.e., $R[h_1+i, h_1+j]- = R_3[i,j]$
  Create a $h_1 \times h_1$ matrix M_4, with $M_4[i,j] = M[h_1 + i, j]$
  Create a $h_1 \times h_1$ matrix N_4, with $N_4[i,j] = N[i,j] + N[h_1 + i, j]$
  Recursively compute $R_4 = M_4 \cdot N_4$
  Subtract R_4 from upper left quadrant of R, i.e., $R[i,j]- = R_4[i,j]$
  Add R_4 to lower left quadrant of R, i.e., $R[h_1 + i, j]+ = R_4[i,j]$
  Create a $h_1 \times h_1$ matrix M_5, with $M_5[i,j] = M[i, h_1 + j]$
  Create a $h_1 \times h_1$ matrix N_5, with $N_5[i,j] = N[i, h_1 + j] + N[h_1 + i, h_1 + j]$
  Recursively compute $R_5 = M_5 \cdot N_5$
  Add R_5 to upper right quadrant of R, i.e., $R[i, h_1 + j]+ = R_5[i,j]$
  Subtract R_5 from lower right quadrant of R, i.e., $R[h_1+i, h_1+j]- = R_5[i,j]$
  Create a $h_1 \times h_1$ matrix M_6, with $M_6[i,j] = M[h_1 + i, j] - M[h_1 + i, h_1 + j]$
  Create a $h_1 \times h_1$ matrix N_6, with $N_6[i,j] = N[h_1 + i, j]$
  Recursively compute $R_6 = M_6 \cdot N_6$
  Add R_6 to lower right quadrant of R, i.e., $R[h_1 + i, h_1 + j]+ = R_6[i,j]$
  Subtract R_6 from lower left quadrant of R, i.e., $R[h_1 + i, j]- = R_6[i,j]$
  Create a $h_1 \times h_1$ matrix M_7, with $M_7[i,j] = M[i,j] - M[i, h_1 + j]$
  Create a $h_1 \times h_1$ matrix N_7, with $N_7[i,j] = N[i, h_1 + j]$
  Recursively compute $R_7 = M_7 \cdot N_7$
  Add R_7 to upper right quadrant of R, i.e., $R[i, h_1 + j]+ = R_7[i,j]$
  Subtract R_6 from upper left quadrant of R, i.e., $R[i,j]- = R_7[i,j]$
  Return the upper left $n \times n$ submatrix of R

Algorithm 3.3 Strassen matrix multiplication (rounding down)

Require: Input $n \times n$ matrices M and N
  if $n = 1$ **then**
    Create R a 1×1 matrix with entry $M[1,1] \cdot N[1,1]$
    Return R
  end if
  Let $h_1 \longleftarrow \lfloor n/2 \rfloor$
  Let $n_1 \longleftarrow 2h_1$
  Create R a $n_1 \times n_1$ matrix with zero entries
  Prepare, perform and post-process 7 recursive calls as in Algorithm 3.2
  if n is odd **then**
    Redefine R as $n \times n$.
    for i from 1 to n_1 **do**
      for j from 1 to n_1 **do**
        Let $R[i,j] \longleftarrow R_[i,j] + M[i,n] \cdot N[n,j]$
      end for
    end for
    for i from 1 to n_1 **do**
      Let $R[i,n] \longleftarrow \sum_{k=1}^{n} M[i,k] \cdot N[k,n]$
      Let $R[n,i] \longleftarrow \sum_{k=1}^{n} M[n,k] \cdot N[k,i]$
    end for
    Let $R[n,n] \longleftarrow \sum_{k=1}^{n} M[n,k] \cdot N[k,n]$
  end if
  Return $n \times n$ matrix of R

Winograd's formulas for multiplying 2×2 matrices also require 7 multiplications, as in Strassen's algorithm, but only 15 additions. Using the same notations, to multiply M by M', the formulas are:

$$S_1 = c + d, \quad S_2 = S_1 - a,$$
$$S_3 = a - c, \quad S_4 = b - S_2,$$
$$S_5 = b' - a', \quad S_6 = d' - S_5,$$
$$S_7 = d' - b', \quad S_8 = S_6 - c',$$
$$Q_1 = S_2 \cdot S_6, \quad Q_2 = a \cdot a',$$
$$Q_3 = b \cdot c', \quad Q_4 = S_3 \cdot S_7,$$
$$Q_5 = S_1 \cdot S_5, \quad Q_6 = S_4 \cdot d',$$
$$Q_7 = d \cdot S_8,$$
$$T_1 = Q_1 + Q_2, \quad T_2 = T_1 + Q_4 \text{ and}$$
$$M \cdot M' = \begin{pmatrix} Q_2 + Q_3 & T_1 + Q_5 + Q_6 \\ T_2 - Q_7 & T_2 + Q_5 \end{pmatrix}.$$

Figure 3.2: Winograd's formulas for matrix multiplication

exhaustive and is here simply to highlight some unexpected technical details that arise when implementing matrix multiplication. We consider two cases of matrix multiplication which are cryptographically relevant. One case uses multiplication of machine size integers and to avoid overflows when multiplying numbers, we chose to work in $\mathbb{F}_p$, where $p = 46337$ is the largest prime such that $2p^2$ fits in an unsigned number on 32 bits. The second case multiplies Boolean matrices whose sizes are multiples of 32. In this second case, elementary block matrices of size 32×32 are multiplied using Program 3.5. Both programs are too long to print here and are available on the book's website.

In both cases, another consideration is to decide whether to use the round-down as in Algorithm 3.2 or the round-up approach of Algorithm 3.3 when implementing Strassen's algorithm. Both approaches work equally well for matrices whose dimensions (or numbers of blocks over $\mathbb{F}_2$) are powers of two. Similarly, both approaches have a worse case for which the other approach would yield a much faster alternative. Namely, using the rounding up approach for a matrix of dimension of the form $2^t + 1$ is a bad idea. Likewise, using the rounding down approach for a matrix of dimension $2^t - 1$ is inadequate. Moreover, rounding up costs more than rounding down on average. Thus, to minimize the adverse consequences of rounding, we implemented both approaches within each of our matrix multiplication codes, choosing the rounding up option only for dimensions congruent to -1 modulo some small power of 2. This choice allows us to use one rounding up instead of four rounding down when possible. It is not optimal, but gives a nice, easy to implement compromise and the resulting running times behave reasonably nicely as a function of the dimension.

In both cases, we compare the running time of our implementation of Strassen's algorithm with an elementary matrix multiplication. Over $\mathbb{F}_2$ this elementary multiplication uses a block by block approach in order to make use of the fast code we have for 32×32 matrices. Programs 3.6 and 3.7 contain the elementary matrix multiplication used as a reference. One drawback of Program 3.7 is that it performs a modular reduction after each multiplication, which is quite costly. A better approach is to perform the modular reduction once, at the end of each scalar product, as in Program 3.8. However, this limits the modulus that can be achieved with a fixed integer size. Similarly, in Program 3.6, instead of using the 32×32 matrix multiplication routine in a black box manner, we can combine the inner optimizations together with the block matrix structure; see Exercise 4. Note that depending on the specific machine or compiler, the programs can sometimes be improved by reversing the order of the two outer loops.

Over $\mathbb{F}_2$, pushing the recursion in Strassen's algorithm all the way down to 32×32 matrices, the resulting running times almost perfectly reflect the asymptotic analysis. The running times are given in seconds in Table 3.2 and shown on a graph in Figure 3.3. Both sets measured running times closely follow the theoretical predictions and can be approximated by curves of the form $t_0 \cdot (x/X)^3$ and $t_1 \cdot (x/X)^{\log_2 7}$, where t_0 and t_1 are the respective running times of the two multiplication programs on the last data point available: X. Moreover, Strassen's algorithm quickly becomes more effective than the ordinary matrix multiplication. It is interesting to remark that near 512 blocks, our rounding strategy does not behave well, which explains the irregular running times. Also note that these codes can still be greatly improved. In particular, enlarging the basic block to 64 or 128 bits would be a very good idea. It is also possible to improve the basic block multiplication with a time-memory tradeoff algorithm called the algorithm of four Russians [ADKF70]. This is especially useful here since each basic block appears in several multiplications. With these improvements, the cutoff point for Strassen's algorithm would be higher.

Over $\mathbb{F}_p$, we experimented several different values of the cutoff parameters, namely 32, 64 and 128, it turns out that, with the simple basic implementation we have, 64 seems to be a reasonable choice for the cutoff value. The running times are shown on a graph in Figure 3.4. Note that the basic implementation we are using is not optimized, as a consequence, by writting it more carefully, it would be possible to speed up the matrix multiplication in $\mathbb{F}_p$ by at least an order of magnitude. The reader can experiment this on the various computer algebra systems listed in the Preface.

With our implementation, Strassen's algorithm over $\mathbb{F}_p$ also behaves as ex-

Program 3.6 C code for elementary $32n \times 32n$ matrix multiplication over $\mathbb{F}_2$

```c
#include <stdio.h>
#include <stdlib.h>

#define WORD unsigned int

#define access(M,i,j,bsize) (&M[((i)*(bsize)*DIM)+((j)*DIM)])

/* External procedure for 32x32 boolean matrix multiplication */
extern void Mul(WORD res[DIM], WORD mat1[DIM], WORD mat2[DIM]);

void matmul(WORD * A, WORD * B, WORD *Res, int bsize) {
  int i,j,k,l;
  WORD tmp[DIM];

  for(i=0;i<DIM*bsize*bsize;i++)
    Res[i]=0;

  for(i=0;i<bsize;i++)
    for(j=0;j<bsize;j++) {
      for(k=0;k<bsize;k++) {
        Mul(tmp, access(A,i,k,bsize), access(B,k,j,bsize));
        for(l=0;l<DIM;l++) {
          access(Res,i,j,bsize)[l]^=tmp[l];
        }
      }
    }
}
```

Num. of blocks (n)	Ordinary mult.	Strassen's mult.
16	0.01	< 0.01
24	0.02	0.01
32	0.04	0.03
48	0.13	0.08
64	0.29	0.20
96	0.98	0.60
128	2.32	1.41
192	7.88	4.22
256	18.79	9.94
384	63.97	29.75
511	147.88	69.94

Table 3.2: Times for $(32n) \times (32n)$ **Boolean matrix multiplication**

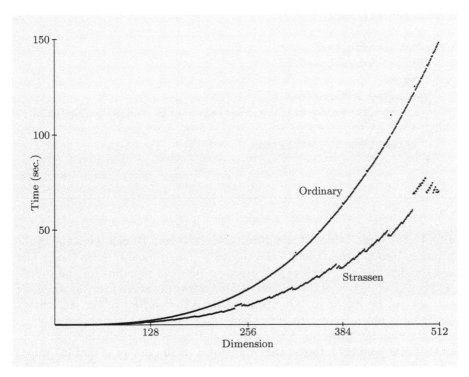

Figure 3.3: Performance of Strassen's multiplication over $\mathbb{F}_2$

Program 3.7 C code for elementary matrix multiplication over $\mathbb{F}_p$

```
#include <stdio.h>
#include <stdlib.h>

#define TYPE unsigned short
#define MODULO 46337 /* Also works with primes up to 65521*/
#define access(M,i,j,size) M[(i)+((j)*(size))]

void matmul(TYPE * A, TYPE * B, TYPE *Res, int size) {
  int i,j,k;
  unsigned int tmp;

  for(i=0;i<size;i++)
    for(j=0;j<size;j++) {
      tmp=0;
      for(k=0;k<size;k++) {
        tmp=(tmp+access(A,i,k,size)*access(B,k,j,size))%MODULO;
      }
      access(Res,i,j,size)=tmp;
    }
}
```

---

pected. For large values of the dimension, however, we can see some cyclic variations around the theoretical behavior. This probably reflects the fact that the difference in terms of performances between rounding up and rounding down is a function of the dimension. It shows once again that the rounding strategy can be improved. On the contrary, the behavior of the elementary algorithm is much more surprising. Instead of following a simple cubic equation, the running times seem to follow some cubic equation, then switch to another and possibly a third. Moreover, for a small number of specific values of the dimension, the behavior is even worse. This mysterious behavior calls for an explanation. In fact, it is due to cache effects when reading the matrices from memory during multiplication. These effects are a side-effect of cache mecanisms, a frequently used tool in computer science. The common basic idea of cache mecanism is to keep local copies of data elements in order to avoid fetching them again when they are requested a second time. This can be used to speed up all kind of data accesses. Frequently encountered applications are disk or webpage accesses. More information about the use of cache in modern processors can be found on Figure 3.5.

To understand why cache effects arise in our implementation of elementary multiplication, we can remark that the two innermost loops on j and k completely read the matrix B and scan the same line of A over and over. On the one hand, the total size of B expressed in bytes is $2n^2$ for a $n \times n$ matrix and for values of n within a reasonable range, it is comparable to the size of either the first (up to $n \approx 150$) or the second level (up to $n \approx 1000$) of memory cache. On the other hand, the size of a single line of A is $2n$ and thus should comfortably fit into the first level of cache. To explain the misbehaviors of the code, we should look for transitions of B from the first level of cache to the second level or from the second level of cache to the main memory. This first cache effect corresponds to the progressive slow-down of the code as the dimension increases. To explain the second effect, we should also look for exceptional cases where – for technical reasons – a line of A cannot be kept in the first level of cache. This happens when many addresses with a large number of equal low order bits are required at the same time. This phenomenon is due to the low level details of the implementation of the cache mechanism. In fact, each memory address can only be stored in a small number of possible different locations in the cache. These authorized locations are determined by looking at a number of low order bits (say 20 or a bit less). When the dimension of the matrix is divisible by a large power of 2, due to the memory representation we are using, several memory locations may compete for the same cache locations. In this case, the same line of A needs to be fetched over and over again from memory. Moreover, reading A in this context is even slower than reading B from main memory, because B takes advantage of automatic prefetching mechanisms.

These cache effects are more visible on the faster implementation that uses fewer modular reductions. The reason is that the memory accesses in this implementation use up a larger fraction of the running time. Note that the effect of cache misses greatly depends on the considered computer architecture. More recent computers seem to perform better in this respect.

All these cache misbehaviors, or at least most of them, can be avoided by reorganizing the loops and thus the memory accesses. However, this is not straightforward and it can take a long time for a programmer to find a working approach. It is extremely interesting to remark that thanks to its use of a divide-and-conquer approach, Strassen's algorithm neatly avoids this problem.

3.2.2 Asymptotically fast matrix multiplication

After the discovery of Strassen's algorithm, the question of finding an optimal asymptotic algorithm for matrix multiplication became an essential issue. Great advances were made and the exponent of matrix multiplication was lowered a lot. These improvements rely on increasingly complex formulas for multiplying various tensor forms. In particular, the use of approximate formulas was essential to lower the number of necessary multiplications. For example, a method of Bini, Capovani, Romani and Lotti [BCRL79] allows to multiply a 3 by 2 matrix and a 2 by 3 matrix using only 10 multiplications of low degree polynomials and leads to a complexity $O(n^{2.7799})$. These methods are discussed in detail in a book by Pan [Pan84]. The best current asymp-

Program 3.8 C code for matrix mult. over $\mathbb{F}_p$ with fewer modular reductions

```
#include <stdio.h>
#include <stdlib.h>

#define TYPE unsigned short
#define MODULO 46337
#define access(M,i,j,size) M[(i)+((j)*(size))]

void matmul(TYPE * A, TYPE * B, TYPE *Res, int size) {
  int i,j,k;
  unsigned int tmp;

  for(i=0;i<size;i++)
    for(j=0;j<size;j++) {
      tmp=0;
      for(k=0;k<size;k++) {
        tmp=(tmp+access(A,i,k,size)*access(B,k,j,size));
        if (tmp>=(MODULO*MODULO)) tmp-=MODULO*MODULO;
      }
      access(Res,i,j,size)=tmp%MODULO;
    }
}
```

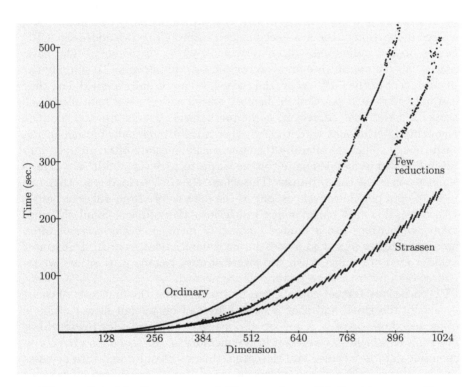

Figure 3.4: Performance of Strassen's multiplication over $\mathbb{F}_p$

In modern processors, a need for cache mechanisms arised from the fact that processors speeds have been progressing much faster than memory speeds. In this context, a CPU cache is a piece of fast memory located as close to the processor unit as possible. Of course, the size of caches is much smaller than the size of the main memory of the computer. As a consequence, it is necessary to devise efficient heuristic techniques to make sure that in many applications the cache will hold the necessary piece of data when it is required. This good event is called a **cache hit**. The bad case, where a required piece of data is not present in the cache, is called a **cache miss**. Each entry in a CPU cache contains both a copy of a data piece from the main memory and the address of this data in the memory. To avoid using a large proportion of the fast cache memory simply to store addresses, CPU caches do not individually store very small pieces of memory. The basic unit is called a **cache line** and can contain several integers. To simplify the management of the CPU cache, the cache lines are usually aligned, i.e., they can only start at a position in memory whose address is a multiple of the size of a cache line. Moreover, cache mechanisms cannot use too complex algorithmic techniques to detect whether a data is already present in the cache or not. Thus, to simplify the issue, cache lines are often grouped into several smaller caches depending on the bit patterns of their addresses. These smaller caches are called **banks**. This allows the CPU to load more than one piece of data per clock cycle as long as the data comes from different banks. Otherwise, there is a bank conflict which slows things down. Similarly, since many applications use a limited amount of memory, cache accesses often ignore the higher bits of addresses during the initial data search. This causes another kind of conflict when two pieces of data located at positions whose addresses differ by a large power of 2 are needed.

CPU caches are further complicated by the fact that the first level of cache is usually too small and that a second level of bigger but slower cache is often added on top of it. They are also usually combined with **prefetching** mechanisms which try to predict future memory accesses and load the corresponding data in advance into the cache, this works quite well with accesses to regularly spaced memory addresses. In multitasking operating systems, the conversion of the virtual memory addresses seen by an individual task into physical memory addresses can become extremely slow if the tables needed for the conversion cannot be held into first level of cache. Another important issue is that cache mechanisms may induce security concerns and allow an adversarial program to learn secret data, such as cryptographic keys, which it should not be able to access [AS08, AScKK07, OST06].

Algorithms and computer programs that take cache mechanisms into account in order to achieve better performance are called **cache-oblivious** [FLPR99].

Figure 3.5: Principles of cached memory in processors

totic result is the method of Coppersmith and Winograd [CW90] which yields a matrix multiplication algorithm running in time $O(n^{2.376})$. This algorithm has recently been reformulated in a group theoretic setting by Cohn, Kleinberg, Szegedy and Umans [CKSU05]. However, no recent advances have been made concerning the exponent of the complexity of matrix multiplication. A widely held opinion is that the correct exponent is 2 and that the asymptotic complexity of matrix multiplication probably is $O(n^2 \log(n))^t$ for some small integer t.

From a practical point-of-view, these algorithms with an asymptotic complexity better than Strassen's are not applicable. Indeed, they involve extremely large constant overheads, due to the highly complex basic formulas that are used. In practice, none of the algorithms with asymptotic complexity better than Strassen's has been reported as useful. Thus, as far as practical aspects are considered, and in particular for cryptanalysis, the best currently achievable exponent for matrix multiplication is $\log_2 7$.

3.2.3 Relation to other linear algebra problems

Since all linear algebra problems can be solved using matrix multiplication, addition and inverse, in order to relate the general complexity of linear algebra to the complexity of matrix multiplication, it suffices to determine the complexity of inversion from the complexity of matrix multiplication. Indeed, addition of $n \times n$ matrices can be done in time n^2, and thus it cannot increase the exponent in the complexity of linear algebra problems. In this section, we assume that we are given a matrix multiplication algorithm with complexity $O(n^\omega)$, for some constant $2 < \omega \leq 3$ and we would like to show that matrix inversion can also be performed in time $O(n^\omega)$. As Strassen's algorithm, this can be obtained by working with 2×2 block matrices. Write:

$$M = \begin{pmatrix} A & B \\ C & D \end{pmatrix}, \tag{3.2}$$

and remark that when D is invertible then multiplying M on the right by

$$N = \begin{pmatrix} Id & 0 \\ -D^{-1}C & Id \end{pmatrix} \tag{3.3}$$

yields:

$$MN = \begin{pmatrix} A - BD^{-1}C & B \\ 0 & D \end{pmatrix}. \tag{3.4}$$

In this equation, $A - BD^{-1}C$ is called the Schur complement of M, in the sequel, we denote it by S. Since a triangular matrix is easy to invert, indeed:

$$\begin{pmatrix} U & V \\ 0 & W \end{pmatrix}^{-1} = \begin{pmatrix} U^{-1} & -U^{-1}VW^{-1} \\ 0 & W^{-1} \end{pmatrix}, \tag{3.5}$$

we can compute the inverse of M as:

$$M^{-1} = N \begin{pmatrix} S & B \\ 0 & D \end{pmatrix}^{-1}. \tag{3.6}$$

This equation only requires two inversions of half-size matrices (D and S) and some matrix multiplications and additions. As a consequence, it can be used as a basis for a recursive matrix inversion algorithm with asymptotic running time of the same form as the underlying matrix multiplication algorithm, i.e., $O(n^\omega)$.

The use of the Schur complement is a powerful tool for numerical analysis, many of its applications in this context are described in Zhang's book [Zha05].

It is also possible to use the Schur complement to invert matrices over finite fields. Two methods are possible. The first option is to lift the matrix to a larger field. For example, a matrix modulo a prime p can be lifted to an integer matrix. However, this approach is costly, since the inverse of the integer matrix may involve large denominators. The other option is to compute Schur complement directly in the finite field. In this case, a specific difficulty arises: we need to make sure that D is invertible at each step of the algorithm. Indeed, over the finite field $\mathbb{F}_q$, even a random matrix is not guaranteed to have full rank. This is most critical when considering matrices over $\mathbb{F}_2$, because a random square matrix of dimension n over $\mathbb{F}_2$ has full rank with probability $\prod_{i=1}^{n}(1 - 2^{-i})$, which is close to 0.29 when n becomes large.

3.3 Gaussian elimination algorithms

When working with matrices, frequently encountered problems are to solve a linear system of equations or to invert a matrix. In Section 3.2.3 we described a recursive approach to matrix inversion. Here, we consider a more direct approach: Gaussian elimination algorithms. Since the problems of solving systems of equations and of inverting matrices are related, it is convenient to start by the simpler problem: solving linear system of equations.

Gaussian elimination works in two phases. During the first phase, called Gauss's pivoting, we progressively modify the original system of equations using reversible transforms, in order to turn it into a triangular system of equations. During the second phase, thanks to the triangular form, it becomes possible to determine, one at a time, the value of each unknown.

Gaussian elimination is quite simple to describe; however, when writing down a complete algorithm, some technicalities arise in order to avoid any division by zero. To avoid these technicalities, we initially write down slightly incorrect versions of the two phases in order to outline the basic idea. To simplify the presentation and avoid dealing with the representation of the linear equations themselves, we simply assume that each equation is given as an equality between a linear combination of variables x_i given on the left-hand side and a constant value given on the right-hand side. Moreover, we assume that we have access to elementary routines that compute the addition or subtraction of two equations, multiply an equation by a given scalar and access individual coefficients within an equation. The first phase in simplified form is given as Algorithm 3.4 and the second phase as Algorithm 3.5. The action of the ingredient of Algorithm 3.5, pivoting, on the matrix of coefficients of a linear system is illustrated in Figure 3.6.

Another application of Gaussian elimination is to compute the determinant of a matrix. Indeed, the first phase does not modify the determinant and once we have a triangular matrix, it suffices to compute the product of diagonal element to obtain the determinant.

Algorithm 3.4 Triangularization of a linear system (simplified, **incorrect**)

Require: Input linear system n equations E_i in n unknowns x_i
  for i from 1 to $n-1$ **do**
    Let P be the **non-zero** coefficient of x_i in E_i
    for j from $i+1$ to n **do**
      Let C be the coefficient of x_i in E_j
      $E_j \longleftarrow E_j - (C/P) \cdot E_i$
    end for
  end for
  Output modified upper triangular system of n equations E_i

Algorithm 3.5 Backtracking to solve a triangular system

Require: Input triangular linear system n equations E_i in n unknowns x_i
  Create an array X of n elements to store the values of variables x_i
  for i from n down to 1 **do**
    Let V be the constant term in equation E_i
    for j from $i+1$ to n **do**
      Let C be the coefficient of x_j in E_i
      $V \longleftarrow V - C \cdot X[j]$
    end for
    Let P be the **non-zero** coefficient of x_i in E_i
    Let $X[i] \longleftarrow V/P$
  end for
  Output array of determined values X

$$\begin{pmatrix} a_{1,1} & \times & \times & \cdots & \times & \times & \cdots & \times \\ 0 & a_{2,2} & \times & \cdots & \times & \times & \cdots & \times \\ 0 & 0 & a_{3,3} & \cdots & \times & \times & \cdots & \times \\ \vdots & \vdots & \vdots & \ddots & \vdots & \times & \cdots & \times \\ 0 & 0 & 0 & \cdots & \mathbf{a_{i,i}} & \times & \cdots & \times \\ 0 & 0 & 0 & \cdots & \mathbf{a_{i+1,i}} & \times & \cdots & \times \\ \vdots & \vdots & \vdots & \ddots & \vdots & \vdots & \ddots & \vdots \\ 0 & 0 & 0 & \cdots & \mathbf{a_{n,i}} & \times & \cdots & \times \end{pmatrix} \Rightarrow \begin{pmatrix} a_{1,1} & \times & \times & \cdots & \times & \times & \cdots & \times \\ 0 & a_{2,2} & \times & \cdots & \times & \times & \cdots & \times \\ 0 & 0 & a_{3,3} & \cdots & \times & \times & \cdots & \times \\ \vdots & \vdots & \vdots & \ddots & \vdots & \times & \cdots & \times \\ 0 & 0 & 0 & \cdots & \mathbf{a_{i,i}} & \times & \cdots & \times \\ 0 & 0 & 0 & \cdots & \mathbf{0} & \times & \cdots & \times \\ \vdots & \vdots & \vdots & \ddots & \vdots & \text{updated} & \vdots \\ 0 & 0 & 0 & \cdots & \mathbf{0} & \times & \cdots & \times \end{pmatrix}$$

Figure 3.6: Effect of a pivoting step

Our simplified algorithms are incorrect in general. They work correctly under several important assumptions. The first of these assumptions is that the input system has n equations in n unknowns and a unique solution. In general, this need not be the case, linear systems may have more unknowns than equations or the reverse. And, as recalled in Chapter 2, even systems of n equations in n unknowns do not necessarily have a unique solution. If their kernel is non-zero, they may have none or infinitely many. The next assumption is that whenever we request the coefficient of x_i in equation E_i at some step of the algorithm, this value is non-zero. This assumption is highlighted in boldface type in the algorithms.

When the first assumption does not hold, i.e., when the system of equations does not have full rank, printing out an error message may, at least for now, be an acceptable solution. However, the second assumption that no zeros are encountered as denominators throughout the algorithm, is a technical assumption that does not reflect any underlying mathematical invariant of the linear system. In order to better understand its role, let us discuss the inner working of Algorithm 3.4. This algorithm has a main loop where variable i is used to successively go through all equations. At each step, we take the coefficient of x_i in the current version of equation E_i. Thanks to the work already performed by the algorithm, all coefficients of previous variables x_1 up to x_{i-1} are zero in E_i. Assuming that the coefficient of x_i is non-zero, the inner loop indexed by j then modifies the system to make x_i vanish in all subsequent equations. It is important to note that this transformation, called pivoting, is reversible and thus does not change the set of solutions of the linear system. As a consequence of the successive pivoting steps, after the algorithm execution, the linear system becomes upper triangular. The non-zero coefficient of x_i in E_i used for pivoting is called the **pivot**. Taking the i-th pivot from equation E_i comes naturally when writing down the algorithm, but it is unessential to its inner working. In truth, for each pivoting step it suffices to find a pivot, i.e., a non-zero coefficient, for x_i in any equation that has not yet been used for this purpose. In fact, as long as the system of equations is invertible, at each step of the algorithm it is always possible to find at least one equation E_j with $j \geq i$ such that the coefficient of x_i in E_j is non-zero. Thus, by reordering the equations, we can modify Algorithm 3.4 into Algorithm 3.6. When working over finite fields, this change suffices to obtain a fully function Gaussian elimination algorithm. Over the real or complex fields, this is not the end of story. Since this is outside of the scope of this book, let us simply state that in this case the pivot should be chosen with even more care in order to minimize the loss of accuracy during the algorithm execution.

Algorithm 3.6 Triangularization of a linear system

Require: Input linear system n equations E_i in n unknowns x_i

  for i from 1 to $n - 1$ **do**

    for j from i to n **do**

      Let P be the coefficient of x_i in E_j

      if $P \neq 0$ **then**

        Exit Loop

      end if

    end for

    if $P = 0$ **then**

      Output 'Non-invertible system', **Exit Algorithm**

    end if

    Exchange equations E_i and E_j

    Let P be the **non-zero** coefficient of x_i in E_i

    for j from $i + 1$ to n **do**

      Let C be the coefficient of x_i in E_j

      $E_j \longleftarrow E_j - (C/P) \cdot E_i$

    end for

  end for

  Output modified system of n equations E_i

Once the linear system is transformed into triangular form, finding the value of each variable within the unique solution is a simple matter. Due to the triangular form of the transformed system, the last equation relates the value of the single variable, the last one, to a constant. Thus dividing the constant in this equation by the coefficient of the last variable x_n, we recover its value. Once the value of x_n is known, we can substitute this value in the previous equation which only contains x_n and x_{n-1}. After the substitution, we can recover the value of x_{n-1}. Following this approach, we successively determine x_{n-2} and all other variables. Algorithm 3.5 uses this approach.

3.3.1 Matrix inversion

With Gaussian elimination as a tool to solve invertible linear systems of equations, we can now turn to the problem of computing the inverse of a matrix. Remember that M is invertible, if and only if, there exists a matrix N such that: $MN = Id$. Here, Id denotes the identity matrix, with a diagonal of ones and zeros everywhere else, which is the neutral element in the multiplicative group of $n \times n$ matrices. Viewing the matrix N as n column vectors $N^{(i)}$, each such vector satisfies a linear equation: $M \cdot N^{(i)} = \Delta^{(i)}$, where $\Delta^{(i)}$ is the vector whose coordinates are 0 everywhere except on line i where the coefficient is 1. Solving these n linear systems of equations, we thus invert M. However, from a complexity point-of-view this approach is not satisfying. Instead, it is preferable to use the similarity between all these systems and solve them all at once. Indeed, the sequence of operations in Algorithms 3.5 and 3.6 does not depend on the constant side of the equations. This allows us to perform these operations in parallel on these n related systems.

To illustrate these ideas in the context of 32×32 Boolean matrices as in Section 3.1, we give in Program 3.9 a C code compatible with the representation, input and output routines given in the matrix multiplication Program 3.2.

3.3.2 Non-invertible matrices

Linear systems of equations encountered in cryptography do not necessarily have full rank. When faced with such a system, Gaussian elimination is also very useful. However, we need to change and complete our algorithms. Our goal is to find whether a given system has solutions and, at least, to compute a single solution. More generally, for non-invertible square matrix M, we aim at computing its kernel $\text{Ker}(M)$ and image $\text{Im}(M)$.

Since these sets are linear subspaces, they can be very large, or even infinite when the base field is infinite. Thus, it is not conceivable to represent them by a list of elements; instead, we would like to obtain efficient representations such as linear bases for these subspaces.

Remember that a linear system of equations, written in matrix form as $M\vec{x} = \vec{y}$, with M non-invertible, has the following properties:

- If $\vec{y}$ does not belong to the image of M, then the equation has no solution.

- If $\vec{y}$ belongs to the image of M, then the set of solutions is an affine subspace.

- More precisely, starting from an arbitrary solution $\vec{x}_0$ of the system, the set of solutions is $\vec{x}_0 + \text{Ker}(M)$.

When solving a single system for a non-invertible matrix M, it is convenient to write a variation of Algorithms 3.5 and 3.6 to solve the system. When faced with many systems of equations involving a single matrix M, it is preferable to first compute global information about the matrix, find its rank and describe

---

Algorithm 3.7 Matrix inversion

Require: Input $n \times n$ matrix M
  Initialize matrix N to identity
  for i from 1 to $n-1$ **do**
    for j from i to n **do**
      Let $P = M_{j,i}$
      if $P \neq 0$ **then**
        Exit Loop
      end if
    end for
    if $P = 0$ **then**
      Output 'Non-invertible matrix', **Exit Algorithm**
    end if
    Exchange lines i and j in M
    Exchange lines i and j in N
    Divide lines i of M and N by P
    for j from $i+1$ to n **do**
      Let $C = M_{j,i}$
      Subtract C times line i of M from line j of M
      Subtract C times line i of N from line j of N
    end for
  end for
  Assert: M is upper triangular with a diagonal of 1
  for i from n down to 1 **do**
    for j from 1 to $i-1$ **do**
      Let $C = M_{j,i}$
      Subtract C times line i of M from line j of M
      Subtract C times line i of N from line j of N
    end for
  end for
  Assert: M is the identity matrix
  Output N the inverse of input matrix M

Program 3.9 Inversion of 32×32 matrix over $\mathbb{F}_2$

```
/* Warning: mat is transformed during MatInv */
int MatInv(WORD mat[DIM], WORD inv[DIM])
{
  int piv,l,c,k;
  WORD val,vali,mask;
  for(piv=0,mask=1;piv<DIM;piv++,mask<<=1)
    inv[piv]=mask;
  for(piv=0,mask=1;piv<DIM;piv++,mask<<=1) {
    for (c=piv;c<DIM;c++) if (mask&mat[c]) break;
    if (c>=DIM) return(FALSE);
    val=mat[c];mat[c]=mat[piv];mat[piv]=val;
    vali=inv[c];inv[c]=inv[piv];inv[piv]=vali;
    for(c=0;c<DIM;c++) if ((c!=piv)&&(mask&mat[c])) {
      mat[c]^=val;inv[c]^=vali; }}
  return(TRUE);
}
```

its kernel and image. This global information then allows fast resolution of the systems involving M. This approach is similar to the computation of a matrix inverse with Algorithm 3.7.

Let us start with the single system case. Since the system is non-invertible, we know that running Algorithm 3.6 on it produces the error message "Non-invertible system." This occurs when no pivot can be found for a variable x_i. However, at that point, since x_i has a zero coefficient on all lines from i to n, the triangularization is in some sense already complete for line i, except that there is a zero on the diagonal at this position, instead of a 1. Thus, instead of aborting, we could simply skip to the next iteration of the main loop. However, if we do not make sure that the current line is going to be considered when looking for the next pivot, then the backtracking phase may incorrectly claim that the system does not have a solution. To avoid this, the simplest approach is to renumber the variables, pushing those without a pivot toward the end. With this simple modification, all systems of equations can be written, after renumbering the variables, in a nice triangular form, possibly with a bunch of zeros at the lower right end of the diagonal. Once such a system goes as input to Algorithm 3.5, it is easy to modify this algorithm to find solutions. For all the trailing zeros on the diagonal, if the corresponding equation has a non-zero constant, the system does not have any solution. Otherwise, the corresponding variable may be set at random. For the non-zero diagonal entries, the backtracking algorithm resumes its former behavior. The modified triangularization and backtracking algorithms (for a single solution) are given as Algorithms 3.8 and 3.9.

Let us now consider the case of multiple systems involving the same matrix

---

Algorithm 3.8 Triangularization of a possibly non-invertible system

---

Require: Input linear system n equations E_i in n unknowns x_i

   Create an array Π of n entries, initialized to $\Pi[i] = i$.

   Let $Last = n$

   for i from 1 to $Last$ **do**

      for j from i to n **do**

         Let P be the coefficient of $x_{\Pi[i]}$ in E_j

         if $P \neq 0$ **then**

            Exit Loop

         end if

      end for

      if $P \neq 0$ **then**

         Exchange equations E_i and E_j

         Let P be the **non-zero** coefficient of $x_{\Pi[i]}$ in E_i

         for j from $i + 1$ to n **do**

            Let C be the coefficient of $x_{\Pi[i]}$ in E_j

            $E_j \longleftarrow E_j - (C/P) \cdot E_i$

         end for

      else

         Exchange $\Pi[i]$ and $\Pi[Last]$

         Decrement $Last$

         Decrement i (to re-run the loop on the same i value)

      end if

   end for

   Output modified system of n equations E_i and renumbering Π of the variables x

---

Algorithm 3.9 Backtracking of a possibly non-invertible triangular system

---

Require: Input triangular linear system n equations E_i in n unknowns x_i
Require: Renumbering of the variables Π
  Create an array X of n elements to store the values of variables x_i
  for i from n down to 1 **do**
    Let V be the constant term in equation E_i
    Let P be the coefficient of $x_{\Pi[i]}$ in E_i
    if $P = 0$ **then**
      if $V \neq 0$ **then**
        Output 'No solution exists', **Exit Algorithm**
      else
        Assign random value to $X[\Pi[i]]$
      end if
    else
      for j from $i+1$ to n **do**
        Let C be the coefficient of $x_{\Pi[j]}$ in E_i
        $V \longleftarrow V - C \cdot X[\Pi[j]]$
      end for
      Let $X[\Pi[i]] \longleftarrow V/P$
    end if
  end for
  Output array of determined values X

M. In that case, we need to compute several objects. More precisely, we need a basis of the kernel of M to recover all solutions of the system, we need an efficient way to test whether a vector $\vec{y}$ belongs to the image of M and a procedure to compute a solution $\vec{x}_0$ for the system.

The easiest approach is to start by modifying Algorithm 3.7 in order to avoid any division by zero, using the same renumbering idea as in Algorithms 3.8 and 3.9. At the end of the modified algorithm, the matrix M is transformed, in renumbered form into a matrix:

$$H = \left(\begin{array}{c|c} Id & K \\ \hline 0 & 0 \end{array} \right). \tag{3.7}$$

At the same point, matrix N contains a transformation matrix such that $H = NMP_{\Pi}$, where P_{Π} is the permutation matrix representing the permutation Π. Alternatively, reversing the roles of rows and columns, we can similarly write $H' = P'_{\Pi}MN'$, where:

$$H = \left(\begin{array}{c|c} Id & 0 \\ \hline K' & 0 \end{array} \right). \tag{3.8}$$

In that case, the matrix H' is a basis for the image of M, with permuted coordinates, and the final columns N' corresponding to null columns after multiplications by M form a basis of the kernel of M.

3.3.3 Hermite normal forms

In all the Gaussian elimination algorithms we presented in this section, we encounter divisions by diagonal elements. When the considered matrix or system of equations is defined over a field, all the divisions, by non-zero elements, can be performed without trouble. However, if the matrix is defined over a ring, we may have difficulties. The typical example is the case of a system of equations defined over the integers. Such a system does not necessarily have an integer solution. In that case, it would be nice to compute the triangularization of the matrix in a way that avoids all non-exact divisions. When the matrix is invertible this is related to the computation of a Hermite normal form.

DEFINITION 3.1 *An invertible square matrix M with integer coefficients is said to be in* **Hermite normal form** *if it is upper triangular, with positive elements on its diagonal and if furthermore all non-diagonal elements are non-negative and smaller than the diagonal element in their column. Equivalently:*

$$\forall i > j : M_{i,j} = 0 \tag{3.9}$$

$$\forall i : M_{i,i} > 0 \tag{3.10}$$

$$\forall i < j : 0 \le M_{i,j} < M_{j,j} \tag{3.11}$$

Computing Hermite normal forms is reminiscent of both Gaussian elimination and GCD computations. During the first phase of Gaussian elimination, at each pivoting step, we choose in the current column the smallest non-zero entry (in absolute value) as pivot. We move the corresponding row in order to have this pivot on the diagonal, potentially replace it by its opposite to get a positive pivot, and remove integer multiples of this row from all rows below it. We choose the multiples to make sure that the coefficients in the current column become non-negative and smaller than the coefficient of the pivot. If all coefficients of transformed rows in this column are zeros, we proceed to the next column, otherwise we repeat the same computation, choosing a smaller pivot. During the second phase of Gaussian elimination, we remove multiples of each row from the rows above it, to make all non-diagonal entries non-negative and smaller than the diagonal entry in their column. This is described as Algorithm 3.10. This algorithm can easily be modified to keep track of the transformations in an auxiliary matrix as in the matrix inversion Algorithm 3.7. Note that a similar algorithm can be used to transform a non-invertible matrix into something called a **row echelon form**. The main difference with the Hermite normal form is that in columns that do not have any non-zero pivot, the size of the entries in previous rows cannot be reduced.

Algorithm 3.10 Hermite normal forms

Require: Input invertible $n \times n$ matrix M with integer entries
  for i from 1 to $n - 1$ **do**
    Let done $\longleftarrow$ false
    while done $\neq$ true **do**
      Let $P \longleftarrow 0$
      for k from i to n **do**
        if $M_{i,k} \neq 0$ **then**
          if $P = 0$ or $|M_{i,k}| < P$ **then**
            Let $j \longleftarrow k$; let $P \longleftarrow M_{i,j}$
          end if
        end if
      end for
      if $P = 0$ **then**
        Output 'Non-invertible system', **Exit Algorithm**
      end if
      Let done $\longleftarrow$ true
      Exchange rows M_i and M_j
      if $P < 0$ **then**
        Let row $M_i \longleftarrow -M_i$; let $P \longleftarrow -P$
      end if
      for j from $i + 1$ to n **do**
        Let $C \longleftarrow M_{j,i}$
        Let row $M_j \longleftarrow M_j - \lfloor C/P \rfloor \cdot M_i$
        if $M_{j,i} \neq 0$ **then**
          Let done $\longleftarrow$ false
        end if
      end for
    end while
  end for
  if $M_{n,n} = 0$ **then**
    Output 'Non-invertible system', **Exit Algorithm**
  end if
  if $M_{n,n} < 0$ **then**
    Let row $M_n \longleftarrow -M_n$
  end if
  for i from 2 to n **do**
    Let $P \longleftarrow M_{i,i}$
    for j from 1 to $i - 1$ **do**
      Let $C \longleftarrow M_{j,i}$
      Let row $M_j \longleftarrow M_j - \lfloor C/P \rfloor \cdot M_i$
    end for
  end for
  Output Hermite normal form M

3.3.3.1 Linear algebra modulo composites and prime powers

When performing linear algebra modulo a composite number N, several cases can be encountered. If the determinant of the matrix that defines the system is invertible modulo N, then the solution is unique and can be derived using Gaussian elimination as in Section 3.3. However, if the determinant is not invertible modulo N, then Gaussian elimination fails. In fact, without loss of generality, we can limit ourselves to considering the case where N is prime power. Indeed, otherwise we can use the Chinese remainder theorem. More precisely, when $N = N_1 \cdot N_2$ and N_1 and N_2 are coprime, then any solution of the linear system modulo N can be obtained by pasting together a solution modulo N_1 and a solution modulo N_2.

When N is a prime power, say $N = p^e$, we can generalize Gaussian elimination and solve the linear system anyway. The basic idea, remains the same, first we triangularize the linear system, choosing for each pivot a value which is "as invertible as possible." Formally, this means that if we define the valuation $v(x)$ of an element x in $\mathbb{Z}/p^e\mathbb{Z}$ as the multiplicity of p in any representative for x, we should always choose as pivot the value in a given column with the smallest valuation. During the second stage, when creating the list of solutions, we can divide a value y by a pivot x if and only if $v(y) \geq v(x)$. When possible, such a division yields $p^{v(x)}$ different solutions.

3.4 Sparse linear algebra

All the linear algebra algorithms that we have presented up to now deal with dense matrices represented by their complete lists of entries. However, there are many applications, both in scientific computing and in cryptography, where sparse matrices are involved. A sparse matrix is a matrix that contains a relatively small number of non-zero entries. Very frequently, it takes the form of a matrix in which each line (or column) only contains a small number of non-zero entries, compared to the dimension of the matrix. With sparse matrices, it is possible to represent in computer memory much larger matrices, by giving for each line (resp. column) the list of positions containing a non-zero coefficient, together with the value of the coefficient. Indeed, assuming an average of l entries per line, storing a $n \times n$ matrix requires about $2ln$ numbers instead of n^2. When dealing with a sparse linear system of equations, using plain Gaussian elimination is often a bad idea. Each pivoting step during Gaussian elimination increases the number of entries in the matrix and after a relatively small number of steps, the matrix can no longer be considered as sparse. As a consequence, if the dimension of the initial matrix was large, Gaussian elimination quickly overflows the available memory. In order to deal with sparse systems, a different approach is required. Moreover, sparsity is not a well-behaved mathematical property. In particular, the inverse of a sparse invertible matrix is not necessarily sparse. As a consequence, the best we can hope for is an efficient, sparsity preserving algorithm to solve a single linear system of equations.

Two main families of algorithms have been devised for sparse systems. One family called structured Gaussian elimination contains variations on the ordinary Gaussian elimination that chooses pivots in order to minimize the loss of sparsity. The other family uses a totally different approach; it does not try to modify the input matrix but instead aims at directly finding a solution of the linear system using only matrix by vector multiplications. In this family, we find the Lanczos's and the Wiedemann's algorithms.

3.4.1 Iterative algorithms

3.4.1.1 Lanczos's algorithm

Lanczos's algorithm is a famous algorithm which has been devised to find solutions of linear algebra systems of real or complex numbers. It is much easier to describe when we can rely on a notion of convergence. Thus to explain this algorithm, we temporarily leave our usual setting, forget about finite field and consider a linear equation $M\vec{y} = \vec{x}$ over the real numbers. For simplicity, we assume that M is square and invertible. Moreover, without loss of generality, we may assume that M is symmetric. Indeed, multiplying the initial equation by the transpose of M, it can be transformed to an equation $(^{\top}MM)\vec{y} = (^{\top}M\vec{x})$, where $(^{\top}MM)$ is a symmetric matrix. When a square $n \times n$ matrix M is real, symmetric and invertible, it induces a scalar product $(\cdot|\cdot)_M$ on the vector space $\mathbb{R}^n$, defined from the usual scalar product by:

$$(\vec{u}|\vec{v})_M = (\vec{u}|M\vec{v}) = (M\vec{u}|\vec{v}). \tag{3.12}$$

This scalar product induces a norm $\|\cdot\|_M$ defined as:

$$\|\vec{u}\|_M = \sqrt{(\vec{u}|\vec{u})_M} \tag{3.13}$$

Over the field of real numbers, Lanczos's algorithm works by first constructing an orthonormal basis of $\mathbb{R}^n$ for the scalar product $(\cdot|\cdot)_M$, i.e., a family of vectors $\vec{v}_1, \vec{v}_2, \ldots, \vec{v}_n$ such that:

$$(\vec{v}_i|\vec{v}_i)_M = 1 \quad \forall i \in [1 \cdots n] \text{ and} \tag{3.14}$$
$$(\vec{v}_i|\vec{v}_j)_M = 0 \quad \forall i \neq j.$$

Then it decomposes the solution $\vec{y}$ of $M\vec{y} = \vec{x}$ over the basis $(v_i)_{i \in [1 \cdots n]}$ using the decomposition formula:

$$\vec{y} = \sum_{i=1}^{n} (\vec{y}|\vec{v}_i)_M \vec{v}_i. \tag{3.15}$$

This decomposition formula can be used to solve the equation because the coefficients $(\vec{y}|\vec{v}_i)_M$ can be computed from $\vec{x}$ without previous knowledge of $\vec{y}$ by remarking that:

$$(\vec{y}|\vec{v}_i)_M = (M\vec{y}|\vec{v}_i) = (\vec{x}|\vec{v}_i). \tag{3.16}$$

Since the decomposition formula does not cost too much in terms of running time, because it only requires n scalar products, this gives an efficient algorithm if and only if the orthonormal basis can be constructed efficiently. A nice approach works as follows, first choose a random vector $\vec{w}_1$, then compute $\vec{v}_1$ as:

$$\vec{v}_1 = \frac{\vec{w}_1}{\|\vec{w}_1\|_M}. \tag{3.17}$$

From $\vec{v}_1$, we construct the orthonormal basis iteratively, computing for each i, $\vec{w}_i = M \cdot \vec{v}_i$ and letting $\vec{v}_i$ by the vector obtained by orthonormalizing $\vec{w}_i$, i.e., letting:

$$\vec{w}_i' = \vec{w}_i - \sum_{j=1}^{i-1} (\vec{w}_i|\vec{v}_j)_M \vec{v}_j \quad \text{and} \tag{3.18}$$

$$\vec{v}_i = \frac{\vec{w}_i'}{\|\vec{w}_i'\|_M}. \tag{3.19}$$

At first, this orthonormalization process does not seem efficient, because a naive implementation requires i scalar products at each step and $O(n^2)$ scalar products for the complete algorithm. However, it is easy to modify the computation and perform two scalar products at each step. The reason is that for $i > j + 1$, the scalar product $(\vec{w}_i|\vec{v}_j)_M$ is already 0, as a consequence we can rewrite the computation of $\vec{w}_i'$ as:

$$\vec{w}_i' = \vec{w}_i - (\vec{w}_i|\vec{v}_{i-1})_M \vec{v}_{i-1} - (\vec{w}_i|\vec{v}_{i-2})_M \vec{v}_{i-2} \tag{3.20}$$

The reason for this simplification is that:

$$(\vec{w}_i|\vec{v}_j)_M = (M\vec{v}_i|\vec{v}_j)_M = (\vec{v}_i|M\vec{v}_j)_M = (\vec{v}_i|\vec{w}_j)_M. \tag{3.21}$$

Since $\vec{w}_j$ is in the vector space spanned by $\vec{v}_1, \ldots, \vec{v}_{j+1}$, whenever $i > j + 1$, this coefficient is already 0.

We stop the process, when the orthogonalized vector is equal to $\vec{0}$. Clearly, the sequence of vectors $\vec{v}_1, \ldots, \vec{v}_k$ that is generated is an orthonormal family. However, is it a basis? The answer is yes if the family is large enough, more precisely, if $k = n$. Due to the initial random choice of $\vec{v}_1$, this is the most frequent case.

Moreover, with most real matrices, Lanczos's algorithm has a very useful property of convergence; even partial decompositions, which are obtained by truncating the sum in Equation (3.15), quickly give very good approximations of $\vec{y}$.

Very surprisingly, Lanczos's algorithm can also be applied to linear system defined over finite fields. Of course, in finite fields, we cannot rely on convergence arguments and need to run the algorithm till the end. However, if the sequence $(\vec{v}_i)$ really forms a basis of the image vector space of M, everything remains fine. Thus, Lanczos's algorithm works over finite field, as long as the construction of the sequence $(\vec{v}_i)$ does not abort prematurely. It can abort for two main reasons. First, as in the real field case, some vector $\vec{w}_k$ may already belong to the vector space spanned by the previously obtained vectors in $(\vec{v}_i)$. Second, in finite field, it may happen that $\|\vec{w}_i'\|_M = 0$ with $\vec{w}_i' \neq \vec{0}$. This comes from the fact that in a finite field, all computations need to be performed modulo the characteristic of the field. To illustrate the problem, take the row vector $\vec{x} = (1, 1)$, over the real field, its Euclidean norm is $\|\vec{x}\| = 2$; however, over $\mathbb{F}_2$ the norm is taken modulo 2 and thus equal to 0. When the characteristic is small, this problem occurs frequently and Lanczos's algorithm needs to be modified into a block algorithm (see Section 3.4.1.3). However, when the characteristic is a large prime p, this is a very rare event. More precisely, due to the initial randomization, we can heuristically estimate the probability of error at each step of the algorithm as $1/p$. As a consequence, the overall probability of failure is roughly n/p. Thus, for large values of p, this probability is negligible.

When implementing Lanczos's algorithm over finite fields, in order to avoid the computation of the square roots that appear in the computation of norms, it is preferable to avoid normalizing the vectors $\vec{v}_i$ and instead to divide by their norms where necessary. This is described in Algorithm 3.11.

3.4.1.2 Wiedemann's algorithm

Wiedemann's algorithm is another approach to find solutions of linear systems using matrix-vector products. However, instead of computing an orthogonal family of vectors, it aims at reconstructing a minimal polynomial. Before presenting this algorithm, we need to recall a few facts about square matrices and minimal polynomials.

3.4.1.2.1 Minimal polynomials of matrices
Given a square matrix A over a field $\mathbb{K}$ and a univariate polynomial f in $\mathbb{K}[X]$, it is clear that we can evaluate f at A, thus computing another matrix $f(A)$ of the same dimension as f. When $f(A)$ is the zero matrix, we say that f annihilates A. Let I_A be the set of polynomials of $\mathbb{K}[X]$ that annihilate A. In fact, this set I_A is an ideal of $\mathbb{K}[X]$. Indeed, if f and g both annihilate A, then for all polynomial α and β, we see that $\alpha f + \beta g$ also annihilates A. Since I_A is an ideal of univariate polynomials, if I_A is different from the zero ideal, there exists a polynomial f_A, unique up to multiplication by a non-zero element of $\mathbb{K}$, that generates I_A. This polynomial f_A is called the minimal polynomial of A.

It remains to show that I_A is not the zero ideal. This is a simple corollary of the Cayley-Hamilton theorem:

Algorithm 3.11 Lanczos's algorithm over finite fields

Require: Input vector $\vec{x}$ and routines for multiplications by M and $^\top M$

Let $\vec{X} \longleftarrow {}^\top M \cdot \vec{x}$

Initialize vector $\vec{y}$ to zero

Initialize vector $\vec{v_1}$ to random

Let $\vec{w_1} \longleftarrow {}^\top M M \vec{v_1}$

Let $N_1 \longleftarrow (\vec{w_1} | \vec{v_1})$

Let $\vec{y} \longleftarrow (\vec{X} | \vec{w_1}) \vec{v_1} / N_1$

Let $\vec{v_2} \longleftarrow \vec{w_1} - (\vec{w_1} | \vec{w_1}) \vec{v_1} / N_1$

Let $\vec{w_2} \longleftarrow {}^\top M M \vec{v_2}$

Let $N_2 \longleftarrow (\vec{w_2} | \vec{v_2})$

Let $\vec{y} \longleftarrow \vec{y} + (\vec{X} | \vec{w_2}) \vec{v_2} / N_2$

for i from 3 to n **do**

    Let $\vec{v_3} \longleftarrow \vec{w_2} - (\vec{w_2} | \vec{w_1}) \vec{v_1} / N_1 - (\vec{w_2} | \vec{w_2}) \vec{v_2} / N_2$

    Let $\vec{w_3} \longleftarrow {}^\top M M \vec{v_3}$

    Let $N_3 \longleftarrow (\vec{w_3} | \vec{v_3})$

    if $N_3 = 0$ **then**

        Exit Loop.

    end if

    Let $\vec{y} \longleftarrow \vec{y} + (\vec{X} | \vec{w_3}) \vec{v_3} / N_3$

    Let $\vec{v_1} \longleftarrow \vec{v_2}, \ \vec{w_1} \longleftarrow \vec{w_2}, \ N_1 \longleftarrow N_2.$

    Let $\vec{v_2} \longleftarrow \vec{v_3}, \ \vec{w_2} \longleftarrow \vec{w_3}, \ N_2 \longleftarrow N_3.$

end for

Let $\vec{z} \longleftarrow M\vec{y}$

if $\vec{z} = \vec{x}$ **then**

    Output: $\vec{y}$ is a solution of the system.

else

    Let $\vec{Z} \longleftarrow {}^\top M \vec{z}$

    if $\vec{Z} = \vec{X}$ **then**

        Output: $\vec{z} - \vec{x}$ is in the kernel of $^\top M$.

    else

        Output: Something wrong occurred.

    end if

end if

THEOREM 3.1
For any square matrix A, let F_A, the characteristic polynomial of A, be defined as $F_A(X) = \det(A - X \cdot Id)$. Then, F_A annihilates A.

PROOF   See [Lan05].   ☐

3.4.1.2.2   Application to linear systems   Writing down the minimal polynomial of A as:

$$f_A(X) = \sum_{i=0}^{d} \alpha_i X^i, \tag{3.22}$$

we can rewrite $f_A(A)0$ as:

$$\sum_{i=0}^{d} \alpha_i A^i = 0. \tag{3.23}$$

As a consequence, for any vector $\vec{b}$, we find:

$$\sum_{i=0}^{d} \alpha_i (A^i \cdot \vec{b}) = \vec{0}. \tag{3.24}$$

This implies that the sequence $\vec{B}$ defined as $\vec{B}_i = A^i \cdot \vec{b}$ satisfies a relation of linear recurrence:

$$\vec{B_{i+d}} = -\frac{1}{\alpha_d} \sum_{j=0}^{d-1} \alpha_j \vec{B_{i+j}}. \tag{3.25}$$

If $\alpha_0 \neq 0$, then this linear recurrence can be used backward and in particular, we may write:

$$\vec{B_0} = -\frac{1}{\alpha_0} \sum_{i=1}^{d} \alpha_i \vec{B_i}. \tag{3.26}$$

Note that when $\alpha_0 = 0$, we can factor X out of the minimal polynomial f_A and writing $A \cdot (f_a/X)(A) = 0$ conclude that A is non-invertible. Indeed, otherwise f_A/X would annihilate A, which by minimality of f_A is not possible.

From this remark, we can derive the basic idea of Wiedemann's algorithm:

- To solve $A\vec{x} = \vec{y}$, build a sequence $\vec{y}, A\vec{y}, \ldots, A^i\vec{y}, \ldots$

- Find a recurrence relation in the above sequence.

- Remark that the sequence can be expanded on its left by adding $\vec{x}$ and use the recurrence backward to recover $\vec{x}$.

However, this basic idea, as we just presented, suffers from a major obstruction. In order to determine this recurrence relation, we need a number of vectors at least equal to the length of the relation. Storing all these vectors requires roughly the same amount of memory as storing a dense matrix of the same dimensions as A. Of course, this is not acceptable for a sparse linear algebra algorithm. To avoid this problem, we do not store all the vectors in the sequence. Instead, we only keep their scalar product with some fixed vector. Of course, the current vector needs to be kept in memory when computing the next one, it is only erased after this computation. After computing enough of these scalar products, roughly twice the dimension of A, we can use Berlekamp–Massey algorithm from Chapter 2 to recover the minimal recurrence relation satisfied by the sequence of scalar product. Expressed as a polynomial, this recurrence relation divides the minimal polynomial f_A. In order to get rid of the bad case where we have a proper divisor, it is useful to study more carefully the relation between these polynomials.

For this analysis, let us look at the following polynomials:

- f_A the minimal polynomial of A.

- $f_A^{\vec{b}}$ the minimal polynomial of the sequence $S^{\vec{b}}$ of vectors $\vec{b}$, $A\vec{b}$, ..., $A^i\vec{b}$, ...

- $f_A^{\vec{b},\vec{u}}$ the minimal polynomial of the sequence $T^{\vec{b},\vec{u}}$ of vectors $(\vec{b}|\vec{u})$, $(A\vec{b}|\vec{u})$, ..., $(A^i\vec{b}|\vec{u})$, ...

It is clear that f_A annihilates $S^{\vec{b}}$ and that $f_A^{\vec{b}}$ annihilates $T^{\vec{b},\vec{u}}$. As a consequence, $f_A^{\vec{b},\vec{u}}$ divides $f_A^{\vec{b}}$ and $f_A^{\vec{b}}$ divides f_A. Note that when the constant term of $f_A^{\vec{b},\vec{u}}$ or of $f_A^{\vec{b}}$ is zero, so is the constant term of f_A and A is non-invertible. The reverse is false in general. In fact, it is even possible to have $f_A^{\vec{b},\vec{u}} = 1$, for example when $\vec{u} = \vec{0}$.

To solve a system $A\vec{y} = \vec{x}$, it suffices to compute $f_A^{\vec{x}}$. Indeed, assume that $f_A^{\vec{x}} = \sum_{i=0}^d \alpha_i X^i$ with $\alpha_0 \neq 0$ and consider the vector:

$$\vec{y} = -\frac{1}{\alpha_0} \sum_{i=1}^d \alpha_i A^{i-1}\vec{x}. \qquad (3.27)$$

Multiplying by A, we find that:

$$A\vec{y} = -\frac{1}{\alpha_0} \sum_{i=1}^d \alpha_i A^i \vec{x} = \vec{x}, \qquad (3.28)$$

by definition of the polynomial $f_A^{\vec{x}}$.

However, as said above, directly computing $f_A^{\vec{x}}$ would require to store the sequence $S^{\vec{b}}$ and would be too costly in terms of memory. As a consequence, Wiedemann's algorithm focuses on the computation of $f_A^{\vec{b},\vec{u}}$ for a random (non-zero) vector $\vec{u}$. Moreover, it uses probabilistic arguments to show that computing $f_A^{\vec{b},\vec{u}}$ for at most a few vectors $\vec{u}$ suffices to recover $f_A^{\vec{b}}$ with a good enough probability. For example, when $f_A^{\vec{b}}$ is an irreducible polynomial over the finite field we are considering, we necessarily have $f_A^{\vec{b}} = f_A^{\vec{b},\vec{u}}$ when $\vec{u} \neq \vec{0}$. Even when $f_A^{\vec{b}}$ is composite, a fraction of the vectors $\vec{u}$ satisfies $f_A^{\vec{b}} = f_A^{\vec{b},\vec{u}}$. Moreover, we can recover $f_A^{\vec{b}}$ by taking the lowest common multiple of a few polynomials of the form $f_A^{\vec{b},\vec{u}}$.

Thus, to complete the description of Wiedemann's algorithm, we need to efficiently compute $f_A^{\vec{b},\vec{u}}$ from the sequence $T^{\vec{b},\vec{u}}$. This can be done using Berlekamp-Massey algorithm, as described in Chapter 2. Since this algorithm recovers the minimal polynomial with degree at most d from a sequence of $2d$ elements over a finite field, it can be used in the context of Wiedemann's algorithm by letting d be the dimension of the matrix A. Note that when A is non-square, the smallest dimension suffices.

3.4.1.3   Block iterative algorithms

In Section 3.4.1.1, we have seen that over small finite fields, Lanczos's algorithm may fail by encountering self-orthogonal vectors during its computation. We recall that a self-orthogonal vector $\vec{x}$ whose coordinates lie in a finite field $\mathbb{F}_q$ is self-orthogonal when $(\vec{x}|\vec{x}) = 0$ (in $\mathbb{F}_q$). Heuristically, this bad event occurs with probability near $1/q$ for each scalar product occurring in the algorithm. This implies that for small finite fields, Lanczos's algorithm is likely to fail. When this happens, there is a simple way to avoid the problem: instead of using Equation (3.15) with an orthogonal basis $(\vec{v}_i)$ defined over $\mathbb{F}_q$, we consider an orthogonal basis defined over an extension field $\mathbb{F}_Q$, with Q a large enough power of q. With this simple change, the individual probability of failure of each scalar product becomes $1/Q$, assuming that the initial vector $\vec{v}_1$ is chosen at random in this large field. The fact that we search a solution for the linear system in $\mathbb{F}_q$ does not affect these probabilities, since it only impacts the scalar products with $\vec{y}$, which are not used as denominators. Thus, this simple change removes the bad behavior of Lanczos's algorithm. However, this change is very costly because each computation in $\mathbb{F}_q$ is replaced by a computation in $\mathbb{F}_Q$, without lowering the number of computations that need to be performed.

Another approach is the block Lanczos algorithm proposed in [Mon95], which starts from a block of several initial random vectors. At each round, this algorithm computes a new block of vectors by multiplying the previous block by the matrix and by performing orthogonalization with respect to the previous blocks. The details are similar to those of plain Lanczos's; however, there are some additional steps. First, the orthogonalization process requires linear algebra on some matrices, whose dimensions are the number of blocks. Second, some self-orthogonal vectors may be encountered during the computation and in order to deal with them, they are removed from the current block and added to the next one. This implies that block Lanczos algorithm needs to deal with block of (slightly) varying size.

In addition to the possibility of working with small finite fields, block Lanczos has another advantage compared to ordinary Lanczos: it is much better suited for parallel computer architecture. As a consequence, it is frequently used for computations in large dimension. Note that Wiedemann's algorithm also has a block version [Cop94], which is also well adapted to parallel or even distributed computation (see [AFK+07] for an example).

3.4.2 Structured Gaussian elimination

Structured Gaussian elimination was first introduced in the context of index calculus algorithms by Odlyzko in [Odl85] (see also Lamacchia and Odlyzko [LO91]). Its goal is, starting from a large sparse system of linear equations, to reduce it to a smaller system, while maintaining a relative sparsity. It has been especially devised for the kind of linear system typically encountered in index calculus algorithms. In these systems, some variables occur very frequently in equations while others are quite rare. Moreover, with enough sieving effort, these systems can be largely overdefined. These two specific properties are essential to the behavior of Lamacchia and Odlyzko's algorithm. Note that this algorithm is heuristic and that no definite analysis of its complexity is available. However, this is not essential in practice. Indeed, structured Gaussian elimination is not used alone, but always in conjunction with an iterative algorithm, as a practical method to reduce the cost of this iterative algorithm. Moreover, using such structured Gaussian elimination does not improve the asymptotic complexity of index calculus methods as described in Chapter 15.

In the sequel, we are given a sparse linear system of equations and we try to simplify the system, reducing the number of equations and unknowns, without losing too much sparsity. Note that it is useful to start with an overdetermined system, i.e., with more equations than unknowns. This allows us to discard some equations along the way, when they become too dense. It is also convenient to assume that most coefficients that appear in the linear system are either 1 or −1 and that an overwhelming majority of coefficients are small enough to fit in a single computer word or even in a single byte.

To introduce the ideas used in structured Gaussian elimination, we start by considering some special cases of interest. The simplest transform that can be applied to the initial system of equations is the removal of singletons. Scan the system of equations and if any unknown only appears in a single equation, remove both the unknown and the equation from the system. This transform is very efficient since it lowers the number of equations and unknowns and also reduces the total memory size of the linear system description. The next transform that we can apply looks either for equations containing two unknowns only or for unknowns that appear in two equations only. In both cases, it is convenient to add the additional constraint that the two coefficients involved are both either 1 or -1. In this case, an equation containing two unknowns can be rewritten as $y = \pm x + a$, where x and y are the involved unknowns and a is a constant; thus y can be replaced everywhere by either x or $-x$, with a corresponding change to the right-hand side constants. Similarly, if x only appears in two equations E and E', we can replace these two equations by either $E - E'$ or $E + E'$ and remove x from the list of unknowns. Note that these two transforms are in some sense dual. Applying one to the matrix of a linear system corresponds to applying the other one to the transposed matrix. These two transforms remove one equation and one unknown, while leaving the memory size of the system essentially constant. In fact, the amount of memory usually decreases by 2 coefficients.

These two simple transforms can be applied iteratively; indeed, removing or merging equations may create new singletons or doubletons. Once this is done, we need to consider heavier equations and/or unknowns that appear more often. Typically, if an equation with t unknowns exists and if, at least, one of its coefficients is 1 or -1, then the corresponding unknown can be replaced by an affine combination of the others everywhere in the linear system. Similarly, if an unknown appears ℓ times, at least once with coefficient 1 or -1, it can be eliminated by subtracting the corresponding equation from each of the $\ell - 1$ others. In fact, if unknown x appears in equation E with coefficient 1 or -1, with x appearing a total of t times and if E has weight ℓ, this transform removes one unknown and one equation; moreover, assuming that no other cancellation occurs, the memory size is increased by $(t-1)(\ell-1) - t - \ell + 1 = (t-2)(\ell-2) - 2$.

Since we took care to start from an overdefined system of equations, another basic transform is available; it consists of simply removing some equations when the system becomes too large to fit into memory. Of course, during this step, it is preferable to remove the equation(s) with the largest weight.

All existing variations of structured Gaussian elimination are essentially heuristic methods that aim at using these transforms or close variations in an efficient way.

3.4.2.1 Odlyzko's method

In the original method of Odlyzko, the geometry of the specific systems of equations encountered during index calculus algorithms[1] is also taken into account. With these systems, some unknowns, corresponding to small primes, occur frequently, while others are much rarer. This dissymmetry is reflected in the structured Gaussian elimination by distinguishing between heavy and light unknowns. The transforms that are applied are adapted to take this distinction into account. More precisely, transforms in the following list are applied repeatedly:

---

[1]See Chapter 15.

- Find all light unknowns which appear only once and remove the corresponding rows and columns.

- Find all light unknowns which appear alone, with coefficient 1 or -1, in some row, replace the unknown everywhere by its expression in terms of heavy unknowns and remove the corresponding row.

- Enlarge the set of heavy unknowns, declaring as heavy some of the previously light unknowns. These unknowns are selected among the more frequently occurring.

- Remove some heavy rows.

The above transforms eliminate singletons, with a modified definition of weight that ignores the contribution of heavy unknowns, i.e., of unknowns which appear too frequently. In [LO91], the elimination of doubletons was also considered; however, experiments showed that in this case, the overall weight grows too quickly.

This approach of Odlyzko works well for linear systems of equations that appear during index calculus algorithms for factoring or computing discrete logarithms. Moreover, when implementing the method it is possible to remove heavy unknowns from memory and to recompute them afterwards by journaling the history of the computation. However, the approach also presents some drawbacks:

- The approach is not well adapted to the large primes variation of index calculus. To deal with such systems, it is preferable to work with the graph method presented below.

- Linear systems arising during the computation of discrete logarithms in finite fields of small characteristics behave badly with this algorithm. The problem is that instead of having unknowns associated with prime numbers, whose frequencies slowly decrease as the primes become larger, we have unknowns that correspond to irreducible polynomials and the frequency is a function of the degree of this polynomial. As a consequence, when modifying the set of heavy unknowns, any reasonable strategy makes jumps and adds a whole family of unknowns, corresponding to the same degree. In this context, the overall behavior is quite bad. There, the greedy method below is preferable.

- Finally, despite the relative simplicity of the method, programming this approach is quite tricky and there is no easy way to check that the structured Gaussian elimination did not subtly alter the system due to an undetected bug. When faced with this concern, using the lightweight approach of Section 3.4.2.3 is a good idea.

3.4.2.2   Markowitz pivoting

Instead of partitioning the unknowns between the heavy and light categories, we can use Markowitz pivoting, a greedy method that simply tries
to apply the basic transforms we described in the introduction of this section. We already know that performing Gaussian elimination to remove an
unknown that appears in ℓ different equations and in particular in an equation
of weight t, with a coefficient 1 or -1, usually increases the total number of
entries in the linear system by $(t-2)(\ell-2)-2$. Markowitz pivoting selects
at each step of the computation to perform the elimination corresponding to
the smallest possible increases of the linear system size. When applying this
strategy, the size of the system slowly increases; when it reaches a predeter-
mined threshold, a fraction of the equations with highest weight is removed.
The program stops when the number of remaining equations becomes smaller
than the number of remaining unknowns plus a small fixed security margin,
e.g., 100 extra equations. Another option for halting structured Gaussian
elimination is to compute at regular intervals an estimate of the complexity
of running an iterative solver on the current system of equations. At the
beginning of structured elimination, this estimate quickly decreases, at some
point of the elimination it reaches a minimum, when this happens, stopping
the structured elimination is a good idea. All in all, this often gives a better
result than pushing the elimination too far, which results in a denser matrix.

The principle of the greedy approach is very simple and was first described
in [Mar57]. However, the bookkeeping required by this approach is quite
massive. The key questions are to efficiently locate at each step the pair
unknown/equation that corresponds to the lowest value of $(t-2)(\ell-2)-2$ and
to efficiently perform the Gaussian elimination steps. Clearly, recomputing
$(t-2)(\ell-2)-2$ for all pairs at each step of the computation to locate the
minimum value is going to be too costly. Similarly, in order to perform the
Gaussian elimination or pivoting step, we cannot afford to scan the complete
linear system in order to locate the unknown that needs to be eliminated.
Thankfully, these two tasks can be managed efficiently by using adequate
data structures.

3.4.2.2.1   Structure for efficient pivoting   Given a pair (E, x) where E
is an equation of weight t and x an unknown that appear in E with coefficient
1 or -1 and ℓ is the total number of equations that contains x, we need to
efficiently locate all equations that contain x and add/subtract some multiple
of E from these equations. A neat solution to perform this task is to store
not only the sparse matrix corresponding to the system of equations but also
its transpose. Note that when storing this transpose, it is possible to take
advantage of the heavy column idea of Odlyzko's original approach. Indeed,
unknowns with a very large value for ℓ are unlikely to minimize $W(\ell, t) =
(\ell-2)(t-2)-2$. As a consequence, it suffices to store a fraction of the
transpose. During pivoting, both the linear system and the transposed matrix

need to be updated. However, since the linear system is sparse, the update task only affects a fraction of the unknowns and equations.

3.4.2.2.2 Structure for pivot selection To select a pivot, we need to find the pair (E, x) that minimizes the quantity $W(\ell, t)$. A first remark is that for a given unknown x, $W(\ell, t)$ can be computed by scanning the column of x using the transposed copy of the linear system and by remembering the equation where x appears with coefficient 1 or -1 that has the lowest weight. Thus, for every unknown x, we can define a best possible pair (E, x) and a corresponding value $W(\ell, t)$. During a pivoting step, thanks to sparsity, only a small number of unknows are subject to change and need an update of their $W(\ell, t)$ value. The only remaining question is to find a data structure that allows to efficiently locate the unknown with the smallest value at each step and that also allows efficient updates. This can for example be achieved using self-balancing tree techniques (see Chapter 6).

3.4.2.3 A lightweight approach

One notable disadvantage of using Markowitz pivoting is that since efficient implementations of this algorithm require a complex data structure, they may contain subtle, hard to detect bugs. In the context of index calculus, such a bug may cause failure of the sparse linear algebra algorithm used after structure Gaussian elimination. However, such a failure could also arise due to a bug in sieving or even due to a hardware failure at some point during the process that would modify a single equation somewhere. When this happens, it may become necessary to use simpler algorithms. For this reason, we now propose a lightweight approach to structured Gaussian elimination. After an initialization step, this approach works in several rounds, each round being broken into two phases. During initialization, we count the total number of occurrence of each unknown and memorize the list of light unknowns which occur less than some fixed threshold. The first phase of each performs statistics on the sparse system of equations, stored in memory. More precisely, it sorts the equations by increasing values of the weight, or rather estimated weight. When an equation contains a single light (active) unknown, this unknown is remembered as a pivot for the equation. In case of conflict between two equations, the lighter one is given precedence to keep the pivot. The second phase performs the pivoting step in a lazy way. It simply adds to each equation containing one or more unknowns declared as pivot a list of entries that remember that a multiple of the equation corresponding to the pivot should be added to it. The addition is not performed, however, the estimated weight of the equation is modified, by adding to it the weight of the equation being added minus 1. After that, all pivots, together with their corresponding equations, are declared unactive unknowns and we proceed to the next round.

In a final phase, the algorithm evaluates for each equation the sum of the initial entry and of the contribution of all equations that have been lazily added. Note, that this needs to be done in a recursive manner since the added equations may also contain further lazy additions. Any equation that becomes too large during this process is simply discarded. Note that for debugging purposes, only this part of the algorithm need to be checked in details. Indeed, by construction, this part can only output linear combinations of the original equations, which by definition are valid equations. Any bug in the previous parts of the program might affect the level of sparsity of the output system but not its correctness.

3.4.2.4   Graph method

In the so-called large prime variation of index calculus algorithms, we encounter systems of equations with some additional structure. More precisely, the set of unknowns is divided in two parts: unknowns corresponding to regular primes and unknowns corresponding to large primes. The main difference is that on average, each large prime appears less than once in the linear system, while regular primes appear at least a few times on average. Moreover, large prime unknowns always have coefficients 1 or -1. Of course, an unknown that appears only once can be safely removed, with the corresponding equation from the linear systems. Thus, the average large prime essentially yields a useless equation. However, due to birthday paradox effects (see Chapter 6), a small fraction of large primes appears at least twice. If there is at most one large prime in each equation, it suffices to merge the two equations where a given large prime occurs as in basic structured Gaussian elimination. The resulting equation has double weight, thus remaining reasonably sparse. However, with current index calculus algorithms, we may have up to four large primes per equation.

Another specificity of the large prime variation is that since a large fraction of large primes does not appear, a large fraction of the collected linear equations is lost. As a consequence, we need to consider a huge system, much bigger than systems arising from index calculus without large primes. For this reason, it is important to use algorithms that do not require loading into memory the complete system of equations. Instead, the algorithms should work using only the large prime parts of equations. This approach called filtering is in particular discussed in [Cav00]. Of course, in a final stage, it is necessary to compute the resulting linear system after large prime elimination. However, this final stage only involves the small fraction of equations that remains and is not a real problem.

With this idea in mind, specific algorithms were proposed. These algorithms treat Gaussian elimination as a graph problem. To explain how, let us first consider the restricted case with at most two large primes per equation. In that case, we can view each unknown of the large type as a node in the graph and each equation as an edge between the two large primes that appear in the equation. When a given large prime appears alone in an equation, we construct an edge between the large prime and a special node no-unknown.

This representation is helpful, because any linear combination (with ±1 coefficients) of equations where all large prime unknowns cancel out induces a cycle in the above graph. Moreover, the converse is "almost" true. If the linear system is considered modulo 2 as when factoring, since any node in a cycle appears an even number of times, usually twice, summing all equations that correspond to the edges of the cycle, the contribution of the large primes vanishes. When the linear system is modulo a larger prime, it is possible to adjust the sign of the contribution of each equation in the cycle to make sure that all unknowns but one vanish. If the special node no-unknown occurs in the cycle, we can make the contribution of all other nodes vanish and thus obtain a new equation without large primes. If it does not occur, we need to rely on luck for eliminating the final large prime unknown and we succeed with probability $1/2$. In truth, the whole approach is more complicated than that, because it does not suffice to find a single cycle. To make the most out of the available equations, we need to construct a set called a cycle base which generates all possible cycles with redundancies.

3.4.2.5 Final backtracking

After applying any of the above structured Gaussian elimination techniques, one obtains a reduced sparse system of equations which can be solved using iterative techniques. However, after that, it is often necessary to turn back to the initial linear system. Two cases arise depending on the target application. For factoring, the goal of the linear algebra step is to find an element of the kernel of the sparse matrix. For discrete logarithms, the goal is to find values for the unknowns.

Let us start by considering the case of discrete logarithms. After applying the iterative solver, we have values only for the unknowns that remain in the linear system after structured Gaussian elimination. We would like to complete this result and recover a value for all unknowns present in the original system. This can be done using a very simple backtracking algorithm. In fact, it suffices to scan the initial system of equations and, whenever an equation with a single undetermined unknown is encountered, to compute a value for this unknown. Repeating this step until no new value can be added yields a complete (or almost complete) table of values for the unknowns of the initial system. Indeed, since structured Gaussian elimination eliminates one unknown at a time, this backtracking method can recover the missing values one by one.

The case of factoring is slightly more complicated. Here, we are given a kernel element as a linear combination of equations in the transformed system and we need to express it as a linear combination of equations in the original system. The simplest way to do this is to use journaling in the structured Gaussian elimination algorithm. In the journal, we give labels to all initial equations and to all intermediate equations that arise during Gaussian elimination. Note that, each intermediate equation is a linear combination of two preexisting equations. We now write a kernel element as a linear combination of equations, allowing all initial, intermediate or final equations in the expression. Such a expression can be iteratively transformed using the journal by replacing each equation by its expression in terms of its immediate predecessors. At the end, we obtain a linear equation that only involves the initial equations. Since the linear combinations that represent kernel elements are not sparse, we can only perform backtracking on a small number of kernel elements. However, this is enough for factoring purposes.

Exercises

1[h]. Construct a linear system of 4 equations in 4 unknowns, without a solution. Another one, over the field of reals, with infinitely many solutions. Write a linear system of 4 equations in 4 unknowns, with a unique solution, such that Algorithm 3.4 fails. Choose your example in a way that ensures that if the order of the equations in the system is reversed, Algorithm 3.4 succeeds.

2. Put together the code fragments in Programs 3.4 and 3.5 to realize a complete multiplication of 32×32 Boolean matrix. Extend this to 64×64 and 128×128. Profile the execution of the resulting code on your computer, using the available compiler(s), identify the critical parts of the program. Without changing the algorithm, rewrite the critical parts of the code to improve the overall performance.

3[h]. Assume that we would like to multiply one fixed square Boolean matrix of small dimension by many matrices of the same size. To improve the performance, we consider the algorithm of four Russians which precomputes and stores the results all the possible multiplication for rectangular slices of the original matrix. Assuming that the number of multiplications is very large, we can ignore the time of this precomputation. Implement and optimize this on your computer.

4. Program 3.6 uses the 32×32 matrix multiplication as a black box, in order to multiply $32n \times 32n$ Boolean matrices. This can be improved in many ways. A first option is to simply share the fold operations rather than doing them independently for each small multiplication. A second option is to use the algorithm of four Russians from the previous exercise.

5[h]. In order to make Strassen's multiplication algorithm running time behave more smoothly as a function of the dimension, we should change the rounding rule and the decision to cut off the algorithm. One simple but time-consuming option is to run the program for each size using three possibilities (or two for even dimensions), Strassen with rounding up, Strassen with rounding down or basic multiplication. After this, save the most efficient option for this dimension in a table. Then, when the multiplication is called recursively from a larger size, reuse the saved option. Note that this table of decision needs to be constructed by starting from small sizes and going up.

6. Reprogram a recursive multiplication algorithm using Winograd's formulas instead of Strassen's.

7^h. Consider Schur's complement method for matrix multiplication. How would you bypass the problem caused by the fact that D might be non-invertible? The goal is to achieve asymptotic complexity with exponent $\log_2 7$.

8. Gaussian elimination can also be used for matrices defined over a ring. For example, consider the issue of solving linear systems of equations modulo 256 when the determinant of the system is a multiple of 256 (see Section 3.3.3.1). What new difficulties do appear? How can you solve them?

9^h. Implement the computation of Hermite normal forms over the integers.

Some of the methods presented in this chapter can also be a basis for programming projects. Implementations of structured Gaussian elimination or iterative algorithms are very interesting in this respect, especially in conjunction with the applications presented in Chapters 11 and 15.

Chapter 4

Sieve algorithms

Many modern algorithms for factoring and computing discrete logarithms in finite fields heavily rely on sieving in practical implementations[1]. These index calculus algorithms themselves are described in the applications part of this book, more precisely in Chapter 15. We chose to present the sieve algorithms in the present chapter in order to focus on the variety of possible implementations and available trade-offs.

4.1 Introductory example: Eratosthenes's sieve

Eratosthenes (276–194 BC) was an ancient Greek mathematician. Among his many contributions, he invented one of the first algorithms known. This algorithm, called Eratosthenes's sieve, is an efficient method for computing all prime numbers up to a given bound. In 2004, Atkin and Bernstein [AB04] found a faster algorithm for achieving this task. However, before that, Eratosthenes's sieve had essentially been unsurpassed for more than two millennia. In this section, in order to introduce sieve algorithms, we show how to efficiently implement Eratosthenes's sieve on modern computers.

4.1.1 Overview of Eratosthenes's sieve

The basic idea of Eratosthenes's sieve is to start from a list of integers from 1 up to some initial limit and to remove all non-primes from the list, thus ending with a list of primes. The first step is to cross 1, which is a unit, thus not a prime, in the ring of integers. After that, the smallest remaining number, 2, is a prime. As a consequence, all multiples of 2 are composite numbers and can be crossed from the list. The next smallest remaining number is 3, which is prime and its multiples need to be crossed. After that, we find in turn the prime numbers 5, 7, 11 and so on, cross their multiples and continue until

---

[1] As we will see in Chapter 15, sieving is only an implementation aspect of these algorithms and can be ignored where asymptotic analysis is concerned. Nonetheless, it is a very important practical aspect.

we reach the end of our list. To make the description more precise, we write down the pseudo-code in Algorithm 4.1.

Algorithm 4.1 Eratosthenes's sieve

Require: Initial limit Lim

  Create array $IsPrime$ indexed from 1 to Lim, initialized to **true**.

  $IsPrime[1] \longleftarrow$ **false**

  for i from 2 to Lim **do**

    if $IsPrime[i]$ is **true** **then**

      $j \longleftarrow 2i$

      repeat

        $IsPrime[j] \longleftarrow$ **false**; $j \longleftarrow j + i$

      until $j >$ Lim

    end if

  end for

  for i from 2 to Lim **do**

    if $IsPrime[i]$ is **true** **then**

      Display(i,'is prime')

    end if

  end for

Before going through the process of programming and optimizing this algorithm, let us analyze its main properties. First of all, the algorithm needs to work with Boolean values. From a theoretical point-of-view, this is extremely nice. However, the practical use of Booleans in the C language is not completely straightforward. Indeed, no Boolean type was defined in early C89, and while an extension was added in C99, it is still not supported by all compilers. Thus, as in Chapter 3 we use the trick of packing several Booleans into the machine representation of a single (unsigned) integer. This allows us to represent the required array of Lim Booleans into Lim/8 bytes of memory.

From the point-of-view of correctness, let us show that the algorithm works correctly. It is necessary to prove that any prime number has **true** stored in the corresponding entry of table $IsPrime$ and that any composite number C has **false** stored. Clearly, since the table is initialized with **true** and since only 1 and proper multiples of accepted numbers are crossed, no prime number may be crossed from the list. Thus, we need to check that all composites are effectively removed. More precisely, this event happens during the first i loop, before i reaches value C. Indeed, if C is composite, it has at least one prime divisor, say $P < C$. Since P is prime, it remains in the final list and thus all its multiples, including C are crossed. Thus the algorithm is correct.

As the runtime analysis goes, the algorithm spends $\lceil \text{Lim}/p \rceil$ steps in the inner loop to remove the multiples of the prime p (when i is equal to p).

Thus, the total running time is bounded by:

$$\sum_{p=2}^{\text{Lim}} \left\lceil \frac{\text{Lim}}{p} \right\rceil. \tag{4.1}$$

Looking more precisely at the algorithm, we see that every composite is crossed more than once. This opens the way for a slight improvement, it suffices to cross multiples of p starting from p^2. Indeed, each composite C has at least one prime divisor p such that $p^2 \leq C$. This reduces the number of iterations of the outside loop down to $\sqrt{\text{Lim}}$ and improves the running time. However, asymptotically, the gain is negligible. The main reason is that small primes have many multiples and sieving over these primes is the costly part of the algorithm. In the next section, we present possible improvements in order to speed up practical implementation of Eratosthenes's sieve. Our starting reference implementation is the C code of Program 4.1.

4.1.2 Improvements to Eratosthenes's sieve

Looking at our reference Program 4.1, we quickly discover two main limitations on its range of applicability. The first limitation, which is easily overcome, is the fact that with many compilers, the integer type `int` represents signed 32-bit integers which cannot count above $2^{31} - 1$. To remove this limitation, it suffices to define the loop variables i and j as 64-bit integers. The next limitation is the need to store the Boolean array containing one bit for each number up to the considered limit. Even with a 1 Gbyte memory, this prevents us from building primes further than 2^{33}.

4.1.2.1 Wheel factorization

Our first improvement pushes this limit further by focusing on the regularity of the distribution of multiples of small primes in order to use less memory. The most basic example of this idea is to remove all even numbers from our table in memory. This is safe since the only even prime is 2. Moreover, this can be done without any essential change to the sieving process. Indeed, if x is an odd multiple of p, the next odd multiple of p is $x + 2p$ whose representation is located p steps further than the representation of x in memory. As a consequence, this improvement also allows us to run Eratosthenes's sieve twice as far for the same cost in terms of both time and memory. To push this idea further, let us consider the three primes 2, 3 and 5. Between, 1 and 30, only eight numbers, namely 1, 7, 11, 13, 17, 19, 23 and 29, are not multiples of either 2, 3 or 5. Moreover, adding 30 or a multiple of 30 to any number preserves the properties of being multiple of 2, 3 or 5. As a consequence, in any interval of 30 consecutive integers, we know that Eratosthenes's sieve already removes 22 numbers, simply by sieving over the small primes 2, 3 and 5. We can use this fact and restrict our table of Booleans in memory

Program 4.1 Basic C code for Eratosthenes's sieve

```c
#include <stdio.h>
#include <stdlib.h>

typedef unsigned int packedbool;
packedbool *_IsPrime;

#define NpckBool (8*sizeof(packedbool))
#define GetIsPrime(x) \
   (_IsPrime[(x-1)/NpckBool]>>((x-1)%NpckBool))&1
#define SetCompo(x) \
   _IsPrime[(x-1)/NpckBool]&=~(1UL<<((x-1)%NpckBool))

SievePrimes(int Limit)
{
  int i,j, tabLimit;

  tabLimit=(Limit+NpckBool-1)/NpckBool;
  for (i=0;i<tabLimit;i++) _IsPrime[i]=~0;
  for (i=2;i*i<=Limit;i++){
    if (GetIsPrime(i)) {
      for (j=i*i;j<=Limit;j+=i){
        SetCompo(j); } } }

  printf("List of primes up to %d:\n",Limit);
  for (i=2;i<=Limit;i++){
    if (GetIsPrime(i)) {
      printf("%d\n",i); } } }

main()
{
  int Limit;
  printf("Enter limit ?");
  scanf("%d",&Limit);
  _IsPrime=(packedbool *)malloc((Limit+NpckBool-1)/8);
  SievePrimes(Limit);
  free(_IsPrime);
}
```

to consider only numbers of the form $30k + \delta$, for all integers k and for δ chosen among 1, 7, 11, 13, 17, 19, 23 and 29. This allows us to represent more than three times more numbers than the initial sieve into the same amount of memory. In order to visualize the algorithmic principle, it is useful to represent the numbers on a modular wheel of perimeter 30 as in Figure 4.1. For this reason, this particular approach to Eratosthenes's sieve is often called wheel factorization.

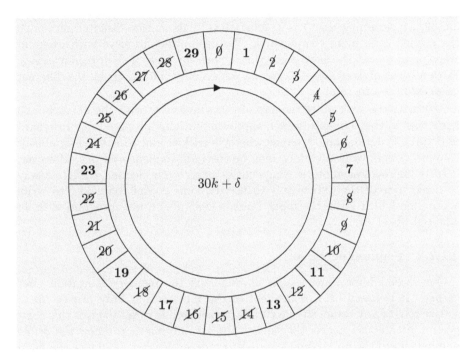

Figure 4.1: Schematic picture of wheel factorization

Of course, we can even add more small primes into the picture to further improve the ratio, going up to the small prime 11, we restrict ourselves to 480 numbers out of 2310, thus gaining a factor of almost five. Using wheel factorization also speeds up the computation. Indeed, it removes the need to sieve over the smallest, more costly, primes. However, the approach has the drawback of complicating the matter of sieving over larger primes, since the locations of multiples of these primes are no longer evenly distributed in memory. To show the difficulty, let us consider the distribution of the multiples of 7 in a wheel of perimeter 30. We already know that our memory only holds numbers which are multiples of neither 2, 3 nor 5. We now need to locate multiples of 7 among these numbers. They are obtained by multiplying together 7 with a number in the wheel, since multiplication by 7 does not change the divisibility by 2, 3 or 5. Thus, we need to mark all numbers of the form: $7 \cdot (30k + \delta)$ for all k and for δ in 1, 7, 11, 13, 17, 19, 23 and 29. These numbers form eight series, one for each value of δ and each series consists in evenly distributed numbers in memory, separated by a distance of 7×8 as shown in Figure 4.2, with one series in each column. However, the figure also shows that the starting points of the different series are not evenly spaced. In our example, the starting points correspond to the numbers 7, 49, 77, 91, 119, 133, 161 and 203 and their respective locations in memory are 1, 13, 20, 24, 31, 35, 42, 54; assuming that 1 is at location 0. This uneven distribution makes the sieving code more complicated. We need either to sieve with irregular steps or to repeat the sieve eight times, once for each of the different series. With a wheel of fixed size, this can be achieved by embedding all the different cases into the program.

Alternatively, we can let the size of the wheel grow with the upper bound Lim, this is the dynamic wheel approach initially proposed by Pritchard in [Pri81]. Programming dynamic wheels is trickier and adds some significant computational overhead, which may be costly in implementations. However, it truly improves asymptotic complexities as shown in Section 4.1.2.4. A way to avoid part of the difficulty is to use dynamic compilation, i.e., to write a program which given the input bounds generates a new program with an adapted wheel size.

4.1.2.2   Segmented sieve

Our second improvement aims at removing the memory limitation altogether. It is based on a refinement of the remark that only primes up to square root of any composite need to be considered in order to cross this composite. As a consequence, the sieving process only needs to know the prime numbers up to the initial limit in order to proceed further up. Thus, the main part of the computation only depends on a short initial segment of Eratosthenes's sieve. This implies that the computation can be split into several smaller computations, each within an interval of length equal to the square root of the initial limit. Each of these smaller computations is essentially independent of the previous ones, the notable exception being the computation of the primes in the first of these intervals. Doing this, we can reduce the amount of required memory drastically, representing only short intervals of square root length. However, sieving each interval independently adds some computational overhead. Indeed, we need to compute the location of the first multiple of each prime in each interval. In fact, it is better to reuse some of the memory we just saved in order to remember this location from one interval to the next. This is illustrated by the C code of Program 4.2.

This approach, which works with each segment in turn, is called segmented sieve. From a complexity point-of-view, it greatly reduces the amount of required memory. However, the number of operations to perform remains

Program 4.2 Small memory code for Eratosthenes's sieve

```c
#include <stdio.h>
#include <stdlib.h>
typedef unsigned int packedbool; typedef long long int int64;
packedbool *_IsBPrime; packedbool *_IsPrime; int64 *Offset;
#define Npck (8*sizeof(packedbool))
#define GetBasePrime(x) (_IsBPrime[(x)/Npck]>>((x)%Npck))&1
#define SetBaseCompo(x) _IsBPrime[(x)/Npck]&=~(1UL<<((x)%Npck))
#define GetIsPrime(x) (_IsPrime[(x)/Npck]>>((x)%Npck))&1
#define SetCompo(x) _IsPrime[(x)/Npck]&=~(1UL<<((x)%Npck))
#define Action(p) printf("%lld\n",(int64)(p))

int InitialSievePrimes(int Limit) {
  int i, j, count, tabLimit;
  count=0; tabLimit=(Limit+Npck-1)/Npck;
  for (i=0;i<tabLimit;i++) _IsBPrime[i]=~0;
  for (i=2;i*i<Limit;i++){
    if (GetBasePrime(i)) { count++;
      for (j=i*i;j<Limit;j+=i) SetBaseCompo(j); } }
  for (;i<Limit;i++){if (GetBasePrime(i)) count++;}
  Offset=(int64 *)malloc(count*sizeof(int64)); count=0;
  for (i=2;i<Limit;i++){
    if (GetBasePrime(i)) { Action(i);
      j=Limit%i; if (j!=0) j=i-j; Offset[count]=j;
      count++; } } }

int SievePrimesInterval(int64 offset, int Length) {
  int i, j, count, tabLimit;
  count=0;  tabLimit=(Length+Npck-1)/Npck;
  for (i=0;i<tabLimit;i++) _IsPrime[i]=~0;
  for (i=2;i<Length;i++) {
    if (GetBasePrime(i)) {
      for (j=Offset[count];j<Length;j+=i) SetCompo(j);
      Offset[count]=j-Length; count++; } }
  for (i=0;i<Length;i++) {if (GetIsPrime(i)) Action(offset+i); }
}
```

Program 4.2 Small memory code for Eratosthenes's sieve (continued)

```
main() {
  int i,j; int64 Limit, tmp; int sqrt;

  printf("Enter limit ?"); scanf("%lld",&Limit);

  for(tmp=0;tmp*tmp<=Limit;tmp++); sqrt=tmp;

  _IsBPrime=(packedbool *)malloc((sqrt+Npck-1)/8);
  _IsPrime=(packedbool *)malloc((sqrt+Npck-1)/8);
  InitialSievePrimes(sqrt);

  for(tmp=sqrt;tmp<Limit-sqrt;tmp+=sqrt)
    SievePrimesInterval(tmp,sqrt);
  SievePrimesInterval(tmp,Limit-tmp);

  free(_IsPrime);free(_IsBPrime);
}
```

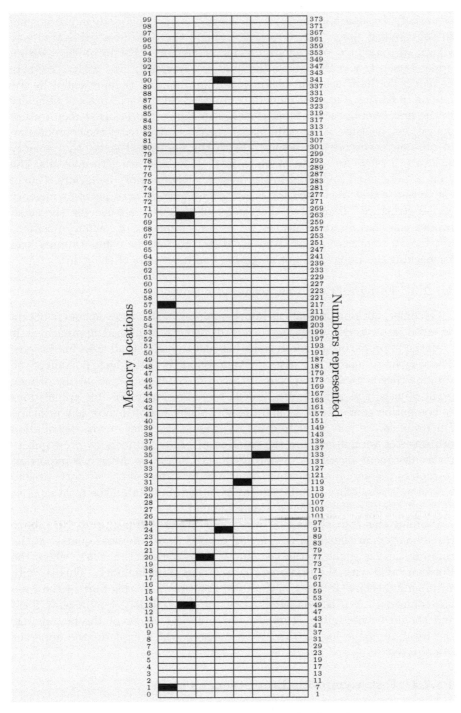

Figure 4.2: Multiples of 7 in a wheel of perimeter 30

essentially the same and the approach does not seem to offer any improvements to the running time. Yet, in practice, doing the sieve in a segmented way allows to work in a very small amount of memory. For example, with an upper-bound Lim $= 2^{34}$, above the limit we previously had with a 1 Gbyte memory, the basic segment contains 2^{17} integers and is represented by 2^{17} bits or 16 Kbytes. On most modern computers, this fits without any difficulty in the first level of cache. Since sieving is a simple process that requires few operations, the time of the main memory accesses dominates the computation of our reference code for sieving. With the segmented sieve, the memory accesses in cache are about twice as fast. It is interesting to note that the gain is quite small compared to the most extreme memory behaviors. One of the reason is that the basic sieve is already quite good in terms of its memory access structure. Remember that at lot of time is spent sieving the small primes and that for small primes we access a sequence of memory positions not too far from each other. Another reason is that the computational load for packing the Boolean is not negligible and hides part of the gain.

4.1.2.3   Fast practical Eratosthenes's sieve

Of course, nothing prevents us from combining both improvements to obtain an even faster code. However, before devising a combined approach, it is essential to profile our previous codes in order to understand their limitations. The very first remark that arises during profiling is that a huge portion of the running time is used to simply print the prime numbers. From an algorithmic point-of-view, this is irrelevant, indeed there are many possible applications to computing prime numbers and many of them do not involve any printing. For example, we may want to compute general statistics on the distribution of primes or we may require the creation of prime numbers as pre-requisite for a subsequent algorithm. As a consequence, before profiling it is important to replace the action of printing the primes by a different, less costly action. One simple possibility is to increment a counter and obtain the total number of prime numbers below the considered bound.

Profiling this code shows that when building the primes up to 10^9, about three quarters are spent in the sieving process itself, the final quarter of the running time is spent in the last line of SievePrimesInterval, where the Boolean table with the prime status is reread from memory. This is quite surprising, because from an asymptotic point-of-view, this part of the code has complexity $O(\text{Lim})$ compared to $O(\text{Lim} \cdot \log \log(\text{Lim}))$ for the sieve itself and should be negligible. However, for practical values of the bound Lim, the resulting value $\log \log(\text{Lim})$ is very small, which explains the apparent discrepancy.

4.1.2.4   Fast asymptotic Eratosthenes's sieve

From an asymptotic point-of-view, there are several different optimums when looking at the sieve of Eratosthenes, depending on the relative contribution of time and space to the complexity. In particular, in [Sor98], Sorenson presented four different combinations with different balance of time and space.

We can either try to optimize the asymptotic running time or take into account both the running time and the memory use. When considering the asymptotic running time only, the best known variation on Eratosthenes's sieve is an algorithm by Pritchard, described in [Pri81]. This algorithm uses wheel factorization with the largest possible wheel applicable to the upper sieve bound `Lim` given in input. Clearly, the wheel size is a function of `Lim` and this approach needs to deal with the additional technicalities of using wheel of varying size. The resulting algorithm is sublinear in time but requires a large amount of space.

Another approach proposed by Pritchard [Pri83] is using both wheels and a segmented sieve. It only achieves linear time but reduces the memory requirements down to the square root of the bound `Lim`.

4.1.3 Finding primes faster: Atkin and Bernstein's sieve

As we saw above, asymptotically, the sieve of Eratosthenes can be sublinear; it can also be performed using a small amount of memory. However, it is not known how to achieve both properties simultaneously for this sieve. This has been an open problem for a long time. In 1999, Atkin and Bernstein proposed in [AB04] a new algorithm, not based on Eratosthenes's sieve, that solves this problem. This algorithm is constructed from 3 subroutines each addressing a subset of the prime numbers, using a characterization of primes which does not make use of divisibility. The first algorithm can be used to find primes congruent to 1 modulo 4, the second finds primes congruent to 1 modulo 6 and the third finds primes congruent to 11 modulo 12. Since all primes, but 2 and 3, are of these forms these three subroutines suffice. For primes congruent to 1 modulo 12, we need to choose either the first or second subroutine.

Let us start with the case of primes congruent to 1 modulo 4. The algorithm is based on the following theorem:

THEOREM 4.1 (Th. 6.1 in [AB04])
Let n be a square-free positive integer with $n \equiv 1$ (mod 4). Then n is prime if and only if the cardinality of the set $\{(x,y)|x > 0, y > 0, 4x^2 + y^2 = n\}$ (or equivalently of the set $\{(x,y)|x > y > 0, x^2 + y^2 = n\}$) is odd.

The first subroutine follows quite naturally from this theorem. It contains two phases: the first phase computes the cardinality of the set given in Theorem 4.1, the second phase removes non-square-free integers. This results in Algorithm 4.2. For more efficiency, this algorithm can be combined with a wheel. For example, in [AB04] it is presented (as Algorithm 3.1) using a wheel of 30. Note that such a wheel is easily combined with the congruence condition modulo 4, thus resulting in an extended wheel of length 60 containing 8 different numbers.

Algorithm 4.2 Sieve of Atkin and Bernstein for primes $\equiv 1 \pmod 4$

Require: Initial range $\texttt{Lim} \ldots \texttt{Lim} + 4Count$ with $\texttt{Lim} \equiv 1 \pmod 4$

  Create array $IsOdd$ indexed from 0 to $Count$, initialized to $\texttt{false}$.

  for all (x, y, i) with $x > 0$, $y > 0$, $i \leq Count$, $4x^2 + y^2 = \texttt{Lim} + 4i$ **do**

    Negate the Boolean value of $IsOdd[i]$

  end for

  for all prime q with $q^2 \leq \texttt{Lim} + 4Count$ **do**

    for all $\texttt{Lim} + 4i$ multiples of q^2 **do**

      Set $IsOdd[i]$ to $\texttt{false}$

    end for

  end for

The other subroutines derive from two other Theorems 4.2 and 4.3. They can also be implemented using extended wheels of 60, respectively containing eight and four numbers.

THEOREM 4.2 (Th. 6.2 in [AB04])

Let n be a square-free positive integer with $n \equiv 1 \pmod 6$. Then n is prime if and only if the cardinality of the set $\{(x, y) | x > 0, y > 0, 3x^2 + y^2 = n\}$ is odd.

THEOREM 4.3 (Th. 6.3 in [AB04])

Let n be a square-free positive integer with $n \equiv 11 \pmod{12}$. Then n is prime if and only if the cardinality of the set $\{(x, y) | x > y > 0, 3x^2 - y^2 = n\}$ is odd.

Putting together the three subroutines implemented with an extended wheel of 60 gives a practically efficient code. A package called $\texttt{primegen}$ written by Bernstein and distributed on his web pages gives an efficient implementation of this algorithm. Moreover, from an asymptotic point-of-view, this new algorithm improves upon the best algorithms based on Eratosthenes's sieve. In this case, we need to use a wheel containing all the primes up to $\sqrt{\log \texttt{Lim}}$ and achieve running time $O(\texttt{Lim}/\log\log(\texttt{Lim}))$ using $\texttt{Lim}^{1/2 + o(1)}$ bits of memory.

4.1.3.1   Further improvements of Atkin and Bernstein's sieve

While more efficient than algorithms based on the sieve of Eratosthenes, Atkin-Bernstein's sieve share a drawback with it. Assume that we do not want all the primes up to $\texttt{Lim}$, but only the primes in some interval $[\texttt{Lim}_1, \texttt{Lim}_2]$. Then, we see that there is a fixed start-up cost $O(\sqrt{\texttt{Lim}_2})$ which is independent of the length of the interval. With the sieve of Eratosthenes, this cost comes from creating the primes up to $\sqrt{\text{Lim}_2}$ and finding their smallest multiples in the interval. For Atkin and Bernstein's sieve, the cost comes from the number of pairs (x, y) we need to enumerate. Let us consider the first case that covers primes congruent to one modulo 4. There, we need to consider all pairs (x, y) with $\text{Lim}_1 \leq 4x^2 + y^2 \leq \text{Lim}_2$. Equivalently, we need to enumerate all pairs (x', y) with $\text{Lim}_1 \leq x'^2 + y^2 \leq \text{Lim}_2$ and x' even. This can be done by finding all points with integer coordinates located between two concentric circles of diameter $\sqrt{\text{Lim}_1}$ and $\sqrt{\text{Lim}_2}$. When Lim_1 is near from Lim_2, it seems natural to expect that this task requires a walk inside the perimeter of the outer circle and costs $O(\sqrt{\text{Lim}_2})$. This is indeed the behavior of the algorithm proposed by Atkin and Bernstein in [AB04].

Surprisingly, it is possible to enumerate all integer points between two nearby concentric circles in a more efficient manner. The key idea comes from a theorem of van der Corput and it has been applied to the algorithm of Atkin and Bernstein by Galway in his PhD thesis. Using this technique, he devised a "dissected" sieve based on Atkin-Bernstein's whose overhead cost is only $O(\sqrt[3]{\text{Lim}_2})$. Using this algorithm, he could enumerate primes in small intervals for much larger values of Lim_2.

At the time of writing, this dissected sieve idea is not used in any cryptographic application. Thus, we do not describe it in detail and refer the interested reader to Galway's thesis [Gal04].

4.2 Sieving for smooth composites

In the previous section, we used sieve algorithms to find prime numbers. However, another very frequent application of sieve algorithms in cryptography searches for composite numbers (resp. polynomials) with many small enough factors, which are usually called smooth numbers (resp. polynomials). Applications are discussed in Chapter 15. In the present chapter, we simply focus on the sieve algorithms themselves and show the many possible variants, together with their performance issues. When sieving for composites instead of primes, there is a first, essential difference. In order to locate primes, a Boolean table is enough, indeed, whenever a number has a divisor, it is composite and can be removed. When locating smooth numbers, this is no longer true, indeed, we need to check that each number has enough divisors and thus we should be able to count higher than 1.

Another important difference between the sieve algorithms in Sections 4.1 and 4.1.3 and the algorithms in this section is that instead of simply searching for numbers in an interval we may have an additional degree of freedom and search in higher dimensional sets. The most frequent case is to consider two dimensional spaces. In this context, there are several possible strategies to cover the sieving zone. We present two classical options: line and lattice sieving.

Finally, when using the algorithms presented in this section in applications, we need to find many smooth objects but it is usually not essential to retrieve all of them. We can overlook some without any problems as long as we find many quickly enough.

Due to these differences, to optimize the sieve algorithms in this section, we cannot follow the same approaches as before and we need to develop specific strategies.

4.2.1   General setting

The most frequent case for sieve algorithms, especially in index calculus methods (see Chapter 15), is to consider algebraic objects of the form $a + b\alpha$ where α is fixed and a and b vary. Without going into the mathematical details, which are given in Chapter 15, we need to explain what "smoothness" means in this context. In fact, there are two different mathematical settings to consider. The varying elements a and b may be either numbers (integers) or univariate polynomials over a small finite field. The two settings are very similar from a high level point-of-view, but the implementation details vary a lot between the two. Moreover, the mathematical language used to described both cases is slightly different. For now, we give an overall description using the language corresponding to the case of numbers. Our first remark is that we only consider pairs (a, b) where a and b are coprime. Otherwise, we would be considering multiples of previously seen objects. In the context of index calculus, this would result in equivalent equations and would not be of any help. Given such a pair (a, b), we need to consider $a + b\alpha$ modulo a prime p and test whether we find 0 or not. To do this, we need to define the image of α modulo p. In general, this image need not be unique, for some primes α has a single image $\alpha_p^{(1)}$, for some it has two $\alpha_p^{(1)}$ and $\alpha_p^{(2)}$, and so on ...

There are also primes with no image at all for α, these primes can never occur among the factors of $a + b\alpha$. For both theoretical and practical reasons, it is best to consider each image of α for the same p as a different factor. With this convention, we say that a pair $(p, \alpha_p^{(i)})$ divides $a + b\alpha$ when $a + b\alpha_p^{(i)} \equiv 0$ (mod p). From a geometric point-of-view, the pairs (a, b) associated to objects of the form $a + b\alpha$ divisible by $(p, \alpha_p^{(i)})$ are nicely distributed in the Euclidean plane. More precisely, they form a 2 dimensional lattice. This is easily shown, let $a + b\alpha$ be divisible by $(p, \alpha_p^{(i)})$, then $a + b\alpha_p^{(i)} \equiv 0$ (mod p). Thus there exists an integer λ such that $a = -b\alpha_p^{(i)} + \lambda p$. This implies that:

$$\begin{pmatrix} a \\ b \end{pmatrix} = b \begin{pmatrix} -\alpha_p^{(i)} \\ 1 \end{pmatrix} + \lambda \begin{pmatrix} p \\ 0 \end{pmatrix}. \tag{4.2}$$

Thus, a point associated to an element $a + b\alpha$ divisible by $(p, \alpha_p^{(i)})$ belongs to the 2-dimensional lattice generated by the two vectors appearing in the right-hand side of Equation (4.2). Conversely, we can check that both generating vectors satisfy the divisibility equation, since $-\alpha_p^{(i)} + 1 \cdot \alpha_p^{(i)} \equiv 0 \pmod{p}$ and $p + 0 \cdot \alpha_p^{(i)} \equiv 0 \pmod{p}$. By linearity, any integer linear combination of the two vectors also satisfies the divisibility equation. As a consequence, the set of points associated to elements $a + b\alpha$ divisible by $(p, \alpha_p^{(i)})$ is precisely this 2-dimensional lattice.

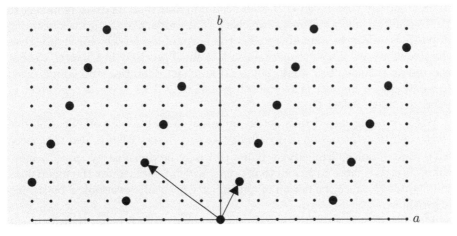

Figure 4.3: Set of points $a + b\alpha$ divisible by $(11, 5)$

To illustrate this geometric point-of-view, we choose the following simple example: let us take the prime $p = 11$ and an image for α: $\alpha_{11}^{(1)} = 5$. In this case, the set of points divisible by $(11, 5)$ is represented by Figure 4.3. In this example, we see that the generating vectors we gave for this lattice are not the best possible choice. The two dotted vectors in Figure 4.3 give a better description of the lattice. Since the lattice is two-dimensional, it is not necessary to use the general purpose algorithms described in Chapter 10 to obtain this good representation. We can either use Gauss reduction as described in Chapter 10 or adapt a continued fraction algorithm to our purpose. For now, we assume when necessary that each pair $(p, \alpha_p^{(i)})$ is given together with a good basis describing the corresponding lattice.

Adaptation to polynomials We mentioned at the beginning of the present section that we adopt here the language of numbers. In the case of polynomials, the necessary adaptations are quite natural. When considering $a + b\alpha$, a and b become univariate polynomials over a finite field, say $a(t)$ and $b(t)$. The primes p are replaced by irreducible polynomials $p(t)$ over the considered finite field. The images α_p also become polynomials and we say that a pair $(p(t), \alpha_p^{(i)}(t))$ divides $a(t) + b(t)\alpha$ when $a(t) + b(t)\alpha_p^{(i)}(t)$ is a multiple of $p(t)$ over the finite field.

4.2.1.1   Smoothness

Now that we know how to find divisors of an element $a + b\alpha$, we are ready to define the notion of smoothness. We start from a smoothness basis $\mathbb{B}$ containing all pairs $(p, \alpha_p^{(i)})$ with $p \leq B$, where B is called the smoothness bound. We say that $a + b\alpha$ is B–smooth if and only if all its divisors belong to $\mathbb{B}$. To detect a smooth element, we check that product of the primes occuring in its divisors is near a number called the norm of $a + b\alpha$ which roughly gives the size of $a + b\alpha$. This notion of norm is defined precisely in Chapter 15. For the rest of this chapter, we do not need an exact definition and simply need to know that a number is B–smooth if and only if its norm is a product of prime numbers below B.

Without going into details, let us mention an additional complication due to the fact that the notion of divisibility we presented earlier is also associated with a notion of multiplicity: a pair (a, b) may be divisible by $(p, \alpha_p^{(i)})$ more than once. We chose to ignore that point, which is needed for the algorithms in Chapter 15, since the notion of multiplicity is usually overlooked by sieving algorithms.

This means that when computing the product of the primes occuring in the divisors of $a + b\alpha$, we do not correctly account for the contribution of multiple factors. Thus, there is no guarantee that the product is equal to the norm, but on average we expect it to be close to it. Indeed, we expect multiple factors to be of small size and the overall multiple contribution to remain small. Of course, we could also try to account for divisors that occur multiple times; however, the additional computational cost makes this approach much less practical. Note that in practice, instead of computing the above product, it is more convenient to compute an additive equivalent by adding the logarithms of the primes occuring in divisors. An even simpler alternative is to compute the numbers of different divisors, if it is high enough, we expect $a + b\alpha$ to be smooth quite often. Of course, using these simplifications, i.e., ignoring multiplicities and counting the numbers of divisors only, we may well miss some smooth elements. However, since we only need to find many, not all, this is not a real problem. A very important parameter when detecting smooth candidates is the detection threshold. It should be carefully chosen by balancing the rate of false candidates and the rate of good candidates being overlooked.

4.2.1.2   Basic Lattice Algorithm

With all these elements in mind, we are now ready to give our first algorithm to sieve for smooth elements $a + b\alpha$ with $|a|$ and $|b|$ smaller than sieve limits

S_a and S_b. As said before, we are only interested in pairs (a, b) where a and b are coprime. Moreover, we may remark that it is useless to consider both $a + b\alpha$ and its opposite. As a consequence, we may assume that $a \geq 0$. We need to explore as sieving zone a rectangle of S_a by $2S_b - 1$ points. Using a direct approach to explore the whole sieving zone results in Algorithm 4.3.

Algorithm 4.3 Two-dimensional sieving for smooth numbers

Require: Smoothness basis $\mathbb{B}$
Require: Sieve limits S_a and S_b
Require: Expected value $LogVal$ for smooth candidates
    Create two-dimensional array $LogSum$ indexed by $[O \cdots S_a - 1] \times [-(S_b - 1) \ldots S_b - 1]$ initialized to zero
    for all $\mathbb{P} = (p, \alpha_p^{(i)})$ in $\mathbb{B}$ **do**
        for all $(a, b) \in [O \cdots S_a - 1] \times [-(S_b - 1) \ldots S_b - 1]$ with $a + b\alpha$ divisible by $\mathbb{P}$ **do**
            Add $\lfloor \log p \rceil$ to $LogSum[a, b]$
        end for
    end for
    for all $(a, b) \in [O \cdots S_a - 1] \times [-(S_b - 1) \ldots S_b - 1]$ **do**
        if $LogSum[a, b] \geq LogVal$ **then**
            Check whether $a + b\alpha$ is really smooth
            If so Display$((a, b),$'corresponds to a smooth value')
        end if
    end for

As stated, this algorithm is not complete; we need to know how to walk over multiples of $(p, \alpha_p^{(i)})$ and also how to check whether a candidate $a + b\alpha$ is really smooth or not. Moreover, even in this basic form, some potential improvements are already worth considering: Can we expect to go faster than this basic algorithm? Can we use less memory and work with a smaller array? In this section, we now give general answers to these questions. More detailed and specific answers are discussed in Sections 4.2.1.3, 4.2.1.4 and 4.2.1.5.

4.2.1.2.1 Walking the multiples. Given an element $(p, \alpha_p^{(i)})$ of the smoothness basis $\mathbb{B}$, we need a good presentation of the corresponding lattice given by two basis vectors $\vec{u}$ and $\vec{v}$. After that, we search for the lattice elements within the total sieve area. This is a reasonably simple task performed by starting from a known lattice element, usually the point at origin and by adding multiples of the two basis vectors in order to find new points. This is usually done by nesting two loops, the inner loop goes through the sieve area by adding multiples of one basis vector, say $\vec{u}$, thus drawing a line. The outer loop adds multiples of the other basis vector v. This basic idea needs to be adapted to each specific setting and this can be done quite efficiently for all of our special cases.

4.2.1.2.2 Speeding up the approach.

One basic technique to speed up the sieve algorithm is to remark that an element of $\mathbb{B}$ associated with a prime p corresponds to a walk over a $1/p$ fraction of the sieving zone. As a consequence, elements associated to small primes contribute more to the runtime. Moreover, these elements offer a low added value: knowing that the norm of an element is divisible by a small prime is a smaller step toward smoothness. To overcome these two drawbacks, one classical technique is to remove from the sieve algorithm all elements of the smoothness basis below a given bound.

4.2.1.2.3 Checking for smoothness.

The easiest way to check for smoothness is simply to compute the norm of $a + b\alpha$ and then to factor this norm using a fast algorithm. At this point, the cases of numbers and polynomials are quite different. Indeed, as recalled in Chapter 2, factoring polynomials can be done in polynomial time by efficient and practical algorithms, thus this approach seems to be a good start in the case of polynomials. In the case of numbers, among known algorithm, the better suited for factoring numbers with many small factors is the elliptic curve factoring method (ECM, see Chapter 14). As a consequence, in this case, checking smoothness seems to be much more difficult and other methods need to be devised to remove this practical obstruction. Note that from a complexity theoretic point-of-view, using ECM is not a real problem and does not change the overall complexity. However, in practice, getting rid of ECM, wherever it is possible, greatly improves the efficiency. Interestingly, these methods can even be used in the case of polynomials to speed up things a little.

The best approach is to view the problem of checking for smoothness as a specific factoring task. We need to factor a list of candidates in order to keep the smooth ones. As with ordinary factoring tasks, it is a good idea to start by using trial division, thus checking for elements of the smoothness basis associated to small primes, whether they divide each candidate. It is essential to follow this approach for the very small primes that were omitted from the initial sieving phase. When a candidate is divisible by such an element we divide its (precomputed) norm by the corresponding prime, taking the multiplicity into account. As a consequence, after going through all the very small primes, we are left for each candidate with a reduced norm without very small factors. Once this is done, some candidates have a large reduced norm and others a smaller one. Clearly, candidates with a large reduced norm have a smaller probability of being smooth. As a consequence, to speed up things, it is possible to filter the candidates, removing those with a large norm. After that, some additional trial division tests can be performed for slightly larger primes. This idea of filtering, i.e., of aborting early for bad candidates was analyzed precisely by Pomerance in [Pom82] for Dixon's factoring algorithm and allowed improvements of its asymptotic time complexity.

Once trial division is finished, we could move to other generic factoring algorithms such as Pollard's Rho or ECM to factor the reduced norms; however, at this point, it is tempting to use the extra structure shared by our candidates. All of them are of the form $a + b\alpha$ with (a, b) in the sieve zone. A good trick in order to find among the candidates those which are divisible by an element of $\mathbb{B}$ is to sieve again. More precisely, we initialize the sieve zone to zero except in locations corresponding to candidates, where we write a identifier for the candidate. During this second sieve, for each point encountered during the lattice walk, we do nothing if the point's location contains a zero and otherwise we divide the reduced norm of the corresponding candidate, taking multiplicity into account. This approach allows us to factor out all elements of $\mathbb{B}$ corresponding to medium-sized primes.

Finally, after another filtering pass, where we keep small enough reduced norms, we complete the factorization using primality tests and either Pollard's Rho (see Chapter 7) or ECM depending on our choice of factor base $\mathbb{B}$.

Note that, while checking for smoothness, we obtained the factorization of the norm of each candidate. In order to avoid redoing this job later, it is preferable in most cases to store this factorization together with the smooth candidate. This stored factorization is used in the subsequent phases of index calculus method described in Chapter 15.

4.2.1.2.4 Working with smaller arrays.
In order to reduce the required amount of memory, a natural approach is to mimic the idea behind wheel factorization described in Section 4.1.2.1. The basic idea behind wheel factorization is to remove from the memory representation all elements which are known to be useless. With the current problem, we known in advance that a pair (a, b) is useless whenever the GCD of a and b is not one. Thus a good way to reduce the memory usage is to filter out such pairs. Of course, the filtering should not be too complicated to avoid high runtime penalties. The simplest approach is to remove all pairs where both a and b are even, this reduces the amount of memory by $1/4$ without involving any complicated computation. In fact, it is similar to the removal of even integers in Eratosthenes's sieve. Of course, one could push the approach further and remove all pairs with 2, 3, 5, ... as a common factor between a and b. However, this is much less interesting than wheel factorization. Indeed, in Eratosthenes's sieve, the proportion of integers removed by adding p to the wheel is roughly $1/p$, here it is $1/p^2$. This new ratio is not so good, since the sum of the series formed of the squared inverses of primes is converging.

4.2.1.3 The case of polynomials over $\mathbb{F}_2$

Some index calculus algorithms need to sieve over pairs (a, b) where a and b are univariate polynomials. In this section, we address the frequent special case where these polynomials are defined over $\mathbb{F}_2$. Using polynomials implies a slight rewriting of some of the mathematical objects involved, for example we need to replace prime numbers by irreducible polynomials. However, these changes are unessential where the sieving process is concerned. Here, we are interested by the direct impact of using polynomials on the sieve algorithm. Interestingly, using polynomials over $\mathbb{F}_2$ greatly simplifies the task of sieving. This is the reason why we start with this case. The first remark concerns the sieving zone. Since we are working over $\mathbb{F}_2$ any polynomial is its own opposite. This additional symmetry leads us to reduce the sieve area to a square zone a and b should both belong to the same set of small polynomials. Moreover, it is extremely convenient to define this set by giving a upper bound on the degrees of a and b. However, this greatly limits the flexibility of the memory usage. Taking all polynomials of degree $< D$, we have 2^{2D} possible pairs (a, b). Writing a and b as polynomials in x, it is easy to remove from the memory representation pairs (a, b) where a and b are both multiples of x. Similarly, it is easy to remove pairs (a, b) where a and b are both multiples of $x + 1$. As soon as $D \geq 2$ removing these pairs lowers the required amount of memory to 9/16 of the initial amount 2^{2D}.

From a practical point-of-view, polynomials over $\mathbb{F}_2$ can be efficiently represented by numbers encoded in binary, as long as their degree is smaller than the available number of bits. Assuming that the bits are numbered from right to left starting with bit 0, we use bit i to represent the coefficient of x^i in the polynomial. For example, the low order bit (rightmost) is used to encode the constant term. With this convention, the polynomial $x^3 + x + 1$ is encoded by the number 11. With this representation, polynomials can be added using the XOR operation on numbers, they can be multiplied by x using a simple left shift (multiplication by 2 of the corresponding number). Moreover, within this representation, multiples of x and $x + 1$ are easily recognized. A multiple of x corresponds to an even number and a multiple of $x + 1$ corresponds to a number whose Hamming weight is even, or equivalently to a number that contains an even number of 1s in its representation.

Despite its drawback of limiting the choice in terms of used memory, choosing a sieving zone comprising all polynomials of degree $< D$ is extremely convenient for sieving. First, when representing polynomials over $\mathbb{F}_2$ as an integer, testing whether the degree is $< D$ can be done by comparing the polynomial's representation with 2^D. Since integer comparison is a native operation on modern computers, this is extremely efficient. Second, when sieving, we need to construct all multiples of a given polynomial up to degree D. If we start with a polynomial of degree d, this is achieved by multiplying it by all polynomials up to degree $D - d$. With another choice of sieving zone, we cannot use the additivity of degrees during multiplication and thus need to use a more complicated and less efficient approach. When constructing all polynomials up to degree D, we can even speed up the construction of the multiples by using a Gray code to enumerate them quickly.

To make this more precise, assume that we want to mark all pairs of

Binary Gray codes offer a way of enumerating all vectors on n-bits, while changing only a single bit between one vector and the next. For example, on 4 bits, we have the following code:

$$
\begin{array}{cccc}
0 & 0 & 0 & 0 \\
0 & 0 & 0 & 1 \\
0 & 0 & 1 & 1 \\
0 & 0 & 1 & 0 \\
0 & 1 & 1 & 0 \\
0 & 1 & 1 & 1 \\
0 & 1 & 0 & 1 \\
0 & 1 & 0 & 0 \\
1 & 1 & 0 & 0 \\
1 & 1 & 0 & 1 \\
1 & 1 & 1 & 1 \\
1 & 1 & 1 & 0 \\
1 & 0 & 1 & 0 \\
1 & 0 & 1 & 1 \\
1 & 0 & 0 & 1 \\
1 & 0 & 0 & 0
\end{array}
$$

which contains the sixteen possibilities. Gray codes are useful whenever manipulating complex objects where changes are costly. In this case, reducing the total number of changes is clearly worthy. In particular, it is possible to use Gray codes to efficiently produce all multiples of a binary polynomial $a(x)$ by all polynomial of degree at most $n-1$ by using a Gray code on n bits. We start from an initial multiple equal to 0, i.e., $0 \cdot a(x)$, assuming that bits are numbered from the right starting at 0, a change on bit c in the Gray code is converted into an addition of $x^c a(x)$ to the current multiple.

Programming Gray codes on n is a simple matter. It suffices to run an arithmetic counter i from 1 to $2^n - 1$ that contains the number of the current change. The position of the bit that needs to be changed during the i-th is the position of the rightmost bit equal to 1 in the binary representation of i. The position of the rightmost bit is easily obtained by shifting a copy of i to the right, until the resulting number becomes odd. Note that Gray codes can easily be generalized to vectors of n elements in a small finite field $\mathbb{Z}/p\mathbb{Z}$. For example, for vectors of 3 ternary digits, we obtain the following Gray code:

0 0 0	1 0 1
0 0 1	1 1 1
0 0 2	1 1 2
0 1 2	1 1 0
0 1 0	2 1 0
0 1 1	2 1 1
0 2 1	2 1 2
0 2 2	2 2 2
0 2 0	2 2 0
1 2 0	2 2 1
1 2 1	2 0 1
1 2 2	2 0 2
1 0 2	2 0 0
1 0 0	

Ternary Gray codes are slightly less suited to the architecture of present computer, because to implement them, it is preferable to know the decomposition of the change counter i in base 3. As a consequence, they use many reductions modulo 3 and division by 3.

Figure 4.4: **Gray codes**

polynomials $(a(x), b(x))$ divisible by some element of the smoothness basis: $(p(x), \alpha_p(x))$, where $p(x)$ is an irreducible polynomial. These pairs satisfy the equation:

$$a(x) + b(x)\alpha_p(x) \equiv 0 \pmod{p(x)}. \tag{4.3}$$

Using a partial run of Euclid's extended GCD Algorithm 2.2 on the pair of polynomials $(\alpha_p(x), p(x))$, we can find two different small solutions, (a_1, b_1) and (a_2, b_2), where the degrees of a_1, a_2, b_1 and b_2 are near the half-degree of $p(x)$. Moreover, any solution (a, b) can be written as a linear combination of these two initial solutions, i.e., there exist two polynomials $\lambda(x)$ and $\mu(x)$ such that:

$$a(x) = \lambda(x)a_1(x) + \mu(x)a_2(x) \quad \text{and} \tag{4.4}$$
$$b(x) = \lambda(x)b_1(x) + \mu(x)b_2(x).$$

In order to exhibit all solutions, it suffices to consider all pairs (λ, μ) with small enough degree. As a consequence, we simply need to nest two independent Gray code enumerations to efficiently enumerate all the (a, b) pairs. This is much simpler than the corresponding algorithm in the case of numbers, as can be seen on Algorithm 4.4. This algorithm also tests whether both coordinates in each pair are multiples of x or $x + 1$ before marking the pair.

4.2.1.4   The case of numbers

Sieving with numbers is more complicated than sieving with polynomials over $\mathbb{F}_2$; we must take into account the geometry of both the sieving zone and the lattice to be sieved. Due to the variety of relative positions of the basis vectors $(\vec{u}, \vec{v})$, many different cases are possible. Assuming that the lattice's basis is Gauss's reduced as described in Chapter 10, we know that $\vec{u}$ is (one of) the shortest in the lattice and that $\vec{v}$ is (one of) the shortest among the lattice's vectors linearly independent from $\vec{u}$. Moreover, $|(\vec{u}|\vec{v})|$ is smaller than $\|\vec{u}\|/2$. Thus, $\vec{u}$ and $\vec{v}$ are both short and they are nearly orthogonal to each other. This means that considering vectors of the form $c_u\vec{u} + c_v\vec{v}$ for c_u and c_v in given intervals allows us to cover an almost rectangular looking parallelogram. However, the axes of these parallelogram have no reason whatsoever to be aligned with the axis of the sieving zone. We describe one possible way of marking the multiples in Algorithm 4.5. In this code, we denote by w_a and w_b the coordinates of a vector $\vec{w}$. A specific instance of the algorithm is illustrated in Figure 4.5.

4.2.1.5   The case of polynomials in odd characteristic

Finally, the case of polynomials in odd characteristic is a mix of the two previous cases. Either the characteristic is small and the degree of polynomials quite large and we proceed as for polynomials over $\mathbb{F}_2$, with the small caveat that it is no longer possible to represent polynomials simply as bit vectors. Or the characteristic is larger and we consider only polynomials of small degree,

---

Algorithm 4.4 Walking the multiples with polynomials

Require: Element of $\mathbb{B}$: $(p(x), \alpha_p(x))$
Require: Maximal degrees of $a(x)$ and $b(x)$ in the sieving zone, d_a and d_b
  Create initial solutions $(a_1(x), b_1(x))$ and $(a_2(x), b_2(x))$
  Let $d_1 \longleftarrow \min(d_a - \deg a_1, d_b - \deg b_1)$
  Let $d_2 \longleftarrow \min(d_a - \deg a_2, d_b - \deg b_2)$
  Let $(\alpha(x), \beta(x)) \longleftarrow (0, 0)$
  for i from 0 to $2^{d_1} - 1$ **do**
    if $i \neq 0$ **then**
      Let $i' \longleftarrow i$
      Let $c \longleftarrow 0$
      while i' is even **do**
        Let $i' \longleftarrow i'/2$
        Let $c \longleftarrow c + 1$
      end while
      Let $\alpha(x) \longleftarrow \alpha(x) \oplus x^c a_1(x)$
      Let $\beta(x) \longleftarrow \beta(x) \oplus x^c b_1(x)$
    end if
    for j from 0 to $2^{d_2} - 1$ **do**
      if $j \neq 0$ **then**
        Let $j' \longleftarrow j$
        Let $c \longleftarrow 0$
        while j' is even **do**
          Let $j' \longleftarrow j'/2$
          Let $c \longleftarrow c + 1$
        end while
        Let $\alpha(x) \longleftarrow \alpha(x) \oplus x^c a_2(x)$
        Let $\beta(x) \longleftarrow \beta(x) \oplus x^c b_2(x)$
      end if
      if $\alpha(0) \neq 0$ or $\beta(0) \neq 0$ **then**
        if $\alpha(1) \neq 0$ or $\beta(1) \neq 0$ **then**
          Mark $(\alpha(x), \beta(x))$
        end if
      end if
    end for
  end for

Algorithm 4.5 Walking the multiples with numbers

Require: Element of $\mathbb{B}$: (p, α_p)
Require: Sieve limits S_a and S_b
 Create reduced basis $(\vec{u}, \vec{v})$ for lattice
 Modify basis to ensure $0 \leq u_a < v_a$ and $u_b \cdot v_b \leq 0$
 Assume $u_b < 0 \leq v_b$ (otherwise adapt boundary condition in the sequel)
 Let $\vec{w} \longleftarrow \begin{pmatrix} 0 \\ 0 \end{pmatrix}$
 repeat
  Let $\vec{z} \longleftarrow \vec{w}$
  Let $\vec{w} \longleftarrow \vec{w} + \vec{u}$
  if $w_a \geq v_a$ **then**
   Let $\vec{w} \longleftarrow \vec{w} - \vec{v}$
  end if
 until $w_b \leq -S_b$
 Let $\vec{w} \longleftarrow \vec{z}$ {Here $\vec{w}$ is in lower left corner}
 Let $\vec{z} \longleftarrow \vec{w}$
 repeat
  Let $\vec{t} \longleftarrow \vec{z}$ {Start first sieving zone}
  repeat
   if t_a is odd or t_b is odd **then**
    Mark point corresponding to $\vec{t}$
   end if
   Let $\vec{t} \longleftarrow \vec{t} + \vec{u}$
  until $t_a \geq S_a$ or $t_b \leq -S_b$
  Let $\vec{z} \longleftarrow \vec{z} + \vec{v}$
 until $z_a \geq S_a$ or $z_b \geq S_b$
 Let $\vec{w} \longleftarrow \vec{w} - \vec{u}$
 if $w_a < 0$ **then**
  Let $\vec{w} \longleftarrow \vec{w} + \vec{v}$
 end if
 repeat
  Let $\vec{z} \longleftarrow \vec{w}$ {Second sieving zone}
  repeat
   if z_a is odd or z_b is odd **then**
    Mark point corresponding to $\vec{z}$
   end if
   Let $\vec{z} \longleftarrow \vec{z} + \vec{v}$
  until $z_a \geq S_a$ or $z_b \geq S_b$
  Let $\vec{w} \longleftarrow \vec{w} - \vec{u}$
  if $w_a < 0$ **then**
   Let $\vec{w} \longleftarrow \vec{w} + \vec{v}$
  end if
 until $w_b \geq S_b$

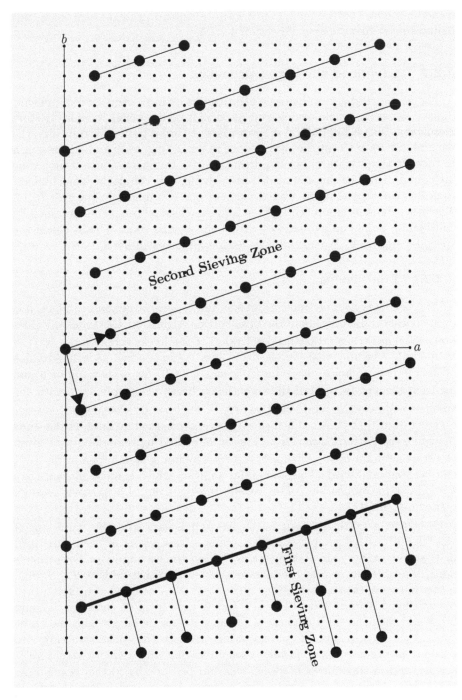

Figure 4.5: Illustration of Algorithm 4.5

sometimes linear polynomials. Since we are sieving over large coefficients, this behaves more like the case of Section 4.2.1.4.

4.2.2   Advanced sieving approaches

The main drawback of the basic algorithm is its huge memory requirement. Since it sieves the whole sieving zone at once, it needs enough memory to store this zone, which is impractical in most cases. In order to lift this difficulty, the natural answer is to cut the sieving zone in small fragments and address each in turn. In the literature, two important ways of fragmenting the sieving zone are considered. The simplest, called line sieving, cuts the sieving zone in lines, each line corresponding to a fixed value of one coordinate in the pair (a, b). The other frequent approach looks at two-dimensional fragments obtained by considering points on sub-lattices within the sieving zone. This is called the special-q approach. We now discuss both approaches in detail.

4.2.2.1   Line sieving

Line sieving is a simple and elegant solution to sieve large zones. In fact, it is even conceptually easier than the basic lattice Algorithm 4.3. Indeed, it does not even require a good lattice description of the linear equation $a + b\alpha_p^{(i)} \equiv 0 \pmod{p}$. Assume that we sieve by lines corresponding to a fixed value of b. Since our choice of sieving zone has $a \geq 0$, for each value of b and for each element $(p, \alpha_p^{(i)})$ of the smoothness basis, we simply start with the smallest non-negative representative of $b\alpha_p^{(i)} \bmod p$ and tick every p-th integer starting from that point. When sieving several consecutive lines on the same machine, the performance can be greatly improved by remarking that given the starting point of $(p, \alpha_p^{(i)})$ for b, the corresponding starting point for $b + 1$ can be computed with a single modular addition which is more efficient than a modular multiplication. This results in Algorithm 4.6 to sieve a set of consecutive lines.

Of course, as with the basic lattice sieve, it is easy to derive variations of this algorithm that simply count the number of divisors of each pair (a, b) or that remove very small elements of $\mathbb{B}$ from the sieving loop. Once this is done, we can observe that the innermost operation that adds to $LogSum[a]$ is the most important as running time goes. As a consequence, it is worth optimizing this inner loop. The key fact is that in most cases, S_a is quite large and thus the total amount of memory we use is larger than the cache size. As a consequence, it is a good idea to take the cache structure into account in order to sieve faster. Assuming that the memory structure is simple, with a single level of cache, this can be done by further segmenting the representation of the a values in memory into several blocks that fit in the cache. For elements of $\mathbb{B}$ with a p value smaller than the blocks' size, we can sieve a given block while staying within the cache. For larger values of p,

---

Algorithm 4.6 Basic line sieve

Require: Smoothness basis $\mathbb{B}$
Require: Sieve limit S_a
Require: Range for b: $b_{min} \cdots b_{max}$
Require: Expected value $LogVal$ for smooth candidates
Create single dimensional array $LogSum$ indexed by $[O \cdots S_a - 1]$
Create single dimensional array $StartPos$ indexed by $\mathbb{B}$
for all $\mathbb{P} = (p, \alpha_p^{(i)})$ in $\mathbb{B}$ **do**
   $StartPos[\mathbb{P}] \longleftarrow b_{min}\alpha_p^{(i)} \pmod{p}$
end for
for b from b_{min} to b_{max} **do**
  Initialize $LogSum$ to zero
  for all $\mathbb{P} = (p, \alpha_p^{(i)})$ in $\mathbb{B}$ **do**
     $a \longleftarrow StartPos[\mathbb{P}]$
     while $a \leq S_a$ **do**
       Add $\lfloor \log p \rceil$ to $LogSum[a]$
       $a \longleftarrow a + p$
     end while
  end for
  for a from 0 to S_a **do**
    if $LogSum[a] \geq LogVal$ **then**
      Check whether $a + b\alpha$ is really smooth
      If so Display$((a, b),$'corresponds to a smooth value'$)$
    end if
  end for
  for all $\mathbb{P} = (p, \alpha_p^{(i)})$ in $\mathbb{B}$ **do**
    $StartPos[\mathbb{P}] \longleftarrow StartPos[\mathbb{P}] + alpha_p^{(i)} \pmod{p}$
  end for
end for

---

unfortunately, we need to jump between blocks and do not gain. Luckily, the small values of p require more operations than the small ones, which makes this optimization very useful. We illustrate this idea with two levels of cache by the C code given in Program 4.3.

4.2.2.2   Special-q lattice sieving

The other frequent way of splitting a sieve into smaller is more technically involved than line sieving. However, it presents the very useful property of guaranteeing that all the pairs being considered are already divisible by a common element. This common element, called the special q is usually chosen among the element $\mathbb{B}$ associated to a large prime value. Each piece of the sieve corresponds to a different special-q. One technicality is the fact that the various pieces are not necessarily disjoint. As a consequence, when using this algorithm for sieving, some solutions are duplicated. These duplicates should be removed using a filtering program. Given a special q written as (q, α_q), the special-q lattice sieving algorithm focuses on pairs (a, b) divisible by the special and thus satisfying the linear condition $a + b\alpha_q \equiv 0 \pmod{q}$. As explained in the case of the basic lattice sieving, this corresponds to pairs (a, b) on some integer lattice, which can be described by a reduced basis $(\vec{u_q}, \vec{v_q})$. In most cases, the largest coordinate of each vector in the reduced basis is in absolute value not too far from $\sqrt{q}$. When this is not the case, the norm of $\vec{u_q}$ is much smaller than the norm $\vec{v_q}$ and the basis is unbalanced. With such an unbalanced basis, special q lattice sieving is much less convenient. As a consequence, we omit special q's with unbalanced basis and can assume from now on that the reduced basis is balanced. More precisely, we fix some constant Λ and keep a given special-q only when the coordinates of the two basis vectors are bounded (in absolute value) by $\Lambda\sqrt{q}$. Using this choice allows us to choose the sieving zone within the sublattice independently of the corresponding special q. We consider pairs (a, b) written as $c_u \vec{u_q} + c_v \vec{v_q}$ with (c_u, c_v) in a fixed sieving zone $[0 \cdots S_u] \times [-S_v \cdots S_v]$. A direct consequence of this choice is that sieving such a sublattice can be done using the basic lattice sieve Algorithm 4.3. The only condition is that all elements of $\mathbb{B}$ should be expressed in terms of the new coordinates (c_u, c_v).

In order to express an element $(p, \alpha_p^{(i)})$ of $\mathbb{B}$ in terms of the new coordinates, let us recall that $a + b\alpha$ belongs to the corresponding lattice whenever the linear condition $a + b\alpha_p^{(i)} \equiv 0 \pmod{p}$ is satisfied. Replacing a and b by their expression in c_u and c_v we find the condition:

$$(c_u u_q^a + c_v v_q^a) + (c_u u_q^b + c_v v_q^b)\alpha_p^{(i)} \equiv 0 \pmod{p} \quad \text{or} \qquad (4.5)$$
$$c_u(u_q^a + u_q^b \alpha_p^{(i)}) + c_v(v_q^a + v_q^b \alpha_p^{(i)}). \equiv 0 \pmod{p}$$

This is a linear equation in c_u and c_v modulo p. As a consequence, we can follow the same approach as in the case of the original equation in a and b

Program 4.3 Line sieving with two levels of cache

```
#define SIZE 510000      /* Max. Number of prime pairs */
#define ZONE 9000000     /* Size of each zone, mult of CACHE2 */
#define CACHE1 25000     /* Approximate size of L1 CACHE */
#define CACHE2 750000    /* Approximate size of L2 CACHE */
#define START_ZONE -500  /* Number of first sieve zone */
#define END_ZONE 500     /* Number of last sieve zone */
#define THRESHOLD 12     /* Num. of hits per smooth candidate */
#define START 30         /* Prime pairs below START not sieved */
int root[SIZE], mod[SIZE]; /* Arrays for prime pairs */
char Zone[ZONE];           /* Array for sieving */
int Size;          /*True number of pairs */
int Size1, Size2; /*Limits to fit in CACHE1 and CACHE2 */

main()
{ int i,j,offset;  int limit2,limit1;
  Read();/*Get prime pairs, set Size1/2 to fit CACHE1/2*/
  for(i=START;i<Size;i++) {
    int r,q; /* Loop shifts primes pairs to first zone */
    q=mod[i]; offset=MulMod(START_ZONE,ZONE,q);
    r=root[i]-offset; if (r<0) r+=q; root[i]=r; }

  for (j=START_ZONE;j<END_ZONE;j++){
    for(i=0;i<ZONE;i++) Zone[i]=0; /*Reset sieve zone */
    for(limit1=limit2=0;limit2<=ZONE;limit2+=CACHE2) {
      for(;limit1<=limit2;limit1+=CACHE1) {
        for(i=START;i<Size1;i++) { /* Sieve small*/
          int r,m;  r=root[i]; m=mod[i];
          while (r<limit1) {Zone[r]++; r+=m;}
          root[i]=r;}}
      for(i=Size1;i<Size2;i++) { /* Sieve medium*/
        int r,m; r=root[i]; m=mod[i];
        while (r<limit2) { Zone[r]++; r+=m;}
        root[i]=r;}}
    for(i=START;i<Size2;i++) { /* Shift to next zone */
      root[i]=root[i]-ZONE;}
    for(i=Size2;i<Size;i++) { /* Sieve large */
      int r,m; r=root[i]; m=mod[i];
      while (r<ZONE) {Zone[r]++; r+=m;}
      root[i]=r-ZONE;}
    for(i=0;i<ZONE;i++){ /* Detect and print smooth candidates*/
      if (Zone[i]>=THRESHOLD) {printf("F(%d*SZ+%d);\n",j,i);
fflush(stdout); }}}}
```

in order to write down a good representation of the lattice comprising all the coordinates (c_u, c_v) of divisible elements.

Finally, we see that lattice sieving with special-q does not differ much from basic lattice sieving. We need to reinitialize the representations of elements of $\mathbb{B}$ for each special-q and change the accepting bound on the number of divisors or on the sum of their logarithm to account for the additional special-q divisors and their influence on the value of norms. Except for this minor change, everything else is a verbatim copy of our previous approach. However, the fact that we need to sieve on a relatively small zone for each special-q greatly enhances the sieve performance.

4.2.3   Sieving without sieving

Before closing this chapter on sieving, it is important to mention an alternative algorithm proposed by D. Bernstein in [Ber00] that allows the implementation of index calculus algorithms without using sieving at all. Instead, his algorithm addresses the problem of finding smooth enough numbers within a long list of numbers. A nice property of this algorithm is that contrary to sieve algorithms, it does not require additional structure[2] on the numbers themselves. This interesting property could be especially useful for some specific applications of index calculus such as Coppersmith factorization factory. Note that this specific algorithm is almost dedicated to index calculus algorithms that use numbers, when polynomials come in play, it is easier to use a fast polynomial factorization algorithm. However, Bernstein notes in his paper that a careful comparison is required in the case of polynomials.

The key ingredient in Bernstein's algorithm is to consider a large set of numbers at the same time, to identify the set of primes belonging to the smoothness basis that occur in the decomposition of the product of these numbers and using a divide and conquer approach to distribute these primes to smaller and smaller sets until the decomposition of each number is known.

The main difficulty of the algorithm is the fact that it requires a two-way divide and conquer approach, to find out which primes are occuring and in which numbers. The algorithm proposed in [Ber00] nests two divide and conquer approaches: the inner one is used to determine the set of primes involved in a particular set of numbers and the outer one to focus on sets on decreasing size. In order to make the inner loop efficient, the given set of primes is preprocessed and represented as a product tree. The product tree is a binary tree whose leaves are labelled by the given primes. Each inner node of the tree is labelled by the product of the label of its two children. As a consequence, the root of the product tree contains the product of all primes in the given set.

---

[2]With sieve algorithms, the additional structure comes from the fact that the numbers are all obtained by evaluating a fixed polynomial at many regularly spaced points.

Given a set of primes P and a set of numbers N, the outer divide and conquer step needs to determine the subset P' of primes that occurs in the decomposition of numbers in N. First, it computes the product Π_N of all numbers in N and a product tree T_P for P. Then it reduces Π_N modulo the label of the root of the product tree T_P. After that, it pushes this reduced value down the product tree, computing Π_N modulo each label by reducing the modular value of the father node in order to compute the reduced value at each node. When the leaves are reached, it is easy to see that the prime label of a leaf occurs in N if and only if the modular value of Π_N at this leaf is zero. Thus, P' is simply the subset of P corresponding where the modular value of Π_N vanishes.

In order to refine the decomposition, we simply need to split N in two (or more) subsets and recursively apply the same algorithm on each subset together with the prime set P'. At the very end of the algorithm, we know the list of primes for each numbers and can easily check for smoothness.

In order to improve the underlying number arithmetic, Bernstein proposes a slightly different method of computing the modular values at each leaf of T_P. His approach assumes that 2 does not belong to the initial list of primes, which is not a real constraint since trial dividing by 2 is so easy. Then, when computing the reduced value of a node from the value of the father, instead of doing an exact computation, he computes the modular value up to some power of 2. This improves the arithmetic and of course does not change the final criteria for odd prime: a prime p occurs in the decomposition of N if and only if Π_N reduced modulo p is zero. Since 2 is invertible modulo the odd prime p, multiplying the value Π_N by a power of 2 preserves the final zero. This modification changes non-zero values but this is irrelevant for our purpose.

A fair conclusion to this section is to remark that determining the impact of Bernstein's algorithm on index calculus methods is currently an open problem. At the time being, no large index calculus publicly announced computation has ever made use of this algorithm to improve running times. However, studying its practical impact is an interesting research problem.

Exercises

1. Program a basic implementation of Eratosthenes's sieve using a wheel of perimeter 30.

2. Modify the implementation of Exercise 1 into a segmented sieve, similar to Program 4.2.

3[h]. Using your answer to Exercise 4.2 as a template, write a program that generates implementations of Eratosthenes's sieve with larger wheels of fixed perimeter. Adding the primes 7, 11 and 13, write a code for wheels of perimeter 210, 2310 and 30030. Time these programs and compare their respective performances.

4. Write a version of Eratosthenes's sieve to deal with wheels of varying perimeter.

5. Combine wheels of varying perimeter with a segmented sieve.

6[h]. Create a variant of Eratosthenes's sieve for finding irreducible polynomials over $\mathbb{F}_2$. Make sure that you are using Gray codes and write a segmented version.

7. An amusing variation of sieving process aims at constructing "Lucky" numbers. It starts with the list of odd integers, consider the second number '3' and cross every third number, i.e., 5, 11, ... In this new list, the third number is '7', thus we cross every seventh number in the list, starting with '19'. Write a basic sieve program for computing lucky numbers. Which optimizations of Eratosthenes's sieve can you adapt to this case?

8[h]. Choose a prime p and a polynomial P of degree d. Compare the following approaches to determine all the polynomials of the form $P + a$ for a in $\mathbb{F}_p$ that have d different roots.

   (a) Sieving on a line.

   (b) Individual factorization of each $P + a$.

   (c) Adapted variant of Bernstein's algorithm.

Chapter 5

Brute force cryptanalysis

5.1 Introductory example: Dictionary attacks

Brute force is one of the less subtle[1] ways of attacking cryptosystems; it simply consists of trying all possible values for the secret to recover, until the correct one is found. Yet, this approach is more tricky than it may seem. To illustrate this point, we start this chapter by a special case of brute force attacks: dictionary attacks. Typically, a dictionary attack is used to find out the password that gives access to a user's account on an information system. Since passwords are used for a large number of purposes, dictionary attacks can be used in a large variety of contexts.

At the first level of dictionary attacks, we find simple password guessing, as often encountered in books and movies. The attacker simply sits at his target's desk, thinks deeply for a little while and, after a couple of unsuccessful trials, comes up with the correct password.

A slightly more sophisticated version is to automate the process and have a tool for submitting many passwords until the correct one is found. For example, this can be easily done when trying to connect to a remote server. This kind of dictionary attack is called an online dictionary attack. For each new tested password, the attacker needs to try to open a new session. One problem (or advantage, depending on the point-of-view) of online dictionary attacks is that they can be easily detected. For example, smart cards are often protected using a very short 4-digit password, also called PIN code[2]. Thus, any attacker could easily discover the correct PIN code, simply by trying out the 10,000 different possibilities. Assuming a slow rate of one trial per second, this could be done in a couple of hours. To avoid this attack, smart cards are programmed to detect sequences of consecutive failures. After three failures, they usually stop accepting PIN codes and put themselves in a locked state. To unlock the card, the user needs a different, much longer password.

Offline attacks are a much stronger form of dictionary attacks. In these attacks, the attacker first recovers some side information and then using this

---

[1] Of course, pointing a gun at the legitimate user and asking for the secret is even less subtle.
[2] Where PIN stands for Personal Identification Number.

information is able to test on his own whether a given password is correct or not. With offline attacks, the attacker is only limited by the raw computing he has access to. The side information used by the attacker can be of several different types: it can be the transcript of a protocol execution where the password was used, it can be an entry from a password file, or even some side-channel information obtained from the system during password testing.

For our illustration, we focus on the simple case of password files entries. Password files are used by operating systems or applications that need to check user passwords in order to store side information that can be used to verify correctness of passwords. In their simplest form, password files simply store plaintext copies of passwords. Of course, this is a bad idea, since in that case passwords are directly accessible to an attacker that obtains the password file, regardless of the passwords' length. An equally bad idea is to store encrypted passwords using a secret key algorithm, indeed, regardless of the strength of the algorithm, the secret key needs to be stored somewhere to allow password checking and if the attacker finds this secret key, he can decrypt all passwords in the file. To avoid these straightforward attacks, the password file can contain either passwords encrypted with a public key algorithm or hashed passwords. It also needs to contain auxiliary information that allows passwords to be tested against the password file. For instance, when using a randomized public encryption, the random value used for encryption needs to be stored. This allows at a later point to reencrypt a given password and to check equality between this new encryption and the stored encryption.

The most common approach is to use hashed passwords, indeed, it is much faster than using public key encryption. However, there is more to it than simply running passwords through a standard hash function. Indeed, in that case, attackers could prepare in advance a dictionary of hashed passwords and store them in sorted form (see Chapter 6). This would allow using a simple search to discover hashed passwords from the dictionary extremely quickly. To avoid this, the hash function needs to be randomized. A simple approach is to store a hash value of the string formed by concatenating a reasonable short random string together with a user's password. To allow verification, the random string is stored together with the corresponding hash value in the password file. When the random value is quite short and used as a side input that modifies the computation of hash values, it is often called a salt. For example, for decades, Unix-based operating systems used a salted variation of the DES encryption algorithm to hash passwords. In order to avoid the possibility of decryption, the password was used as key and served to encrypt a fixed message. Even with a relatively short salt, storing hashed dictionaries is highly impractical. Indeed, since you need enough dictionaries to cover a noticeable fraction of all possible salts, the storage requirements quickly become impractical.

More recently, passwords have been used, in conjunction with public key mechanisms, in order to protect cryptographic protocols between users which only share a short password (see [GL03]). In this context, the goal of the designers is to make make sure that the only efficient attacks need one online attempt to test one possible password and that, under some security hypothesis, no offline attack exists.

In addition to brute force, dictionary attacks may also rely on time-memory tradeoffs. A typical example is Hellman's time memory tradeoff, described in Chapter 7. Where brute force attacks are concerned, the key ingredient is to combine an efficient dictionary of potential passwords with a very fast implementation of the underlying cryptographic algorithm. In the next section, we illustrate this issue of fast implementations using the famous DES algorithm.

5.2 Brute force and the DES algorithm

The DES, Data Encryption Algorithm [DES77], was the first modern block cipher available to a wide audience. It was standardized by the NBS, National Bureau of Standards, in 1977. This block cipher encrypts 64-bit blocks using 56-bit keys. At the time of DES standardization, this keysize was judged to be too small and potentially vulnerable to brute force attacks. However, no successful brute force attack against DES was publicly reported until 1997. At that time, RSA Inc. launched three successive challenges, asking challengers to decrypt cyphertexts encrypted with the DES. The final challenge in 1999 was broken in less than a day, using a mix of hardware and software.

When software brute force attacks against DES are considered, it is essential to use DES implementations that are very efficient. However, using implementations that have been written with fast message encryption in mind are far from optimal when used for cryptanalytic purposes. The main reason is that to optimize the encryption speed, it is often a good idea to assume that keys change once in a while and that message blocks change rapidly. For brute force cryptanalysis, it is exactly the contrary, the known plaintext/ciphertext pairs (or the available ciphertext for ciphertext only attacks) are fixed, and the test key changes extremely quickly. As a consequence, optimizing encryption algorithms with brute force in mind is a very specific task. In the rest of this section, we describe in detail how this can be done in the case of DES. For the sake of completeness, we start by recalling the description of the DES algorithm.

5.2.1 The DES algorithm

DES is based on the Feistel construction and uses 16 consecutive rounds of Feistel. Before applying these rounds, we find an unkeyed initial permutation which moves bits around. After the Feistel rounds, we find an unkeyed final permutation which is the inverse of the initial permutation. In the standard

58	50	42	34	26	18	10	2
60	52	44	36	28	20	12	4
62	54	46	38	30	22	14	6
64	56	48	40	32	24	16	8
57	49	41	33	25	17	9	1
59	51	43	35	27	19	11	3
61	53	45	37	29	21	13	5
63	55	47	39	31	23	15	7

Table 5.1: DES initial permutation

40	8	48	16	56	24	64	32
39	7	47	15	55	23	63	31
38	6	46	14	54	22	62	30
37	5	45	13	53	21	61	29
36	4	44	12	52	20	60	28
35	3	43	11	51	19	59	27
34	2	42	10	50	18	58	26
33	1	41	9	49	17	57	25

Table 5.2: DES final permutation

DES description, individual bits within blocks are numbered from left to right, from 1 to 64 for full blocks, from 1 to 32 for half blocks or from 1 to t for t-bit subblocks. Using this convention, the initial and final permutations of DES are given in Tables 5.1 and 5.2. It says that the first bit of the permuted input is bit 58 of the original input, that the second bit is bit 50, the ninth bit is bit 52 and so on. It is worth noting that if the inputs are viewed as a concatenation of eight bytes, then the fifth permuted byte is the concatenation of the first bits of all original bytes, the first permuted byte the concatenation of second bits and similarly for other bytes. Thus, implementing the input permutation in hardware costs virtually nothing when data is entered on an 8-bit bus. It suffices to consider that each bit of the bus is a serial input and to put together each series of incoming bits in a corresponding byte.

After the initial permutation, the 64 bits of plaintext are split into two 32-bit halves L_0 and R_0. The left half L_0 is formed of bits 1 to 32 of the permuted input. The right half R_0 is formed of bits 33 to 64 of the permuted input. For each round of Feistel numbered from 1 to 16, DES generates a specific subkey: K_i for round i. At round i, the algorithm computes:

$$L_i = R_{i-1} \quad \text{and} \quad R_i = L_{i-1} \oplus f(R_{i-1}, K_i), \tag{5.1}$$

where f is the round function keyed by K_i. After round 16, the concatenation of R_{16} and L_{16} is passed through the final permutation. Note that on output,

16	7	20	21	29	12	28	17
1	15	23	26	5	18	31	10
2	8	24	14	32	27	3	9
19	13	30	6	22	11	4	25

Table 5.3: Permutation of the round function

the left and right halves are reversed. This is an important feature; thanks to it DES (with round keys in inverse order) is its own inverse.

To make our description of DES complete, we now need to describe the round function f and the key expansion process which is used to derive the round keys from the DES 56-bit key. The round function can be written as:

$$f(x, k) = P \circ S(E(x) \oplus k), \qquad (5.2)$$

where E is a linear expansion from 32 to 48 bits, S a non-linear transform consisting of 8 S-boxes from 6 to 4 bits each and P is a bit permutation on 32 bits. The linear expansion E expands its 32 bits of input by duplicating some of them. Using the same bit numbering convention as before, the expansion E is described in Table 5.4. After expansion, the block is xored with the round subkey, then split into 8 slices of 6 bits each. The first slice, consisting of bits 1 to 6, is used as input for the first S-box, usually called S1. Similarly, the other slices are used as input to the other S-boxes, S2 to S8. The outputs of these S-boxes are then concatenated back into a 32-bit word, starting by S1 and ending with S8. The S-boxes S1 to S8 are described in Tables 5.5 to 5.12. The reader unaccustomed to the description of DES, should take great care with these tables. Indeed, they are presented in a way that emphasizes the presence of four permutations on the set of integers from 0 to 15, embedded within each S-box. However, they are not used in a straightforward manner. Each S-box takes 6 bits of the expanded input xored with the round subkey. Due to the structure of the expansion, the middle two bits in each set of six are only used once, while the other four are also used in neighbour sets. The S-boxes are used in a way that reflects this important fact. When computing S_i on a set of six bits $b_1 b_2 b_3 b_4 b_5 b_6$, the number obtained from $b_1 b_6$ (i.e., $2b_1 + b_6$) is used to determine the row number from 0 to 3 and the number obtained from $b_2 b_3 b_4 b_5$ is used to determine the column number from 0 to 15. The output value of the S-box is then interpreted as a bitstring $c_1 c_2 c_3 c_4$ corresponding to the number $8c_1 + 4c_2 + 2c_3 + c_4$. After concatenation, the output bits of S_1 are numbered from 1 to 4, while the output bits of S_8 are numbered from 29 to 32. Using this way to read the S-boxes makes sure that bits duplicated during the expansion are used once as row indices and once as column indices, while non-duplicated bits are only used as column indices.

32	1	2	3	4	5	4	5	6	7	8	9	8	9	10	11
12	13	12	13	14	15	16	17	16	17	18	19	20	21	20	21
22	23	24	25	24	25	26	27	28	29	28	29	30	31	32	1

Table 5.4: Expansion of the round function

14	4	13	1	2	15	11	8	3	10	6	12	5	9	0	7
0	15	7	4	14	2	13	1	10	6	12	11	9	5	3	8
4	1	14	8	13	6	2	11	15	12	9	7	3	10	5	0
15	12	8	2	4	9	1	7	5	11	3	14	10	0	6	13

Table 5.5: S-box S1

15	1	8	14	6	11	3	4	9	7	2	13	12	0	5	10
3	13	4	7	15	2	8	14	12	0	1	10	6	9	11	5
0	14	7	11	10	4	13	1	5	8	12	6	9	3	2	15
13	8	10	1	3	15	4	2	11	6	7	12	0	5	14	9

Table 5.6: S-box S2

10	0	9	14	6	3	15	5	1	13	12	7	11	4	2	8
13	7	0	9	3	4	6	10	2	8	5	14	12	11	15	1
13	6	4	9	8	15	3	0	11	1	2	12	5	10	14	7
1	10	13	0	6	9	8	7	4	15	14	3	11	5	2	12

Table 5.7: S-box S3

7	13	14	3	0	6	9	10	1	2	8	5	11	12	4	15
13	8	11	5	6	15	0	3	4	7	2	12	1	10	14	9
10	6	9	0	12	11	7	13	15	1	3	14	5	2	8	4
3	15	0	6	10	1	13	8	9	4	5	11	12	7	2	14

Table 5.8: S-box S4

2	12	4	1	7	10	11	6	8	5	3	15	13	0	14	9
14	11	2	12	4	7	13	1	5	0	15	10	3	9	8	6
4	2	1	11	10	13	7	8	15	9	12	5	6	3	0	14
11	8	12	7	1	14	2	13	6	15	0	9	10	4	5	3

Table 5.9: S-box S5

12	1	10	15	9	2	6	8	0	13	3	4	14	7	5	11
10	15	4	2	7	12	9	5	6	1	13	14	0	11	3	8
9	14	15	5	2	8	12	3	7	0	4	10	1	13	11	6
4	3	2	12	9	5	15	10	11	14	1	7	6	0	8	13

Table 5.10: S-box S6

4	11	2	14	15	0	8	13	3	12	9	7	5	10	6	1
13	0	11	7	4	9	1	10	14	3	5	12	2	15	8	6
1	4	11	13	12	3	7	14	10	15	6	8	0	5	9	2
6	11	13	8	1	4	10	7	9	5	0	15	14	2	3	12

Table 5.11: S-box S7

5.2.1.1 Round key expansion

In the DES algorithm, the 56-bit key is extracted from a 64-bit word with bits numbered from 1 to 64, by ignoring the high order bit of each byte, i.e., ignoring bits 8, 16, ..., 64. The remaining bits are split in two halves after a permutation called PC-1 described in Table 5.13. The bits obtained from the first part of the table are stored in a register C_0 and the bits obtained from the second part of the table in a register D_0. At round i, registers C_{i-1} and D_{i-1} are transformed into two new values C_i and D_i using left rotations. The 48-bit round key K_i is obtained by concatenating 24 bits from C_i and 24 bits from D_i the selected bits are obtained using PC-2 from Table 5.14, under the convention that bits from C_i are numbered from 1 to 28 and bits from D_i from 29 to 56. To obtain C_i and D_i from C_{i-1} and D_{i-1} the left rotations rotate two bits to the left except when i is 1, 2, 9 or 16 where they are rotated by a single bit. As a consequence, the accumulated amount of rotation during the 16 rounds is 28 bits, thus $C_{16} = C_0$ and $D_{16} = D_0$.

5.2.2 Brute force on DES

First, we can note that during known plaintext brute force attacks, the initial and final permutations of DES can safely be ignored since the permutations can be applied beforehand to plaintext/ciphertext pairs. For ciphertext

13	2	8	4	6	15	11	1	10	9	3	14	5	0	12	7
1	15	13	8	10	3	7	4	12	5	6	11	0	14	9	2
7	11	4	1	9	12	14	2	0	6	10	13	15	3	5	8
2	1	14	7	4	10	8	13	15	12	9	0	3	5	6	11

Table 5.12: S-box S8

Algorithmic Cryptanalysis

57	49	41	33	25	17	9
1	58	50	42	34	26	18
10	2	59	51	43	35	27
19	11	3	60	52	44	36
63	55	47	39	31	23	15
7	62	54	46	38	30	22
14	6	61	53	45	37	29
21	13	5	28	20	12	4

Table 5.13: Permutation PC-1 of the key bits

14	17	11	24	1	5
3	28	15	6	21	10
23	19	12	4	26	8
16	7	27	20	13	2
41	52	31	37	47	55
30	40	51	45	33	48
44	49	39	56	34	53
46	42	50	36	29	32

Table 5.14: Table PC-2 to extract K_i from C_i and D_i

only attacks, the input permutation often allows us to express our knowledge about the plaintext distribution in a more compact form. For example, if we know that the plaintext consists of printable ASCII characters, we know that all high order bits are zeros. After the initial permutation, this implies that bits 25 to 32 form a block of eight consecutive zeros. As a consequence, an implementation of DES dedicated to brute force attacks does not need to implement the initial and final permutations. Similarly, several other minor improvements are possible. For testing several keys that differ only on a few bits, it is not necessary to recompute the first rounds of decryption completely. It is also not necessary to decrypt the whole block of message for all keys. Indeed, if we compute a few bits of the decrypted message and see that they are wrong, the rest of the bits are useless.

The most efficient idea to attack DES by brute force is probably the bitslicing technique proposed by Biham in [Bih97]. With this technique, we view the available computer as a parallel computer that operates in parallel on many single bit registers. At first, this approach may seem surprising; however, it allows us to replace many slow operations such as bit permutations and S-box accesses by much faster operation. Depending on the specific computer used for the computation, we may obtain 32, 64 or even 128 bit operations in parallel. This level of parallelism is used to try several keys simultaneously. As a consequence, to get a fast bitslice implementation of DES, it essentially suffices to express DES efficiently using bit operations. We now study in more details how the various basic operations of DES may be implemented using bit operations:

- **Bit permutations:** Permuting bits on a single-bit computer essentially comes for free. Instead of moving values around, it suffices to wait until the time where the permuted values are used and to directly access the corresponding non-permuted register.

- **Expansion:** Since the DES expansion is a simple copy of bits, instead of doing the copy, we proceed as we explain above for bit permutations and directly read the required bit when needed.

- **Xor:** Since xoring is a bit by bit operation, it can clearly be done without any change at the bit level.

- **S-boxes:** In fact, S-boxes access are the only operations which are not straightforward on a single bit computer. Each S-box is a function from 6 bits to 4 bits. Alternatively, it can be viewed as four functions from 6 bits to 1 bit. Moreover, it is well known that any bit valued function can be written using only XOR and Logical-AND operations. Thus, S-boxes can be computed on single bit computers without any real difficulty. To optimize bitslice implementations of DES, it is important to express each S-box, using as few bit operations as possible. Currently, the best available expressions are those described by Matthew Kwan in [Kwa00].

In addition to this bitslicing technique, one should also remark that at the very end of the computation the internal structure of the DES allows us to stop almost two rounds early. On a total of 16 rounds, this permits to save another 10% of the running time. First, the final round can be avoided in most cases, indeed, once the penultimate round has been performed, half of the ciphertext values for the set of keys being tested are already known. Most of the time, these values do not match the expected ciphertext and the computation can be aborted. Similarly, thanks to the S-box structure, this penultimate round can be computed by small 4-bit chunk. As soon as one chunk does not match the expected value, the corresponding key can safely be forgotten. This approach is called an early abort strategy for the brute force computation.

5.3 Brute force as a security mechanism

As most cryptanalysis tools, brute force can also be used to add security features in computerized systems. The basic idea is to remark that in many computer-assisted tasks, the amount of computation required from a legitimate user is often extremely small or even negligible. For example, testing the correctness of a password by hashing at in Section 5.1 takes an extremely small amount of time. Especially when this amount of time is compared to the inherent delay of the human-computer interface involved. Indeed, no user really cares whether the password verification can be done within a millisecond or a tenth of a second. For practical purposes, this is as good as immediate. Of course, longer delays of the order of the second might be perceived as unbearable. Similarly, when sending out emails, a small delay would be perfectly acceptable. This simple practical remark is the key to the material in this section. During the short delay induced by the user, the computer could perform some real work instead of simply waiting. For password testing for example, one can easily conceive a slow hash function that would take a full tenth of a second to test a password. No user would see the difference. However, from an attacker point-of-view any slow down in the password testing routine is going to slow any attempt to dictionary attack by the same factor. Testing a password in a tenth of a second instead of a millisecond is not an issue for the legitimate user, but it costs the attacker a factor of a hundred, either in terms of running time or of computing power.

However, implementing this simple idea is tricky. It is not enough to make the password testing routine slow. For example, in the early days of computer science, wait loops were added to the password testing routines. However, since an attacker only need to remove the wait loops to make his code for dictionary attacks faster, this approach is essentially useless. Similarly, using unoptimized code for a password testing routine is not a solution. Any motivated attacker can sit down and optimize the code on his own. Instead, one should make sure to use a deeply optimized code implementing an inherently slow function. These functions have been proposed in the literature under the name of **moderately hard-function**. They were initially introduced in [DN93], an interesting variation making use of memory accesses to slow the computation even on fast computers was proposed in [ABW03].

5.4   Brute force steps in advanced cryptanalysis

Brute force is not necessarily a stand-alone cryptanalytic technique, it can also occur as the final step of a larger cryptanalytic effort. To illustrate this, we propose to look at the differential cryptanalysis of the hash function SHA-0 and its practical implementation. The hash function SHA-0 was proposed in 1993 by NIST, under the name SHA. It was modified in 1995 and the new version is called SHA-1. We use the name SHA-0 to emphasize that we are considering the early version of 1993. In this section, we describe SHA-0, show how to devise a differential attack to find collisions in this hash function. Finally, we show that the implementation of this differential attack amounts to sampling using brute force a large enough set of message pairs.

5.4.1 Description of the SHA hash function family

Since SHA-0 and SHA-1 are very similar, they can easily share a common description. For this common description, we refer to both algorithms using the generic name of SHA. The general architecture of SHA is very similar to other existing hash functions. It consists of two parts, a low-level compression function which underlies the hash function and a mode of operation which turns it in a full-fledged hash function. The low level compression function takes as input two values, one on 160 bits and one on 512 bits and outputs a 160 bits value. It is built from an ad-hoc block cipher with 160-bit blocks and 512-bit keys. The block cipher is transformed into a function, rather than a permutation, using a variation of the Davies-Meyer transform. The mode of operation used SHA is the Merkle-Damgård construction with final strengthening. The idea of the construction is to process the message to be hashed iteratively by individual 512-bit blocks. Starting from a fixed initial value IV on 160 bits and the first block of message, the compression function produces a new 160-bit value. This value is compressed together with the second block of message and so on ...

To make the description more precise, let us see in details how a message M is processed. The first step is to make sure that M is encoded into a multiple of the block size in a non-ambiguous fashion. This requirement is very important. Indeed, if two different messages M and M' can be encoded into the same sequence of blocks, then they necessarily have the same hash value and the hash function cannot be collision resistant. With SHA, a simple encoding is used. The bitstring representation of M is padded with a single bit set to 1, a sequence of bits set to 0 and finally a 64-bit representation of the bitlength of the unpadded message M. The length of sequence of zeros is the smallest possible number to have a multiple of 512 as length of the padded message. Thus, the number of zeros is contained in the interval between 0 and 511. It is worth noting that with this padding the length of the unpadded message is encoded twice, once explicitly by the 64-bit number at the end of the padding and once implicitly, since the original message can be recovered by removing from the padded message the 64 final bits, all trailing zeros and a single one. This double encoding of the length was introduced as a strengthening of the Merkle-Damgård construction. It is usually called MD-strengthening.

After padding, the message can be divided into 512-bit blocks, numbered $M_1, M_2, \ldots, M_\ell$, where ℓ is the number of blocks. Denoting by F the compression function, the hash value of M is computed as follows:

$$h_0 = IV \tag{5.3}$$
$$h_i = F(h_{i-1}, M_i) \text{ for all } i \text{ from 1 to } \ell.$$

The last value h_ℓ is the hash value of M. The first value h_0 is the fixed initial value. For convenience, the other values h_i are called intermediate hash values. The compression function F of SHA is described in Section 5.4.1.2.

5.4.1.1 Security properties of the Merkle-Damgård construction

As discussed in Chapter 1, when considering hash functions, we are interested by three security properties: collision resistance, preimage resistance and second preimage resistance.

When considering the Merkle-Damgård construction, it is possible to show that the security of the hash function H can be derived from the security of the compression function H. For example, given a collision for H, it is easy to produce a collision for the compression function F. First, if M and M' have different lengths, their final blocks M_ℓ and $M'_{\ell'}$ are different. If $H(M) = H(M')$, we necessarily have:

$$F(h_{\ell-1}, M_\ell) = F(h'_{\ell'-1}, M'_{\ell'}), \tag{5.4}$$

which is a collision for F. Otherwise, M and M' are different messages of the same length and we look at the largest block index i in the interval $[1, \ell]$ such that either $M_i \neq M'_i$ or $h_{i-1} \neq h'_{i-1}$. Of course, $h_i = h'_i$ and

$$F(h_{i-1}, M_i) = F(h'_{i-1}, M'_i) \tag{5.5}$$

is a collision for F. It is important to remark that in general we obtain collisions for F with different values on the h component. As a consequence, proving that $F(h, \cdot)$ is collision resistant for all fixed values of h is not sufficient to establish collision resistance of H. In relation to this comment, note that some specific differential attacks, called multiblock attacks, require several consecutive calls to F before producing a collision for H. From this multiblock attack, one can derive a collision on F but usually not on $F(h, \cdot)$.

Using a very similar analysis, we can prove that if F is preimage resistant then H is preimage resistant. Indeed, if we can find a preimage M of a target value h, we see that $F(h_{\ell-1}, M_\ell) = h$ and obtain a preimage of h for the compression function F.

However, where second preimage resistance is concerned, we obtain a slightly weaker result. Any second preimage attack on H can be used to obtain a collision attack on F, but not necessarily a second preimage attack on F.

5.4.1.2 Compression function of SHA

At the core of the compression function of SHA, we find a specific block cipher that encrypts 160-bit blocks using a 512-bit key. In order to be efficient in software, this specific block cipher is mostly based on 32-bit unsigned arithmetic. As a consequence, each 160-bit block is decomposed in five 32-bit values. The input to the block cipher is thus denoted by the vector $\langle A^{(0)}, B^{(0)}, C^{(0)}, D^{(0)}, E^{(0)} \rangle$. The block cipher itself is a generalized Feistel

Round i	Function $f^{(i)}(x, y, z)$		Constant $K^{(i)}$
	Name	Definition	
0–19	IF	$(x \wedge y) \vee (\bar{x} \wedge z)$	0x5A827999
20–39	XOR	$(x \oplus y \oplus z)$	0x6ED9EBA1
40–59	MAJ	$(x \wedge y) \vee (x \wedge z) \vee (y \wedge z)$	0x8F1BBCDC
60–79	XOR	$(x \oplus y \oplus z)$	0xCA62C1D6

Table 5.15: Definition of the round functions and constants

scheme with 80 rounds. At round i, we denote the inner state of the encryption algorithm by $\langle A^{(i)}, B^{(i)}, C^{(i)}, D^{(i)}, E^{(i)} \rangle$. To update the inner state between round i and round $i+1$, a new value is computed for $A^{(i+1)}$ by mixing together the previous values of A, B, C, D, E and some key material. At the same time, $B^{(i+1)}$, $C^{(i+1)}$, $D^{(i+1)}$ and $E^{(i+1)}$ are obtained as (rotated) copies of $A^{(i)}$, $B^{(i)}$, $C^{(i)}$ and $D^{(i)}$. More precisely, for each round, with i varying from 0 to 79, the registers A, B, C, D and E are updated according to the following formulas:

$$A^{(i+1)} = \text{ROL}_5\left(A^{(i)}\right) + f^{(i)}(B^{(i)}, C^{(i)}, D^{(i)}) + E^{(i)} + W^{(i)} + K^{(i)},$$

$$B^{(i+1)} = A^{(i)},$$

$$C^{(i+1)} = \text{ROL}_{30}\left(B^{(i)}\right), \tag{5.6}$$

$$D^{(i+1)} = C^{(i)} \text{ and}$$

$$E^{(i+1)} = D^{(i)}.$$

In these equations, addition is performed modulo 2^{32} and $\text{ROL}_b X$ denotes the value obtained by rotating b bits to the left the binary representation of the 32-bit unsigned integer X. Moreover, $f^{(i)}$ is a round function and $K^{(i)}$ a round constant, they both depend on i. They are specified in Table 5.15. Finally, $W^{(i)}$ is the contribution of the 512-bit key to round i. The 16 values from $W^{(0)}$ to $W^{(15)}$ are simply obtained by decomposing the 512-bit input into 16 consecutive 32-bit words. The next 64 values $W^{(16)}$ to $W^{(79)}$ are obtained as the result of an expansion process. In SHA-0 the expansion is defined by:

$$\forall i: \ 16 \leq i < 80, \quad W^{(i)} = W^{(i-3)} \oplus W^{(i-8)} \oplus W^{(i-14)} \oplus W^{(i-16)}. \tag{5.7}$$

Note that this only difference between SHA-0 and SHA-1 concerns this expansion. In SHA-1, the expansion becomes:

$$W^{(i)} = \text{ROL}_1\left(W^{(i-3)} \oplus W^{(i-8)} \oplus W^{(i-14)} \oplus W^{(i-16)}\right). \tag{5.8}$$

After 80 rounds, the block cipher output is $\langle A^{(80)}, B^{(80)}, C^{(80)}, D^{(80)}, E^{(80)} \rangle$. However, despite the fact that SHA is based on the block cipher as mentioned at the beginning of this section, using the Merkle-Damgård construction directly with a block cipher is not a good idea; see Exercise 5 in Chapter 6. Instead, an additional step, similar to the Davies-Meyer construction, is used to transform this block cipher into a function, rather than a permutation. Recall that the Davies-Meyer construction is a very useful way to turn a cryptographic permutation π into a cryptographic function $x \rightarrow \pi(x) \oplus x$. With SHA, a variation based on 32-bit addition in place of the exclusive or is used. With this transform, the final output of the compression function is:

$$\left\langle A^{(80)} + A^{(0)}, B^{(80)} + B^{(0)}, C^{(80)} + C^{(0)}, D^{(80)} + D^{(0)}, E^{(80)} + D^{(0)} \right\rangle.$$

5.4.2 A linear model of SHA-0

In order to study the propagation of differences in SHA-0 and to learn how to control this propagation well enough to construct collisions, it is very useful to start by looking at a fully linearized variation of SHA-0. First, remark that in SHA-0, there are two sources of non-linearity: the $f^{(i)}$ functions in rounds 0 to 19 and in rounds 40 to 59 (respectively IF and MAJ) and the addition modulo 2^{32}. Indeed, addition modulo 2^{32} may involve propagation of carries, and as a consequence it is not linear over $\mathbb{F}_2$. The linear model of SHA-0 is obtained by replacing the additions by XOR operations on 32-bit words and by using the XOR function as round function in every round. We also replace addition by XOR in the final Davies-Meyer transform.

With this change, we obtained a completely linear compression function. As a consequence, finding collisions becomes a straightforward task. It suffices using linear algebra algorithms to construct an element of the kernel of this function. However, this approach does not shed any light on SHA-0 itself. Instead, we are now going to write this linear algebra attack using a specific sparse approach. The advantage is that this approach can then be transformed into a probabilistic attack of SHA-0. The specific approach splits the problem of collision finding into two easier tasks. The first task concentrates on a small number of consecutive rounds ignoring the expansion of the vector W, yielding local collisions. The second task is to merge together several local collisions while remaining consistent with the expansion.

5.4.2.1 Local collisions in linearized SHA-0

In order to construct local collisions in the linearized version of SHA-0, it suffices to introduce a change on a single bit of W, to follow the propagation of this change in the subsequent rounds and using adequate corrections to prevent the initial change from impacting too many bits. More precisely, starting from a common inner state $\left\langle A^{(i)}, B^{(i)}, C^{(i)}, D^{(i)}, E^{(i)} \right\rangle$, assume that we perform in parallel two partial hash computations. In the first computation, we input a message block $W_1^{(i)}$ and in the second computation a message block

$W_2^{(i)}$. We call $\delta = W_1^{(i)} \oplus W_2^{(i)}$ the difference[3] of the two message blocks. To make the analysis simpler, it is useful to assume that the difference δ is on a single bit.

Clearly, by linearity, the difference δ directly propagates in the computation of $A^{(i+1)}$ and we find that $A_1^{(i+1)} \oplus A_2^{(i+1)} = \delta$. In the subsequent steps, this initial difference on $A^{(i+1)}$ moves to $B^{(i+2)}$, $C^{(i+3)}$, $D^{(i+4)}$ and $E^{(i+5)}$. However, if we do not make any other changes in the message blocks, secondary changes also occur in $A^{(i+2)}$, $A^{(i+3)}$, ... To control the propagation of the differences, the key idea is to make other changes in W in order to prevent any secondary change in A. We refer to these additional changes as corrections. The first correction is to prevent the difference in $A^{(i+1)}$ to propagate to $A^{(i+2)}$. The reader can easily check that, by linearity, this is effected by choosing $W_1^{(i+1)} \oplus W_2^{(i+1)} = \text{ROL}_5(\delta)$. The second correction prevents the difference in $B^{(i+2)}$ to affect $A^{(i+3)}$, it consists of choosing $W_1^{(i+2)} \oplus W_2^{(i+2)} = \delta$. The third correction prevents the difference in $C^{(i+3)}$ to affect $A^{(i+4)}$, since $C^{(i+3)}$ is a rotated copy of $B^{(i+2)}$, the correction requires $W_1^{(i+3)} \oplus W_2^{(i+3)} = \text{ROL}_{30}(\delta)$. Finally, the fourth and fifth corrections account for $D^{(i+4)}$ and $E^{(i+5)}$ using $W_1^{(i+4)} \oplus W_2^{(i+4)} = \text{ROL}_{30}(\delta)$ and $W_1^{(i+5)} \oplus W_2^{(i+5)} = \text{ROL}_{30}(\delta)$.

After that fifth correction, we see that the initial difference vanishes and thus the two parallel computations yield the same inner state

$$\left\langle A^{(i+6)}, B^{(i+6)}, C^{(i+6)}, D^{(i+6)}, E^{(i+6)} \right\rangle.$$

This propagation of differences is summarized in Figure 5.1. In this figure, $A^{(i)}[j]$ denotes the j-th bit of $A^{(i)}$.

5.4.2.2 Combining local collisions with the message expansion

In order to combine local collisions with the message expansion, it suffices to remark that thanks to linearity, xoring together the changes and corrections of several local collisions yields a possible pattern of collision. As a consequence, it suffices to put together several local collisions in a way that guarantees that the XOR of all involved changes and corrections is a valid output of the message expansion. With SHA-0, the fact that the expansion is not only linear but also applies in parallel to individual bits of the words W is quite helpful. Indeed, any valid expanded sequences of words W_0, ..., W_{79} can be constructed by pasting together 32 expanded sequences of bits. Since the expansion computes 80 expanded bits from 16 initial bits, it is even easy to exhaustively compute the 2^{16} different expanded sequences on a single bit position.

The key idea is to insert several local collisions in a way that follows such an expanded sequence of bits. In particular, this means that all changes of

---

[3]This is a standard name in differential cryptanalysis, reminiscent from the fact that in $\mathbb{F}_2$, $a - b$ is the same thing as $a \oplus b$.

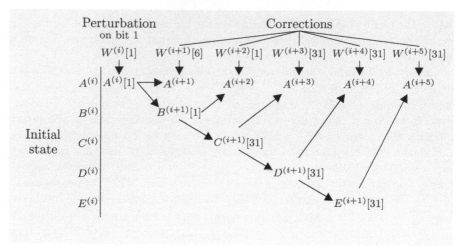

Figure 5.1: Propagation of differences in a local collision of SHA

these local collisions act on the same bit position with the words W. Let j denote this bit position. Clearly, according to our previous description of the local collisions, all in all, three bit positions are involved in the changes and correction, namely j, $j+5$ and $j-2$, where bit positions are numbered from 0 to 31 and considered modulo 32. Due to the parallel nature of the expansion, an expanded bit sequence can be moved around from one bit position to another without any difficulty. However, the construction of local collisions does not only require to move bit sequence between bit position but also to shift them between word position. For example, a change on $W^{(i)}$ leads to a first correction on $W^{(i+1)}$. Here, we encounter a possible difficulty. Indeed, nothing guarantees that an expanded sequence of bits remains a valid expanded sequence if we shift it in this way. For example, the sequence:

$$0000000000000001$$
$$0010010110010010$$
$$0110110000110001$$
$$1101011111011010$$
$$0001000010110100$$

cannot be shifted in that way. Indeed, in order to obey the expansion formula, it would need to become:

$$1000000000000000$$
$$1001001011001001$$
$$0011011000011000$$
$$1110101111101101$$
$$0000100001011010$$

with a one at the beginning. However, half of the possible sequences can be shifted without problems. In fact, in order to perform the five necessary corrections, the chosen sequence of changes needs to be shifted 5 times. This is possible for a fraction 1/32 of expanded bit sequences. An easy way to formalize this is to run the expansion backward and to require that the five bits in the negative position are zeros. Equivalently, this adds five linear conditions on the 16 input bits of the expansion.

For any such sequence of bits, putting together the corresponding local collisions gives an interesting differential pattern for linearized SHA-0. However, this pattern need not be a collision. Indeed, if a local collision starts in one of the 5 last rounds, it cannot be canceled in time. In that case, we only obtain a **pseudo-collision**. To get a collision for linearized SHA-0, an additional constraint is needed. There should be no local collision remaining active in the final round, thus we need to choose the local collision positions according to a sequence of bits with five zeros in the five last positions. To satisfy this, we need to add five more linear conditions.

When we enforce the conditions of having five zeros in the negative positions and five in the final positions, there remain 63 non-zero bit sequences that can be used to introduce local collisions in a way that is consistent with the message expansion. Some of these possible sequences are described in Table 5.16; the complete table is available on the book's website.

Let δ be any sequence on 80 bits from Table 5.16. Denote by $\delta^{(i)}$ the i-th bit of δ, with the convention that the sequence in the table are numbered from 0 to 79. In addition, when i is a negative integer from -5 to -1, let $\delta^{(i)} = 0$. Given δ together with a bit position j, we can construct an 16-word vector Δ in the kernel of the linearized SHA-0 by letting:

$$\Delta^{(i)} = \text{ROL}_j(\delta^{(i)}) \oplus \text{ROL}_{j+5}(\delta^{(i-1)}) \oplus \text{ROL}_j(\delta^{(i-2)}) \oplus \text{ROL}_{j-2}(\delta^{(i-3)}) \oplus$$
$$\text{ROL}_{j-2}(\delta^{(i-4)}) \oplus \text{ROL}_{j-2}(\delta^{(i-5)}), \quad \text{for all } 0 \leq i \leq 15. \tag{5.9}$$

Due to the care we took to construct Δ, Equation 5.9 also holds after expansion for $\Delta^{(i)}$ with $16 \leq i \leq 79$. It is clear that for any 16-word message W, the linearized SHA-0 hashes of W and $W \oplus \Delta$ form a collision.

5.4.3 Adding non-linearity

Once the linear algebra collision attack on linearized SHA-0 is written in the sparse form of Section 5.4.2, we can try to use it as a probabilistic collision attack on the real SHA-0 hash algorithm. As before, given a sequence of 80 bits δ from Table 5.16 and a bit position j, we construct a vector Δ and then consider pairs of messages $(W, W \oplus \Delta)$. For such a pair, a collision may occur in two different ways. It may occur by wild chance after two extremely different computations or it may result from a nice case where the two computations follow essentially the same relationship as in the linearized case. The first type of collision is not interesting for us because we do not get any control

```
000100000001001000000001000011011011111101101001000010101001010100010111001100000
001000100000000101111011000111000000010100010001001001001110110011000011111000000
001100100000100001111010000100011011010101001011010000110100110010010000111100000
101001010000011111001111001100011111110110111100001100010101011010010100100000000
101101010001010110011010010101010000011011011100010010001111101011001000110000000
100001110000010100111001000010011110100111110001010001011100100010001011000000000
100101110001011100111011000100101001011100101010101011011111001110011010111110000
011000001001001100010011101110111001010001001000100111000110010010000100111000000
011100001000000100010001101000001110101010011010100010110001100010101010100000000
010000101001001111100101100000111000000000001100000011011000000011000101101100000
010100101000001111100111100110001111111011011110000100001010101110100101000000000
110001011001010011011100100010100110100111101001010101111101100101110011011110000
110101011000011011011110101010001000101110010011010111010010011111110000010000000
111001111001011000010101010110010011111011011100000011110011010110110000010110000
111101111000010000101000101001000000011011000100010100111111100111011111000000000
100011000100011000111100111111011000001011010011001110100000001010011000100000000
100111000101010000111110111001101111110000000010010100000101000101011011011100000
101011100010001001100101011000101100101101000101111010111010110001100101110000000
101111100101011011001000110110111101010001000101101101110011011110111011100101100000
001010010100000111110011110011000111111011011110000100001001010111010010100000000
001110010101001111110001110101110000000110111101000010010111111111111110011100000
000010110100001100000010111110100011010110010101110011111111001101101101101100000
000110110101000100000111111011110001010111111100110000101011001100111100110100000
111011001101010100101111010001100001011010011011101000110011000000001110001100000
111111001100011100101101010110101011010101010000100100110110001101000011001000000
110011101101011111011100101111110000000010110111110010110110001000010001001111100000
110111101100010111011011011001010111100000011010010101101010010011110110000000
010010011101001011000000111011111101011001001110010010010011001110101011001100000
010110011100000011100010011011001001010111110101100001110100110101110000000000
011010111101000000010110010011111111111101100011000000011101010001011001111100000
011110111100001000010100010010001001101100010011011011111110011100010111010111100000
011001000010000111100010000100110101010001011010000101001100100010000111100000
011101000011001111101010010111011010101011111111100011000000110000110110110100000
010001100010001100011110011111101100000101010011001110100000010100110001000000
010101100011000100011100011001011011111110111011100010111010101101100010000100000
110000010010011000100110111011110010100010010010011110001100101000010011000000
110100010011010000100101011011000101100001000100100100110100110100111001110000
111000110010010011010001010011110011110010101010101111101011000000011001000000
111100110011011011010011010101000100010000011101101011111110000010100000100000
000001001011001011110111111111010100000101100101100100110000000011000111001001000000
000101001010000011111001110011000111111011011110000110001010101110100101000000
001001101011000000011011100010101010101001000010000000011001111100100010100000
001101101010001000001111110111100010101111100110001010110011001110011011000000
```

⋮

```
000111111110001111111100000100100101010010011100000110011100110000110100010000000
```

Table 5.16: Possible expanded bit sequences for local collisions

compared to a random independent message pair. The second type of collision is much nicer and we now want to study this case.

In order to find a colliding pair $(W, W \oplus \Delta)$, we proceed in two phases. First, we choose a good value for the difference Δ in order to maximize the probability of success. After that, we search for a good value of W. In order to choose a good difference Δ, we need to analyze the sources of non-linearity in SHA-0 and consider how to minimize their impact. There are two sources of non-linearity in SHA-0, additions modulo 2^{32} and the functions from Table 5.15. Clearly, the impact of the functions does not depend on the bit position j, since they operate in parallel on the bit positions. On the contrary, for the modular addition, not all bit positions are equivalent. Indeed, when adding two numbers X and Y, it is easy to see that changes on the high order bits of both numbers never affect the value of the sum $X + Y \pmod{2^{32}}$ in a non-linear manner. On the other hand, when changing the value of some lower order bits of X and Y, the corresponding bit of $X + Y \pmod{2^{32}}$ is affected in a linear way, but the next bits of the sum may be modified non-linearly. To give an example, when we modify the low order bit of X, we either add or subtract 1 from X. When changing this bit both in X and Y, if one change corresponds to an addition and the other to a subtraction, $X + Y$ does not change. Otherwise, the value of $X + Y$ changes by either 2 or -2. This carry to the next bit value is a non-linear change. If in addition the carry propagates, other bits may also be affected in the binary representation of $X + Y$.

As a consequence, to build a good value for the difference Δ, we roughly need to minimize the total number of individual bit changes occurring during the computation, in order to limit the risk of encountering carries. We also need to have a large fraction of the differences located on the high order bit in order to take advantage of the linearity of the high order bit in additions. To satisfy these two conditions, the basic idea is to choose a low weight element in Table 5.16 and use it as a locator to insert local collisions on bit position 1. Indeed, we know that such local collisions are going to involve changes on bit positions 1, 6 and 31. Moreover, bit position 31, which corresponds to high order bits occurs more frequently than the others. Indeed, a change on bit position 1 at round i requires one correction on each of the bit positions 1 and 6, but three corrections on bit 31. Note the weight of the elements in Table 5.16 is only a rough indicator. In order to choose the best possible locations to insert the local collisions, it is necessary to analyze the propagation of changes within SHA-0 computations in more details. We now give this detailed analysis.

5.4.3.1 Propagation of changes in SHA-0

In order to study the propagation of changes throughout SHA-0, we are going to analyze each of the non-linear functions that occur here. There are three such functions to consider: addition modulo 2^{32} on a bit position

< 31 and the IF and MAJ functions. Indeed, addition on the high order bit and the XOR function both behave linearly. For each function, we not only need to consider the presence or not of a change on each input bit but also the direction of the change. In the tables given in this section, we use the following notations, 0 represents a bit equal to 0 that does not change, 1 represents a bit equal to 1 that does not change, ↑ represents of bit that changes from 0 to 1 and ↓ represents of bit that changes from 1 to 0.

With this notation, describing the case by case behavior for each function on three bits is quite simple. They are given in Tables 5.17, 5.19 and 5.20. In the case of addition, we only give the details for an addition of three summands but we also indicate the behavior of the carry bit. It is, of course, possible to perform the same analysis for more summands. However, in the case of SHA-0, this simplified analysis is sufficient to understand the probabilistic behavior of local collisions.

With these three tables in mind, it is easy to understand the behavior of a single, stand-alone local collision. The first action of this local collision is to insert a difference on bit 1 of $W^{(i)}$. To simplify the analysis, it is useful to assume that we know the direction of the bit change. For example, let us assume that the changing bit goes from 0 to 1. Following the notation of Tables 5.17 to 5.20, it is a ↑ transition. As in the linear case described in Section 5.4.2.1, this change directly affects bit 1 of $A^{(i+1)}$. Looking at Table 5.20, we see that two main cases are possible. If there are no unwanted carries, then bit 1 of $A^{(i+1)}$ is also a ↑ and all other bits of $A^{(i+1)}$ are constants. Otherwise, bit 1 of $A^{(i+1)}$ is a ↓ and bit 2 no longer is a constant. Note that the carry effect may propagate further and affect more bits. Assuming that the inputs of the addition are random values, we can see that the nice case where no bad carry occurs happens with probability $1/2$. Once the ↑ change is in $A^{(i+1)}$, it is involved (after rotation) in the computation of $A^{(i+2)}$. Since we are considering a local collision, another change is also present on bit 6 of $W^{(i+2)}$. Looking again at the table, we see that when both changes are ↑, they never cancel each other. Thus, we need to make sure that the correction present on bit 6 of $W^{(i+2)}$ is a ↓, in which case it always correctly performs the correction. In the next round, the ↑ change is on bit 1 $B^{(i+2)}$ and involved in the computation of $A^{(i+3)}$, together with the correction located on bit 1 of $W^{(i+2)}$. This change undergoes two non-linear processes; first it goes through one of the three functions IF, XOR or MAJ, then it enters the addition. Looking at the three functions, we see that a total of three different behaviors may occur. The change may vanish, it may remain unchanged or it may be reversed into a ↓. Note that with XOR, the change never vanishes. With MAJ, it is never reversed. With IF, all three behaviors are possible. If the change vanishes, then the correction never occurs correctly and the local collision fails. With IF or MAJ, this occurs with probability $1/2$. Concerning the direction of the change, we need to consider two cases. When going through MAJ, the ↑ is never reversed, thus if we make sure that the correction on bit 1 of $W^{(i+2)}$ is a ↓, the addition also behaves nicely, assuming that the

$x=0$	$y=0$	$y=1$	$y=\uparrow$	$y=\downarrow$	$x=1$	$y=0$	$y=1$	$y=\uparrow$	$y=\downarrow$
$z=0$	0	0	0	0	$z=0$	0	1	$\uparrow$	$\downarrow$
$z=1$	0	1	$\uparrow$	$\downarrow$	$z=1$	1	1	1	1
$z=\uparrow$	0	$\uparrow$	$\uparrow$	0	$z=\uparrow$	$\uparrow$	1	$\uparrow$	1
$z=\downarrow$	0	$\downarrow$	0	$\downarrow$	$z=\downarrow$	$\downarrow$	1	1	$\downarrow$
$x=\uparrow$	$y=0$	$y=1$	$y=\uparrow$	$y=\downarrow$	$x=\downarrow$	$y=0$	$y=1$	$y=\uparrow$	$y=\downarrow$
$z=0$	0	$\uparrow$	$\uparrow$	0	$z=0$	0	$\downarrow$	0	$\downarrow$
$z=1$	$\uparrow$	1	$\uparrow$	1	$z=1$	$\downarrow$	1	1	$\downarrow$
$z=\uparrow$	$\uparrow$	$\uparrow$	$\uparrow$	$\uparrow$	$z=\uparrow$	0	1	$\uparrow$	$\downarrow$
$z=\downarrow$	0	1	$\uparrow$	$\downarrow$	$z=\downarrow$	$\downarrow$	$\downarrow$	$\downarrow$	$\downarrow$

Table 5.17: Case by case behavior of $\mathbf{MAJ}(x, y, z)$

$x=0$	$y=0$	$y=1$	$y=\uparrow$	$y=\downarrow$	$x=1$	$y=0$	$y=1$	$y=\uparrow$	$y=\downarrow$
$z=0$	0	1	$\uparrow$	$\downarrow$	$z=0$	1	0	$\downarrow$	$\uparrow$
$z=1$	1	0	$\uparrow$	$\downarrow$	$z=1$	0	1	$\downarrow$	$\uparrow$
$z=\uparrow$	$\uparrow$	$\downarrow$	0	1	$z=\uparrow$	$\downarrow$	$\uparrow$	1	0
$z=\downarrow$	$\downarrow$	$\uparrow$	1	0	$z=\downarrow$	$\uparrow$	$\downarrow$	0	1
$x=\uparrow$	$y=0$	$y=1$	$y=\uparrow$	$y=\downarrow$	$x=\downarrow$	$y=0$	$y=1$	$y=\uparrow$	$y=\downarrow$
$z=0$	$\uparrow$	$\downarrow$	0	1	$z=0$	$\downarrow$	$\uparrow$	1	0
$z=1$	$\downarrow$	$\uparrow$	1	0	$z=1$	$\uparrow$	$\downarrow$	0	1
$z=\uparrow$	0	1	$\uparrow$	$\downarrow$	$z=\uparrow$	1	0	$\downarrow$	$\uparrow$
$z=\downarrow$	1	0	$\downarrow$	$\uparrow$	$z=\downarrow$	0	1	$\uparrow$	$\downarrow$

Table 5.18: Case by case behavior of $\mathbf{XOR}(x, y, z)$

$x=0$	$y=0$	$y=1$	$y=\uparrow$	$y=\downarrow$	$x=1$	$y=0$	$y=1$	$y=\uparrow$	$y=\downarrow$
$z=0$	0	0	0	0	$z=0$	0	1	$\uparrow$	$\downarrow$
$z=1$	1	1	1	1	$z=1$	0	1	$\uparrow$	$\downarrow$
$z=\uparrow$	$\uparrow$	$\uparrow$	$\uparrow$	$\uparrow$	$z=\uparrow$	0	1	$\uparrow$	$\downarrow$
$z=\downarrow$	$\downarrow$	$\downarrow$	$\downarrow$	$\downarrow$	$z=\downarrow$	0	1	$\uparrow$	$\downarrow$
$x=\uparrow$	$y=0$	$y=1$	$y=\uparrow$	$y=\downarrow$	$x=\downarrow$	$y=0$	$y=1$	$y=\uparrow$	$y=\downarrow$
$z=0$	0	$\uparrow$	$\uparrow$	0	$z=0$	0	$\downarrow$	0	$\downarrow$
$z=1$	$\downarrow$	1	1	$\downarrow$	$z=1$	$\uparrow$	1	$\uparrow$	1
$z=\uparrow$	0	$\uparrow$	$\uparrow$	0	$z=\uparrow$	$\uparrow$	1	$\uparrow$	1
$z=\downarrow$	$\downarrow$	1	1	$\downarrow$	$z=\downarrow$	0	$\downarrow$	0	$\downarrow$

Table 5.19: Case by case behavior of $\mathbf{IF}(x, y, z)$

$x = 0$	$y = 0$	$y = 1$	$y =\uparrow$	$y =\downarrow$	$x = 1$	$y = 0$	$y = 1$	$y =\uparrow$	$y =\downarrow$
$z = 0$	00	01	$0\uparrow$	$0\downarrow$	$z = 0$	01	10	$\uparrow\downarrow$	$\downarrow\uparrow$
$z = 1$	01	10	$\uparrow\downarrow$	$\downarrow\uparrow$	$z = 1$	10	11	$1\uparrow$	$1\downarrow$
$z =\uparrow$	$0\uparrow$	$\uparrow\downarrow$	$\uparrow 0$	01	$z =\uparrow$	$\uparrow\downarrow$	$1\uparrow$	$\uparrow 1$	10
$z =\downarrow$	$0\downarrow$	$\downarrow\uparrow$	01	$\downarrow 0$	$z =\downarrow$	$\downarrow\uparrow$	$1\downarrow$	10	$\downarrow 1$
$x =\uparrow$	$y = 0$	$y = 1$	$y =\uparrow$	$y =\downarrow$	$x =\downarrow$	$y = 0$	$y = 1$	$y =\uparrow$	$y =\downarrow$
$z = 0$	$0\uparrow$	$\uparrow\downarrow$	$\uparrow 0$	01	$z = 0$	$\downarrow$	$\downarrow\uparrow$	01	$\downarrow 0$
$z = 1$	$\uparrow\downarrow$	$1\uparrow$	$\uparrow 1$	10	$z = 1$	$\downarrow\uparrow$	$1\downarrow$	10	$\downarrow 1$
$z =\uparrow$	$0\uparrow 0$	$\uparrow 1$	$\uparrow\uparrow$	$\uparrow\downarrow$	$z =\uparrow$	01	10	$\uparrow\downarrow$	$\downarrow\uparrow$
$z =\downarrow$	01	10	$\uparrow\downarrow$	$\downarrow\uparrow$	$z =\downarrow$	$\downarrow 0$	$\downarrow 1$	$\downarrow\uparrow$	$\downarrow\downarrow$

Table 5.20: Case by case behavior of ADD(x, y, z) (carry bit on left)

change does not vanish. With IF or XOR, we do not know in advance if the change remains a $\uparrow$ or becomes a $\downarrow$, thus we do not care about the direction of the change on bit 1 of $W^{(i+2)}$, the addition cancels the two with probability $1/2$. Summing up, we see that the computation of $A^{(i+3)}$ works out correctly with probability $1/4$ when the function is an IF, with probability $1/2$ due to the possibility of vanishing changes when it is a MAJ and with probability $1/2$ due to the possibility of reversing changes when it is a XOR. The next correction concerns bit 31 of $A^{(i+4)}$. Since it involves the high order bit of addition, no carry propagation is possible. Thus, at this point, we do not care whether the change is preserved or reversed, as long as it does not vanish. Thus, there is a probability of good correction of $1/2$ with IF and MAJ, while the correction is always correct with XOR. The fourth correction on bit 31 of $A^{(i+5)}$ behaves exactly in the same way. Finally, the final correction on bit 31 of $A^{(i+6)}$ is always correctly cancelled by the correction bit present on bit 31 of $W^{(i+5)}$.

Thanks to this analysis, it is possible to compute for any given round position i, the probability of successfully applying a single local collision at this round. However, this calls for several comments. First, we need to make sure in advance that where needed the correction bits are indeed of type $\downarrow$ to compensate for the initial change $\uparrow$. Second, this probability is correct when the computation of the local collision is uncontrolled, i.e., when all values that enter the computation, with the exception of the correction bits, are considered to be random. As we will see later, when it is possible to choose (some of) the input values, then it is possible to greatly improve this probability.

5.4.3.1.1 Superposing several local collisions The above analysis is not complete because it only considers isolated local collisions that do not interfere with each other. However, looking at Table 5.16, we see that none of the candidates for placing the local collisions can guarantee that they are always isolated from each other. As a consequence, we need to understand the interaction between interleaved local collisions. Two cases need to be considered. On one hand, when two such local collisions only act on different bits, it is safe to assume that the individual probability of success are preserved and simply need to be multiplied together to obtain the total probability. On the other hand, when two local collisions impact the same bit of some word in A, the exact behavior is quite complicated and each of the functions IF, XOR and MAJ is a specific case.

To analyze these interactions, let us determine the relative positions of interfering pairs of local collisions. Assume we are superposing two local collisions respectively starting on $A^{(i_1)}$ and $A^{(i_2)}$, with $i_1 < i_2$. Note that when $i_2 > i_1 + 5$, the local collisions are clearly not overlapping. Thus, there are five cases to consider. For each case, there is an interaction for each round where the two local collisions act on the same bit. For example, when $i_2 = i_1 + 1$, we see that at round $i_1 + 1$ the first collision acts on bit 6, the second on bit 1 and that they do not interfere. At round $i_1 + 2$, they respectively act on bits 1 and 6. At round $i_1 + 3$, they act on bits 31 and 1. At rounds $i_1 + 4$ and $i_1 + 5$, they both act on bits 31 and do interfere. We summarize all the possible interactions in Table 5.21. We see in this table that there are only four possible cases of interaction:

- When $i_2 = i_1 + 2$, there is an interaction on bit 1 in round i_2. On message word $W^{(i_2)}$, we see that the two local collisions both modify bit 1, as a consequence, their interaction leaves this bit fixed. Thus, we need to consider the consequence of a single difference on bit 1 of $B^{(i_2)}$. If this difference does not propagate through the function $f^{(i_2)}$, then the second local collision is not correctly inserted in round i_2 and the corresponding corrections in later rounds fail. When $f^{(i_2)}$ is a MAJ function, the direction $\uparrow$ or $\downarrow$ of the change is preserved. This information should be used when determining the direction of the corrections for the second local collision. With IF or XOR, the direction may change and we can choose between two sets of possibilities for the corrections. Of course, once we have chosen to consider that the direction is preserved (or changed), any message pair that behaves differently encounters miscorrection at some later round.

We can now compare the probability of success with two interacting local collisions and the probability of two independent local collisions one in round i_1 and one in round i_2. With two independent collisions, the first succeeds when the perturbation correctly goes through f, i.e., with probability $1/2$ with XOR or MAJ and $1/4$ with IF; the second succeeds if there is no carry, i.e., with probability $1/2$. With two interacting collisions, the probabilities are allocated slightly differently, but the overall contribution is left unchanged. Thus, this type of interaction does not deeply affect the construction of a collision path.

- When $i_2 = i_1 + 1$, there is an interaction on bit 31 in round $i_1 + 4 = i_2 + 3$. On message word $W^{(i_1+4)}$, the corrections cancel each other. Thus,

$i_2 - i_1$	Round number				
	$i_1 + 1$	$i_1 + 2$	$i_1 + 3$	$i_1 + 4$	$i_1 + 5$
1	$6 \neq 1$	$1 \neq 6$	$31 \neq 1$	**31**	**31**
2	—	1	$31 \neq 6$	$31 \neq 1$	**31**
3	—	—	$31 \neq 1$	$31 \neq 6$	$31 \neq 1$
4	—	—	—	$31 \neq 1$	$31 \neq 6$
5	—	—	—	—	$31 \neq 1$

Table 5.21: Interferences of overlapping local collisions

we have two changes on bit 31, one in $D^{(i_1+4)}$, the other in $C^{(i_2+3)}$. Note that the two exponents are equal; they are written differently to emphasize that the first change corresponds to the first collision and the second to the second one. With the XOR function, these two changes cancel each other, which is the desired behavior. With MAJ, the two changes cancel if and only if one is a $\uparrow$ and the other a $\downarrow$. Since this can be controlled in advance when choosing the direction of changes in the collision path, this improves the probability of success compared to independent collisions. With IF, since bit 31 of B cannot belong to any local collision, it should be seen as a constant. This constant controls which of the two changes is going through the IF function, but in any case, one change always goes through. Thus, with two adjacent local collisions that share an IF function, it is never possible to construct a correct collision path that follows our linear model. This imposes a new constraint on the construction of collision paths.

- When $i_2 = i_1 + 1$, there is a possibility of interaction on bit 31 in round $i_1 + 5 = i_2 + 4$. In fact, this is not a real interaction, because the function f is only applied to one of the two changes and the changes are then linearly combined by addition on the high order bit. Once again, the global cost of the interacting collisions is the same as the cost of two independent local collisions.

- When $i_2 = i_1 + 2$, there is an interaction on bit 31 in round $i_1 + 5 = i_2 + 3$. As in the previous case, this interaction does not modify the probability of success. Note that this case may be combined with the second case to yield an interaction of three consecutive local collisions with $i_2 = i_1 + 1$ and $i_3 = i_1 + 2$ on bit 31 of round $i_1 + 5$. In that case, the probability of success is modified as in the second case where the interaction between the second and third cases is concerned, while the first local collision is essentially independent of the other two at this round. Remember that with such a triple, the first two local collisions also interact at round $i_1 + 4$.

Taking all these considerations into account, we find that two differential paths for SHA-0 which are almost equally good, these two paths are described by the following disturbance vectors:

00010000000100100000001000011011011111110
110100100001010100101010100010111001100000

and

00100010000000101111011000111000000010100
010001001001001110110011000011110000000

In order to compute the exact cost of the cryptanalytic attack associated with these disturbance vectors and to determine which of these two vectors is the best one, it is necessary to explain with more details how these vectors are used in practical attacks. This is the goal of the next section.

5.4.4 Searching for collision instances

The above analysis shows that there exists a large number of message block pairs that collide after following our differential path. However, this does not suffice to build a collision search attack. In addition, we need to find an efficient way to construct such message pairs. This is done using a guided brute force approach. This approach is a kind of brute force, because we try message pairs again and again until a collision is found, it is guided because we use the known properties of the collision path to make the search more efficient and faster. Several techniques are used for this purpose. Two basic techniques were described in [CJ98], others were introduced later in [WYY05a], [WYY05b], [BC04] and [BCJ+05].

5.4.4.1 Early abort

In a basic implementation of the attack, a possible approach would be for each message pair to compute the compression function completely for both messages. Of course, this would work but requires two complete evaluations of the compression function for each message pair. To improve this, we can use our extensive knowledge of the differential path. For each intermediate step in the computation of the compression function, the differential path essentially specifies the value of the XOR of the computed $A^{(i)}$ for each message. In this context, an early abort strategy consists of checking at each step of the computation that the effective value of the XOR is compatible with the differential path. If this is not the case, it does not mean that the two message blocks cannot collide, only that if they collide, they do so in a non-controllable manner. As a consequence, the probability of collision for the pair is very low and it is not worth pursuing the computation. Thus, as soon as the message blocks stray from their intended path, we abort the computation.

5.4.4.2   Partial backtrack

One of the most important optimizations to search for a differential collision in the SHA-0 family is to remark that constructing messages conformant
with the differential path during the first 16 rounds essentially comes for free.
Indeed, during these rounds, if the current message pair is incompatible with
the differential path, instead of restarting from scratch, we can instead backtrack for a few rounds in order to modify the register whose value prevents
the conditions of the differential path to be satisfied. Using this approach,
it is easy to find message pairs which conform to the differential up to the
start of round 15. In fact, remarking that both disturbance vectors given at
the end of Section 5.4.3 have a difference inserted in round 14, we can make
the analysis more precise. For this difference to be corrected properly, the
following conditions need to be satisfied:

- The difference in round 14 should be inserted without carry. This can
  be easily satisfied by backtracking by a single round.

- The second correction in round 16 should be performed correctly. Since
  the non-linear function in round 16 is an IF, this implies that the bits in
  position 1 of $C^{(16)}$ and $D^{(16)}$ should be opposite, to make sure that the
  difference propagates through the IF function. Moreover, we can make
  sure that it propagates in the right direction by enforcing specific values
  for these bits. Note that $C^{(16)}$ and $D^{(16)}$ are rotated copies of $A^{(14)}$ and
  $A^{(13)}$. As a consequence, the condition can be ensured before choosing
  $W^{(14)}$.

- To perform the third correction in round 17, we need to make sure that
  bit 31 of $B^{(17)}$ is a 1. This can be ensured by a correct choice of the
  high order bit of $W^{(15)}$.

- With this basic analysis, the fourth correction in round 18 is the first
  one that cannot be simply enforced in advanced. Indeed this condition
  is controlled by bit 31 of $B^{(18)}$, i.e., of $A^{(17)}$. As a consequence, the
  probability associated with the difference in round 14 is only $1/2$. Using
  a slightly more complicated analysis, we see that, with high probability[4],
  this bit can be changed to the correct value by modifying bit 26 of $W^{(15)}$,
  thus affecting $A^{(16)}$ and, indirectly, $A^{(17)}$.

The second disturbance vector also has differences inserted in rounds 16 to 19.
It is clear that an analysis similar to the one we performed for the difference of
round 14 allows us to prearrange a fraction of these conditions. Due to these
prearranged conditions, the second disturbance vector becomes more efficient
than the first one for the differential attack on SHA-0.

---

[4]This may fail when changing bit 26 induces a carry which propagates up to round 31.

5.4.4.3 Neutral bits and message modifications

Looking at the difference paths induced by a disturbance vector in SHA-0 and at the conditions that are necessary for these paths, we see that most bits of the inner state, i.e., most bits of A, are not directly involved in these conditions. For example, if we modify a single bit in the middle of register A at some round, this change is not going to affect the differential path immediately. Of course, after enough rounds, the change propagates throughout the registers and affects the path. The basic idea of neutral bits [BCJ⁺05] is, given a message pair compatible with a fixed differential path up to some round, to identify bit positions where A can be modified without immediately affecting the differential path. Since A can be modified by changing the corresponding message word, it is possible to perform this change by flipping a bit at the same position in the message word. Such bit positions are called neutral bits for the message pair. We speak of a neutral bit up to round r when we want to emphasize the range of rounds we are considering. There are two kinds of neutral bits that can be encountered, simple neutral bits and composite neutral bits. A simple neutral bit corresponds to a single bit position, usually located in one of the final rounds, say between round 12 and round 15. A composite neutral bit consists of several bit positions which need to be flipped together in order to preserve conformance. A more careful analysis shows that a composite neutral bit behaves like a kind of local collision, inserts a change and corrects it, often in a non-linear way, for a limited number of rounds.

Another important property of neutral bits is obtained when considering several neutral bits at the same time. Two neutral bits, either simple or composite, are said to be pairwise independent when the four pairs of messages consisting of an original pair, the pair with one neutral bit flipped, the pair with the other bit flipped and the pair with both neutral bit flipped all conform to the differential path up to round r. This property of pairwise independence reflects the fact that the changes effected by flipping the two neutral bits involve bit position which can only interact through rare long range carries. An important heuristic fact is that given n neutral bits, such that each pair of these neutral bits are pairwise independent up to round r, a large fraction of the 2^n pairs of messages obtained by flipping an arbitrary subset of the neutral bits also conform to the differential path up to round r. In practice, this works very well up to r around 22.

To summarize, the raw effect of large set of pairwise independent neutral bits is to give a shortcut during the search for a valid message pair. Thanks to this shortcut, the number of message pairs to be tested can be estimated by counting the probability of success from round 22–24 instead of round 19.

Boomerang attacks

Neutral bits, as defined above, occur naturally in message pairs. From time to time, some composite neutral bits with a very long range are encountered. However, this event is unfrequent and we cannot count on the natural occurrence of these long range neutral bits. The idea of the boomerang[5] attack [JP07] for hash functions is to construct specific differential paths that embed a few of these long range neutral bits. This idea requires the use of sophisticated differential paths relying on non-linear behavior during the IF rounds, as introduced in [WYY05a, WYY05b]. It was used in [MP08] to derive the fastest known differential attack against SHA-0.

Message modifications

Another approach to bypass the probabilistic cost of a few rounds after round 16 when searching for a message pair is the use of message modification and advanced message modification techniques introduced in [WYY05a, WYY05b]. The basic idea is given a message pair to identify the first unsatisfied condition which prevents the message pair from conforming to the differential path. Once this condition is found, the message modification technique consists of flipping a few bits of messages. These bits are chosen to make sure that, with high probability, flipping them preserves conformance up to the first failure and reverse the value of the first non-conforming bit. Very roughly, it can be seen as a kind of neutral bit, whose first non-neutral effect is used as a correction.

5.5   Brute force and parallel computers

Brute force attacks can be very easily implemented using parallel computers. Indeed, they are of the embarrassingly parallel kind: it suffices to launch n copies of the same program, each searching in a fraction of the space of possible keys (or message pairs for the differential attack on SHA) to speed up the computation by a factor of n. Despite its extreme simplicity, this basic remark should never be forgotten when evaluating the security of a cryptosystem. A brute force attack whose computational cost is at the boundary of feasible computations is much easier to implement and run than a sophisticated computation with roughly the same cost.

A new trend in parallel computation is the use of powerful graphic processor units for general purpose computations. It is interesting to note that brute force computations can easily be implemented on such cards. As a consequence, reports of practical brute force attacks on graphic processors should be expected in the near future. In the same vein, people are also considering the use of video game consoles, since they offer very good performance/price ratios.

---

[5]This name is due to similarity with the boomerang attack for block ciphers introduced by Wagner [Wag99].

Exercises

1. To better understand the figures involved in brute force cryptanalysis, compute, assuming that trying one key costs one CPU cycle, the total number of keys that can be tried during a year using your personal computer, using all computers in a university or company, and using all computers on earth.

2. Using the Moebius transform of Chapter 9, compute expressions for each bit of each S-box of DES as polynomials over $\mathbb{F}_2$. Count the number of operations (XOR and AND) needed to compute each S-box. Compare with the number of operations given in [Kwa00].

3. Historically, monoalphabetic substitution ciphers have frequently been encounted. Assuming an alphabet with 26 letters, count the number of keys. Is a monoalphabetic substitution cipher vulnerable to a simple minded brute force attack which simply tries all possible keys.

4[h]. When considering brute force attacks, the rapid increase of computing power through time needs to be taken into account. Assume that you are given five years to attack through brute force an instance of a cryptosystem on 64 bits. How you would proceed to minimize the cost? Assume that the cost of computing power follows a geometric evolution which divides the cost of trying a single key by two during a one-year period.

5[h]. One important difficulty with automated brute force attacks is the need for a stopping condition, i.e., for a criterion which can be used to distinguish the correct key. Find a possible stopping condition to decrypt a DES ciphertext, assuming an english plaintext written using the ASCII code.

6[h]. Assume that you need to write an encryption program but are, for external reasons, limited to small encryption keys, say on 40 bits. How would you proceed to build a system with as much security as possible? Do not forget that the key limit only applies to keys, not to any auxiliary data used during the encryption, as long as this data can be recovered from the ciphertext.

7. Write an exhaustive search algorithm for placing n tokens on a $n \times n$ grid and make sure that there never are two tokens on the same row, column or diagonal. What is the cost of your approach as a function of n? What values are achievable?

   - Improve your algorithm to make sure before placing a new token that it does not conflict with previous tokens.

This chapter can also be a source for projects, a few examples are:

i. Program a bitslice implementation of DES, dedicated to brute force. Compare the speed of your implementation with available libraries that contain fast implementations of DES encryption. Port this program to a graphic processor unit.

ii. Write an implementation of the basic SHA-0 attack described in this chapter. Improve your implementation using advanced attacks available in the literature.

iii. Write a toolkit for brute force cryptanalysis against historical cryptosystem.

Chapter 6

The birthday paradox: Sorting or not?

The birthday paradox is a ubiquitous paradigm in cryptography. In some sense, it can be seen as an efficient probabilistic variation on the pigeonhole principle. Recall that the pigeonhole principle is a very general rule, that says that, given n disjoint classes of objects and a set of N objects, if $N > n$ then there is at least one class containing two or more objects. In this context, we say that the two objects collide in the same class or simply that we have a collision. Using this terminology, the pigeonhole principle simply states that given n classes, with at least $n + 1$ objects, a collision **always** occurs. In a probabilistic context, assuming that objects are equiprobably distributed among classes, it is natural to ask: How many objects do we need to have a collision with probability greater than one half?

The answer to this question is the core of the birthday paradox, it is a paradox in the sense that it does not fit the common sense answer. With n classes, it is not necessary to have $n/2$ objects, much less are needed. May be surprisingly, the correct answer is of the order of $\sqrt{n}$.

The consequences of the birthday paradox in cryptography are uncountable; they are encountered in public and secret key cryptography, in cryptanalysis and provable security and impacts all subtopics of cryptography from the study of stream ciphers to number theory. For this reason, we devote this chapter and the next two to the study of the birthday paradox. The three chapters differ by the amount of memory used by their respective algorithms. In the present chapter, we consider algorithms that use a lot of memory, of the same order as their running time. In the next chapter, we focus on algorithms that use a small amount of memory. The third chapter on birthday paradox techniques describes the intermediate case and presents algorithms with medium memory requirements.

6.1 Introductory example: Birthday attacks on modes of operation

To illustrate the importance of the birthday paradox in cryptography and show that it is the key to understanding the security of many cryptographic protocols, we study the case of modes of operations for block ciphers. As explained in Chapter 1, a mode of operation for a block cipher is a way to extend the domain of the block cipher and allow it to encrypt and/or authenticate messages of varying length instead of blocks of fixed length. Another essential feature of modes of operation is that they use block ciphers as black boxes. Thus, a given mode can be used with many different block ciphers. As a consequence, it is useful to study the security of the mode of operation independently of the underlying block cipher. The standard proof technique used to perform this study is to consider the mode of operation when used in conjunction with a truly random permutation. With this idealization, two types of results can be derived. On the one hand, given a mode of operation, we can describe attacks which work with all block ciphers. They are called **generic** attacks and give an upper bound on the security of the mode of operation. On the other hand, it is also possible to give lower bounds by writing security proofs which show that when used in conjunction with a secure block cipher the mode of operation is secure. The study of security proof is outside of the scope of this book. The interested reader can refer to one of the numerous research papers on this topic such as [BR06].

In this section, we are going to consider attacks on two basic modes of operation, Cipher Block Chaining (CBC) including its authentication version called CBC-MAC and Counter mode (CTR). More precisely, we are going to study the security of these modes by describing generic chosen message distinguishing attacks, as described in Chapter 1.

6.1.1 Security of CBC encryption and CBC-MAC

In order to start the study of the security of CBC encryption and CBC-MAC, we first address the motivation behind the slight differences between the two modes and explain why these differences are essential to the security. For this purpose, we need to use the notion of forgery recalled in Chapter 1.

Initial value. First, recall that one of the essential differences between CBC encryption and CBC-MAC is the fact that the former needs an IV, while the latter needs no IV. We already know that without an IV, CBC encryption becomes deterministic and thus cannot be secure against chosen message distinguishers. It is thus natural to ask what happens to CBC-MAC, when used with an IV. Clearly, in order to allow the message recipient to verify the MAC tag, the IV used when creating the authentication tag needs to be transmitted along with the message. Thus, the attacker knows the chosen IV value. Assume that this IV value is used as the starting intermediate ciphertext block $C^{(0)}$. Then, the intermediate block $C^{(1)}$ is obtained from the plaintext block $P^{(1)}$ by encrypting $C^{(0)} \oplus P^{(1)}$. Since both values $C^{(0)}$ and $P^{(1)}$ are known to the attacker, he can easily create a different message with the same authentication tag by xoring any value of his choice with both $C^{(0)}$ and $P^{(1)}$. This allows him to change the first message block to any value of his choice.

If the IV is used as a plaintext block, this attack does not work; however, this simply corresponds to authenticating using the ordinary CBC-MAC without IV, a message extended by adding a random first block. As a consequence, this is more costly than CBC-MAC without an IV and does not add any security.

Intermediate values. Releasing intermediate values during the MAC computation would be both costly and insecure. The cost part is obvious since this modification would make the authentication tag much longer. The insecurity part can be seen using variations on the above attack on initial values.

Final block. In order to see that CBC-MAC is insecure if the final block is not reencrypted, let us consider the following simple attack. The adversary first queries the CBC-MAC of a single block message. Clearly, in that case, assuming that the underlying block cipher is modeled by the random permutation Π, the authentication tag is $t = \Pi(P^{(1)})$. Given this tag, the adversary can immediately assert that the authentication tag of the two-block message $P^{(1)} \| (P^{(1)} \oplus t)$ is also t. We let the reader verify this fact in Exercise 2. This attack and variations on it can be performed only if messages of varying length are accepted by the MAC computations/verifications algorithms. When considering messages of a fixed length, CBC-MAC without a final block becomes secure[1].

6.1.1.1 Birthday attacks on CBC encryption and CBC-MAC

After seeing the simple attacks that explain the differences between CBC encryption and CBC-MAC, we are now ready to consider the security of these modes and its relation with the birthday paradox. For simplicity, it is easier to start with the security of CBC-MAC because this mode is deterministic and the attack is thus easier to follow. Assuming that we play the attacker's role, in order to build a forgery on CBC-MAC, we first ask authentication tag for many (different) two-block messages. Each message M_i is formed of the blocks $M_i^{(1)}$ and $M_i^{(2)}$, the corresponding authentication tag is t_i. Assume

---

[1] Up to the birthday paradox bound.

that making these queries, we find that the tags t_i and t_j of two different messages are equal. At this point we make an additional query, by asking for the authentication tag t of a message obtained by completing M_i by an arbitrary sequel S. We then assert that the authentication tag of M_j completed by the same sequel S is also t. This comes from the fact that CBC-MAC computations propagate equalities between intermediate ciphertext blocks as long as the message blocks are also equal. As a consequence, as soon as two authentication tags are seen to be equal, forgery becomes an easy matter. Since this is exactly the definition of a collision between authentication tags, it shows that the (non) existence of collisions is crucial to the security of CBC-MAC authentication. Such an attack which relies on the existence of collisions is often called a **birthday attack**, by reference to the birthday paradox.

Similarly, CBC encryption is also vulnerable to birthday attacks. More precisely, if an attacker observes a collision between two encrypted blocks, he gains some partial information about the corresponding plaintext blocks. For simplicity of exposition, let us consider a single message M, with blocks $M^{(i)}$. If we have equality between the two blocks of ciphertext $C^{(i)}$ and $C^{(j)}$ we also have equality between the corresponding inputs to the block cipher, i.e., between $M^{(i)} \oplus C^{(i-1)}$ and $M^{(j)} \oplus C^{(j-1)}$. As the consequence, the attacker learns the value of $M^{(i)} \oplus M^{(j)}$. This remark is easily transformed into a chosen message distinguishing attack (see Exercise 3).

6.1.1.2 Birthday attacks and the counter mode

The security of the counter mode is also an interesting example. At first, it seems completely immune to birthday attacks. Indeed, since the pseudo-random sequence is obtained by encrypting different values, no collision may occur. More precisely, starting from an initial value and incrementing it, we may not loop back to the initial value until all possible values of the b-bit block have been considered. Since there are 2^b different values[2], this is well beyond the birthday paradox bound.

However, when considering the encryption of multiple messages, matters become more complicated. Indeed, we should avoid reusing the same counter values between messages. A simple solution is to memorize the current value after encrypting a message and to restart from here for the next message. However, in some practical scenarios, this solution which requires an encrypting device with memory is not possible. In that case, a frequently encountered alternative is to choose a random initial value for each message. Of course, with this solution, collisions between initial values lead to attacks and the birthday paradox comes back into play.

More surprisingly, as distinguishing attacks go, even memorizing the counter value is not enough to avoid the birthday paradox. Again playing the at-

---

[2]For completeness, note that if the incrementation is done using a LFSR, the all zero block should be excluded and there are "only" $2^b - 1$ values to be considered.

tacker's role, we go directly to the challenge step and submit two long random messages M_0 and M_1 of the same length L (measured in blocks). We obtain the encryption C of either M_0 and M_1. By xoring C with M_0 we get a first candidate for the pseudo-random stream used during encryption. By xoring C with M_1 we get a second candidate. By construction, there can be no collision between blocks within the correct candidate. However, nothing prevents random collisions to occur in the incorrect one. As a consequence, if the adversary observes a collision in one of the two pseudo-random candidates, he knows for sure which of M_0 and M_1 was encrypted and he can announce it. Otherwise, he tosses a random coin and announces the corresponding guess. Clearly, if the probability of collision in the incorrect pseudo-random candidate is non-negligible, the adversary gains a non-negligible advantage. It is interesting to note that this distinguisher is not really based on a property of the counter mode itself but rather on the fact that the mode is constructed on a permutation. If the block cipher or random permutation is replaced by a (pseudo) random function the distinguisher vanishes. However, this would not remove the need to memorize counter values. Moreover, in practice, a counter mode based on a pseudo-random function is a rarely seen construction in cryptography. The main reason probably being that block ciphers are standardized and thus more readily available to developers.

6.2 Analysis of birthday paradox bounds

Since the mere existence of collisions can be essential to the security of cryptographic algorithms and protocols, it is important to estimate the probability of having collisions within sets[3] of random objects, before considering the algorithmic issue of effectively and efficiently finding such collisions. Of course, the exact probability depends on the distribution of these random objects. If they are generated from distinct values by application of a random permutation or equivalently if they are randomly drawn without replacement from a set, no collision may occur and the probability of having a collision is zero. On the other hand, the probability for twins to have the same birthday is almost one (see Exercise 4).

The easiest case to analyze is the case of objects taken uniformly at random from a given set (with replacement). In this context, two parameters need to be considered, the number of randomly drawn objects N and the size of the set $\mathcal{N}$. In order to determine the probability of collision, we may first consider a simple heuristic method. Among N objects, we can build

---

[3]Or more precisely multisets: by definition a set may only contain a single copy of a given object. This abuse of language is frequently encountered and we simply follow the tradition.

$N(N-1)/2$ pairs. Moreover, for any pair, the probability of collision is $\mathcal{N}^{-1}$. Assuming independence between pairs, we have an average estimated number of collisions equal to:

$$\frac{N(N-1)}{2\mathcal{N}}.$$

This simple heuristic analysis already points to a threshold for collisions around $N \approx \sqrt{\mathcal{N}}$. However, this analysis is not satisfying for several reasons. First, it incorrectly assumes independence between pairs of objects in the random set. Second, it does not give a probability of collision but instead estimates the expected number of collisions within a multiset.

Our goal in the sequel is to make the analysis more precise. In this analysis, we denote the probability of having at least one collision in a randomly chosen set of N elements among $\mathcal{N}$ by $\mathrm{Coll}_{\mathcal{N}}^{N}$. Let us first start by giving an upper bound. For convenience, we assume that the drawing order of elements in the random set has been kept. In that case, we have a random list and can access each element by its rank from 1 to N. Element i is denoted by X_i. For any pair (i,j), the probability of having $X_i = X_j$ is $\mathcal{N}^{-1}$. If any pair collides, then of course, there is a collision in the set. Since the probability of a union of events is always upper bounded by the sum of the probability of individual events, regardless of independence we find that:

$$\mathrm{Coll}_{\mathcal{N}}^{N} \leq \sum_{\mathrm{Pairs}(i,j)} \frac{1}{\mathcal{N}} = \frac{N \cdot (N-1)}{2\mathcal{N}}. \tag{6.1}$$

This directly implies that when $N \ll \sqrt{\mathcal{N}}$ the probability of collision is low. This fact is routinely used in security proofs, when showing that no birthday attack is applicable.

Should the reader want more than an upper bound, the easiest approach is to compute the probability of the reverse event, i.e., the probability to get no collisions when drawing N elements among $\mathcal{N}$. This probability is exactly:

$$\prod_{i=0}^{N-1} \frac{\mathcal{N}-i}{\mathcal{N}} = \frac{\mathcal{N}!}{\mathcal{N}^N \cdot (\mathcal{N}-N)!}. \tag{6.2}$$

Then, using Stirling formula or even better upper and lower bounds derived from the Stirling series, we can get a very precise estimate for the desired probability.

6.2.1 Generalizations

In some cryptographic contexts, it is interesting to generalize the analysis. One important generalization is the existence of collisions between two different sets. Another frequently encountered issue is the existence of multicollisions, i.e., the existence of 3, 4 or more random elements sharing a common value. For these generalizations, we only make a simplified heuristic analysis based on the expected number of (multi)collisions assuming independence. This simplified analysis can be refined to give upper and lower bounds as in the case of collisions. However, in most cases, this is not necessary and the simple estimate suffices.

Collisions between two sets. In this generalization, we are given two subsets, each obtained by drawing at random without replacement from a large set. We denote by N_1 and N_2 the respective cardinality of the subset and by $\mathcal{N}$ the cardinality of the set. By construction, there is no collision within any single subset and we are interested by collisions between elements of the first set and elements of the second one. Since $N_1 \cdot N_2$ pairs of elements can be constructed, the expected number of collisions is:

$$\frac{N_1 \cdot N_2}{\mathcal{N}}.$$

When the two subsets are of the same size $N_1 = N_2 = N$, this expected number of collisions lies between the expectation we had for a single subset of N elements and the expectation for a subset of size $2N$. Note that considering a subset of size $2N$ is quite natural, since it is the size of the union of the two considered subsets.

Multicollisions. For multicollisions, as for collisions, two subcases may be considered. Either we have a single subset and search for ℓ different elements with the same value. Or we have ℓ different subsets and want an element common to all. In short, a multicollision involving ℓ elements is called a ℓ-multicollision. Making our usual heuristic analysis under an independence hypothesis, we find that the expected number of ℓ-multicollisions in a subset of size N chosen among $\mathcal{N}$ elements is:

$$\frac{\prod_{i=1}^{\ell} N + 1 - i}{\ell! \cdot \mathcal{N}^{\ell-1}} \approx \frac{N^{\ell}}{\ell! \cdot \mathcal{N}^{\ell-1}}. \tag{6.3}$$

For a ℓ-multicollisions between subset of respective sizes $N_1 \cdots N_\ell$ we find an expected number equal to:

$$\frac{\prod_{i=1}^{\ell} N_i}{\mathcal{N}^{\ell-1}}$$

From these expected numbers, assuming that ℓ remains small enough to neglect the $\ell!$ factor, we find that for $N \ll \mathcal{N}^{(\ell-1)/\ell}$ the probability of ℓ-multicollisions within a subset remains small.

Non-uniform statistical distributions. Up to this point, when looking for collisions, we assumed a uniform random distribution for the would-be colliding objects. In practice, this is not always the case and we may in some cases be looking for collisions under other statistical distributions. Assuming that the random objects remain independent from each other, the probability of collisions is always larger with a non-uniform distribution. This comes from the higher probability of collision within a single pair of randomly drawn objects. Indeed, assume that we draw at random among N values and that value i is taken with probability p_i then the probability of collision is:

$$\sum_{i-1}^{N} p_i^2. \qquad (6.4)$$

Under the additional condition $\sum_{i=1}^{N} p_i = 1$ this probability is maximized when all p_i values are equal to $1/N$. A more precise analysis of collisions in hash functions is presented for the unbalanced case in [BK04].

Finally, when the objects are not drawn independently from each other, we cannot say much about collisions. Depending on the precise details, collisions may either be more frequent or vanish completely. In the non-independent case, a specific analysis is required for each specific problem.

6.3   Finding collisions

A natural algorithm question given a list of elements is to check whether this list contains a collision or not. The first idea that comes to mind is to consider all pairs of elements in the list and to test for equality. This approach involves two intricated loops and requires $N \cdot (N-1)/2$ tests. For cryptanalytic purposes, this algorithm is dreadful, because it takes back the edge that the existence of collisions gave us. As a consequence, we need better algorithms for this purpose. Ideally, we would like to obtain a linear time algorithm. However, unless non-standard assumptions are made about the computing device, it is not known how to solve this problem.

Still, there are several algorithmic techniques that allow us to find collisions in quasi-linear time, i.e., with $O(N \log N)$ operations for a list of N elements. These techniques involve sorting, hashing, balanced trees or a mix of those.

Before delving into the algorithmic details, let us discuss why these techniques are helpful to find collisions.

Collisions finding by sorting.   If we can sort efficiently, then the problem of finding collisions or multi-collisions is reduced to the easier problem of finding collisions within a sorted list. Of course, sorting first requires a total order on the values we are considering. However, for the purpose of collision finding, the precise choice of this order is irrelevant, the only property we need is shared by all orders: equal elements are necessarily neighbors in a sorted list. As a consequence, finding a collision in a sorted list is very simple, it suffices to read the list in order and compare each element with its immediate successor. If there are collisions in the list, we necessarily find some. However, as stated the algorithm may miss some collisions. Assume that we have a triple of equal elements. After sorting, they can be found in positions i, $i+1$ and $i+2$. The method described above locates the collision in the pair $(i, i+1)$, it also detects the collision $(i+1, i+2)$. However, it misses the collision $(i, i+2)$. Fixing this problem is easy: when a first collision is detected for some value, say in $(i, i+1)$, it suffices to scan the list forward until a different element is detected. After this scanning, we know that all elements in the range $[i, i+\delta]$ are equal. Generating the corresponding $\delta \cdot (\delta+1)/2$ collisions is easily done. Algorithm 6.1 fully describes the way to generate all collisions in a sorted list.

Algorithm 6.1 Generating all collisions in a sorted list

Require: Input a sorted array X of N elements
  for i from 1 to $n-1$ **do**
    if $X[i] = X[i+1]$ **then**
      Assert: First collision detected for the value $X[i]$
      Let $\delta = 1$
      while $i + \delta < n$ and $X[i] = X[i+\delta+1]$ **do**
        Increment δ
      end while
      for j from 0 to $\delta - 1$ **do**
        for k from $j+1$ to δ **do**
          Print 'Collision between positions $i+j$ and $i+k$'
        end for
      end for
      Let $i \longleftarrow i + \delta$ (to skip over the values equal to $X[i]$)
    end if
  end for

Since the order chosen for sorting does not affect this collision finding algorithm, it is not necessary to have a naturally ordered set to find collisions. If the set is not naturally ordered, a simple way to go is to view the binary representation of the elements in the set as the number encoded by the same binary string and then to sort these numbers. This order is meaningless, in the sense that it is not in any way related to the semantics of the elements. However, it suffices to find collisions.

Collisions finding by sorting one of two lists. When looking for collisions between two lists, i.e., for common elements, several methods are possible. The first idea is to re-use the technique we used in the single list case. Simply merge both lists, while associating to each value a tag that remembers its original list. Then sort and look for collisions in the merge lists. Only keep the collisions involving one element from each original list. Of course, this approach is not optimal and has several drawbacks. In particular, the fact that an additional bit of memory is required to keep track of the original list is extremely cumbersome. Instead, it is much better to devise a specific method. The next idea probably is to sort both lists and then to look for collisions in the sorted lists. To illustrate the technique used to look up the collisions, imagine that the sorted lists are written on two sheets of paper. Put your left index at the top of one list and your right index at the top of the other. At each step of the comparison, move down by one position the finger corresponding to the smallest of the two elements. If at any time the elements below both fingers are equal, this detects a collision. This basic procedure is going to be used as part of the merge sort Algorithm 6.7, so we do not write down for now. Finally, the best way to proceed is to break the symmetry between the two lists, just sorting one of then, preferably the shortest one. Once this is done, take elements of the other lists one by one and try to locate them in the sorted list. If we find any one of them, we have a collision. The advantage of this approach is that we only need to store the shortest list in main memory, the other one can be read from disk during the collision search. The method can even be used when the second list is not available at the start of the computation but produced online after that. In a cryptanalytic setting, this can be extremely useful, for example, the second list could be obtained by observing ciphertext blocks during an encrypted communication. The key fact which makes this technique efficient is that finding an element in a sorted list can be done efficiently using a dichotomy search, given as Algorithm 6.2. This way of searching needs about $\log_2 N$ operations to find an element within a sorted list of N elements. Moreover, if the element is not present in the sorted list, the algorithm can return the position where it would fit: at the start, at the end or at some place between two consecutive elements.

Collisions finding without sorting. In order to better understand why it is possible to detect collisions without sorting, let us perform a thought experiment, using a non-standard computer architecture. This non-standard computer has access to a very large amount of direct access memory, initialized for free to a special value $\perp$ upon start-up. With such a computer, it is possible to have a memory size not bounded by the running time. Note that with Turing machines this is not possible because they do not offer direct access to the memory but only sequential access. With a conventional computer, it is not possible either since the memory initialization never comes for free. On our hypothetical computer, finding collisions is very simple. For each element X_i in the list, interpret X_i as a number, view this number as a memory address,

---

Algorithm 6.2 Dichotomy search

Require: Input an array X of N elements and a value Y

  Let $start \longleftarrow 1$

  Let $end \longleftarrow N$

  if $Y < X[1]$ **then**

    Output 'Y not in list, its place is at the start'

  end if

  if $Y > X[N]$ **then**

    Output 'Y not in list, its place is at the end'

  end if

  if $Y = X[N]$ **then**

    Output 'Y is in list, in position N'

  end if

  while $start + 1 < end$ **do**

    Assert: We have $X[start] \leq Y < X[end]$

    Let $mid \longleftarrow \left\lfloor \frac{start+end}{2} \right\rfloor$

    if $Y < X[mid]$ **then**

      $end \longleftarrow mid$

    else

      $start \longleftarrow mid$

    end if

  end while

  if $Y = X[start]$ **then**

    Output 'Y is in list, in position $start$'

  else

    Output 'Y not in list, its place is right after position $start$'

  end if

---

check that the address is non-initialized and then write i at this address. If an already initialized address is encountered, then we have found a collision between X_i and X_j (assuming that the address contained the value j). Of course, this approach is not feasible; however, sorting through hashing can be seen as a practical way of implementing it. Also, finding collisions with a balanced tree can depending on the point-of-view be considered either as a hidden sort or as another implementation of the large memory machine.

6.3.1   Sort algorithms

Sorting is widely encountered in computer science. In particular, it is one of the favorite problems of computer science teachers. The main reason probably is that good sorting algorithms are more efficient than would be expected by the uninitiated, very elegant and tough to understand. They are also optimal at least when considered as generic algorithms. In this section, we first present an overview of some sort algorithms before considering implementation issues. The overview is further divided into two parts. The first part contains quadratic, asymptotically unefficient, algorithms. The second describes asymptotically fast algorithms. The selection of sort algorithms presented here is somewhat arbitrary and based on simplicity and usefulness criteria. Many more sort algorithms do exist, such as comb sort, or smooth sort. They are not included in this selection.

6.3.1.1   Quadratic algorithms for sorting

6.3.1.1.1   Bubble sort. Bubble sort is probably the conceptually simpler of all sort algorithms. It starts from an unsorted list of elements, looks at the current list to see if any pair of consecutive elements is in an incorrect order and repeats this step until no badly ordered pair can be found. To find out badly ordered pair, the bubble sort simply scans the list of elements starting at the beginning and going straight to the end of the list. This is called a pass on the list. It then performs successive passes, until a full pass without a swap occurs, at which point the algorithm stops. It is easy to check that if a pass has nothing to do, then each element is smaller than the next, thus by transitivity the list is fully ordered. In order to bound the running time of the bubble sort, it is useful to remark that after the first pass, the largest element in the list has been moved into place at the end of the list. After two passes, the largest two elements are in place and so on ... As a consequence, there is a growing tail in each pass that never performs any swap. We can take advantage of this fact and write a slightly improved bubble sort as Algorithm 6.3. The running time of this algorithm measured in number of comparisons is at most $1 + 2 + \cdots + (n-1) = n(n-1)/2$.

6.3.1.1.2   Selection sort. Selection sort is a way of sorting which is built on a simpler procedure: finding a minimal element in a list. From this simple

---

Algorithm 6.3 Bubble sort

Require: Input an array X of N elements
  for *pass* from 1 to $N-1$ **do**
    Let Active ⟵ `false`
    for i from 1 to $N - pass$ **do**
      if $X[i] > X[i+1]$ **then**
        Assert: Incorrectly order pair found
        Let Active ⟵ `true`
        Exchange $X[i]$ and $X[i+1]$
      end if
    end for
    if Active $=$ `false` **then**
      Abort loop
    end if
  end for
  Output sorted array X

Algorithm 6.4 Find minimal element

Require: Input an array $X[start \cdots end]$
  Let *mini* ⟵ *start*
  for i from *start* $+ 1$ to *end* **do**
    if $X[i] < X[mini]$ **then**
      Let *mini* ⟵ i
    end if
  end for
  Output position of smallest element, i.e., *mini*

---

procedure, sorting is easy: the first element in the sorted list is the minimal element and sorting the remaining elements can be done by repeating the same idea. Of course, by symmetry, it is also possible to find a maximal element and sort by placing the largest element at the end of the list. Since the time to find an element is linear in the size of the list N, selection sort, which requires N such passes, is a quadratic algorithm. The search for a minimal element is given as Algorithm 6.4 and the selection sort as Algorithm 6.5.

Algorithm 6.5 Selection sort

Require: Input an array X of N elements
    for *pass* from 1 to $n - 1$ **do**
        Find the position i of the minimum element in $X[pass \cdots n]$.
        Swap $X[i]$ and $X[pass]$
    end for
    Output sorted array X

6.3.1.1.3   Insertion sort.   A well-known sort algorithm in the real life is the way we proceed to sort cards. The basic idea is simple: put the unsorted cards in a stack, draw one card at a time from the stack and place it in your hand. Each time a card is to be placed, locate the correct position among the already sorted cards in your hand and insert it there. In computer science, this is called insertion sort. In order to perform this algorithm, we need to efficiently locate the right position in the already sorted list using the dichotomy search of Algorithm 6.2. We also need to be able to insert the new element in the correct position quickly. Surprisingly, this apparently simple step is the catch which makes insertion sort unefficient. The reason is that in general a new element needs to be inserted somewhere in the middle of the list. Since there is no room at that place, we need to make some, without disordering the list. With cards, it is easy, when you insert one, the others move to let it fit in. With elements in an array, we need to move them one by one to make room. On average, for each insertion, half the elements in the list need to be displaced. Thus each insertion requires linear time and the insertion sort, given by Algorithm 6.6, is quadratic.

6.3.1.2   Fast algorithms for sorting

Beyond their purpose, all fast sort algorithms that we present here but one have a common point. Except heap sort, all of them are recursive algorithms based on the divide and conquer approach. Using a linear time transformation called a stage, they reduce the problem of sorting a list to the problem of sorting two smaller lists of roughly half-size. Of course, once we reach

---

Algorithm 6.6 Insertion sort

Require: Input an array X of N elements
  for i from 2 to N **do**
    Assert: List from 1 to $i-1$ is sorted
    Let $Y \longleftarrow X[i]$
    Using dichotomy, find the correct place k for Y in $X[1 \cdots i-1]$ (The end position is encoded by i).
    for j from i down to $k+1$ **do**
      Let $X[j] \longleftarrow X[j-1]$
    end for
    Let $X[k] \longleftarrow Y$
  end for
  Output sorted array X

---

extremely small lists of zero or one element we stop. As a consequence, the number of stages is limited by the number of integer divisions by two that can be performed, i.e., $\log_2 N$. Since the stages take linear time, this yields algorithms with complexity $O(N \log N)$.

6.3.1.2.1 Merge sort. Our first fast algorithm for sorting, called merge sort, is a straight implementation of the divide and conquer idea. It works by dividing the list in two sublists of half-size, sorting the sublists and merging the two sorted sublists into a single sorted list. The basic idea for merging is already described in Section 6.3 as a way of finding collisions between two sorted lists. Writing down the merge sort algorithm from this description is a simple matter. The main difficulty comes from merging. Indeed, when merging two sorted lists into a single one, it is essential to have some place to store the merged list. Indeed, it is very hard to do the merge in place, reusing the storage of the two sublists. As a consequence, merge sort requires extra memory in addition to the memory used by the input array. With a little care, a single auxiliary array of the same size as the input array suffices. One approach is to copy the original array into the auxiliary array, to sort both halves of the auxiliary array and to merge them back into the original. When sorting the halves of the auxiliary array, the role of original and auxiliary arrays are reversed. Moreover, most of the copy operations can be removed, indeed at each stage after the first one, we may see that copying the current original arrays to the auxiliary ones simply rewrite the same data over and over again. In fact, if the array's size is a power of two, it is even better to avoid the copying altogether and directly start by merging. First merge small lists and proceed upward until the final merging is done. With luck, if the number of stages is even, the sorted list already is in the correct array. Otherwise, copy it once from the auxiliary to the main array. However, when the size of the array is not a power of two, the depth of the computation's tree is not uniform and making an initial copy is a great simplification. To distinguish between the top level where the allocation of an auxiliary array and the initial copy is done from the other levels, we split the algorithm into two parts: a wrapper (Algorithm 6.8) and the main procedure (Algorithm 6.7).

Algorithm 6.7 Merge sort main procedure

Require: Input an array $X[start \cdots end]$ of $end - start + 1$ elements
Require: Input an auxiliary array $Y[start \cdots end]$
  If $start \geq end$, the array is already sorted, **return**
  Let $mid \longleftarrow \lfloor \frac{start+end}{2} \rfloor$
  Recursively merge sort $Y[start \cdots mid]$ with auxiliary array X
  Recursively merge sort $Y[mid + 1 \cdots end]$ with auxiliary array X
  (Note the reversal of main and auxiliary array in the calls)
  Let $start_1 \longleftarrow start$
  Let $start_2 \longleftarrow mid + 1$
  Let $merge \longleftarrow start$
  while $start_1 \leq mid$ and $start_2 \leq end$ **do**
    if $Y[start_1] \leq Y[start_2]$ **then**
      Let $X[merge] \longleftarrow Y[start_1]$
      Increment $start_1$
    else
      Let $X[merge] \longleftarrow Y[start_2]$
      Increment $start_2$
    end if
    Increment $merge$
  end while
  while $start_1 \leq mid$ **do**
    Let $X[merge] \longleftarrow Y[start_1]$
    Increment $start_1$
    Increment $merge$
  end while
  while $start_2 \leq end$ **do**
    Let $X[merge] \longleftarrow Y[start_2]$
    Increment $start_2$
    Increment $merge$
  end while
  Return sorted array $X[start \cdots end]$

Merge sort is a very efficient algorithm, however, since it requires additional memory, it cannot be used when the array to be sorted fills the computer's main memory. In that case, it is better to use a different sorting algorithm. Note that using a more complicated merging method, it is also possible to

---

Algorithm 6.8 Merge sort wrapper

Require: Input an array $X[start \cdots end]$ of $end - start + 1$ elements
  If $start \geq end$, the array is already sorted, **return**
  Create an auxiliary array $Y[start \cdots end]$
  Copy X into Y
  Call merge sort procedure on $X[start \cdots end]$ with auxiliary array Y
  Free the memory occupied by Y
  Return sorted array $X[start \cdots end]$

---

devise an in-place variation of the merge sort algorithm, e.g., see [KPT96].

6.3.1.2.2 Quicksort sort. In order to largely reduce the need for additional memory, quicksort implements the divide and conquer idea in a different way. At each stage, it randomly chooses an element in the list and uses this element called the pivot to divide the list into two sublists. The first sublist contains all elements smaller than the pivot and the second sublist the other elements. During the stage, elements are moved around to put the list in a new order, with the first sublist at the beginning, the pivot in the middle and the second sublist at the end. This can be done in linear time, without requiring extra memory. If the two sublists are then independently sorted, it is clear that the list itself becomes sorted. Despite the simplicity of the basic idea, this algorithm presents several complex issues. First of all, the need to choose the pivot at random is essential. If the pivot is chosen in a deterministic way, then it is possible to conceive examples of adversary lists which would require quadratic time to be sorted. Second, making the complexity analysis precise is a difficult matter because the running time depends on the random choices made by the algorithm and making the bad choice at each stage leads to a quadratic running time. However, with overwhelming probability, the size of the sublists decreases quickly between stages and the running time stays in the range $O(N \log N)$. Finally, to avoid bad lists leading to quadratic running time, it is also necessary to specify more precisely what happens when we encounter an element which is equal to the pivot. If we decide that all such elements go to the first (or second) sublist, we are in trouble. Indeed, in that case, sorting a list containing N copies of the same element takes quadratic time! To avoid this, the simplest approach is to put elements in the first sublist if they occur before the pivot in the original list and to put them in the second sublist otherwise.

Despite these difficulties, quicksort is a very useful practical algorithm. Moreover, it is quite easy to detect bugs which lead to quadratic running time. Indeed, due to the recursive nature of the algorithm, each time the program enters a deeper stage, it adds some elements on the recursivity stack. When a bug leads to quadratic behavior, the size of the stack needs to become linear in N, which quickly leads to an overflow. When such an overflow occurs,

the program aborts and reports an error. This misbehavior usually suffices to
find the bug. Note that the recursive nature of the algorithm also implies that
quicksort still requires more than a constant amount of memory in addition
to the original array. The additional memory needed is proportional to the
depth of the computation tree, i.e., it is of the form $O(\log N)$.

Algorithm 6.9 Quicksort

Require: Input an array $X[start \cdots end]$ of $end - start + 1$ elements
   If $start \geq end$, the array is already sorted, **return**
   Choose a random position pos in $[start \cdots end]$
   Copy $start$ to $start_0$
   Copy end to end_0
   Let $Y \longleftarrow X[pos]$
   Let $X[pos] \longleftarrow X[start]$
   Let $i \longleftarrow start + 1$
   while $i \leq end$ **do**
      if $(X[i] < Y)$ or $(X[i] = Y$ and $i < pos)$ **then**
         Let $X[start] \longleftarrow X[i]$
         Increment $start$
         Increment i
      else
         Exchange $X[end]$ and $X[i]$
         Decrement end
      end if
   end while
   Let $X[start] \longleftarrow Y$
   Recursively apply quicksort to $X[start_0..start - 1]$
   Recursively apply quicksort to $X[end..end_0]$
   Return sorted array $X[start_0 \cdots end_0]$

6.3.1.2.3   Radix sort.   From the descriptions of merge sort and quicksort,
it is natural to ponder whether it is possible to devise a sort algorithm that
requires neither additional memory nor randomness. Radix sort neatly solves
this problem. However, this algorithm is not generic and it requires additional knowledge about the order used for sorting. For many cryptographic
applications, this is not a problem at all, but it makes index sort badly suited
in other cases, especially when sorting complex objects, such as, for example,
database entries. In fact, there are two different version of radix sort, which
sort by considering bits either from left to right or from right to left. Here, we
only consider the left to right version. To explain this version of radix sort,
let us assume that we want to sort an array of N unsigned n-bit numbers. We

know that in the sorted array, the numbers with a high order bit set to 0 occur before the numbers with a high order bit of 1. Similarly, with each of these subarrays, the numbers should be sorted according to the value of the second high order bit. And so on, until the low order bit is reached. The complexity analysis is simple, each stage which consists of moving numbers with some bit equal to 0 before numbers with the bit equal to 1 can be performed in linear time. Since there is one stage per bit, the complexity is $O(nN)$. With very large numbers, when $n \gg \log N$, this is worse than merge sort or quicksort. However, it is not a very frequent case. Moreover, if the numbers are (or look) random, the complexity can be improved with a very simple change: simply abort the recursion when a list of zero or one element is to be sorted. With random numbers, this cuts off the depth of the computation tree around $\log_2 N$, thus improving the complexity as required.

Radix sort can also be used with signed numbers. In that case, we should remember that the high order bit is the sign bit. As a consequence, numbers with a high bit set to 1 need to be at the beginning because they are negative. Thus reversing the role of the first and second sublists for this bit is the only change to deal with signed numbers. For all other bits, we should proceed as we did with unsigned numbers (assuming that negative numbers are represented using the two-complement convention).

Algorithm 6.10 Radix sort

Require: Input an array $X[start \cdots end]$ of $end - start + 1$ unsigned integers
Require: Bit position b to consider in the integers
    If $start \geq end$, the array is already sorted, **return**
    Copy $start$ to $start_0$
    Copy end to end_0
    while $start \leq end$ **do**
      if Bit b of $X[i]$ is 0 **then**
        Increment $start$
      else
        Exchange $X[end]$ and $X[start]$
        Decrement end
      end if
    end while
    Recursively radix sort $X[start_0..start - 1]$ on bit $b - 1$
    Recursively radix sort $X[end + 1..end_0]$ on bit $b - 1$
    Return sorted array $X[start_0 \cdots end_0]$

6.3.1.2.4 Heap sort. From a theoretical point-of-view, heap sort is very interesting. First, it is a non-recursive algorithm and thus it can truly achieve in-place sorting: in addition to the array being sorted, it only requires a constant amount of memory. Second, it uses a very nice implicit tree structure as the basis of the algorithm. This tree structure is called a heap, it is a binary tree where the value of each node is larger[4] than or equal to the values of all nodes in the subtree attached to it. The tree structure is implicit, because it does not use any pointers but instead works with the convention the left and right children of the element in position i are respectively stored[5] in positions $2i + 1$ and $2i + 2$. Whenever the position of a child is beyond the end of the array, it simply means that the child does not exist. For example, when position $2i + 1$ is not in the array, then node i is a leaf. When $2i + 1$ is the last position in the array, then node i has a single child (on the left).

Using this structure, heap sort works in two phases. During the first phase, it takes the input array with the implicit tree structure and reorder it to satisfy the heap property. This is done by making sure that each element is larger than its children. During the second phase, it progressively reduces the size of the tree by taking out the largest element and placing it where it belongs in the sorted array. During this second phase, one part of the array is used to store a heap tree and the other part to store an already sorted part of the array. At the end of the second phase, the tree part is empty and the array is fully sorted.

To write down the heap sort algorithm, we first need a heap insertion procedure which allows us to correctly add elements to an existing heap. This insertion works as follows: assuming that the subtrees starting in position $2i + 1$ and $2i + 2$ are correct heap trees, it inserts an element in position i and if necessary moves it down the tree to create a correct heap tree in position i. During the first phase, this procedure is used iteratively on all positions, starting from the end of the original array. Of course, remarking that subtrees with a single node are always heap trees, there is nothing to be done for the second half of the array and we may start in the middle. During the second phase, we take the element at the root of the heap tree, i.e., the largest element in the tree and put it at the beginning of the sorted part. Since this partially breaks the heap structure, we use the insertion procedure again. More precisely, we move the last element of the previous heap at the root and apply the insertion to guarantee the heap property on a new tree, with one fewer element. Both phases of the heap sort algorithm require N calls to the insertion procedure, whose complexity is bounded by the tree depth. As a consequence, the total runtime is $O(N \log N)$ as with the other fast sort algorithms.

Despite its good complexity analysis, in practice, heap sort does not perform very well, especially on large arrays, because it causes many cache misses when moving elements around during the heap insertion.

---

[4]There is, of course, a variation of the heap tree where each node is smaller than its successors, but this is not the version we use here.

[5]This assumes that the array starts at position 0, for arrays starting with position 1, we should use positions $2i$ and $2i + 1$ instead.

Algorithm 6.11 Heap sort

Require: Input an array $X[0 \cdots end - 1]$ of end elements

  First Phase: Heap creation
  for i from $\lfloor \frac{end-2}{2} \rfloor$ down to 0 **do**
    (Note: Subtrees at positions $2i + 1$ and $2i + 2$ are already correct)
    Call heap insertion on position i with tree size end
  end for
  Second phase: Sorting
  while $end > 1$ **do**
    Put largest element at the end: Exchange $X[0]$ and $X[end - 1]$
    Decrement end
    Call heap insertion on position 0 with (reduced) tree size end
  end while
  Return sorted array X

Algorithm 6.12 Insertion in heap procedure

Require: Input an array $X[0 \cdots end - 1]$ of end elements
Require: Input a position pos where the heap property must be enforced

  while $2 \cdot pos + 1 \leq end - 1$ **do**
    Let $val \longleftarrow X[pos]$
    Let $dest \longleftarrow pos$
    if $X[2 \cdot pos + 1] > val$ **then**
      Let $val \longleftarrow X[2 \cdot pos + 1]$
      Let $dest \longleftarrow 2 \cdot pos + 1$
    end if
    if $2 \cdot pos + 2 \leq end - 1$ and $X[2 \cdot pos + 2] > val$ **then**
      Let $dest \longleftarrow 2 \cdot pos + 2$
    end if
    if $dest \neq pos$ **then**
      Exchange $X[pos]$ and $X[dest]$
      Let $pos \longleftarrow dest$ {loop continues at lower level of the tree}
    else
      Break from loop {The tree is now a heap}
    end if
  end while

6.3.1.3   Optimality of sort algorithms and beating the bound

Fast algorithms are able to sort N elements in $O(N \log N)$ time and there is a trivial lower bound of $O(N)$ on this running time, since sorting an array at least requires reading the N elements in the array and writing the sorted array as output. A natural question is to ask whether it is possible to outperform $O(N \log N)$. In fact, it can be shown that in a generic sense, this is impossible. However, in some special cases, where additional information is known about the array and the elements it contains, it is possible to beat this bound and come up with linear time special purpose sort algorithms.

Let us first explain why $O(N \log N)$ is optimal. For that, assume that we need to sort an array of N distinct elements. We also assume that the array is accessed only in a limited way. Elements can be compared and moved around, nothing else. For our array of N elements, $N!$ different orders are possible. Moreover, given the sorted array that the algorithm outputs and the computation trail, it is possible to reconstruct the initial order of elements. As a consequence, the algorithm has to learn enough information to be able to fully reconstruct the input order. The number of bits necessary to encode this order is $\log_2(N!) \approx N \log_2 N$. Since any comparison reveals at most one bit of information, the algorithm requires on average at least $N \log_2 N$ comparisons. This yields the desired lower bound on the running time.

To show that this bound can be beaten in some special cases, assume that we want to sort an array of N integers in $[1 \cdots \lfloor \lambda N \rfloor]$ for some constant λ. Consider Algorithm 6.13. The initialization of the algorithm costs $O(\lambda N)$, the first loop costs $O(N)$, the second loop costs $O(\lambda N)$ for the control and $O(N)$ for the printing. All in all, since λ is a constant, this algorithm running time is $O(N)$. Clearly, the algorithm outputs numbers in increasing order, since it only writes down the loop variable i. Moreover, it writes i as many times as i appears in X. Thus, the algorithm outputs a sorted copy of array X. It beats the lower bound on sort algorithms, because it is able to obtain information about X through an additional channel. This algorithms does not restrict itself to comparisons; in fact, it does not perform any comparisons at all, only (implicit) equality tests.

Algorithm 6.13 Count sort: Beating the sort lower bound

Require: Input a sorted array X of N elements in $[1 \cdots \lfloor \lambda N \rfloor]$
    Allocate an array M of $\lfloor \lambda N \rfloor$ elements initialized to zero
    for i from 1 to N **do**
        Increment $M[X[i]]$
    end for
    for i from 1 to $\lfloor \lambda N \rfloor$ **do**
        Print $M[i]$ times the number i
    end for

6.3.1.4 Sort algorithms and stability

When using sort algorithms with duplicate elements, we can ask what happens to equal elements: Do they remain in the same relative order or can they be moved around? If a sort algorithm always preserves the relative orders of copies of the same element, it is called a **stable** sort algorithm. Since this can sometimes be an important property, it is useful to know which sort algorithms are stable and which are not. Of course, any sort algorithm can be made stable by simplify adding the original position to any element, using this value to break ties between identical entries. The problem with this approach is that it requires a potential large amount of extra memory to store the initial positions. As a consequence, when needed an intrinsically stable sort is preferable. Among our sort algorithms with quadratic running time it is easy to check that bubble sort and selection sort are stable. For insertion sort, the sort is stable if and only if the dichomoty search we use always return the last position where an element can be inserted. We leave this verification as an exercise.

Of course, it would be interesting to use an algorithm which is both fast and stable. Among our selection of sort algorithms, only merge sort has this extra property. Quicksort and radix sort can reasonably easily be made stable 9 if the sorted elements are written in a copy of the initial array. However, the versions that do not need such a copy move elements around too much to satisfy this property.

6.3.2 Hash tables

In a book about cryptography, when hash functions are discussed, the reader is usually expected to think about cryptographically strong hash functions, without collisions, preimages or other strong properties. This section is an exception to this general rule, here we consider algorithmic hash functions and only require that for ordinary inputs (not selected by an adversary) to these functions, the outputs of the hash functions are well distributed. When looking for collisions, such a non-cryptographic hash function can be used to filter out most non-colliding pairs of elements. The key requirement is that for given two distinct elements, their hash values are rarely equal. Of course, given two equal elements, their hash values are always equal, so using a hash function to filter pairs cannot miss collisions. In practice, depending on the set of inputs, the hash functions we use can be extremely simple. The most extreme case is encountered with random numbers, say on n bits. There, it suffices to truncate the numbers, keeping b bits only to get a useful hash function. For this purpose, we can use the low order bits, the high order bits or anything else that fits well with the particular problem being considered. It is clear that this process cannot qualify as a strong cryptographic hash function, for example, finding collisions in this function is straightforward.

To find collisions among n-bit numbers using a non-cryptographic hash function h, we proceed as follow. Assuming that the hash values are b-bit numbers, we create a large hash table of 2^b numbers, which can be indexed by hash values. This table is initialized to some special value $\perp$. Then we take numbers of the list one by one, and process number x by writing x in the hash table T at position $h(x)$ when the position still contains the special value $\perp$. Otherwise, we test whether $x = T[h(x)]$, if so, since $T[h(x)]$ is a previous element, we have found a collision. A very important remark is that this approach may miss some collisions. For example, assume that the list of elements contains in this order three elements, x, y and z with $h(x) = h(y) = h(z)$, $y = z$ and $x \neq y$. We also assume that no other element in the list has the same hash value $h(x)$. Then during the execution of the algorithm, x is written in $T[h(x)]$ since the position $h(x)$ in table T is empty when x is processed. When we encounter y, we see that $T[h(y)]$ is not empty and that $y \neq T[h(y)](= x)$, so no collision is detected. Similarly, for z we see that $z \neq T[h(z)](= x)$, and do not detect a collision. All in all, we missed the collision between y and z. In order to solve this problem, we should make sure that when a false collision is detected, the corresponding element is not dismissed but stored somewhere. In fact, this is a classical problem when using hash table for other purposes such as database indexing. In these alternative contexts, collision problems between hash values must be solved and specific collision resolution techniques have been devised. However, in our context, it is often easier to choose the parameter b in order to avoid this problem altogether or, at least, with high probability.

In order to use hash tables for collision finding as in Algorithm 6.14, it is important to choose the size of indexes correctly. Assume that we want to find collisions in a list of L elements using a table indexed by b-bits. To avoid collision resolution, we should take care to avoid multicollisions involving 3 or more elements. From the analysis of Section 6.2.1, assuming that the hash values behave as random numbers, such multicollisions appear when:

$$L^3/6 \approx 2^{2b}. \tag{6.5}$$

Forgetting the constant divisor 6, we find that we need to choose $b \approx 1.5 \log_2 L$. Assuming that we search a collision between n-bit numbers and that we have the usual case $L \approx 2^{n/2}$, this yields $b \approx 3n/4$.

Since this value of b is quite large, it is important to refine the analysis, assuming that we are ready to accept some probability of failure. Clearly, if we lower b, 3-multicollisions between hash values are going to occur. These multicollisions make us fail when they hide a real collision on the original numbers as in the x, y, z example we described earlier. Note that we miss the collision if only if the bad element occurs first. Since the numbers essentially appear in a random order, we miss the collision with probability $1/3$. Thus, preventing 3-multicollisions is not essential, it suffices to prevent

4-multicollisions[6]. Redoing the analysis we start with:

$$L^4/24 \approx 2^{3b},\qquad(6.6)$$

and we find $b \approx 2n/3$. Since b is still quite large compared to n, this approach is mostly useful when n is not too large. A typical example is to consider 40-bit numbers and to take $b = 26$ or maybe 27. With these parameters, the hash table can fit in memory, even if we use a wasteful representation and take 64 bits to represent 40-bit number. Indeed, the memory requirements are either 512 Mbytes or 1 Gbyte. A noticeable property of using hash tables compared to sort algorithms is the fact that the list of numbers does not need to be completely available at the start of the algorithm, it can be generated online.

Algorithm 6.14 Collision search using hash tables

Require: Input (potentially one by one) a list X of N numbers on n bits

  Choose b near $2n/3$

  Allocate a hash table $T[0 \cdots 2^b - 1]$ initialized to $\perp$

  for i from 1 to N **do**

    if $T[h(X_i)] = \perp$ **then**

      Let $T[h(X_i)] = X_i$

    else

      if $T[h(X_i)] = X_i$ **then**

        Output 'Collision found for value X_i'

      end if

    end if

  end for

6.3.2.1 Hash tables and cache misses

Using hash tables to search for collisions yields a very simple algorithm which is easily implemented. However, in terms of memory accesses, it has a very bad behavior. Indeed, it reads and writes in a large table in memory at essentially random positions. As we already know, with current memory architecture, this behavior causes many cache misses and slows the program a lot. It would be nice to be able to change the hash table algorithm in order to minimize this difficulty. We now show how this can be done. For this purpose, let us first define b_0 as the largest possible bit size such that a hash table with 2^{b_0} entries fits into the memory cache. Of course, if $b \leq b_0$, Algorithm 6.14

---

[6]We could even go further if we lower again the probability of success, but this particular choice achieves a nice balance.

does not encounter cache misses and we need no improvement. Thus the interesting case happens for $b > b_0$. Since the main memory is usually not arbitrarily larger than the cache memory, it is safe to also assume that $b \leq 2b_0$. For example, with $b_0 = 16$, i.e., a 64-Kbyte cache, this means that the main memory is smaller than 4 Gbytes. Under this assumption, we divide each b-bit hash values into two parts, say the b_0 low order bits on one side and the $b_1 = b - b_0$ other bits on the other side. For simplicity, we decompose the hash function h in the same way and get two smaller hash functions, h_0 on b_0 bits and h_1 on b_1 bits. Then, we proceed as follow: in a first phase, for each incoming value X, we simply throw it on one among 2^{b_1} stacks, more precisely on the stack numbered $h_1(X)$. Once this first phase is finished, the initial list of numbers has been divided into 2^{b_1} stacks. Within each stack, all numbers share the same value for h_1, i.e., they share the top b_1 bits of their h values. If the inputs are random numbers, the stacks are roughly of the same size. In a second phase, we simply run Algorithm 6.14 independently on the content of each stack in order to find collisions within these stacks. Of course, this yields the same result as a direct application of Algorithm 6.14 on the original list of numbers. This approach is described in pseudo-code as Algorithm 6.15. Due to our hypothesis on b_0, we know that no (or few) cache misses occur during this second phase. Of course, we also need to check for cache misses during the first phase. If we represent each stack as an array and write each new element at the end of the corresponding array, we are simply writing elements in memory in an ordered fashion within each array. If the number of arrays is small enough, this does not cause too many cache misses. To estimate the number of arrays, note that we need to store in cache, the current starting position within each stack array, together with a few positions at the current start of each stack. Thus to be safe, the numbers of stacks 2^{b_1} should remain below some fraction of the cache size. In practice, this means that b_1 should be smaller than $b_0 - 5$ or something similar. Since we only assumed $b_1 \leq b_0$, there is a remaining gap. If required, this gap can be filled by using the same idea twice: split the original list into stacks, split each stack into substacks and then search for collisions.

6.3.3   Binary trees

Another interesting method to find collisions is the use of binary search trees. A binary search tree, is a binary tree that satisfies an important additional property: for any node in the tree, all the nodes in its left subtree have smaller (or equal) values and all the nodes in its right subtree have greater values. Binary search trees are represented explicitly, by storing for each node, its value and a pointer to each of its children. Due to this representation, they are quite costly in terms of memory usage. From a runtime point-of-view, all operations are fairly quick. Inserting, removing or searching for an element in the tree costs at most the depth of the tree. Since a tree of depth k may hold $2^{k+1} - 1$ entries, the elementary cost is logarithmic in the size of the tree,

---

Algorithm 6.15 Avoiding cache misses with hash tables

Require: Input a list X of N numbers on n-bits
  Choose b near $2n/3$
  Split b as $b_0 + b_1$
  Create 2^{b_1} stacks of numbers represented by arrays of size $\lceil 1.1N \times 2^{-b_1} \rceil$
  {with a memory margin of 10 percent}
  Allocate a hash table $T[0 \cdots 2_0^b - 1]$
  for i from 1 to N **do**
    Add $X[i]$ to stack number $h_1(X[i])$
  end for
  for k from 0 to $2^{b_1} - 1$ **do**
    If stack k overflowed, **Abort**
    Reinitialize hash table $T[0 \cdots 2^b - 1]$ to $\bot$
    for all x in stack k **do**
      if $T[h_0(x)] = \bot$ **then**
        Let $T[h_0(x)] = x$
      else
        if $T[h_0(x)] = x$ **then**
          Output 'Collision found for value x'
        end if
      end if
    end for
  end for

---

assuming that the tree is reasonably full. As a consequence, the main issue when working with binary trees is to control their depths and to check that they do not grow too much.

The favorable case is when a tree is built by inserting randomly selected elements. In that case, the resulting tree is well balanced, all subtrees are similar and the resulting depth is quite small, i.e, it remains within a factor of the logarithm of the size. The worst case occurs when a sorted sequence of elements is inserted one at a time to build the tree. In this case, the tree degenerates into a list and its depth is equal to its size. As a consequence, searching for an element takes linear time in the size of the tree.

Because of these extreme behaviors, binary search trees are not suited to all applications. For a given application, it is prudent to test the trees by monitoring their depths to see if problems do occur. Thankfully, there exits a nice solution to fix the depth problem with binary search trees, this solution uses AVL trees, named after their inventors Adelson-Velsky and Landis. These trees are also called self-balancing trees. The key idea of AVL trees is for each operation that modifies the tree, insertion or deletion, to check that the tree does not become too unbalanced and if necessary to take corrective action. Note that AVL trees do not try to minimize the tree depth, this would be too costly, instead they simply keep the tree depth within a factor of the optimal value by using a simple criteria. The criteria used by self-balancing trees works by memorizing for each node the difference between the depth of its left and right subtrees. Differences of either 0, 1 or -1 are allowed. Whenever a larger difference appears, the tree is modified to reduce it to an acceptable value. The transformations used to modify the tree are called rotations. Two basic rotations are possible, left and right rotations, they are inverse of each other. Applying the correct rotation maintains the binary search property of the tree while rebalancing the left and right subtrees of the rotated node. Depending on the detailed structures of the subtree, rebalancing steps requires either a single or a double rotation.

Use of binary search trees in cryptography.　Because of their memory costs, direct application of binary trees to collision search is not a good idea. These structures are better suited to maintain collections of data that are modified as time goes. An interesting example that may be encountered is the case where after an initial construction the tree evolves in the following way: at each time remove a tree element, say the smallest one and add in a new element. Such an example is given in Chapter 8 in order to construct an improved birthday paradox attack against knapsack cryptosystems.

6.3.3.1　Detailed description of basic binary search trees

Using a binary search tree requires the application of several algorithms. Each algorithm starts from a correctly constructed tree and modifies into a new, also correct, tree. At the very beginning, the tree is initialized at empty.

To encode a tree, we maintain a pointer which points to the **root node** of the tree. When the tree is empty, the pointer to the root node is the nil pointer. Each node in the tree is a structure which contains the value stored in the node and two pointers, one to the left child and one to the right child. When a pointer to a child is the nil pointer, there is no child on the corresponding side. A node without children is called a **leaf**, any other node is called an **internal node**. Optionally, a node may also contain an additional field to memorize the size of the left subtree connected to this node. With this additional field, it becomes possible to recover the i-th element in the binary search tree in logarithmic time. We describe all algorithms for binary search trees with this optional field in place. In some algorithms, it is necessary to be able to backtrack from an element to its father. This can be done in two different ways, the first one is to maintain for each node an additional pointer that memorizes its father, the second one is when passing an internal node to a tree manipulation algorithm to pass it along with the full path from the root to this node. The first method is simpler to implement but requires even more memory to encode the trees. An alternative method is never to pass an internal node directly, but to let the manipulation algorithm search for it, thus determining the access pass itself. This alternative method may sometimes require the program to perform the same search twice.

Given a binary search tree, the most important algorithm is probably the search algorithm that tries to locate a given value in the tree. This algorithm is quite simple, starting from the root, it compares the reference value to the value stored in the current node. If they are equal, the search is successful, if the reference value is smaller, the algorithm moves to the left child and continues the search, if it is greater, it moves to the right child. Another very important algorithm is the insertion algorithm. Insertion works in two phases, first it searches for the value to be inserted, if the value is already present in the tree, it adds a copy either right after or right before the existing copy. Otherwise, it adds it at the position where the search ended. Since the insertion algorithm contains a search part, we do not describe the search in pseudo-code by itself but include it as part of the insertion Algorithm 6.16. An important part of the insertion algorithm is the ability to find an empty position right before or after an internal node. For an empty position right before a node, move once to the left and then as many times to the right as needed to get to a nil pointer. The nil pointer is the place we looked for. To find the position right after a node, move once right, and continue with left moves until an empty position is reached. While locating the place where the insertion itself is going to happen, each time we move to the left, the field representing the size of the left subtree needs to be incremented in order to remain at the correct value.

Thanks to the optional field that counts the size of the left subtree, we may also search for the i-th element in the tree (we assume as usual that elements are counted from zero). Once again we start at the root and proceed by moving down the tree. If the size of the left subtree for the current node is

---

Algorithm 6.16 Insertion in a binary search tree

Require: Input tree root R, free node N and value to insert X

Set the value of N to X

Set the children of N to nil

Set the left subtree size of N to 0

if R is the nil pointer **then**

    Replace R by N and **return**

end if

Set Current $\longleftarrow R$ {First search for insertion position}

while true do

    if Value of Current node $< X$ **then**

        if Current node has no right child **then**

            Set right child of Current node to N

            Return

        else

            Let Right be the right child of Current node

            Set Current $\longleftarrow$ Right

        end if

    else

        Increment the left subtree size field within Current

        if Current node has no left child **then**

            Set left child of Current node to N

            Return

        else

            Let Left be the left child of Current node

            Set Current $\longleftarrow$ Left

        end if

    end if

end while

---

s, three cases are possible, if $i < s$ then we need to look for the i-th element in the left subtree, if $i = s$ then the current node is the right one, if $i > s$ then look for element $i - s - 1$ in the right subtree.

Finally, we may also delete a given tree node. If the node to delete is a leaf, it is quite easy, it suffices to remove it by replacing the pointer to it in its father node by the nil pointer. After that, we need to backtrack the access path to this node all the way up to the root to correct the size of each left subtree where needed. Similarly, to delete a node with a single child, it suffices to remove the node and link its child directly to its father. If the node to delete has two children, deletion is more tricky. First, we need to locate the immediate predecessor (or successor). Then, we swap the values stored in the node to delete and in its successor. Finally, we delete the successor. Since this successor has at most one child, it can be deleted using the simple approach. Note that during the deletion algorithm, the binary search tree is temporarily invalid. The value to be deleted may be stored in incorrect order compared to the rest of the tree. However, when the value is finally removed, the tree becomes correct once again. The deletion algorithm is given as Algorithm 6.17.

Algorithm 6.17 Deletion in a binary search tree

Require: Input node N to delete, with path P of ancestors to the root
  if N is a leaf or has at most one child C **then**
    Replace pointer to N in its father by a pointer to C or to nil
    for all Ancestors A in path P **do**
      if N is in the left subtree of A **then**
        Decrement the left subtree size field within A
      end if
    end for
    Return
  else
    Let M be the left child of N
    Let path $Q \longleftarrow (P\|N)$
    while M has a right child **do**
      Let path $Q \longleftarrow (Q\|M)$
      Replace M by its right child
    end while
    Exchange the value fields of M and N
    Call deletion algorithm on M with path Q {M has at most one child}
  end if

6.4 Application to discrete logarithms in generic groups

Collisions can be used to computed discrete logarithms in arbitrary groups, using the baby-step, giant-step method. Before presenting this method, it is useful to first show that computing discrete logarithms in a group whose cardinality is known is, assuming that the factorization of the cardinality is given, no harder than computing discrete logarithms in all subgroups of prime order. The Pohlig-Hellman algorithm is a constructive method to compute discrete logarithm in the whole group from a small number of calls to discrete logarithms computations in the subgroups of prime order.

6.4.1 Pohlig-Hellman algorithm

First, let us recall the definition of discrete logarithm. Given a group $\mathbb{G}$, denoted multiplicatively and two elements of $\mathbb{G}$, say g and h, computing the discrete logarithm of h in basis g amounts to finding, if it exists, an integer a, such that $h = g^a$ in $\mathbb{G}$. For discrete logarithms computations are not necessarily possible for all elements h of $\mathbb{G}$. However, if $\mathbb{G}$ is a cyclic group generated by g, then the discrete logarithm is defined for all elements of $\mathbb{G}$. In the sequel, we make this assumption.

Note that, if N denotes the cardinality of $\mathbb{G}$, the discrete logarithm of h in basis g is only determined modulo N, since $g^N = 1$ the identity element in $\mathbb{G}$. An interesting consequence is that any algorithm able to compute discrete logarithm can be used to obtain N. For simplicity, assume that we know the order of magnitude of N and, more precisely, assume that N lies in the range $[N_0 + 1, 2N_0]$. If the discrete logarithm algorithm outputs a normalized value between 0 and $N - 1$, it suffices to ask for the discrete logarithm of g^{2N_0}, say a. Then we know that N divides $2N_0 - a$ and even that $N = 2N_0 - a$. If the discrete logarithm is allowed to output any of the multiple possible values for the discrete logarithm, choose a random integer b between 0 and some multiple of N_0, say $10N_0$. Then ask for a discrete logarithm of g^b and let a denote the resulting value. Since $g^a = g^b$, $|b - a|$ is a small multiple of N, possibly 0. If $a \neq b$, it is a non-zero multiple. Since there are not many divisors of this multiple in the range $[N_0 + 1, 2N_0]$, we can easily recover N. However, we need to make sure that the discrete logarithm algorithm does not systematically output $a = b$. This comes from the fact that we are choosing b at random in a large range. With this strategy, even an adversarial discrete logarithm algorithm cannot systematically determine b and, at some point, it outputs some other value for the discrete logarithm. Finally, even if we do not know a precise range, it is usually possible to find N by computing the GCD of a few multiples obtained as above.

As a consequence, in the context of discrete logarithm computations, it is reasonable to ask for the group cardinality N in advance. We also assume that the factorization of N into prime is known[7], i.e., we have the decomposition:

$$N = \prod_{i=1}^{n} p_i^{e_i}. \tag{6.7}$$

In the cyclic group $\mathbb{G}$, generated by g, there are subgroups $\mathbb{G}_i$ of order p_i for $1 \leq i \leq n$. Moreover, each subgroup $\mathbb{G}_i$ is generated by $G_i = g^{N/p_i}$. In addition, $\mathbb{G}$ also contains subgroups $\mathbb{G}_i'$ of order $p_i^{e_i}$ generated by $G_i' = g^{N/p_i^{e_i}}$. Pohlig-Hellman algorithm works in two steps. First, it uses e_i calls to the discrete logarithm algorithm in $\mathbb{G}_i$ to compute discrete logarithms in $\mathbb{G}_i'$. Second, it pastes these values together to obtain a discrete logarithm in $\mathbb{G}$.

Starting with the second step of the algorithm, recall that we are given g and h in $\mathbb{G}$ and want to find a number a, determined modulo N, such that $h = g^a$. If we raise this equality to the power $N/p_i^{e_i}$ and let $H_i' = h^{N/p_i^{e_i}}$, we obtain a discrete logarithm problem in $\mathbb{G}_i'$, namely $H_i' = (G_i')^a$. Since $\mathbb{G}_i'$ is a group of order $p_i^{e_i}$, this yields the value of a modulo $p_i^{e_i}$. Since the $p_i^{e_i}$ are pairwise coprime, we can use the Chinese remainder theorem and obtain a modulo N.

It remains to see how discrete logarithms in $\mathbb{G}_i'$ can be computed from e_i calls to the discrete logarithm algorithm in $\mathbb{G}_i$. Of course, it is clear for $e_i = 1$. To treat the general case, it is useful to introduce a family of groups: $(\mathbb{G}_i^{(j)})$ with $1 \leq j \leq e_i$. Each group $\mathbb{G}_i^{(j)}$ has cardinality p_i^j and is generated by $g_i^{(j)} = g^{N/p_i^j}$. We have $\mathbb{G}_i^{(1)} = \mathbb{G}_i$ and $\mathbb{G}_i^{(e_i)} = \mathbb{G}_i'$. In order to prove our result, it suffices to show by induction on j that discrete logarithms in each group $\mathbb{G}_i^{(j)}$ can be computed using j calls to discrete logarithms in $\mathbb{G}_i$. It is clear for $j = 1$. To compute a discrete logarithm in $\mathbb{G}_i^{(j+1)}$, or equivalently, given z in this group to compute x such that $z = (g_i^{(j+1)})^x$, we proceed as follows:

- Raise the equation to the power p_i and remark that

$$z^{p_i} = (g_i^{(j+1)})^{p_i x} = (g_i^{(j)})^x$$

is a discrete logarithm problem in $\mathbb{G}_i^{(j)}$. By induction hypothesis, we learn x_0 the value of x modulo p_i^j in j calls to discrete logarithms in $\mathbb{G}_i$.

- Write $x = x_0 + p_i^j x_1$, with $0 \leq x_1 < p_i$ and remark that:

$$\frac{z}{(g_i^{(j+1)})^{x_0}} = (g_i^{(j+1)})^{x-x_0} = (g_i^{(j+1)})^{p_i^j x_1} = (g_i^{(1)})^{x_1}.$$

Thus x_1 can be obtained by an additional discrete logarithm computation in $\mathbb{G}_i$.

---

[7]This is not a real limitation either, because in most practical cases, factoring N is no harder than computing the discrete logarithm.

This concludes the proof. An iterative version of Pohlig-Hellman method is given as Algorithm 6.18.

Algorithm 6.18 Pohlig-Hellman discrete logarithm algorithm

Require: A group $\mathbb{G}$ of cardinality $N = \prod_{i=1}^{n} p_i^{e_i}$.
Require: A group generator g and a group element h
  for i for 1 to n **do**
    Let $a_0 \longleftarrow 0$
    Let $g_0 \longleftarrow g^{N/p_i}$
    for j from 1 to e_i **do**
      Let $h_0 \longleftarrow (hg^{-a_0})^{N/p_i^{j}}$
      Call discrete logarithm algorithm in subgroup of order p_i generated by g_0 for element h_0. Let a denote the result.
      Let $a_0 \longleftarrow a_0 + p_i^{j-1}a$
    end for
    if $i = 1$ **then**
      Let $A \longleftarrow a_0$ and $M \longleftarrow p_1^{e_1}$.
    else
      Compute the Chinese remainder of A modulo M and a_0 modulo $p_i^{e_i}$
      Put the result in A.
      Let $M \longleftarrow Mp_i^{e_i}$.
    end if
  end for

6.4.2 Baby-step, giant-step algorithm

Thanks to Pohlig-Hellman algorithm, we only need to compute discrete logarithms in a group $\mathbb{G}$ of prime order p, generated by g. We are given h and search for an integer a in the range $[0, p-1]$ such that $h = g^a$. Let r be the integer obtained by rounding up the square root of p. It is clearly possible to write a as $a_0 + ra_1$, with $0 \leq a_0 < r$ and $0 \leq a_1 < r$. As a consequence, we have an equality between hg^{-a_0} and g^{ra_1} in the group $\mathbb{G}$. Conversely, any equality of this form yields the discrete logarithm of h.

From an algorithmic point-of-view, this means that we can find the discrete logarithm we seek by searching for a collision between two lists of r elements each. The first list contains all group elements of the form hg^{-a_0}. The second list contains all group elements of the form g^{ra_1}. The name of the algorithm, comes from the fact that elements of the first list are separated by small steps (or baby steps), corresponding to multiplication by g and elements of the second list by large steps (or giant steps), corresponding to multiplication by g^r.

Algorithm 6.19 Baby-step, giant-step discrete logarithm algorithm

---

Require: A group $\mathbb{G}$ of prime order p.
Require: A group generator g and a group element h
  Let $r = \lceil \sqrt{p} \rceil$
  Create T an array of r group elements
  Create I an array of r integers
  Let $k \longleftarrow h$
  for i for 0 to $r - 1$ **do**
    Let $T[i] \longleftarrow k$ and $I[i] \longleftarrow i$
    Let $k \longleftarrow k/g$
  end for
  Sort T and perform all exchanges simultaneously on I.
  Let $k \longleftarrow 1$ (neutral element in $\mathbb{G}$)
  Let $G \longleftarrow g^r$ in $\mathbb{G}$
  for i from 0 to $r - 1$ **do**
    Lookup k in T (using a binary search)
    If k is found at position j, return $ri + I[j]$ as discrete logarithm.
    Let $k \longleftarrow k \cdot G$ in $\mathbb{G}$
  end for

Exercises

1. Neglecting skip years and seasonal birthrate irregularities, compute for sets of ten to thirty individuals, possibly writing a program, the probabilities of birthday collisions.

2. Check that CBC-MAC without final reencryption is insecure by showing that if a message containing a single block $M^{(1)}$ has authentication tag t, then the two-block message $M^{(1)} \| M^{(1)} \oplus t$ has the same authentication tag.

3[h]. Write down an explicit chosen message distinguishing attack on CBC encryption, when used beyond the birthday paradox limit, using the technique described in Section 6.1.1.1.

4. Why is the probability for twins to have the same birthday not equal to one?

5[h]. Consider a hash function obtained by directly applying the Merkle-Damgård construction (Chapter 5, Section 5.4.1) to a permutation family π. The goal of this exercise is to show a weakness of this hash function with respect to the preimage resistance property. This means that starting from an intermediate hash value h and a message block m, the next hash value is $h' = \pi_m(h)$.

   (a) Show that when π^{-1} is available (for example, when considering a block cipher), the hashing process can be reversed, i.e., given a intermediate hash value h' and a message block m, the previous hash value h can be recovered.

   (b) Let h_0 be the initial value of the hash function. Choose a large set containing short sequences of message blocks. For each such sequence M_i, starting from h_0 and considering each block of M_i in turn, we can compute an intermediate hash value which we denote by h_i.

   (c) Let h_F be a target value for the hash function. Similarly choose another large set containing short sequences of message blocks. For each such sequence M'_i, starting from h_F and considering each block of M'_i in turn, we can backward compute an intermediate hash value which we denote by h'_i.

6[h]. Again considering a Merkle-Damgård based hash function H, we now turn to multicollisions. Let $A^{(1)}$ and $A^{(2)}$ be two single-block messages such that $H(A^{(1)}) = H(A^{(2)})$. Let $B^{(1)}$ and $B^{(2)}$ be two message blocks such that $H(A^{(1)} \| B^{(1)}) = H(A^{(2)} \| B^{(2)})$, construct a 4-collision.

(a) More generally, construct a 2^t-collision on messages of t-blocks.

(b) How secure against collision resistance is the hash function obtained by concatenating two Merkle-Damgård hash functions, where each function is on n bits.

7. Check that the dichotomy search given as Algorithm 6.2 always returns the position of the last occurrence of an element. Deduce that the insertion sort Algorithm 6.6 is stable.

8. Any sorting algorithm can be made stable by a generic technique, based on storing pairs containing elements and their initial position. Give the order relation that should be used to achieve this effect.

9^{h}. At first, it may seem that in quicksort and radix sort, the basic partitioning preserves the relative order of elements stored in the stack of small element and reverses the order of elements in the stack of large elements. Explain why it is not the case. Show how to do this with an auxiliary array of N elements. Conclude by writing a stable version of quicksort and radix sort.

Chapter 7

Birthday-based algorithms for functions

In Chapter 6, we introduced several algorithmic techniques, that allow us to apply the birthday paradox to arbitrary lists of randomly distributed objects. These techniques do not take into account the specific details of how these objects are generated. In this chapter, we are going to focus on a more specific problem and look for collisions among objects which are generated by iterating some fixed function. In this specific case, many improvements are possible. This is quite important for cryptanalysis, because there are many applications where this improved setting can be exploited. In particular, there are some very important number theoretic algorithms due to Pollard that use these techniques to factor integers or compute discrete logarithms.

In the most general setting, we are given a set S and a function F from S to itself. Then starting from a point X_0 in S, we generate the sequence defined by:

$$X_{i+1} = F(X_i). \tag{7.1}$$

Our main goal is to find a collision within this sequence, say $X_i = X_j$. Note that due to the deterministic nature of the above computation, from any such collision, we may generate many. Indeed, whenever $X_i = X_j$, we find that $X_{i+1} = F(X_i) = F(X_j) = X_{j+1}$ and obtain a new collision. Of course, iterating this remark, we also have $X_{i+2} = X_{j+2}$ and so on.

In order to study this algorithmic problem, it is important to assume that F behaves more or less randomly. Without this assumption, the birthday paradox has no reason to hold. Furthermore, when omitting the assumption, it is easy to build counterexamples. Take the set $S = [0 \cdots 2^n - 1]$ of integers modulo 2^n and define $F(X) = X + 1 \pmod{2^n}$. Then, for any choice of X_0, we cannot find a collision in the sequence X_i, unless we have considered the full set of 2^n elements. In this case, the birthday paradox would predict a collision after $2^{n/2}$ elements. Thus, we see that it does not apply to this counterexample. In order to theoretically analyze the behavior of such sequences, F is usually modeled as a random function. This model and the corresponding analysis are presented in Section 7.2 below. For now, we simply assume that after some time the sequence X loops back on one of its previous values. Note that since F is a function, not a permutation, nothing guarantees that X loops back on X_0. On the contrary, it usually restarts at a later position in the sequence. A traditional way of describing this fact is to draw a picture

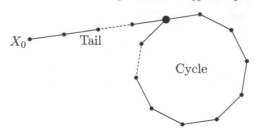

Figure 7.1: Rho shape while iterating a random function

that represents the sequence X by making sure that consecutive elements of the sequence are neighbors on the picture. On the resulting Figure 7.1, we see a shape similar to the letter ρ. This gave its name to Pollard's Rho factoring algorithm. In this figure, we see that the sequence has two distinct parts: a cycle and a tail. If the first collision in X is a collision between X_s and $X_{s+\ell}$, then the tail contains elements of the sequence from X_0 up to X_{s-1} and the cycle starts from X_s and its length is ℓ, the period of X.

7.1 Algorithmic aspects

The key question in this chapter is to find collisions in the sequence X or even better to determine the respective lengths of the cycle and tail of the sequence. A first solution to this problem is to use classical birthday-based algorithms. For example, we can use a binary search tree to store the elements of the sequence X one at a time as long as there is no duplicate in the tree. The first duplicate occurs when $X_{s+\ell}$ is inserted in the tree and we find that this value is already present due to X_s. This method has a running time $O((s+\ell)\log(s+\ell))$ and needs to store $O(s+\ell)$ elements. The running time is essentially optimal, since, unless the function F has very specific properties, we need to generate the sequence up to $X_{s+\ell}$ to detect its cycle. Moreover, this algorithm directly generates both the length of the cycle ℓ and the length of the tail s. However, its storage costs are high. The good news is that there exist alternative algorithms that do not require such a large amount of memory. These algorithms do not directly output s and ℓ but instead find a collision somewhat later in the cycle. However, with a reasonable additional computational cost, it is possible to recover s and ℓ from this late collision.

We start by presenting two simple algorithms that obey an additional restrictive condition. More precisely, these two algorithms access values of the sequence X in a restrictive way, either as inputs to F or in equality tests, but in no other manner. This restrictive use of the sequence's values can be very important because it allows the cycle finding algorithms to be used in a black-box context where the sequence X can only be accessed indirectly. In particular, this restriction is essential to Pollard's Rho factoring algorithm (see Section 7.3.2). Indeed, in Pollard's Rho factoring, equality tests can only be performed indirectly through GCD computations.

7.1.1 Floyd's cycle finding algorithm

In this first algorithm, we start by defining a new sequence Y which is related to X by the equation:

$$Y_i = X_{2i}. \tag{7.2}$$

Alternatively, the sequence Y can be defined directly by the formulas:

$$Y_0 = X_0 \text{ and } Y_{i+1} = F(F(Y_i)). \tag{7.3}$$

Algorithm 7.1 Floyd's cycle detection algorithm

Require: Input initial sequence value X_0, max. iterations M
    Let $x \longleftarrow X_0$
    Let $y \longleftarrow X_0$
    for i from 1 to M **do**
        Let $x \longleftarrow F(x)$
        Let $y \longleftarrow F(f(y))$
        if $x = y$ **then**
            Output 'Collision between i and $2i$'
            Exit
        end if
    end for
    Output `Failed`

Once Y is defined, Floyd's algorithm looks for the first index t such that $Y_t = X_t$. Clearly, this can be done in time $O(t)$ using a very small amount of memory, as shown by the pseudo-code in Algorithm 7.1. Indeed, it suffices to compute the two sequences in parallel until they collide. To analyze the behavior of this algorithm, we need to better understand the relation between the position t of this first collision between X and Y and the parameters s and ℓ of the sequence X.

First, since $X_{2t} = X_t$, we find that t is a period of X and thus a multiple of the period ℓ. Second, since X_t is in the cycle of X, $t \geq s$. In fact, t is the smallest multiple of ℓ greater than (or equal to) s:

$$t = \left\lceil \frac{s}{\ell} \right\rceil \ell. \tag{7.4}$$

Moreover, once t is known, recovering ℓ is easy, it suffices to compute the sequence X_{t+i} until we loop back to X_t. This happens after exactly ℓ steps.

7.1.2 Brent's cycle finding algorithm

The main drawback of Floyd's algorithm is the fact that each step requires three applications of the function F in order to advance both sequences X and Y. Brent's algorithm is an alternative that only requires the computation of a single copy of the original sequence. The starting point of the method is to remark that once we are given a point X_k inside the cycle, we can find the cycle's length in exactly ℓ steps, using the same method as when recovering ℓ after Floyd's algorithm. The difficulty is to find a start-point in the cycle. In Brent's algorithm, this is done by trying several start-points until an adequate one is found. In order to minimize the overall cost, the computation steps used to find a potential cycle are shared and also used to compute the next start-point to be tested. More precisely, we consider the sequence of start-points X_1, X_2, X_4, ..., X_{2^i}, ... indexed by powers of two. For each start-point, X_{2^i}, we compute the sequence of successors X_{2^i+j} until we either find a cycle or reach $X^{2^{i+1}}$. In the former case, we are done, in the latter, we change to the next start-point $X^{2^{i+1}}$. This approach is described in pseudo-code as Algorithm 7.2.

Algorithm 7.2 Brent's cycle detection algorithm

Require: Input initial sequence value X_0, max. iterations M

  Let $x \longleftarrow X_0$

  Let $y \longleftarrow x$

  Let $\mathtt{trap} \longleftarrow 0$

  Let $\mathtt{nexttrap} \longleftarrow 1$

  for i from 1 to M **do**

    Let $x \longleftarrow F(x)$

    if $x = y$ **then**

      Output 'Collision between $\mathtt{trap}$ and i'

      Exit

    end if

    if $i = \mathtt{nexttrap}$ **then**

      Let $\mathtt{trap} \longleftarrow \mathtt{nexttrap}$

      Let $\mathtt{nexttrap} \longleftarrow 2 \cdot \mathtt{trap}$

      Let $y \longleftarrow x$

    end if

  end for

  Output $\mathtt{Failed}$

As before, given the parameters s and ℓ of the sequence X, we can analyze the behavior of Brent's method. It succeeds with starting point X_{2^k}, if and only if, $2^k \geq \max(s, \ell)$. Indeed, to find the cycle with this start-point, the point should be in the cycle, i.e., $2^k \geq s$, and the cycle's length ℓ should be at most 2^k. Clearly, Brent's method finds ℓ with complexity $O(s + \ell)$.

Finally, note that our choice of putting traps in positions of the form 2^k is not the only option. Depending on the specific problem at hand, we could consider another quickly increasing sequence. This could be either a (rounded) geometric progression $\lfloor \alpha^k \rceil$ or something different such as a Fibonacci sequence.

7.1.3 Finding the cycle's start

We have seen that both Floyd's and Brent's algorithms can be used to find ℓ the length of the cycle in sequence X. However, neither produces the length of the tail s or equivalently the position where the cycle starts. From the point-of-view of collision finding, this position is quite important. Indeed, we have:

$$F(X_{s-1}) = X_s = X_{s+\ell} = F(X_{s+\ell-1}) \text{ and } X_{s-1} \neq X_{s+\ell-1}. \qquad (7.5)$$

Thus, finding the entrance of the cycle yields a true collision in F and this is the only place in the sequence X where such a true collision can be found.

After either Floyd's or Brent's algorithm, we know a number $t \leq 2(s + \ell)$ such that X_t is in the cycle of X. Thus, the start of the cycle is an integer s in the interval $[0 \cdots t]$. From this information, it is possible to obtain s in several different ways. We now present two different approaches to solve this problem.

7.1.3.1 Dichotomy search

Once ℓ is known, it is possible for any number σ in the interval from 0 to t, whether $s \leq \sigma$ by testing if $X_\sigma = X_{\sigma+\ell}$. Indeed, if the condition holds, X_σ is in the cycle and $\sigma \geq s$, otherwise, $\sigma < s$. This test can be performed by computing the sequence X up to $\sigma + \ell$, which requires $O(s + \ell)$ steps. Thus we can find s using a dichotomy search with complexity $O((s + \ell) \log(s + \ell))$. It is interesting to remark that this approach has the same runtime complexity as a generic birthday method that relies on a fast sort algorithm.

7.1.3.2 Direct search

The main problem with the above method is that we need to compute several overlapping parts of the same sequence, thus recomputing the same information many times. To avoid this drawback, we can proceed differently. First compute X_ℓ. Then, from X_0 and X_ℓ compute in parallel the two sequences X_i and $X_{\ell+i}$, testing for equality at each step. When the two sequences collide, we have $s = i$. This allows us to recover s and obtain a true collision[1] in time $O(s + \ell)$. This is described as Algorithm 7.3.

Algorithm 7.3 Algorithm for recovering a cycle's start

Require: Start-point X_0, cycle length ℓ, max. iterations M
  Let $x \longleftarrow X_0$
  Let $y \longleftarrow x$
  for i from 1 to ℓ **do**
    Let $y \longleftarrow F(y)$
  end for
  if $x = y$ **then**
    Output 'Cyclic sequence, no real collision'
    Exit
  end if
  for i from 1 to M **do**
    Let $x' \longleftarrow F(x)$
    Let $y' \longleftarrow F(y)$
    if $x' = y'$ **then**
      Output 'Collision between images of x and y'
      Exit
    end if
    Let $x \longleftarrow x'$
    Let $y \longleftarrow y'$
  end for
  Output `Failed`

As a consequence, we see that finding collisions in recursively defined sequences of this type is much more practical than finding collisions in lists of random objects.

7.1.4   Value-dependent cycle finding

As explained above, both Floyd's and Brent's algorithms detect a cycle in a sequence X, while using X only in a restrictive way, either through F or in equality tests on the values taken by the sequence. We know that this additional property is extremely useful in some cases. However, in other cases, the sequence X is directly available. Thus, it is natural to ask whether cycle finding algorithms can be improved using more value-dependent computations. The surprising answer is that it is indeed possible. A first step in this direction can be achieved using a distinguished point method, as proposed for

---

[1]Unless $s = 0$.

example in [QD90]. This consists of memorizing values that satisfy a specific property, called distinguished points, and waiting until they come back. The main difficulty here is to choose the right probability of occurrence for distinguished points. If they are too frequent, then we need to store a huge number of values and the algorithm is inefficient. If they are too rare, then the cycle may not contain any distinguished point at all and the technique fails. We discuss the use of distinguished points in more detail in Section 7.5.2.

A stronger way of using the values taken by the sequence in a cycle detection algorithm was proposed, in 2004, by Nivasch in [Niv04]. The main advantage of his algorithm is that it is guaranteed to stop somewhere between $X_{s+\ell}$ and $X_{s+2\ell-1}$, i.e., during the second iteration of the sequence cycle. This is particularly useful for sequences where ℓ is small compared to s. The key idea is to maintain a (hopefully) short stack of already encountered values in a way that guarantees cycle detection. To build this stack, Nivasch's algorithm relies on the existence of a total ordering on the set of sequence values and focuses on keeping small values taken by the sequence on the stack. The ultimate goal is to make sure that the smallest value occuring in the cycle is always stored on the stack and is always present for detection during the second iteration around the cycle. In order to achieve that, at any point in time, the stack memorizes the minimum value, say m_1, encountered up to that time while computing X, followed by the minimum value (say m_2) encountered after m_1 in the computation, followed by the minimum value (say m_3) encountered after m_2, ... Clearly, the minimum value of the cycle is one of those, say m_k. Indeed, any value of the sequence X smaller that m_k is out of the cycle and thus part of the pre-cycle. Since m_k occurs after these possibly smaller elements, by definition of our stack, it gets to be stored on the stack and remains there afterward.

From an algorithmic point-of-view, building such a stack is an easy matter. Whenever, we compute a new value X_i of the sequence X, we remove all values larger than X_i from the stack and add X_i to the stack. Moreover, it is easy to see that if X_i we add at the end of the stack, then this stack is always sorted in increasing order. Thus, erasing larger values can be done very easily by using a dichotomy search on the pre-existing stack and by truncating the stack after the point where X_i needs to be inserted. It is shown in [Niv04] that the expected size of the stack at time i is $\log i$ and that the size of the stack remains below $O(\log i)$ almost surely. Thus, except in extremely rare cases, the memory requirements of Nivasch's algorithm are small. When considering the time complexity of this algorithm, we need to consider two different kinds of operations: evaluations of the function F and other operations. Indeed, in almost all cases, the computation of values of F dominates the running time. Clearly, since X is computed once for all considered positions, up to at most position $s + 2\ell - 1$, this bounds the number of evaluations of F. The dichotomy search within the stack dominates the other operations, and the cost of each search is logarithmic in the size of the stack. Thus, the time complexity of other operations is bounded by $O((s + 2\ell) \log \log(s + 2\ell))$. At first, this complexity may appear worse than the complexity of the evaluation of F. However, remember that F operates on numbers of at least $O(\log(s+2\ell))$ bits, otherwise the cycle would be shorter. Thus, each evaluation of F costs at least $O(\log(s+2\ell))$ and, when accounting for this unit cost, evaluations of F dominates the complexity. Thus, where applicable, Nivasch's algorithm gives a nice improvement of cycle finding algorithms. We give a full description of it in Algorithm 7.4.

Algorithm 7.4 Nivasch's cycle detection algorithm

Require: Input initial sequence value X_0, max. iterations M

    Create empty stack S

    Let $x \longleftarrow X_0$

    Add pair $(x, 0)$ to the stack

    for i from 1 to M **do**

        Let $x \longleftarrow F(x)$

        Search x by dichotomy in the (sorted) first component

        if (x, j) is found **then**

            Output 'Collision between i and j'

            Exit

        else

            We know that $S_k < x < S_{k+1}$

            Truncate S after S_k

            Add pair (x, i) at the end of S

        end if

    end for

    Output `Failed`

Another advantage of Nivasch's algorithm compared to Floyd's or Brent's algorithms is that it is guaranteed to directly output the cycle length ℓ, rather than a multiple thereof. Thus, for applications where the exact cycle's length is needed, we can obtain an additional gain of ℓ function's evaluations.

7.1.4.1   Nivasch and the cycle's start

After using Nivasch's algorithm, we may still want to find the entrance of the cycle and obtain a real collision. In that case, we can of course use Algorithm 7.3; however, this algorithm can be improved by using the information already collected by Nivasch's algorithm. Indeed, let us consider the state of the stack in Algorithm 7.4 at the point where the collision is detected. At that point in time i, we know that the current value x is equal to a value on the stack $S_k = (x, j)$. Clearly, thanks to the principle of Nivasch's algorithm, x is the minimal value of the whole cycle. As a consequence, unless $k = 0$, we know that S_{k-1} corresponds to a value outside of the cycle. If $S_{k-1} = (X_t, t)$, this value X_t can be used as a replacement for the start-point X_0 in Algorithm 7.3. When $k = 0$, we use the original start-point X_0. On average, this allows us to recover the start of the cycle a bit faster.

7.2 Analysis of random functions

The good behavior of Floyd's algorithm and Brent's algorithm depends on the specific properties of the function F used to define the considered recursive sequences. We already known that some functions such as $F(X) = X + 1$ (mod 2^n) exhibit extremely bad behaviors. In this section, we present background information about the case of random functions, also called random mappings. A very nice survey about this topic is *Random Mapping Statistics* [FO90] by Flajolet and Odlyzko. It derives a large number of useful facts and statistics about random mappings and their cryptographic applications.

All the facts stated in this section are extracted from [FO90] and apply to the graph of a random function on a set of N elements. They are asymptotic results and give asymptotic equivalents for the various parameters as N tends to infinity.

Each random function F on the set S_N of N elements is also viewed as a directed graph G_F whose vertices are the points of S_N and whose directed edges are of the form $(v, F(v))$, for all vertices v in S_N. Each vertex in the graph has out-degree 1, the in-degree is variable, it can be zero or possibly a large number. Of course, the average in-degree is equal to the average out-degree and thus to 1.

7.2.1 Global properties

Connected components. Asymptotically, the number of connected components in the graph G_F, is $\log N/2$.

We recall that each connected component is obtained by grouping together all the points that can be reached from a start-point by following the edges of G_F, either in the forward or the backward direction.

Terminal nodes. A terminal node in G_F is a vertex v of in-degree 0. Equivalently, it is an element v of S_N not in the image of the function F. Asymptotically, the number of terminal nodes is N/e, where $e = \exp(1)$ is the basis of natural logarithms.

Image points. An image point is a point in the image of F, asymptotically, there are $(1 - 1/e)N$ image points in G_F.

Iterate image points. A point v of S_N is a k-th iterate image point, if and only if, there exists an element x of S_N such that $v = F^{(k)}(x)$, where $F^{(k)}$ denotes k consecutive applications of F. Note that a k-th iterate image point also is a $(k-1)$-th iterate image point. Asymptotically, the number of k-th iterate image points is $(1 - \tau_k)N$, where τ_k is defined by the recursion law:

$$\tau_0 = 0 \quad \text{and} \quad \tau_{k+1} = e^{\tau_k - 1}. \tag{7.6}$$

Cyclic nodes. A cycle node v of G_F is a node that belongs to some cycle of G_F, i.e., v can be obtained as a k-th iterate image of itself for some value of k. Asymptotically the average number of cyclic nodes in a random mapping is $\sqrt{\pi N/2}$.

7.2.2 Local properties

By opposition to global properties which only depend on the graph G_F itself, we now consider local properties obtained when looking at G_F from some random start-point. These local properties are extremely important for us, since they mostly dictate the behavior of cycle finding algorithms.

Tail and cycle lengths. As before, the tail length is the distance from the start-point to the cycle obtained when iterating F. Asymptotically, for a random mapping and a random start-point, the average tail length is $\sqrt{\pi N/8}$. In the same conditions, the average cycle length is equal to the average tail length and also is $\sqrt{\pi N/8}$. This implies that on average, the total length on the sequence obtained by recursively applying F to the start-point up to its first repetition is $\sqrt{\pi N/2}$.

Component size and tree size. On average the size of the connected component containing the starting point is $2N/3$. Similarly, if we restrict ourselves to the non-cyclic part of this component and take the tree containing all points which enter, in the same place, the same cycle as our start-point, then the average size of this tree is $N/3$.

Note that these results imply that a large fraction of the points in the graph G_F are grouped in a single connected component. This component is often called the giant component of the random mapping.

Predecessor size. Another question we can ask is the number of points which are iterated preimages of our chosen start-point. On average, the number of such predecessors is $\sqrt{\pi N/8}$.

7.2.3 Extremal properties

Another important class of properties within the graph of random mappings is the class of extremal properties.

Longest paths. The first extremal property we would like to determine is the behavior of a random mapping when choosing the worst possible starting point. How long do cycle finding algorithms take in this worst case? The good news is that the expected length of the longest tail, longest cycle and longest path (including both tail and cycle) are of the same order of magnitude $O(\sqrt{N})$ as in the average case. Only the constants differ.

More precisely, asymptotically the expected length of the longest cycle is $c_1\sqrt{N}$ with $c_1 \approx 0.782$. The expected length of the longest tail is $c_2\sqrt{N}$ with $c_2 = \sqrt{2\pi}\log 2 \approx 1.737$. Finally, the expected length of the longest path is $c_3\sqrt{N}$ with $c_3 \approx 2.415$. It is remarked in [FO90] that quite interestingly, $c_3 < c_1 + c_2$. As a consequence, for a non-negligible fraction of random mappings, the longest tail does not lead to the longest cycle.

Giant component. Other noteworthy properties are the average size of the giant component and of the largest tree in the graph of a random mapping. Due to the results on the average component size and average tree size for a random start-point, we already know that these sizes are respectively larger than $2N/3$ and $N/3$. Asymptotically, the giant component is of size $d_1 N$ and the largest tree of size $d_2 N$, with $d_1 \approx 0.758$ and $d_2 \approx 0.48$.

7.3 Number-theoretic applications

7.3.1 Pollard's Rho factoring algorithm

Pollard's Rho is a very practical factoring algorithm introduced in 1975 by Pollard [Pol75]. This algorithm aims at finding relatively small factors of an integer N. It heavily uses the algorithms of Floyd or Brent in order to find cycles in sequences defined by a recursion formula. With Pollard's Rho, the sequence that we consider needs some additional property. More precisely, if we denote this sequence by X, we require that for any prime factor p of N, the sequence $X \pmod{p}$ is also defined by a recursion formula. This limits the possible choices for the function F used to compute the sequence. To ensure this property and have correctly defined sequences modulo each prime factor, a good approach is to choose for F a polynomial. In fact, the most common choice is to use:

$$F(x) = x^2 + c \pmod{N}, \tag{7.7}$$

for some constant c.

This may seem surprising because the analysis of sequences defined by a recursion formula is done under the assumption that F is a random function, which $x^2 + c$ is not. However, Pollard's Rho algorithm is used routinely and works extremely well. So despite the apparent inconsistency, this choice of

F is a good one. A partial explanation is given in [FO90]. More precisely, they remark that the graph of $F(x) = x^2 + c$ is very specific, each node but one (the value c) has either 0 or 2 predecessors. As a consequence, they also analyze random binary function graphs where each node has in-degree 0 or 2 and find that the multiplicative constants appearing in the graph statistics of the previous section are affected by this change but that their orders of magnitude are preserved.

Once we choose for F a polynomial, we can easily check that for the sequence defined by:

$$X_{i+1} = F(X_i) \pmod{N}, \qquad (7.8)$$

reduction modulo p for any factor p of N gives the same recursion formula (modulo p instead of modulo N) for the reduced sequence.

For a prime divisor p of N, let $X^{(p)}$ denote the sequence $X \pmod{p}$. We know that $X^{(p)}$ satisfies the recursion:

$$X_{i+1}^{(p)} = F(X_i^{(p)}) \pmod{p}. \qquad (7.9)$$

Using the usual analysis for these sequences, we expect that $X^{(p)}$ cycles with parameters s and ℓ of the order of $\sqrt{p}$. In that case, Floyd's or Brent's algorithm find a cycle in time $O(\sqrt{p})$, assuming that we can test for collisions within the sequence $X^{(p)}$. To see that it suffices to remark that:

$$X_i^{(p)} = X_j^{(p)} \Leftrightarrow \mathrm{GCD}(X_i^{(p)} - X_j^{(p)}, N) \neq 1. \qquad (7.10)$$

Thus, we can efficiently test for collisions using GCD computations. Moreover, when a collision occurs modulo p, there is no special reason to have collisions modulo the other factors of N. As a consequence, as soon as a collision occurs, we usually recover a proper factor of N. Pollard's Rho using Brent's collision finding algorithm is given as Algorithm 7.5.

Of course, Pollard's Rho can also be used in conjunction with Floyd's algorithm. In both cases, one costly part of the method is that each equality test requires a GCD computation and, thus, becomes expensive. In practice, this number of GCD computations can be lowered. The idea is to multiply together several consecutive values of the form $X_i^{(p)} - X_j^{(p)}$ corresponding to several equality tests modulo p. Then, we compute a single GCD of the resulting product with N. If this GCD is 1, all the equality tests fail. If the GCD is a proper factor, we are done. However, if the GCD is N, we need to backtrack and redo the equality tests one at a time, hoping for a proper factor of N. The reason behind the need to backtrack is that by grouping several tests together, we increase the probability of having simultaneous collisions for the various factors of N.

Pollard's Rho is a very interesting application of cycle finding algorithms, which call for several specific comments. The first comment is that with Pollard's Rho, there is no need to compute the parameters s and ℓ of the sequence $X^{(p)}$. Instead, it suffices to find a collision by two values $X_i^{(p)}$ and

---

Algorithm 7.5 Pollard's Rho factoring algorithm

Require: Input number to factor N, parameter c , max. iterations M

  Let $x \longleftarrow 0$
  Let $y \longleftarrow x$
  Let trap $\longleftarrow 0$
  Let nexttrap $\longleftarrow 1$
  for i from 1 to M **do**
    Let $x \longleftarrow x^2 + c \bmod N$
    Let $f \longleftarrow GCD(x - y, N)$
    if $f \neq 1$ **then**
      if $f = N$ **then**
        Output 'Fails: collision modulo N'
        Exit
      else
        Output 'Found f as factor of N'
        Exit
      end if
    end if
    if $i =$ nexttrap **then**
      Let trap $\longleftarrow$ nexttrap
      Let nexttrap $\longleftarrow 2 \cdot$ trap
      Let $y \longleftarrow x$
    end if
  end for
  Output Failed

$X_j^{(p)}$ which can be arbitrarily located. In particular, we do not need to use Algorithm 7.3 to find the start of the cycle. A puzzling question concerning Pollard's Rho algorithm is whether we can devise a variation of this algorithm based on a traditional birthday-based algorithm. In other words, given two lists of numbers L_1 and L_2 with comparable sizes and an integer N, can we efficiently discover two numbers x_1 in L_1 and x_2 in L_2 such that $\text{GCD}(x_2 - x_1, N) \neq 1$. This is a difficult problem which was solved by P. Montgomery in his thesis as a tool to improve elliptic curve factorization [Mon92].

7.3.2 Pollard's Rho discrete logarithm algorithm

As many factoring algorithms, Pollard's Rho can be adapted to compute discrete logarithms in arbitrary cyclic groups. Let G be a multiplicative group of prime[2] order p, let g be a generator of G and let h be an element of G. Solving the discrete logarithm problem corresponds to finding an integer α such that $h = g^\alpha$. In order to apply a cycle finding algorithm, we choose a random looking function F on G. To construct the function F, we first partition G into three disjoint subsets G_1, G_2 and G_3, approximately of the same size, preferably with 1 not in G_1. Then, we define F as follows:

$$F(x) = \begin{cases} x^2 & \text{if } x \in G_1, \\ gx & \text{if } x \in G_2, \\ hx & \text{if } x \in G_3. \end{cases} \tag{7.11}$$

Then, we define the recursive sequence X, by letting $X_0 = 1$. Note that if $1 \in G_1$, 1 is a fixed point of F. Thus in that case, we would need to choose a different start-point. This is why it is preferable to have $1 \notin G_1$. As usual, we expect, after $O(\sqrt{p})$ steps, to detect a cycle in F and a collision $X_i = X_j$. By itself, this collision does not suffice to recover α and we need to compute some additional information about the sequence X.

In order to do this, let us consider the function ϕ from $H = \mathbb{Z}/p\mathbb{Z} \times \mathbb{Z}/p\mathbb{Z}$ to G defined by:

$$(a, b) \longrightarrow g^a h^b. \tag{7.12}$$

This function is a surjection onto G and each element of G has p distinct preimages in H. Moreover, given two distinct elements in H with the same image in G, we can compute α. Indeed, if $\phi(a, b) = \phi(a', b')$ then $g^{a+\alpha b} = h^{a'+\alpha b'}$ and $\alpha = \frac{a'-a}{b'-b} \pmod{p}$.

The next step is to lift the sequence X to H and to find a sequence Y of elements in H such that $X = \phi(Y)$. This can be done using the following

---

[2]For a general treatment about discrete logarithms and a justification for restricting ourselves to groups of prime order, see Chapter 6, Section 6.4.1.

function G from H to H:

$$G(a, b) = \begin{cases} (2a, 2b) & \text{if } \phi(a, b) \in G_1, \\ (a+1, b) & \text{if } \phi(a, b) \in G_2, \\ (a, b+1) & \text{if } \phi(a, b) \in G_3, \end{cases} \qquad (7.13)$$

It is easy to verify that $\phi(G(a, b)) = F(\phi(a, b))$ for all elements (a, b) in H. As a consequence, the sequence Y defined by choosing $Y_0 = (0, 0)$ together with the recursion law $Y_{i+1} = G(Y_i)$ satisfies $X = \phi(Y)$.

Of course, Y is a sequence on a much larger set than X, as a consequence, we do not expect to have a collision in the sequence Y as early as in X. Thus, the collision returned for the sequence X by the cycle finding algorithm is usually not a collision for Y. As a consequence, we obtain two different preimages from the same element of G and can compute α.

Note that if we have a collision on Y itself, the discrete logarithm can also be recovered for a slightly higher cost by finding the entrance of the cycle. The only exception occurs if Y collides on its initial value Y_0.

7.3.3 Pollard's kangaroos

Together with his Rho method, Pollard introduced another algorithm for computing discrete logarithms when some additional information is known. More precisely, he considers the problem $g^x = h$ in G and wants to recover x from g and h, if we know in addition that x belongs to some interval $[a, b]$. In this context, using Pollard's Rho in the group G, without taking in this extra information, is not always a good idea. In particular, when $b - a$ is smaller than the square root of the size of G, it is more efficient to recover x by brute force, try all values in $[a, b]$. To deal with this problem, Pollard proposes to use what he calls kangaroos. This method is also known as Pollard's lambda algorithm.

A kangaroo in Pollard's method is a sequence of elements in G whose logarithm increases by successive jumps. In order to apply the method, we need two kangaroos, a tame one, whose start-point is g^a and a wild one, with start-point is h. The first kangaroo is called tame, because the logarithm of the corresponding sequence is always known. To precisely define the kangaroos, we need a jump function j that associates to each element of G a positive number upper bounded by $\sqrt{b-a}$. A classical method is to divide G in k subset of roughly equal sizes $S_0, \ldots S_{k-1}$, with $k = \lfloor \log_2 b - a/2 \rfloor$, and for x in S_i to define $j(x) = 2^i$. Once the jump function j is fixed, we let $F(x) = x \times g^{j(x)}$.

The tame kangaroo is the sequence defined by:

$$T_0 = g^a \qquad (7.14)$$
$$T_{i+1} = F(T_i).$$

The wild kangaroo is the sequence defined by:

$$W_0 = h \qquad\qquad (7.15)$$
$$W_{i+1} = F(W_i).$$

While computing the tame kangaroo sequence, it is important to keep track of the discrete logarithm of T_i, i.e., of the value $a + \sum_{k=0}^{i-1} j(T_k)$. As soon as the discrete logarithm goes beyond b, we abort the sequence T and recall the final position T_n, together with its discrete logarithm l_n, i.e., $T_n = g^{l_n}$. After this final point is known, we start the wild sequence keeping track of the sum of the jumps taken by the wild kangaroo. We abort the wild sequence either when encountering T_n or when the sum of the jumps becomes larger than $b - a$. In the latter case, the algorithm fails. Otherwise, we have found a position W_i with corresponding sum s_i such that $W_i = T_n$, since $W_i = h \times g^{s_i}$ and $T_n = g^{l_n}$ we conclude that:

$$h = g^{l_n - s_i}. \qquad\qquad (7.16)$$

As Pollard's Rho, Pollard's kangaroo method is a generic algorithm. This algorithm has a good probability of success and takes time $O(\sqrt{b-a})$ to compute the discrete logarithm of h; for a rigorous analysis, see [MT09].

7.4 A direct cryptographic application in the context of blockwise security

Blockwise security is a practical concern about the security of modes of operation for block cipher which was discovered independently by two groups of researchers. In [BKN04], it is shown that the practical implementation of secure shell (SSH) which relies on the cipher block chaining (CBC) mode of operation is insecure against a variation on chosen plaintext attacks. At Crypto 2002, in [JMV02], the theoretical idea of blockwise security was introduced. Moreover, in the same paper, it is shown that for several common modes of operation there is a gap between blockwise and ordinary messagewise security. After this discovery, several papers such as [FJMV04], [BT04] or [FJP04] studied this paradigm of blockwise security and came up with blockwise secure modes. The key result is that there exist modes of operation with blockwise security levels essentially equivalent to usual messagewise security levels. More precisely, these modes of operation are secure up to the birthday paradox bound and after that become insecure. In this section, after some brief reminders about the CBC mode of operation and its security both as a messagewise mode and as a blockwise mode, we study the behavior of this mode beyond the birthday paradox bound and show that in the blockwise security model, we can use Brent's algorithm to devise attacks which work more efficiently than their messagewise counterparts.

7.4.1 Blockwise security of CBC encryption

In Section 6.1, we considered the security of CBC encryption, viewed as a messagewise process, with the plaintext provided as a whole and the corresponding ciphertext returned as a whole. However, in practice, implementing this approach may become problematic, since the encryption box needs to be able to internally store whole messages, which can potentially be very long. For this reason, practitioners often use CBC encryption as a blockwise mode. In that case, the plaintext message is transmitted block by block, and each encrypted block is immediately returned. However, as noticed in [BKN04] and [JMV02], in this blockwise model, CBC encryption is no longer secure. Indeed, an active attacker may observe the current ciphertext block $C^{(i)}$ before submitting the next block of plaintext $P^{(i+1)}$. As a consequence, this allows him to choose at will the value v which enters E_K by letting $P^{(i+1)} = v \oplus C^{(i)}$. This property can be used in several ways to show that CBC encryption is not secure as a blockwise mode of operation, for example by creating a simple distinguisher able to make the difference between a CBC ciphertext and a random message of the same length. To create this distinguisher, it suffices to make sure that the same value v is to be encrypted twice in CBC mode. Thus, a CBC encrypted message reflects this equality. Of course, in a random message, the equality has no special reason to hold.

Interestingly, there is a simple countermeasure to fix this problem. This is called Delayed CBC encryption and is proven to be secure in [FMP03]. The basic idea on Delayed CBC is very simple, once a ciphertext block $C^{(i)}$ has been computed, instead of returning it immediately, the encryption device waits until it gets $P^{(i+1)}$ and then returns $C^{(i)}$. For the last encrypted block, the encryption device returns it upon reception of a special block indicating that the message is complete, this special block is not encrypted, it only serves a bookkeeping purpose. The rationale behind Delayed CBC is that the attacker can no longer control the block cipher inputs. Thus, Delayed CBC prevents the simple attack that breaks plain CBC.

7.4.2 CBC encryption beyond the birthday bound

Beyond the security bound of CBC encryption, the birthday paradox comes into play. As a consequence, among the ciphertext blocks, we expect to find at least one collision, say $C^{(i)} = C^{(j)}$. If we now replace each block of ciphertext in this equality by its expression, we find that:

$$E_K(C^{(i-1)} \oplus P^{(i)}) = E_K(C^{(j-1)} \oplus P^{(j)}), \quad \text{and thus} \qquad (7.17)$$
$$C^{(i-1)} \oplus P^{(i)} = C^{(j-1)} \oplus P^{(j)} \quad \text{since } E_K \text{ is a permutation.}$$

We can rewrite this as $P^{(i)} \oplus P^{(j)} = C^{(i-1)} \oplus C^{(j-1)}$. As a consequence, given one collision, we learn the XOR of two blocks of plaintext. As the length of the message grows, more and more collisions are found and thus more and more information is learned about the plaintext message.

For a cryptanalyst to use this attack, he must be able to efficiently detect collisions among the blocks of ciphertext. One very common technique is to proceed as in Chapter 6 and sort the ciphertext blocks in order to detect these collisions. However, this requires either a large amount of main memory or at least a fast auxiliary storage device. For 64-bit block ciphers, the cryptanalyst needs to store and sort about 2^{32} blocks or equivalently 2^{35} bytes. In fact, the situation is slightly worse, since the cryptanalyst also needs to keep track of the initial block positions when sorting in order to know the plaintext blocks involved in the equation derived from the collision. If there is not enough memory to sort a table containing both the ciphertext value and the original position, an alternative is to keep the unsorted values somewhere on an auxiliary storage device, to sort the ciphertext blocks without indices and then to scan the auxiliary device to locate the original position of the colliding blocks once their common value is known.

Even with this approach, we need 2^{35} bytes of fast memory. Despite the quick progress of computer hardware, 32 Gbytes is still a large amount of memory, only available on dedicated computers. Thus, in practice, even when CBC encryption is used with a 64-bit block cipher, say Triple-DES, putting the attack into play requires some important computing effort from the cryptanalyst. For this reason, among others, CBC encryption with Triple-DES is still widely encountered.

7.4.3 Delayed CBC beyond the birthday bound

After considering the security of messagewise CBC beyond the birthday bound, we now turn to blockwise security of Delayed CBC encryption beyond this bound. In order to construct this attack, we are going to use a trick similar to the one used for attacking ordinary blockwise CBC. More precisely, whenever we receive a block of ciphertext, we determine the next block of plaintext as a function of this block and send it back to CBC encryption. Due to the one block delay, we can only start this from the second block of plaintext. For the first plaintext block, we can submit any value of our choice, such as the all-zeros block. Starting from the second block, we choose our next plaintext block as:

$$P^{(i)} = P^{(i-1)} \oplus C^{(i-2)}. \tag{7.18}$$

Even with Delayed CBC, both values P_{i-1} and C_{i-2} are effectively known at the time we need to submit P_i. Now, with this specific choice for plaintext blocks, the ciphertext values are determined by the following equation:

$$C^{(i)} = E_K(C^{(i-1)} \oplus P^{(i-1)} \oplus C^{(i-2)}) \tag{7.19}$$
$$= E_K(E_K(P^{(i-1)} \oplus C^{(i-2)}) \oplus (P^{(i-1)} \oplus C^{(i-2)})).$$

In other words, if we let Z_i denote the input to the block cipher in round i, i.e., $Z_i = P^{(i)} \oplus C^{(i-1)}$, we see that Z_i satisfies a recursion law:

$$Z_i = E_K(Z_{i-1}) \oplus Z_{i-1}. \tag{7.20}$$

Quite interestingly, xoring the input and the output to a block cipher is a pretty standard construction used to build a pseudo-random function from a pseudo-random permutation, known as Davies-Meyer construction. This Davies-Meyer construction is often used in hash functions. For example, it is encountered in SHA-1. If we define the function F by $F(x) = x \oplus E_K(x)$, the recursion law becomes $Z_i = F(Z_{i-1})$.

Using the analysis of Section 7.2, we know that the sequence Z loops back to itself after about $O(2^{n/2})$ steps. Moreover, we can consider using a cycle detection algorithm to discover this loop.

7.4.3.1 Floyd's algorithm

Remember that to use Floyd's algorithm, we need to compute in parallel the sequence Z_i and the sequence $W_i = Z_{2i}$ in order to find an equality $Z_i = W_i$. Without going further into the details, we see that this is troublesome in our context of blockwise CBC encryption. Indeed, since we do not control the initial values C_0 used during encryption, there is no way to compute both sequences Z_i and W_i at the same time. Of course, we can always store Z and compare Z_i with Z_{2i} at each step, but this requires a large amount of memory, which is exactly what we are trying to avoid. Still, note that this approach would be better than sorting and could even work with slow memory or disk storage.

7.4.3.2 Brent's algorithm

On the contrary, with Brent's algorithm, we immediately see that the loop we induced in Delayed CBC encryption can easily be detected. Following our description from Section 7.1.2, whenever the index i is a power of 2, we store this value Z_{2^t} and then compare it to the subsequent values of Z_i up to the next power of two. This can be done without any need to compute the same sequence twice or to control initial values.

In some sense, detecting a loop is already an attack. Indeed, this can be seen as a real or random distinguisher. If the output is real, i.e., comes from a blockwise CBC encryption, we find a cycle in time $O(2^{n/2})$ with high probability. If the output is random, then Brent's algorithm almost never finds a new value equal to the stored value. However, we can strengthen the attack and do more than this basic distinguisher. In fact, we can use the cycle to obtain the encryption of any value v of our choice under the underlying block cipher E_K. For this, when storing Z_{2^t}, which is a block cipher input, we also store the corresponding output $C^{(2^t)}$. Due to the one block delay, $C^{(2^t)}$ is not available immediately, but we store it whenever we receive it. When the collision is detected, we have a new input Z_i equal to Z_{2^t}. Of course, this equality between inputs to the block cipher implies equality between the output. So we know in advance that $C^{(i)} = C^{(2^t)}$. Thanks to this knowledge, we can set $P^{(i+1)} = C^{(i)} \oplus v$ and thus obtain the encrypted value $E_K(v)$.

This shows that thanks to Brent's algorithm, Delayed CBC encryption as a blockwise mode of operation is (slightly) more vulnerable to attacks beyond the birthday paradox bound than ordinary CBC encryption used as a messagewise mode of operation.

7.4.3.3 Nivasch's algorithm

In this context of blockwise attacks, Nivasch's algorithm can be very useful. Indeed, it allows us to reduce the amount of required ciphertext before the cycle is detected. In fact, using Algorithm 7.4 as presented in Section 7.1, we are guaranteed to require at most $s + 2\ell$ blocks of ciphertext. Of course, it would be interesting to further reduce this number. Clearly, the first collision occurs at position $s+\ell$ and we cannot use fewer than $s+\ell$ blocks of ciphertexts to detect a collision. Following Nivasch in [Niv04], it is possible to devise a variation of Algorithm 7.4 that uses some additional memory and can detect the cycle with $s + (1 + \alpha)\ell$ blocks, for arbitrary small values of α. This algorithm uses a technique called partitioning and its memory requirements increase as α decreases.

7.5 Collisions in hash functions

Another application of cycle detection algorithms is the search for collisions in hash functions, especially when their output length is too short. For example, given a hash function with a 128-bit output, thanks to the birthday paradox, we can find a collision with roughly 2^{64} evaluations of the hash function. However, it is clearly infeasible with current computers to store and sort the corresponding list of 2^{64} hash values in order to discover the collision. Thus, it is natural to ask how to use cycle detection algorithms to obtain a collision. The first answer is to simply invoke the cycle detection algorithm on a recursive sequence, where each value is obtained by hashing the previous value. Yet, this answer is not satisfactory for several reasons. The first reason is that this approach can only produce a meaningless collision and output two random looking messages with the same hash. Of course, it would be much preferable for a cryptanalyst to obtain two meaningful messages with different meanings and the same hash. The second reason is that the cycle detection algorithms presented in Section 7.1 are inherently sequential. And while performing 2^{64} operations is feasible with today's computers, performing 2^{64} sequential operations on a single computer is a completely different matter.

7.5.1 Collisions between meaningful messages

Finding a collision between meaningful messages is a now classical trick, first introduced in [Yuv79]. The idea is to start by choosing two messages M and M' with different meanings. Then in each message, we need to find a number of places where safe modifications can be performed. Here, we call safe modification any change which does not alter the meaning of the message. There are usually many possible safe modifications in a human readable message. We could replace a space by a line feed, a non-breaking space or even two spaces. Similarly, at some points in a message, punctuation can be added or removed; upper-case and lower-case letters can be exchanged without altering the message's meaning. Given t places of safe modification for a message M, it is possible to construct a set of 2^t different messages with the same meaning. Note that for the hash function the 2^t messages are genuinely different and we obtain 2^t random looking[3] hash values. If we proceed in the same way, we also obtain 2^t different copies of M'. Clearly, for a n-bit hash function, with $t > n/2$ we expect the existence of a collision between a copy of M and a copy of M'.

The same idea of safe modifications can also be used in conjunction with a cycle finding technique. However, we need some extra care to avoid uninteresting collisions between two copies of M or two copies of M'. Given t points of safe modification of M, for any t-bit number i, we denote by M_i the copy obtained by setting the specific value of each point of modification according to the value of a corresponding bit in the binary expansion of i. Similarly, we denote by M'_i the i-th copy of M'. Using this notation, we can construct a function F from $t + 1$ to n bits as follows:

$$F(v, i) = \begin{cases} M_i & \text{if } v = 0, \\ M'_i & \text{if } v = 1. \end{cases} \tag{7.21}$$

Adding a truncation from n bits to $t + 1$, we can obtain a sequence of hash values of meaningful messages and look for a collision. However, this direct approach with $t \approx n/2$ as above usually yields a trivial collision. Indeed, if two hash values collide after truncation to $t+1$, at the next iteration we hash a previously encountered message and enter a cycle with a collision between the hash values of twice the same message. To avoid this bad case, we need more points of safe modification, namely we need to choose $t = n - 1$. With this choice, the entrance of the cycle corresponds to a genuine collision. Moreover, with probability $1/2$ this genuine collision is between a copy of M and a copy of M'. If not, we can simply restart the algorithm with a different start-point.

---

[3]Unless the hash function is badly broken.

7.5.2 Parallelizable collision search

Since cycle finding algorithms are inherently sequential, in order to find collisions using a parallel computation, we need a different approach. This approach cannot look for a cycle within a single recursively defined sequence, but instead needs to rely on the computation of several independent sequences. We already introduced an algorithm of this kind to compute discrete logarithms in a short interval: Pollard's Kangaroo algorithm (see Section 7.3.3). However, to use this idea in a parallelizable way, we need to adapt it. When running on a parallel computer, it is essential to reduce the amount of communication and synchronization between parallel processes. A common approach when looking for a specific value or property among a large set of data is to use a large number of computers to create likely candidates and then to report these candidates to a dedicated server which performs the final checking. For a collision search, we need to report values which are likely to collide. This can be achieved using the distinguished point technique from [QD90] and [vW96]. Note that in distinguished point algorithms, one node in the parallel computer plays a special role. This node receives distinguished points from all other nodes, checks for collisions within the set of distinguished points using a sort based birthday algorithm as in Chapter 6 and pulls back the collisions from this set of distinguished points to the real values. It is very important to devise parallel collision search algorithms in a way that minimizes the communication cost, the memory requirement and the computational overhead of this server node.

In this section, we present one of the numerous possible options for parallelizable collision search on a random function F. We assume that F operates on a set of N elements and that the parallel computer consists of P identical parallel nodes and a special server node. We also choose a set of D distinguished points. For algorithmic convenience, the distinguished points should be easy to recognize and easy to generate. A typical example of set of distinguished points is the set of n-bits integers with d leading zeros. This set contains $D = 2^{n-d}$ elements in a larger set of cardinality $N = 2^n$. Each distinguished point x is easily recognized by the property: $0 \leq x < 2^{n-d}$. Moreover, with this definition of distinguished points, it is very easy to generate random distinguished points at will.

With this setting, our algorithm works as follows. First, the server node allocates a (distinct) distinguished point to each computation node. Given this distinguished point, the computation node uses it as start-point for Nivasch's cycle detection Algorithm 7.4, with a small change. Namely, when Nivasch's algorithm encounters a distinguished point, it aborts and sends a report. The report contains three values: the starting distinguished point, the distinguished end-point and the length of the path between the two points. The server node maintains a list of reports and detects collisions between end-points. If a collision occurs between the values taken by F during the global computation, necessarily this implies the existence of either a cycle on one computation node or a collision between two different nodes. Both possibilities are detected by our algorithm, in the first case, Nivasch's algorithm detects the cycle, in the second case, the server node discovers a collision between two end-points. From this collision, between end-points, the real collision is easily recovered. Indeed, assume that there is a path of length L_1 from a distinguished point S_1 to a end-point P and another path of length L_2 from S_2 to P. Without loss of generality, we can further assume that $L_1 \geq L_2$. Then, starting from S_1, the server node walks the first path, taking exactly $L_1 - L_2$ steps. From that point, it now suffices to walk both paths in parallel, until a collision is reached. Note that, it is always a genuine collision. Indeed, since E is the first distinguished point encountered by walking from S_1, S_2 cannot lie on the first path. As a consequence, the two parallel walks start from different places and end at the same position. Moreover, they have the same length and the collision necessarily occurs after taking the same number of steps on both paths.

Since any collision between two of the computed values of F is detected, the number of values that need to be computed remains $O(\sqrt{N})$, as with sequential cycle detection algorithms. As a consequence, with P processors, the parallelization speed-up is essentially optimal. One caveat is that the computed values are not integrated into the global count until the corresponding distinguished point is reported. Thus the number of distinguished points should be adequately chosen. With D distinguished points, the fraction of distinguished points is N/D. Thus the average length of a path between two distinguished points is essentially N/D. Note that for some start-points, we may enter a cycle. In this case, we cannot nicely define the distance between two distinguished points. Ignoring this technical difficulty, we find that, as a consequence, running our algorithm on P processors allows it to compute PN/D points. In order to have $O(\sqrt{N})$ values of F, we need to choose:

$$D = O(P\sqrt{N}). \qquad (7.22)$$

Since the server node needs to store P paths, i.e., one per computation node, the amount of required memory is $O(P)$. Finally, the running time of the code that extracts the collision on F from a collision between distinguished points is upper bounded by the time needed to walk through two paths, i.e., $O(N/D) = O(\sqrt{N}/P)$.

Note that with the above description and analysis, the computation is not perfectly balanced between processors. Indeed, some paths between distinguished points are shorter than others. To improve this imbalance, an easy patch is to serially run several instances of Nivasch's algorithm on each computation node, in order to average the paths' length. This increases the parameter P and as a consequence the memory required on the server node.

Another remark is that when P becomes large, collisions coming from a cycle within a single path become extremely rare. As a consequence, we may replace Nivasch's algorithm by a simpler one, which simply runs from a distinguished point to the next, possibly aborting when it takes too long to find a distinguished end-point. The only drawback of this approach is that the exact analysis of the algorithm's behavior becomes much more cumbersome.

7.5.2.1 Looking for many collisions

Another interesting aspect of the above parallelizable collision search algorithm is that it can be used to efficiently construct a large number of collisions. Note that the sequential cycle finding algorithms are not very efficient for construction multiple collisions. Indeed, to construct k collisions, we essentially need to repeat the cycle finding algorithm k times, for a total cost $O(k\sqrt{N})$. On the contrary, with the above parallel algorithm, the birthday paradox continues to apply and to play for us after the first collision. As a consequence, the overall cost to construct k collisions is reduced to $O(\sqrt{kN})$. However, we should be careful when choosing the set of distinguished points. In particular, it is clear that each stored path can yield at most one collision. As a consequence, the parameter P should be chosen larger than the number of desired collisions k.

An alternative would be to use arbitrary points instead of distinguished points as start-points for the stored path. However, this approach introduces a new drawback. Indeed, nothing would prevent us to compute twice the same path or, more precisely, to compute two paths where one is a subpath of the other. These two paths would not yield any new collision and would increase the computational cost without any gain.

7.6 Hellman's time memory tradeoff

Hellman's time memory tradeoff is a generic method for recovering the key K of a block cipher E. This method assumes that the cryptanalyst can perform a massive precomputation about E beforehand and memorize some information summarizing this precomputation, in order to help in recovering the key K at a later stage. The goal is to achieve a non-trivial time memory tradeoff for the late stage of the attack. Two trivial possibilities are to either memorize nothing and use brute force to recover K, or to memorize a sorted table containing all pairs $(E_K(0^n), K)$ and to perform a table lookup of $E_K(0^n)$ in order to recover K.

Despite the fact the Hellman's time memory tradeoff uses a large amount of memory, it is presented in this chapter for two main reasons. First, the behavior of Hellman's tradeoff is deeply connected with the analysis of random functions. Second, Hellman's tradeoff is extremely useful for extracting a block cipher key after the blockwise attacks of Section 7.4.

Hellman's algorithm is based on the analysis of the function $F(K) =$

$E_K(0^n)$ from the set of possible keys to itself. Note that if the key size and the block size are different, this function needs to be adapted. If the key size is smaller than the blocksize, we need to truncate the result. If the key size is larger than the blocksize, then we concatenate several encryptions (say choosing the constant blocks, 0^n, $0^{n-1}1$, ...) and truncate the concatenation if necessary. Breaking the block cipher E and finding K is clearly equivalent to computing the inverse of F on the point $E_K(0^n)$ obtained from an encryption of 0^n. Thus, Hellman's tradeoff is usually considered to be a chosen plaintext attack. In fact, it is slightly better than that, since we do not insist on the choice of 0^n. The only requirement is to make sure that there is some fixed value v, for which it is easy to obtain an encryption of $E_K(v)$. For example, if the adversary knows that users routinely encrypt messages with some fixed headers, which is a frequent feature of several word processors, then, assuming a deterministic encryption mode, chosen plaintext attacks may not be necessary. Hellman's algorithm gives another reason why using randomized encryption modes is important when using block ciphers. Indeed, with a randomized mode, the adversary can no longer obtain the encryption of his chosen fixed value and Hellman's method can no longer be used.

7.6.1 Simplified case

To understand the key idea behind Hellman's algorithm, we first consider a very special case and assume that F is a cyclic permutation. In real life, this never happens. However, in this case the analysis is much simpler. With this unlikely hypothesis, we can start from an arbitrary initial key K_0 and precompute the complete sequence given by $K_{i+1} = F(K_i)$. This sequence follows the cycle of F and goes in turn through all possible keys for the block cipher, eventually coming back to K_0 after 2^k iterations, where k is the number of key-bits. Every 2^l steps, with $l = \lfloor k/2 \rfloor$, we memorize the pair $(K_{j \cdot 2^l}, K_{(j-1) \cdot 2^l})$. In other words, we remember the current key together with its ancestor located 2^l steps before. Once we have collected the 2^{k-l} possible pairs around the cycle, we sort these pairs by values and keep this sorted array.

Given access to this array of sorted pairs, we can now invert F quite easily in 2^l steps. Starting from $E_K(0^n)$ for an unknown key K, we look up this value in our array. If it is not present, we apply F and look up again. After t steps, we find a value $F^{(t)}(E_K(0^n))$ in the array. For this value, we read the second number in the pair, thus going 2^l steps backward in the cycle of F. Then, we apply F again $2^l - t - 1$ times and recover the (unique) preimage of $E_K(0^n)$, i.e., the key K.

In this simplified case, Hellman's tradeoff needs to memorize 2^{k-l}, i.e., 2^l or 2^{l+1} with our choice for l, pairs and the running time of the attack stage (omitting the precomputation phase) is $l \cdot 2^l$, when counting the cost of the dichotomy lookups in the sorted array. Note that this technique is very similar to the baby step, giant step algorithm described in Section 6.4.2. Reformulating Hellman's algorithm in this context, we see that to take a single baby step backward, we need to take many baby steps forward and a single giant step backward.

7.6.2 General case

In the general case, F is no longer cyclic and no longer a permutation. Moreover, we expect it to behave like a random mapping. Thus, choosing a random start-point and applying F iteratively quickly leads to a cycle of square-root size. As a consequence, if we want to cover more than a negligible fraction of the keys, we clearly need to use many start-points. Another difficulty is that with several start-points, nothing prevents two different sequences to merge at some point. In fact, any two starting points in the giant component lead to merging sequences. At the latest, the sequences merge somewhere in the cycle corresponding to the giant component. Even worse, finding a preimage for $E_K(0^n)$ is not sufficient to recover K, since preimages are no longer unique. In order to bypass all these problems, Hellman's tradeoff requires more time and more memory than in the simplified case. The usual balance (neglecting logarithmic factors) is $2^{2k/3}$ in terms of both time and memory.

The classical analysis of Hellman's tradeoff assumes that we are computing sequences of length t for m random start-points. For each sequence, the start- and end-points are memorized, sorted by end-point values. In order to find a preimage by F, the same idea as in the simplified case is used. From a start-point, F is applied repeatedly until a memorized end-point is reached. Then the algorithm jumps back to the corresponding start-point and moves forward again. Several cases are possible, in the most favorable we discover K, in the other cases, we fail, either by finding a wrong preimage for $E_K(0^n)$, i.e., a different key that yields the same encryption, or by not even coming back to the point $E_K(0^n)$. The probability of success is of the order of $mt/2^k$ as long as mt^2 remains below 2^k. If we go beyond this limit, the number of collisions between chains becomes very large and the tradeoff becomes unworthy. Taking $m = t = 2^{k/3}$, we find a probability of $2^{-k/3}$ with memory requirements of $O(2^{k/3})$ keys and running time $O(k2^{k/3})$. To improve the probability, Hellman proposes to use $2^{k/3}$ different choices for the function F, by slightly changing the way a ciphertext $E_K(0^n)$ is converted back into a key. Assuming independence between the tables generated for each of these functions, the success probability increases to a constant fraction of the key space with time and memory requirements $2^{2k/3}$ (neglecting logarithmic factors).

Exercises

1. At the end of Floyd's algorithm, prove that t is the smallest multiple of ℓ greater than or equal to s.

2[h]. Given a discrete logarithm problem $y = g^x$ with the additional information that every bit in odd position in x is a 0, devise a generic algorithm to efficiently obtain x, assuming that there are n unknown bits in x.

3. In this exercise, we consider the simultaneous use of several copies of Nivasch's algorithm.

   (a) Given a sequence $X_{i+1} = f(X_i)$ and two different order relations on elements X_i, run in parallel two copies of Nivasch's algorithm defining one with each order relation. We detect the loop as soon as one of the two copies does. Assuming that the result of each different comparison between unequal values behaves randomly, when do you expect this detection to happen?

   (b) What can be achieved with t different copies of Nivasch's algorithm?

   (c) Describe an application where this technique offers a noticeable gain.

4[h]. What is the expected complexity of finding the cycle's start given the complete state of Nivasch's algorithm, after detecting a collision? Can this be improved with t copies of the algorithm?

5. Implement the search of hash collisions using a cycle finding algorithm on a reduced-sized hash function. For example, you can use a truncated copy of SHA.

6[h]. Devise a method for searching many collisions (not multicollisions, just several independent collisions) within a hash function, using a cycle finding technique. Compare with seach for many collisions using a birthday paradox approach with memory.

7. **Open problem.** Find a method to construct multicollisions (at least a 3-multicollision) using a cycle finding approach.

Chapter 8

Birthday attacks through quadrisection

In practice, cryptanalytic attacks based on the birthday paradox are often limited by the amount of memory they require. Indeed, the time and memory requirements of birthday paradox based attacks are roughly the same and for practical purposes, memory costs much more than time. Of course, when looking for collisions involving functions, we can use the techniques of Chapter 7, however, their range of applicability is limited. In this chapter, we study other specific instances of the birthday-based attacks for which the techniques of Chapter 7 cannot be applied, but which, nonetheless can be tackled using less memory than the generic methods of Chapter 6. These techniques presented here reduce the required amount of memory, without significantly increasing the running times. In fact, even when there is enough memory available to perform a generic birthday attack, using these variations may improve performance, due to cache effects.

8.1 Introductory example: Subset sum problems

To serve as an introduction to the techniques presented in this chapter, we present an algorithm of Schroeppel and Shamir [SS81] first presented in 1979. This example studies the problem of finding all the solutions to a given subset sum problem. We first recall the definition of a subset sum problem.

DEFINITION 8.1   *A subset sum problem consists, given n positive integers x_i and a target sum s, in finding all the solutions to the equation:*

$$\sum_{i=1}^{n} e_i x_i = s, \tag{8.1}$$

where the e_i values are integers equal to either 0 or 1.

   The subset sum problem also has an associated decision problem where the goal is to say whether a solution exists or not.

The subset sum problem is also called the **knapsack** problem.

8.1.1   Preliminaries

Clearly, the subset sum problem can be broken by brute force. It suffices to compute the 2^n different sums. This requires essentially no memory and has a running time of 2^n evaluations of a sum of n terms. One can do better with a simple application of the birthday paradox: first split the x values into two sets of roughly equal sizes, compute L_1 the list of all possible sums for the first set and L_2 the list obtained by subtracting from s all the possible sums for the second set. Any collision between L_1 and L_2 corresponds to an equation:

$$\sum_{i=1}^{\lfloor n/2 \rfloor} e_i x_i = s - \sum_{i=\lfloor n/2 \rfloor+1}^{n} e_i x_i, \tag{8.2}$$

thus yielding a solution to the knapsack problem. As usual, this birthday-based attack has a runtime $O(n2^{n/2})$ and require to store $2^{\lfloor n/2 \rfloor}$ partial sums in memory. For values of n where the brute force attack is at the edge of possible computations, say $n = 80$ it can be argued whether the birthday-based attack is better than the brute force attack. In practice, it is easier to perform 2^{80} computations in a distributed effort on many small computers or to find a large computer with enough memory to store 2^{40} numbers?

With the results of Chapter 7 in mind, it is natural to try solving such a knapsack problem using a cycle finding algorithm. At first, this approach seems reasonable. Indeed, assuming for simplicity that n is even, we can easily define a function F on $n/2 + 1$ bits that maps values starting by 0 to a sum of the first $n/2$ numbers in the knapsack as in the left-hand side of Equation (8.2). The same function would map values starting by a 1 to the target value s minus a sum of the last $n/2$ numbers, as in the right-hand side of the same equation. Clearly, a random collision on F yields a solution to the knapsack problem whenever the colliding values differ on their first bits, which should heuristically occur with probability roughly $1/2$. As a consequence, it is natural to try using cycle finding algorithms to obtain a collision on F. However, in many cryptographic applications of knapsacks, the values of the knapsack elements are large numbers, of n bits or more. Otherwise, given a possible value for the sum s there would be a huge number of solutions. As a consequence, F is an expanding function which maps $n/2 + 1$ bits to approximately n bits. For this reason, it cannot directly be used repeatedly to construct an ultimately periodic sequence. Instead, we need to use some function G to truncate the n bits back to $n/2 + 1$. Of course, from a starting point y_0 we can easily define a sequence, using the recursion formula:

$$y_{i+1} = G(F(y_i)). \tag{8.3}$$

Clearly, this sequence usually leads to a cycle which can be found using a cycle detection algorithm. Yet, this cycle does not usually produce any real collision and cannot yield a knapsack solution.

The reason for this phenomenon is that by writing down the sequence y, we also implicitly defined another sequence z by:

$$z_0 = F(y_0) \quad \text{and} \quad z_{i+1} = F(G(z_i)). \tag{8.4}$$

This implicit sequence z also yields a cycle. However, z values are short numbers on $n/2 + 1$ bits, while y values are long numbers on n bits. As a consequence, with overwhelming probability the sequences enter a cycle through a z-collision and not through a y-collision. This means that we eventually find an initial collision $z_i = z_j$ which defines the entrance of the cycle. This collision of course leads to a collision on the subsequent y values, i.e., $y_{i+1} = y_{j+1}$. However, both values result from the same sum expression and thus cannot offer a solution to the knapsack problem. It is worth noting that for knapsack problems based on small knapsack elements, G would no longer be necessary and this approach would work.

8.1.2   The algorithm of Shamir and Schroeppel

To overcome this difficulty, Shamir and Schroeppel devised an algorithm which uses less memory than the birthday attack but still requires some non-negligible amount of memory. Before presenting the algorithm of Shamir and Schroeppel, it is useful to first remember from Section 6.3 that the birthday attack can be implemented in two different ways. The first possibility is to store a sorted copy of L_1, then to generate on the fly elements of L_2 and search them by dichotomy in L_1. The second possibility is to sort both L_1 and L_2 then to read both lists sequentially, advancing in turns either in L_1 or in L_2, until collisions are found. In the generic setting, the second possibility is the more costly in terms of memory, because two lists[1] need to be stored. However, in order to implement this option, we do not really need to store L_1 and L_2, instead we only need to be able to go through L_1 (and L_2) step by step in increasing order. The key idea introduced by Shamir and Schroeppel shows that for subset sum problems, we can produce the elements of L_1 in increasing order without fully storing L_1. We also need to produce the elements of L_2 in increasing order. Since L_2 is constructed by subtracting a partial subset sum from s, it means that we need to produce the corresponding partial sums in *decreasing* order. This can be achieved through a straightforward adaptation of the method used with L_1.

Assume for now than we are given a subset of the knapsack elements and denote them by $y_1, \ldots, y_l$. We need a memory efficient subroutine to produce

---

[1]Moreover, when n is odd, the second list is twice as long as the first.

all the sums for this subset in either increasing or decreasing order. For this, we first split l roughly in half, letting $l = l_1 + l_2$ and construct two sets:

$$Y_1 = \left\{ \sum_{i=1}^{l_1} e_i y_i \mid \forall (e_1, \cdots, e_{l_1}) \in \{0,1\}^{l_1} \right\} \quad \text{and} \quad (8.5)$$

$$Y_2 = \left\{ \sum_{i=l_1+1}^{l} e_i y_i \mid \forall (e_{l_1+1}, \cdots, e_l) \in \{0,1\}^{l_2} \right\}.$$

Any knapsack element can be uniquely written as the sum of an element from Y_1 and an element from Y_2.

To solve our problem we first ask the following question: "Given two such sums, $\sigma = \gamma_1 + \gamma_2$ and $\sigma' = \gamma_1' + \gamma_2'$ what can we say about the relative orders of σ and σ' from the relative orders between the γ values?" Clearly, if $\gamma_1 \geq \gamma_1'$ and $\gamma_2 \geq \gamma_2'$ then $\sigma \geq \sigma'$. Similarly, when both γs are smaller, then σ is smaller. However, when if $\gamma_1 \geq \gamma_1'$ and $\gamma_2 \leq \gamma_2'$, prediction is not easy.

To make good use of this partial information, we first sort Y_2, then for each number in Y_1, we add it to the first element of Y_2 and memorize the sum together with its decomposition. For efficiency, we memorize these sums in a binary search tree. This allows us to retrieve these sums in sorted order. After this initialization, we proceed as follows, at each step take the smallest sum from the tree and produce it as the next element of $Y_1 + Y_2$. After that, we look at the decomposition, update the sum and put it back in the tree. The update is done by keeping the same number from Y_1 and by moving to the next number in Y_2 (which is easy, since Y_2 is sorted). After the update, we put the new value back into the tree, unless the end of Y_2 was reached. Using a self-balancing tree, since the size of tree is the size of Y_1, the cost of insertion and deletion is guaranteed to be $O(\log |Y_1|)$.

To generate the values in decreasing order, two minor changes[2] are required, start from the end of Y_2 instead of the beginning and take the last tree element at each step instead of the first one. In both versions of the algorithm, each element of Y_1 is paired with each element of Y_2 and thus the set $Y_1 + Y_2$ is completely constructed. Putting together one copy of algorithm going in increasing order for $Y_1 + Y_2$ with another copy going in decreasing order for $Y_3 + Y_4$ (and thus in increasing order for $s - Y_3 - Y_4$) allows us to find all occurrences of s in $Y_1 + Y_2 + Y_3 + Y_4$. From a complexity point-of-view, this requires a running time $O(n2^{n/2})$ and a memory of $O(2^{n/4})$ integers. The running time is the same as the running time of the generic birthday attack, but the required memory is much lower. Note that the constants implicit in the O notations depend on the value of n modulo 4. The best case occurs when n is a multiple of 4.

---

[2] Alternatively, we can simply use the exact same algorithm as before, using the reversed order instead of the natural order on integers.

Algorithm 8.1 Initialization of Shamir and Schroeppel algorithm

---

Require: Input ℓ knapsack elements $y_1, \ldots, y_\ell$

Let $\ell_1 \longleftarrow \lfloor \ell/2 \rfloor$ and $\ell_2 \longleftarrow \ell - \ell_1$

Construct an array Y_2 containing the 2^{ℓ_2} possible sums of $y_{\ell_1+1}, \ldots, y_\ell$.

Sort the array Y_2

Create an empty AVL-tree T and a set a 2^{ℓ_1} free nodes for this tree structure

for $(e_1, \cdots, e_{\ell_1})$ in $\{0, 1\}^{\ell_1}$ **do**

    Compute $Y_1 \longleftarrow \sum_{i=1}^{\ell_1} e_i y_i$

    Take the next free node N

    Let $N_{\text{value}} \longleftarrow Y_1$, $N_{\text{index}} \longleftarrow 0$ and $N_{\text{sortKey}} \longleftarrow Y_1 + Y_2[0]$ {Insert into the fields of N the value Y_1, the index 0 and the sorting key $Y_1+Y_2[0]$}

    Insert N into T {according to the value of the sorting key}

end for

Output T and Y_2 for use in Algorithm 8.2

Algorithm 8.2 Get next knapsack sum with Shamir and Schroeppel algorithm

---

Require: Input array Y_2 and current state of tree T

If T is empty, Output 'Finished'

Remove the smallest node of T, let N be this node

Let $S \longleftarrow N_{\text{sortKey}}$

Let $N_{\text{index}} \longleftarrow N_{\text{index}} + 1$

if $N_{\text{index}} < 2^{\ell_2}$ **then**

    Let $N_{\text{sortKey}} \longleftarrow N_{\text{value}} + Y_2[N_{\text{index}}]$

    Insert modified node N back into the tree T

end if

Output sum S and updated tree T

Clearly, the algorithm of Shamir and Schroeppel does not make use of the strong internal structure of the list Y_1, Y_2, Y_3 and Y_4 which are built by adding together knapsack elements. However, taking advantage of this extra structure to either speed up the search for a knapsack solution or further reduce the required memory is a nice open problem. Not using the internal structure of Y_1, Y_2, Y_3 and Y_4 also means that the algorithm of Shamir and Schroeppel has a large range of applicability. We study it in the sequel of this chapter together with other algorithms which present the same kind of tradeoff between time and memory.

8.2 General setting for reduced memory birthday attacks

The idea of splitting a problem in four parts instead of two for the usual birthday attacks, may be used in a variety of applications. In the general setting, we are given four sets L_1, L_2, L_3 and L_4 of elements from a group G with group operation $\odot$, together with a target group value g. The problem is then to find all solutions $(g_1, g_2, g_3, g_4) \in L_1 \times L_2 \times L_3 \times L_4$ of the equation:

$$g_1 \odot g_2 \odot g_3 \odot g_4 = g. \tag{8.6}$$

Assuming that the sets have sizes N_1, N_2, N_3 and N_4 of the same order of magnitude, a generic birthday-based attack exists for this problem. This generic attack first constructs the two sets:

$$\begin{aligned}\mathcal{L} &= \{g_1 \odot g_2 \mid \forall \, (g_1, g_2) \in L_1 \times L_2\} \quad \text{and} \\ \mathcal{L}' &= \{g \odot (g_3 \odot g_4)^{-1} \mid \forall \, (g_3, g_4) \in L_3 \times L_4\},\end{aligned} \tag{8.7}$$

where h^{-1} denotes the inverse of h in the group G. Then the attack searches for collisions between $\mathcal{L}$ and $\mathcal{L}'$. Each collision yields a solution of Equation (8.6).

The size of $\mathcal{L}$ is $N_1 N_2$ and the size of $\mathcal{L}'$ is $N_3 N_4$. Assuming without loss of generality that $N_1 N_2 \leq N_3 N_4$, the time complexity of the generic attack is $O(max(N_1 N_2 \log(N_1 N_2), N_3 N_4))$ and the memory complexity measured in number of group elements is $N_1 N_2$. The knapsack example from Section 8.1 shows that for the additive group of integers $(\mathbb{Z}, +)$, the memory complexity can be lowered without increasing the time complexity. More generally, we can ask for a list of groups where this improvement can be achieved. In the sequel, we give a list of specific examples where the improvement can be achieved, followed by a more general treatment and examples of groups where no known method exists.

Note that without loss of generality, we can always consider that the target value g is the group identity element. Indeed composing Equation (8.6) with

g^{-1} under the group operation, we find:

$$g_1 \odot g_2 \odot g_3 \odot (g_4 \odot g^{-1}) = 1_G. \tag{8.8}$$

Thus it suffices to transform L_4 by replacing each element g_4 by $g_4 \odot g^{-1}$ to make g disappear from Equation (8.6).

8.2.1 Xoring bit strings

A simple and natural group to consider in computer science and cryptography is the group of n-bit strings, together with the group operation obtained by bitwise xoring strings. Mathematically, this group can be viewed in many different ways, the simplest is to consider it as the n-th fold direct product $\mathbb{G}_2^n$ where $\mathbb{G}_2$ is $\{0,1\}$ with addition modulo 2. Alternatively, adding some more structure, it can also be viewed as the additive group of the finite field $\mathbb{F}_{2^n}$.

In this group, our generic problem reduces itself to find all quadruples of values whose XOR is equal to the specified target. As usual, we can assume without loss of generality that the specified target is the group identity element, i.e., the string 0^n.

Clearly, with this group, the approach used by Shamir and Schroeppel for subset sum problems cannot work. Indeed, this approach heavily relies on the existence of an order $\leq$ on $\mathbb{Z}$, compatible with the group operation. In $\mathbb{G}_2^n$, no such order exists. Instead, we need to use a very different method introduced in [CJM02].

The basic idea of the method starts by selecting a subset of t bits among the n available bits. For simplicity, we choose the t high bits of the n-bit numbers that encode elements of $\mathbb{G}_2^n$. For each value x in $\mathbb{G}_2^n$, we denote by $[x]_t$ the restriction of x to its t high bits. This restriction is compatible with the XOR operation, more precisely we have:

$$[x \oplus y]_t = [x]_t \oplus [y]_t. \tag{8.9}$$

Thus for any solution (x_1, x_2, x_3, x_4) of Equation (8.6) rewritten in $\mathbb{G}_2^n$ as:

$$x_1 \oplus x_2 \oplus x_3 \oplus x_4 = 0^n \tag{8.10}$$

we necessarily have:

$$[x_1 \oplus x_2]_t = [x_3 \oplus x_4]_t. \tag{8.11}$$

In other words, the value of the t high bits is the same when xoring the leftmost elements x_1 and x_2, or when xoring the rightmost elements x_3 and x_4. Of course, we do not know in advance this middle value of the t high bits; however, we know for sure that it is shared between both sides. As a consequence, denoting by M_t this middle value, we can enumerate all possibilities for M_t and for each possibility create the list $\mathcal{L}_{M_t}$ of all pairs (x_1, x_2) with $[x_1 \oplus x_2]_t = M_t$

and the list $\mathcal{L}'_{M_t}$ of pairs (x_3, x_4) with $[x_3 \oplus x_4]_t = M_t$. Any collision between these lists for any value of M_t yields a solution of Equation (8.10).

For a fixed value of M_t, the expected number of pairs in the first of these two lists is $N_1 N_2 \cdot 2^{-t}$ and the expected numbers of pairs in the second list is $N_3 N_4 \cdot 2^{-t}$. For typical applications, we have $N_1 \approx N_2 \approx N_3 \approx N_4 \approx 2^{n/4}$. Thus, to keep the memory requirements around $O(2^{n/4})$ it is natural to choose $t \approx n/4$. Indeed, with this choice the two middle lists are of approximate size $2^{n/4} \cdot 2^{n/4} \cdot 2^{-t} \approx 2^{n/4}$. Moreover, for each value of M_t finding collisions between these lists costs $O((n/2 - t) 2^{n/2-t})$, i.e., $O(n\, 2^{n/4})$ for our choice of t. Thus, taking into account the loop on the 2^t possibilities for M_t, the total time for finding collisions is $O(n\, 2^{n/2})$.

To make sure that the approach is sound, we need to make sure that $\mathcal{L}_{M_t}$ and $\mathcal{L}'_{M_t}$ can be constructed efficiently. Assuming that L_2 is already sorted with respect to the value of the t high bits of each element, all solutions of $[x_1 \oplus x_2]_t = M_t$ can be found in time $O(N_1 \log N_2)$ using a simple variation of the collision search method with a sorted list. For each element x_1 of L_1, compute $[x_1]_t \oplus M_t$ and search this value by dichotomy in the high bits of L_2. When $N_1 \approx N_2 \approx N_3 \approx N_4 \approx 2^{n/4}$ and $t \approx n/4$ the time and memory complexity for building all lists $\mathcal{L}_{M_t}$ and $\mathcal{L}'_{M_t}$ in turn are $O(n2^{n/2})$ and $O(2^{n/4})$. Note that it is essential to remember that we only need to store the lists $\mathcal{L}_{M_t}$ and $\mathcal{L}'_{M_t}$ corresponding to the current value of M_t. The corresponding pseudo-code for this approach is given as Algorithm 8.3 or alternatively as Algorithm 8.4

This method is quite different from the method of Shamir and Schroeppel but it gives similar performances. Moreover, it removes the need to deal with balanced trees altogether and, thus, it is simpler to implement.

8.2.2 Generalization to different groups

In the general setting, the group G from which elements of our four lists are drawn is an arbitrary group. For now, we assume that this group is abelian. In that case, group classification tells us that G is isomorphic to an additive group:

$$\mathbb{Z}/n_1\mathbb{Z} \times \mathbb{Z}/n_2\mathbb{Z} \times \cdots \times \mathbb{Z}/n_t\mathbb{Z},$$

where each n_i divides n_{i-1}. As a consequence, it is natural to first consider the case where G has this specific form. In the previous section, we dealt with one extreme example of this case: $(\mathbb{Z}/2\mathbb{Z})^n$.

Here, the size of the group G is N, the product of the n_i values. As usual, we expect a solution when the size of the lists is of the order of $N^{1/4}$. Depending on the distribution of the n_i values, several approaches are possible. The first approach works when some partial product of the n_i values is of the order of $N^{1/4}$. In that case, we can write G as a direct product $G_1 \times G_2$ by grouping all the contributions to this partial product in G_1 and the rest in G_2. In this direct product, G_1 is of size around $N^{1/4}$ and G_2 of size around $N^{3/4}$. Due

---

Algorithm 8.3 Generating all solutions to Equation (8.10)

Require: Input four lists L_1, L_2, L_3, L_4 of $\approx 2^{n/4}$ n-bit values each
Require: Parameter t
  Sort L_2 and L_4
  Allocate arrays L and L' of size $2^{n/2-t}$ plus some margin
  for M_t from 0 to $2^t - 1$ **do**
    Reinitialize L and L' to empty arrays
    for all $x_1 \in L_1$ **do**
      Search first occurrence x_2 of $([x_1]_t \oplus M_t)$ in L_2 high bits
      while $[x_2]_t = [x_1]_t \oplus M_t$ **do**
        Add $x_1 \oplus x_2$ to list L {Keeping track of x_1 and x_2 values}
        Skip to next x_2 in L_2
      end while
    end for
    for all $x_3 \in L_3$ **do**
      Search first occurrence x_4 of $[x_3]_t \oplus M_t$ in L_4 high bits
      while $[x_4]_t = [x_3]_t \oplus M_t$ **do**
        Add $x_3 \oplus x_4$ to list L' {Keeping track of x_3 and x_4 values}
        Skip to next x_4 in L_4
      end while
    end for
    Sort L and L'
    Find and output all collisions between L and L' as quadruples (x_1, x_2, x_3, x_4).
  end for

Algorithm 8.4 Alternative option to Algorithm 8.3

Require: Input four lists L_1, L_2, L_3, L_4 of $\approx 2^{n/4}$ n-bit values each
Require: Parameter t
  Sort L_2 and L_4
  Allocate arrays L and L' of size $2^{n/2-t}$ plus some margin
  for M_t from 0 to $2^t - 1$ **do**
    Reinitialize L and L' to empty arrays
    for all $x_1 \in L_1$ **do**
      Search first occurrence x_2 of $([x_1]_t \oplus M_t)$ in L_2 high bits
      while $[x_2]_t = [x_1]_t \oplus M_t$ **do**
        Add $x_1 \oplus x_2$ to list L {Keeping track of x_1 and x_2 values}
        Skip to next x_2 in L_2
      end while
    end for
    Sort L
    for all $x_3 \in L_3$ **do**
      Search first occurrence x_4 of $[x_3]_t \oplus M_t$ in L_4 high bits
      while $[x_4]_t = [x_3]_t \oplus M_t$ **do**
        Search $x_3 \oplus x_4$ by dichotomy in L
        if $x_3 \oplus x_4$ found **then**
          Output collision as a quadruple (x_1, x_2, x_3, x_4).
        end if
        Skip to next x_4 in L_4
      end while
    end for
  end for

---

to this repartition, a straightforward adaptation of Algorithm 8.3 gives the desired result. It suffices to take as middle value M_t the projection of group elements on the first coordinate G_1. Then, for each element x_1 of the first list L_1, with first coordinate $[x_1]_{G_1}$ it is easy to find all elements x_2 from L_2 with a G_1-coordinate that satisfies:

$$[x_1]_{G_1} + [x_2]_{G_1} = M_t. \tag{8.12}$$

Similarly, we find pairs (x_3, x_4) with:

$$[x_3]_{G_1} + [x_4]_{G_1} = -M_t. \tag{8.13}$$

To finalize the computation, it suffices to search for collisions between G_2 coordinates of the values of $x_1 + x_2$ and $-(x_3 + x_4)$ occurring in the middle lists.

When it is not possible to write G as a direct product with a correct size repartition, the four-list problem needs a different approach. To illustrate this, let us consider $G = \mathbb{Z}/p\mathbb{Z}$ for a prime p. Here, G cannot be written as a proper direct product. Instead, a simple alternative is to lift the problem to $\mathbb{Z}$ and remark that for any solution:

$$x_1 + x_2 + x_3 + x_4 = 0 \pmod{p}, \tag{8.14}$$

the integer $x_1 + x_2 + x_3 + x_4$ is either 0, p, $2p$ or $3p$, assuming that each representative x_i is chosen in $[0, p-1]$. Moreover, the value 0 can only be reached when $x_1 = x_2 = x_3 = x_4 = 0$. As a consequence, over $\mathbb{Z}/p\mathbb{Z}$ we can use three consecutive applications of the approach of Shamir and Schroeppel described in Section 8.1 and solve the four-list problem without using too much memory.

The same approach is easily generalized to a direct product with a bad repartition. In that case, we write $G = \mathbb{Z}/N_1\mathbb{Z} \times G_2$ where N_1 is an integer greater than the size of G to the power $3/4$ and G_2 is the remaining part of the group structure. Denoting elements of G as pairs (x, y) and using the approach of Shamir and Schroeppel, it is easy to create the sublist of values $x_1 + x_2$ in increasing order over the integers and the sublist of $x_3 + x_4$ in decreasing values. From these lists, we extract partial solutions with x-coordinates summing up to either N_1, $2N_1$ or $3N_1$. Checking that in addition the y coordinates also sum to 0 modulo N_2 is a simple matter.

8.2.2.1 Badly presented groups

When the group G is not directly given as in additive form as

$$\mathbb{Z}/n_1\mathbb{Z} \times \mathbb{Z}/n_2\mathbb{Z} \times \cdots \times \mathbb{Z}/n_t\mathbb{Z},$$

it is not completely clear whether and how a reduced memory birthday paradox approach can be followed. Once again, several cases need to be considered. If the cardinality of G is known and if its factorization is accessible, the additive representation of G can easily be derived. If, furthermore, an iso-morphism from G to its additive representation can be computed efficiently enough, we are clearly done. Similarly, when G can be expressed as a direct product $G_1 \times G_2$ with the right repartition of sizes, we can directly use the first approach we used for groups in additive representation, even when G_1 and G_2 are given in arbitrary form. Note that when G has a small enough subgroup (smaller than the square root of the size of G), then it is possible to tabulate the discrete logarithms in this subgroup in order to use an additive representation for this part.

Nevertheless, in our context, there remains a few bad group presentations which prevent us from using a birthday approach with reduced memory re-quirement. A typical example is the group G of an elliptic curve over a finite field (see Chapter 14). If the order of the group is prime (or even a product of two primes of comparable sizes), then there is no known algorithms to solve the problem of four lists with memory requirement lower than the generic birthday-based attack. Another natural example to consider is the case of multiplicative subgroups in finite fields. Of course, if we consider the full mul-tiplicative group and look for naturally occurring random solutions, the order of the group cannot be too large and we are thus in the case where discrete logarithms can be computed efficiently. Indeed, with a group of order N, the subexponential complexity of discrete logarithm computation, even repeated $N^{1/4}$ times to deal with all the elements in the four lists is negligible compared to the birthday running time of the algorithm $N^{1/2}$. On the other hand, with a moderately large subgroup in a large finite field, computing $N^{1/4}$ discrete logarithms with Pollard's Rho is not an option. Similarly, if we consider small sets of the full multiplicative group and know that a solution has somehow been prearranged, we do not know how to apply the four-list approach. We illustrate this in Section 8.4.2 where we study possible cryptanalysis of plain RSA and plain ElGamal encryptions.

8.2.3   Working with more lists

Clearly, Equation (8.8) can easily be generalized to a different number of lists. With two lists, the algorithmic issue is simple since the ordinary birthday-based approach is almost optimal. With three lists of roughly equal sizes $N^{1/3}$ in a group of N elements, the best known algorithm is a simple adaptation of the basic birthday approach. Sort the first list L_1 and then for all pairs (g_2, g_3) of elements in $L_2 \times L_3$, search for $(g_2 \odot g_3)^{-1}$ in L_1. The required memory is the optimal value $N^{1/3}$ but the running time is $O(N^{2/3} \log N)$.

With more than four lists, the best known algorithm is to transform the problem into a four-list version and proceed. Assuming t lists of roughly equal sizes, we first group the lists into packets of approximately $t/4$ lists. When t is a multiple of four, we group the $t/4$ first lists and put all the corresponding sums (or group compositions) into the first list L_1 of the target four-lists problem. We group the next packet of $t/4$ lists into L_2, L_3 and L_4. Clearly, any solution of the transformed four-list problem is a solution of the original problem with t lists. When t is not a multiple of four, we group the lists in order to balance the size of the lists as far as possible. However, in that case, the resulting time-memory tradeoff is not as good.

8.3   Extensions of the technique

8.3.1   Multiple targets

A natural generalization is to consider the case where instead of a single acceptable target, there are many. Clearly, if we are given a list L' of possible target, a very easy solution is to move the target taken from the list L' from the right-hand side to the left-hand side and obtain a new instance of the original problem with a single target. This instance has one more list but we already know how to address this case.

However, if the set of targets has some extra structure and can be described more efficiently than by a simple list, it is possible to obtain better approaches. To illustrate this, we start with a simple example. Assume that we are considering a four-list problem with lists L_1, L_2, L_3 and L_4 of elements in a group $\mathbb{G}_1 \times \mathbb{G}_2$ and that the list of allowed targets is a direct product $L' = L'_1 \times L'_2$. In other words, a pair (g_1, g_2) is a valid target, if and only if, $g_1 \in L'_1$ and $g_2 \in L'_2$. Moreover, assume that $\mathbb{G}_1$ and $\mathbb{G}_2$ have similar sizes, of the order of 2^n, that each of the lists L_1 to L_4 have size $2^{n/4}$ and that each of the lists L'_1 and L'_2 have size $2^{n/2}$. With these parameters, the expected number of solutions is 1.

With this specific instance, a first approach is to rewrite the problem as a a problem of finding a fixed target within a sum of six lists. In approach, we view each element g_1 of L'_1 as a representative for $(g_1, 0)$ in $\mathbb{G}_1 \times \mathbb{G}_2$ and each element g_2 of L'_2 as a representative for $(0, g_2)$. If, in addition, we construct the two sums of lists $L_1 + L_2$ and $L_3 + L_4$, we can directly use the ordinary case of four sets, grouping $L_1 + L_2$ and L'_1 on one size, $L_3 + L_4$ and L'_2 on the other. We obtain a running time of 2^n and a memory requirement of $2^{n/2}$. However, this approach does not take advantage of the extra structure within L', or equivalently of the fact that L'_1 only contains elements of $\mathbb{G}_1$ and L'_2 elements of $\mathbb{G}_2$.

In order to improve this case, we now consider a different approach. Here, we first focus on the partial solutions, with a correct value in $\mathbb{G}_1$ but not necessarily in $\mathbb{G}_2$. Clearly, L'_2 is not needed to construct these partial solutions, since any element of L'_2 has a zero contribution on the $\mathbb{G}_1$ component of sums. The expected number of partial solutions is $2^{n/2}$; moreover, using the basic algorithm with one group containing L_1, L'_1 and the other group containing $L_2 + L_3$, L_4 we can construct all these solutions in time $2^{3n/4}$. In our precise case, the memory requirement is $2^{n/2}$, due to the need to store L'_1. Once a partial solution is constructed, it is easy to evaluate the $\mathbb{G}_2$ component of the corresponding sum. Then, to check whether the partial solution is also correct on $\mathbb{G}_2$, it suffices to test if the opposite of the $\mathbb{G}_2$ component belongs to L'_2, using a dichotomy search. As a consequence, with this specific set of parameters, using the idea of partial solutions reduces the running time, on a group $\mathbb{G}_1 \times \mathbb{G}_2$ of size 2^{2n}, from 2^n down to $2^{3n/4}$.

8.3.2 Wagner's extension

Up to now, we have concerned ourselves with finding all solutions to Equation (8.8), usually in the case where a small number of solutions exist. David Wagner [Wag02] asks a slightly different question and looks for a single solution to this equation, especially in cases where a large number of solutions is expected. The surprising conclusion, in this case, is that the algorithms can be adapted to yield a completely different and much more efficient time/memory tradeoff. A precursor of this idea was already present in [CP91].

Let us start by considering the problem of xoring four bitstrings in this new context. We are given four lists L_1, L_2, L_3 and L_4, each containing $2^{\beta n}$ n-bit numbers for some parameter $\beta \geq 1/4$. On average, we expect $2^{4\beta-1}$ solutions. The key idea of Wagner, is to restrict the search and only look for solutions (x_1, x_2, x_3, x_4) where the restriction of $x_1 \oplus x_2$ (and of $x_3 \oplus x_4$) to a fraction τn of the bits is zero. The expected number of pairs $x_1 \oplus x_2$ such that $x_1 \oplus x_2$ is zero on these τn bits is $2^{2\beta-\tau}$. From this, we deduce that the expected number of solutions to Equation (8.8) that satisfy the additional condition is $2^{4\beta-\tau-1}$. A natural choice of parameters is $\tau = \beta = 1/3$. In that case, we expect to find one solution (among many) to the four-list problem, using time and memory $O(n \cdot 2^{n/3})$. From a running time point-of-view, this is more efficient than the usual $O(n \cdot 2^{n/2})$ of birthday-based attacks. On the other hand, it only works when much larger lists of numbers are available.

Thus, we already see with four lists that asking for a single solution among a large set leads to a better time/memory tradeoff. However, the real power of this technique appears when we consider a larger number of lists. To simplify the exposition of the method, let us assume that we are given 2^t lists of $2^{\alpha n}$ numbers on n bits. The simplification comes from the fact that when the number of lists is a power of 2, we can more easily organize the computation, since we can group lists by packets of 2, 4, ... Within this list, we want to find elements $x_1, x_2, \ldots, x_t$, with each x_i in the corresponding list L_i, whose sum is zero (using a XOR sum). We proceed by using a succession of steps. In each step, we regroup lists two by two, thus halving the number of lists. Moreover, at each step, we cancel a fraction of the high bits. To balance the computation, it is best to preserve the size of the lists considered at each step, except in the final one, where we are happy to obtain a single solution. Since the original lists contain $2^{\alpha n}$ elements, it is possible to cancel αn bits at each step while preserving the sizes of the lists. Indeed, combining two lists together yield $2^{2\alpha n}$ possibilities and forcing αn bits to be zeros reduces this number by a factor of $2^{\alpha n}$. At the final step, we use the birthday paradox and expect a collision on $2\alpha n$ bits to occur for lists of $2^{\alpha n}$ elements. With 2^t lists, we perform t steps and we can work with $\alpha \approx 1/(t+1)$. Both from a time and memory point-of-view, the complexity of this approach is $2^{t+\alpha n}$. If we are free to optimize this complexity by choosing t, then using $t \approx \sqrt{n}$ yields a subexponential algorithm with complexity $2^{2\sqrt{n}}$.

With this extension, using bitstrings is a very comfortable option, indeed, once a fraction of the bits has been set to zero in all groups at some steps, this fraction clearly remains zero throughout the end of the algorithm. As already noticed in [Wag02], this can be generalized to many other groups. However, when generalizing Wagner's extension to different groups, we need to take care to avoid the resurgence of non-zero values in unwanted places. Over a product of small groups, things are also well-behaved, it suffices to cancel some of the small groups at each step to be sure that they do not reappear. For integers modulo 2^n, the best option is to cancel low order bits at each steps, since carries only propagate from right to left. Over the integers, this also is the best approach. Note that, in that case we encounter a new problem: the sum of 2^t numbers on n bits is a number on $n+t$ bits. Luckily, t is small compared to n and this problem is minor. As a consequence, we see that Wagner's extension has essentially the same spectrum of applicability as the original reduced memory birthday paradox attacks.

8.3.3 Related open problems

When searching collisions between two lists, the available algorithms are essentially optimal. Indeed, it is clear that the running time is bounded by the time required to read the input lists. With lists formed of $2^{n/2}$ elements, we get a lower bound of $2^{n/2}$ operations to compare with the running time $O(n2^{n/2})$ of a fast sort algorithm. On the contrary, with the algorithms presented in this chapter, we have no such guarantee. As a consequence, it is worth pondering whether these algorithms can be improved. As a special case, it is interesting to consider three lists L_1, L_2 and L_3 of respective sizes N_1, N_2 and N_3. How much does it cost to find a solution of the equation:

$$x_1 \oplus x_2 \oplus x_3 = 0, \quad \text{with } x_1 \in L_1, x_2 \in L_2 \text{ and } x_3 \in L_3. \tag{8.15}$$

Clearly, the running time should be expressed as a function of $N_1 N_2 N_3$. Indeed, with random lists of n-bit numbers, we expect a solution when $N_1 N_2 N_3 = 2^n$, independently of the individual values of N_1, N_2 and N_3. With some specific choices for the sizes of the lists, we can achieve a complexity $O(n2^{n/2})$. For example, when $N_3 = 1$ and $N_1 = N_2 = 2^{n/2}$, a plain collision search approach suffices. Similarly, when $N_1 N_2 = 2^{n/2}$ and $N_3 = 2^{n/2}$, the same approach works. A much more problematic case is $N_1 = N_2 = N_3 = 2^{n/3}$. In this case, we can sort one list, say L_3 and then for each pair $(x_1, x_2) \in L_1 \times L_2$

search for $x_1 \oplus x_2$ in L_3. However, this solution yields a running time larger than $2^{2n/3}$. To reduce the running time, we can follow the approach of Wagner and enlarge the lists L_1, L_2 and L_3. For lists of size, $2^{\alpha n}$ with $1/3 \le \alpha \le 1/2$, this leads to a running time $2^{(1-\alpha)n}$. However, the best we can do is $2^{n/2}$ with $N_1 = N_2 = N_3 = 2^{n/2}$, which is worse than the basic birthday algorithm.

This specific example emphasizes two more general open problems with generalized birthday algorithms. The first open problem is to devise a new algorithm that searches a unique solution with a better tradeoff between time and memory. Typically, the current best algorithms have time $2^{n/2}$ and memory $2^{n/4}$, so we could afford to use some extra memory in order to go faster. The second open problem concerns Wagner's algorithm to search for one solution among many. In its present form, this algorithm works extremely well when the number of lists is a power of two. However, when this is not the case, its behavior is obtained by rounding the number of lists down to a power of two. In the worst case, when the number of lists is a power of two minus one, this is highly unsatisfactory.

When considering the XOR operation with a large number of lists, it was noted in [BM97] that Gaussian elimination can be used to solve the problem.

8.3.3.1   An incremental improvement

Looking again at the specific example of three lists, it is possible to slightly improve the basic birthday algorithm when using a specific configuration of the sizes of the lists. Instead of working with $L_1 = L_2 = 2^{n/2}$ and $L_3 = 1$, we now use $L_1 = L_2 = 2^{n/2}/r$ and $L_3 = n/2$, with $r = \sqrt{n/2}$. With this choice, it is possible to rearrange the computations in order to scan the lists L_1 and L_2 only once, simultaneously testing each candidate against the list L_3. In order to do that, we first need to perform some elementary linear algebra on the elements of the lists. Viewing each element of the three lists as a vector in $\mathbb{F}_2^n$, it is clear that a solution $x_1 \oplus x_2 \oplus x_3 = 0$ also satisfies $M x_1 \oplus M x_2 \oplus M x_3 = 0$, for any $n \times n$ Boolean matrix. Moreover, if M is invertible, any solution of the second equation is a solution to the original problem. As a consequence, for any invertible matrix M, we may transform the three lists by applying M to all their elements, without changing the set of solutions.

Now, since L_3 contains $n/2$ vectors, it spans a linear subspace of $\mathbb{F}_{2^n}$ of dimension at most $n/2$. We can easily choose a basis $(b_1, b_2, \cdots, b_n)$ of $\mathbb{F}_{2^n}$ such that each vector in L_3 belongs to the subspace generated by $(b_1, b_2, \cdots, b_{n/2})$. We now choose for M the matrix that transforms each vector of $\mathbb{F}_{2^n}$ to the basis $(b_1, b_2, \cdots, b_n)$. Clearly, each element x_3 of L_3 is such that $M x_3$ has $n/2$ trailing zeros in position from $n/2$ to n. Moreover, L_3 usually spans a subspace of rank exactly $n/2$. In that case, b_1, b_2, ..., $b_{n/2}$ can simply be chosen as the elements of L_3. In that case, any vector $M x_3$ is zero everywhere except in a single position between 1 and $n/2$. Once M is chosen, we can use it to transform L_1 and L_2. After this transformation, it suffices to search for pairs of elements (x_1, x_2) such that Mx_1 and Mx_2 collide on the last $n/2$ bits in the new representation. As usual, we expect that a fraction $2^{n/2}$ of all pairs satisfies this property. For each such candidate pair, it then suffices to check that the corresponding sum belongs to L_3. Since there are $2^{n/2+1}/n$ such pairs and since it suffices to check that sum contains a single one, this can be done efficiently. All in all, we gain a factor r or $\sqrt{n/2}$ compared to the basic birthday algorithm.

8.4 Some direct applications

8.4.1 Noisy Chinese remainder reconstruction

The Chinese remainder theorem, as recalled in Chapter 2, allows to reconstruct an integer x in a given interval $[B_l, B_h]$ from a list of modular values $x \bmod p_i$, as long as the values p_i are mutually coprime, assuming that their product is at least $B_h - B_l + 1$.

The notion of "noisy" Chinese remainder reconstruction covers several possible extensions of this basic problem. One simple extension is presented as Exercise 1. In this section, we present a different extension, which arises from the problem of counting points on elliptic curves. This problem, together with a birthday approach with reduced memory to solve it, was first presented in [JL01].

The goal of the problem is to recover a unique integer x in a range $[B_l, B_h]$ from modular values modulo coprime numbers. The only difference with the basic Chinese remainder theorem is that, instead of having a single candidate for each modular value, we have a set of candidates. Some of these sets may still contain a single element. Note that, by using a preliminary application of the ordinary Chinese remainder theorem, it is possible to regroup all the modulus with a single candidate into their product. Thus, without loss of generality, we assume that we have a list of coprime numbers $p_0, p_1, \ldots, p_k$ together with sets S_i of allowed modular values modulo p_i. The set S_0 contains a single element, all the other sets contain at least 2. If for some reason, none of the modulus corresponds to a unique modular value, we simply set $p_0 = 1$ (and $s_0 = \{0\}$).

To solve this problem in the context of point counting algorithms, the original method was to use Atkin's match and sort algorithm. With this method, we use the fact that the elliptic curve whose cardinality is desired can be used to test each candidate and check its validity. This algorithm is a birthday paradox based algorithm of the type discussed in Chapter 6. We now recall a brief description of the algorithm, since this is useful to present the noisy Chinese remainder variation. The first step of the match and sort algorithm is, excluding p_O to sort the coprime modulus by increasing values of $|S_i|/p_i$.

After doing this, we consider the smallest possible values of ℓ such that the product of all coprime values from p_0 up to p_ℓ is larger than the length of the interval $B_h - B_l + 1$. Combining all the allowed modular values for each p_i we obtain $N = \prod_{i=1}^{\ell} |S_i|$ candidates. Note that N is almost minimal, but depending on the exact values of the modulus p_i and of the cardinality $|S_i|$ it can in some cases be improved, especially when there is a gap between the length of the interval and the product of the modulus.

After choosing a set of modulus, it is possible to obtain the cardinality of the elliptic curve by brute force. It suffices, given an element Q in the additive group of the elliptic curve to check, for each of the N candidates c_i, whether $c_i Q$ is the neutral element in the group. Atkin's match and sort refines this into a birthday-based attack, by splitting the product of modulus in two parts P and P'. To each subproduct, we associate a set of allowed values, respectively S and S'. The decomposition of the product is chosen to balance the sizes of these two sets as much as possible. At this point, it is useful to note that by shifting everything by a well-chosen constant, we can replace the interval $[B_l, B_h]$ by another interval centered around 0, say $[-B, B]$, with $B \leq PP'/2$. Once this is done, we can make sure that each candidate value s in S is replaced by a new value $\tilde{s}$ such that: $\tilde{s} = s \pmod{P}$, $\tilde{s} = 0 \pmod{P'}$ and $\tilde{s} \in]-PP'/2, PP'/2]$. Similarly, each s' in S' is replaced by a new value $\tilde{s}'$ such that: $\tilde{s}' = s' \pmod{P'}$, $\tilde{s}' = 0 \pmod{P}$ and $\tilde{s}' \in [-PP'/2, PP'/2[$.

Making these replacements, we can rewrite each of the N possible candidates as $\tilde{s} + \tilde{s}' + \lambda PP'$ for $\tilde{s}$ and $\tilde{s}'$ in the modified sets and $\lambda \in \{-1, 0, 1\}$. Finally, it suffices to search for collisions between the sets, $\tilde{s}Q$ and $-(\tilde{s}' + \lambda PP')Q$ on the elliptic curve.

The approach presented in [JL01] no longer makes use of the group of the elliptic curve. Instead, it relies on the fact that the modular information that is obtained at the beginning of point counting algorithms is redundant and tries to make use of this redundancy to recover the cardinality. Like Atkin's match and sort, it uses a centered interval $[-B, B]$ and it starts by determining a product of modulus larger than the interval length that generates the smallest amount of candidates. Instead of splitting this product in two, we now split it in four factors P_1, P_2, P_3 and P_4, chosen to balance the sizes of the lists of allowed modular values for each P_i. To each factor P_i, we associate a set S_i that contains each of the candidate values modulo P_i. The values stored in S_i are normalized as follows: for each s_i in S_i, we make sure that $s_i = 0 \pmod{P_j}$ for $j \neq i$ and that $v \in [-P_1 P_2 P_3 P_4/2, P_1 P_2 P_3 P_4/2]$. As a consequence, each of the N candidates modulo $P_1 P_2 P_3 P_4$ can be written as $s_1 + s_2 + s_3 + s_4 + \lambda \cdot P_1 P_2 P_3 P_4$, with each s_i in the corresponding set S_i and $\lambda \in \{-2, -1, 0, 1, 2\}$.

Let $q_1, \ldots, q_\ell$ be the remaining modulus values, i.e., the values among the initial $p_1, \ldots, p_k$ which do not appear in the product $P_1 P_2 P_3 P_4$. To each element s_i in one of the sets S_1, S_2 or S_3, we can associate a vector $V(s_i)$ with ℓ coordinates, where the j-th coordinate is obtained by considering s_i modulo q_j. Similarly, assuming that we have guessed the value of λ, we can construct $V(s_4 + \lambda \cdot P_1 P_2 P_3 P_4)$ for each s_4 in S_4. It now remains to solve a sum-of-four problem with multiple targets as in Section 8.3.1. More precisely, we are looking for (s_1, s_2, s_3, s_4) such that the sum

$$V(s_1) + V(s_2) + V(s_3) + V(s_4 + \lambda \cdot P_1 P_2 P_3 P_4) \qquad (8.16)$$

has an acceptable value modulo each modulus q_i. Thus, the set of allowed targets is structured as a direct product. We refer the reader to [JL01] for more details.

8.4.2 Plain RSA and plain ElGamal encryptions

In this section, we describe some birthday-based attacks from [BJN00] against some variations of RSA and ElGamal encryption which do not follow the state-of-the-art recommendations for secure encryption schemes. We call these variations plain RSA and plain ElGamal. It should be clear from the coming discussion that these simplified cryptosystems are insecure and should not be used. We start by defining the simplified systems that we are going to consider.

Plain RSA. Given an RSA public key (N, e), we use this key to directly encrypt random session keys for a block cipher. Let K be such a random key, the corresponding encrypted key simply is $K^e \pmod{N}$.

Plain ElGamal. Given a large prime p, an element g of order q modulo p, with q much smaller than p, more precisely, with $(p-1)/q \geq 2^m$, and an ElGamal public key $y = g^x$ (where x is the corresponding secret key). We encrypt a randomly chosen session key K of at most m bits by choosing a random number r and forming the ciphertext $(g^r, K \times y^r)$.

When encrypting a random key K with either plain RSA or plain ElGamal, we do not check for any specific properties of this key. In particular, it is conceivable that K viewed as an integer is a product of two integers K_1 and K_2 of roughly equal size. The birthday paradox based attacks we present here work in this specific case. As a consequence, they are probabilistic attacks which only succeed when a bad key is chosen. We give some data about the probability of this bad event occurring at the end of the section and show that both asymptotically and in practice, this approach works for a non-negligible fraction of the keys.

8.4.3 Birthday attack on plain RSA

With plain RSA, if the key K can be written as $K = K_1 K_2$, the multiplicativity of RSA implies that:

$$K^e = K_1^e K_2^e \pmod{N}. \qquad (8.17)$$

Given the encrypted key K^e, it is possible to obtain K_1 and K_2 from a list $\mathcal{L}$ of encrypted small key (k^e, k) sorted by increasing order of the first component. Indeed, it suffices for all small keys K_1 in $\mathcal{L}$ to search for K^e/K_1^e in the first component of $\mathcal{L}$. If the search is successful, we clearly obtain an equality

$$K^e/K_1^e = K_2^e \pmod{N}, \tag{8.18}$$

for two small keys, from which we deduce that $K = K_1 K_2$.

Forgetting about logarithmic factors, we can recover a n-bit key in time $2^{n/2}$ instead of 2^n. This clearly shows that plain RSA is not a secure cryptosystem. However, this attack is costly in terms of memory and we can ask whether it is possible to lower these memory requirements. The natural option would be to look for K as a product of four smaller keys $K = K_1 K_2 K_3 K_4$; of course, this covers a smaller fraction of the keys. In this case, we would like to use the algorithms from the present chapter. However, here the group order is unknown and discrete logarithm computations are not feasible, because N is too large to factor. As a consequence, we find ourselves in one of the bad cases presented in Section 8.2.2.1. A nice open problem is to adapt the algorithms to this specific case and recover K using less memory.

8.4.4   Birthday attack on plain ElGamal

With plain ElGamal, instead of directly using multiplicative properties, we first remark that both g and y belong to the (usually unique) subgroup of order q in $\mathbb{F}_p$. Thanks to this fact, we are able to derandomize the encryption scheme and get a new value which is a function of the encrypted key only. This value is:

$$(y^r K)^{\frac{p-1}{q}} = (y^{\frac{p-1}{q}})^r K^{\frac{p-1}{q}} = K^{\frac{p-1}{q}} \pmod{p}. \tag{8.19}$$

Once again, if $K = K_1 K_2$ then

$$K_1^{\frac{p-1}{q}} K_2^{\frac{p-1}{q}} = K^{\frac{p-1}{q}}, \tag{8.20}$$

thus we can attack this scheme exactly as we did in the case of plain RSA.

The key difference here is that the order of the group where we want to solve this multiplicative equation is no longer an unknown, indeed it is equal to $(p-1)/q$. Assuming that this number can be factored, we now know the decomposition of the group as discussed in Section 8.2.2.1. From this discussion, it follows that for n-bit keys, if K can be written as $K = K_1 K_2 K_3 K_4$ a product of four numbers of about $n/4$ bits, we can use the quadrisection approach if there is a small subgroup of size approximately $2^{n/4}$. As stated in Section 8.2.2.1, we do not even need to compute logarithms and put the problem into additive form in order to solve it.

However, computing logarithms is a good idea, since it allows us to solve the problem efficiently even when the available subgroup is too large, up to

$2^{n/2}$ elements. Indeed, we have a list of $2^{n/4}$ small numbers. Thus, we need to compute $2^{n/4}$ discrete logarithms. In our context, the best algorithms available are generic algorithms such as Pollard's rho. With a group of $2^{n/2}$ elements, each computation of a discrete logarithm costs $2^{n/4}$. As a consequence, the total cost is $2^{n/2}$ and remains comparable to the running time of the quadrisection approach. Once we have reduced the problem to its additive form, we can, as in Section 8.2.2, lift the problem to $\mathbb{Z}$ and solve it efficiently.

Exercises

1^{h}. Let N be a product of many small primes (or prime powers). Let X a number in the interval $[0, B]$, where B is a known upper bound. Assume that we are given values for $X \bmod p$ for each prime (power) in N, with the following caveat: most modular values are correct, but some might be incorrect. To parametrize the problem, it is useful to denote by N_I the product of the primes (or prime powers) for which the modular value is incorrect.

- Show that if N_I is given, we can obtain X as long as $N/N_I > B$.

- Assuming that each prime (or prime power) in N has the same order of magnitude P, we now assume that N contains n terms, i.e., $N \approx P^n$ and N_I contains n_I terms. Describe a brute force algorithm that search for X by trying all possibilities for N_I. Under which condition do we obtain a unique solution for X?

- An alternative approach is to first construct a value $\tilde{X}$ modulo N that matches all the given modular values. Show that $\tilde{X} = X$ $(\bmod\ N/N_I)$. Deduce that X/N can be approximated by a fraction δ/N_I. Can this fraction be obtained without exhaustive search?

- Combining exhaustive search together with the above approach, when can X be recovered and at what cost?

2. Implement the birthday attack using quadrisection against RSA (or El-Gamal) described in Section 8.4.2.

3^{h}. The quadrisection algorithms from this chapter have all been presented for commutative groups. Discuss their applicability to the group of 2×2 matrices over $\mathbb{F}_p$.

4. Let $L_1, \ldots, L_n$ be n lists, each containing 2 bitstrings on n bits. Write an efficient method for finding n elements, one from each list, whose sum is zero.

Chapter 9

Fourier and Hadamard-Walsh transforms

9.1 Introductory example: Studying S-boxes

Substitution boxes or S-boxes are very important ingredients in the construction of secret key cryptographic algorithms. Their main role is to break the linearity of these algorithms. Due to the binary nature of computers, most S-boxes act on bits. They take a number of bits as input and produce a number of bits as output. Having S-boxes with good properties is essential to the security of secret-key encryption. For example, it is well known that the security of the DES algorithm drops dramatically when its S-boxes are replaced without care. The cryptanalysis of FEAL in [BS91b, GC91, MY92, TCG92], which is an encryption algorithm similar to DES, illustrates this point. As a consequence, the properties of S-boxes need to be studied with care when choosing them, or when trying to cryptanalyze an encryption algorithm. In particular, the linear and differential properties of S-boxes, which are respectively used by linear [Mat93, Mat94a, Mat94b, TSM94] and differential [BS91a, BS92, BS93] cryptanalysis, should always be determined.

9.1.1 Definitions, notations and basic algorithms

Let S be an S-box with a n-bit input and a t-bit output. The differential characteristics of S are denoted by $\mathcal{D}_\Delta^\delta(S)$ and stand for the number of input pairs (x, y) with input difference Δ and output difference δ, i.e., such that $x \oplus y = \Delta$ and $S(x) \oplus S(y) = \delta$. The linear characteristics of S are denoted by $\mathcal{L}_M^m(S)$ and stand for the difference between the number of n-bit values x such that the two bitwise scalar products $(M|x)$ and $(m|S(x))$ are equal modulo 2 and the number of pairs such that these scalar products are different. The n-bit value M is called the input mask and the t-bit value m the output mask.

Throughout this study, to simplify notations, we identify numbers with the binary bitstrings that represent these numbers. To make the notation explicit, let a and b be two numbers. First, write both a and b in basis 2, i.e., assuming ℓ-bit values as sums $\sum_{i=0}^{\ell-1} a_i 2^i$ and $\sum_{i=0}^{\ell-1} b_i 2^i$, with all values a_i and b_i in $\{0, 1\}$. In this context, we identify a with the bitstring $a_{\ell-1} \cdots a_1 a_0$, we also define the bitwise scalar product of a and b as:

$$(a|b) = \sum_{i=0}^{\ell-1} a_i b_i \bmod 2.$$

From the definitions, we can directly derive two basic algorithms: one for computing differential characteristics (Algorithm 9.1) and the other for linear characteristics (Algorithm 9.2). For linear characteristics, the algorithm simply enumerates all input masks, all output masks and all input values, compute the scalar products, do some counting and derive each characteristic. The runtime is $O(2^{2n+t})$ operations on n-bit counters and in terms of memory it suffices to store S and a couple of additional variables, thus requiring $t \cdot 2^n + O(1)$ bits of memory (assuming that the characteristics are produced but not stored). For differential characteristics, we proceed slightly differently, by first enumerating all values of the input difference Δ. For each possible difference, we initialize to zero a table of 2^t counters (one for each possible output difference). Then, for each pair $(x, x \oplus \Delta)$ with the prescribed input difference, its output difference $S(x) \oplus S(x+\delta)$ is computed and the corresponding counter is incremented. The runtime is $O(2^{2n})$ and the memory is $t \cdot 2^n + n \cdot 2^t + O(1)$ bits to store S and the counters (again assuming that the characteristics are not stored). Note that it is straightforward to gain a factor of two using the fact that due to symmetry, each pair is computed twice, once as $(x, x \oplus \Delta)$ and once as $(x \oplus \Delta, x)$. We also remark that for $\Delta = 0$ the result is known without any computation, since the 2^n pairs with input difference 0 all have output difference 0. We see that, using these basic algorithms, for any choice of input and output sizes for S, it is easier to compute differential characterics than linear characteristics.

Algorithm 9.1 Algorithm for computing differential characteristics
---
Require: Input Table S containing 2^n elements on t bits
    Create table `Count` of 2^t integers
    for Δ from 0 to $2^n - 1$ **do**
        Set table `Count` to zero
        for x from 0 to $2^n - 1$ **do**
            Increment `Count`$[S[x] \oplus S[x \oplus \Delta]]$
        end for
        Print table `Count`: `Count`$[\delta]/2^n$ is the probability of transition from input difference Δ to output difference δ
    end for

Algorithm 9.2 Algorithm for computing linear characteristics

Require: Input Table S containing 2^n elements on t bits
  for M from 0 to $2^n - 1$ **do**
    for m from 0 to $2^t - 1$ **do**
      Let Count $\longleftarrow$ 0
      for x from 0 to $2^n - 1$ **do**
        if $(x|M) = (S[x]|m)$ **then**
          Increment Count
        end if
      end for
      Print Count, where Count$/2^n$ is the probability of the event $(x|M) \oplus (S[x]|m) = 0$
    end for
  end for

9.1.2 Fast linear characteristics using the Walsh transform

The Walsh (or Hadamard-Walsh) transform is a kind of discrete Fourier transform which has been used for a long time in coding theory. It was applied to cryptography by Xiao and Massey in [XM88]. Among its possible applications, we find the fast computation of linear characteristics. The easiest case happens for S-boxes with a single bit of output, i.e., when $t = 1$. In that case, it suffices to consider the case of an output mask m equal to 1. Indeed, when $m = 0$ the result is independent of the S-box and quite uninteresting. From the definition, we have:

$$\mathcal{L}_M^1(S) = \sum_{(x|M)=S(x)} 1 - \sum_{(x|M)\neq S(x)} 1 \qquad (9.1)$$

$$= \sum_x (-1)^{(x|M)} \cdot (-1)^{S(x)} = \sum_x (-1)^{(x|M)} \cdot T(x),$$

where $T(x)$ is defined as $(-1)^{S(x)}$. In this form, it is interesting to split T in two halves T_0 and T_1 with inputs on $n - 1$ bits. More precisely, for all $0 \leq x < 2^{n-1}$, we let:

$$T_0(x) = T(x) \quad \text{and} \quad T_1(x) = T(2^{n-1} + x). \qquad (9.2)$$

Similarly, we split S in two halves S_0 and S_1. Using these notations, we see that for input masks $M < 2^{n-1}$, we have:

$$\mathcal{L}_M^1(S) = \sum_{x<2^{n-1}} (-1)^{(x|M)} \cdot T_0(x) + \sum_{x<2^{n-1}} (-1)^{(x|M)} \cdot T_1(x) \qquad (9.3)$$

$$= \mathcal{L}_M^1(S_0) + \mathcal{L}_M^1(S_1).$$

Likewise, for input masks of the form $2^{n-1} + M$ with $M < 2^{n-1}$, we have:

$$\mathcal{L}^1_{2^{n-1}+M}(S) = \sum_{x < 2^{n-1}} (-1)^{(x|M)} \cdot T_0(x) - \sum_{x < 2^{n-1}} (-1)^{(x|M)} \cdot T_1(x)$$

$$= \mathcal{L}^1_M(S_0) - \mathcal{L}^1_M(S_1). \tag{9.4}$$

Clearly, T_0 and T_1 can also be divided in halves and so on ... until we reach tables of size one where nothing remains to be done. This yields the following algorithm, which receives as input the table T, modifies it in place and outputs a new table, the Walsh transform of T denoted $W(T)$. At position M in this table, we find the value $W_M(T) = \mathcal{L}^1_M(S)$. This algorithm can be written without using recursion, as described in pseudo-code as Algorithm 9.3.

Algorithm 9.3 Walsh transform algorithm

Require: Input Table T containing 2^n elements $-1, 1$
   Comment: Variable Sz is the small table size
   Comment: Variable Pos is the small table position
   for i from 0 to $n-1$ **do**
      Let Sz $\longleftarrow 2^i$, Pos $\longleftarrow 0$
      while (Pos $< 2^n$) **do**
         for j from 0 to Sz $- 1$ **do**
            Let Sum $\longleftarrow T[\text{Pos} + j] + T[\text{Pos} + \text{Sz} + j]$
            Let Diff $\longleftarrow T[\text{Pos} + j] - T[\text{Pos} + \text{Sz} + j]$
            $T[\text{Pos} + j] \longleftarrow$ Sum
            $T[\text{Pos} + \text{Sz} + j] \longleftarrow$ Diff
         end for
         Let Pos $\longleftarrow$ Pos $+ 2 \cdot$ Sz
      end while
   end for
   Output overwritten content of T containing $W(T)$

We see that the time complexity of the Walsh transform is $O(n \cdot 2^n)$ and that the required memory is $O(2^n)$ numbers or $O(n \cdot 2^n)$ bits since the numbers contained in the Walsh transform are integers in the range between -2^n and 2^n. Another important fact about the Walsh transform is that it can easily be inverted by a similar looking algorithm, the inverse Walsh transform $\hat{W}$, described in pseudo-code in Algorithm 9.4.

Note that the only difference between the Walsh transform and its inverse is the fact that in the inverse transform the variables Sum and Diff are divided by 2 before being put back into the table. It is easy to check that each elementary step of size Sz in the inverse Walsh transform is the inverse of the corresponding step in the regular transform. Moreover, the order in which the various elementary steps are performed is irrelevant because each of them

---

Algorithm 9.4 Inverse Walsh transform algorithm

Require: Input Table T containing 2^n elements $-1, 1$
  Comment: Variable Sz is the small table size
  Comment: Variable Pos is the small table position
  for i from 0 to $n - 1$ **do**
    Let Sz $\longleftarrow 2^i$, Pos $\longleftarrow 0$
    while (Pos $< 2^n$) **do**
      for j from 0 to Sz $- 1$ **do**
        Let Sum $\longleftarrow T[\text{Pos} + j] + T[\text{Pos} + \text{Sz} + j]$
        Let Diff $\longleftarrow T[\text{Pos} + j] - T[\text{Pos} + \text{Sz} + j]$
        $T[\text{Pos} + j] \longleftarrow$ Sum$/2$
        $T[\text{Pos} + \text{Sz} + j] \longleftarrow$ Diff$/2$
      end for
      Let Pos $\longleftarrow$ Pos $+ 2 \cdot$ Sz
    end while
  end for
  Output overwritten content of T containing $\hat{W}(T)$

---

commutes with all the others. As a consequence, we deduce that the two algorithms are indeed inverses of each other.

For general S-boxes with t bits of output, it is also possible to speed up the computation using the Walsh transform. It suffices for each output mask m to construct an auxiliary table $S_m(x) = (m|S(x))$ with a single bit of output, and then to proceed as above. This lowers the time complexity down to $O(n \cdot 2^{n+t})$ compared to $O(2^{2n+t})$ for the basic algorithm. The memory requirements are essentially multiplied by $1 + n/t$, since in addition to S we now need to hold in memory the Walsh transform of one S_m. Using this Walsh transform algorithm, the computation of linear characteristics now compare favorably with the computation of differential characteristics, especially when the output size t is much smaller than the input size n. Indeed, we have to compare $O(n \cdot 2^{n+t})$ with $O(2^{2n})$.

9.1.2.1 Basic implementation of the Walsh transform

In this section, we give a basic implementation of Algorithm 9.3 as Program 9.1. One interesting experiment is to measure the speed of this code for tables of increasing sizes. In this experiment, we take all possible sizes from 32 to 2^{28} and, in order to reflect the time per element, we repeat the Walsh transform $2^{28}/S$ times for a table of size S. We give the observed running time in Table 9.1. For very small tables, the running times are high due to function calls overhead. For large tables, we see that there is a progressive slowdown between 2^{11} and 2^{20}, followed by a brutal slowdown after 2^{20}. This is the result of cache effects for large tables. Cache friendly versions of the Walsh transform are given on the book's website.

Program 9.1 C code for Walsh transform

```
#define TYPE int

/*In place Walsh transform*/
void Walsh(TYPE *Tab, TYPE size)
{
  TYPE i,i0,i1;  TYPE step;
  TYPE sum,diff;
  for (step=1;step<size;step<<=1) {
    for (i1=0;i1<size;i1+=2*step) {
      for (i0=0;i0<step;i0++) {
        i=i1+i0;
        sum=Tab[i]+Tab[i+step];diff=Tab[i]-Tab[i+step];
        Tab[i]=sum;Tab[i+step]=diff;
      }
    }
  }
}
```

Size S	Runtime for $2^{28}/S$ executions	Size S	Runtime for $2^{28}/S$ executions
32	16.0 s	64	9.4 s
128	6.5 s	256	4.9 s
512	4.3 s	1024	4.1 s
2048	3.9 s	4096	4.0 s
8192	4.1 s	16384	4.5 s
32768	4.6 s	65536	4.8 s
131072	5.0 s	262144	5.1 s
524288	5.3 s	1048576	6.0 s
2097152	11.9 s	4194304	13.5 s
8388608	14.2 s	16777216	14.9 s
33554432	15.4 s	67108864	16.1 s
134217728	16.7 s	268435456	17.3 s

Table 9.1: Timings on Intel Core 2 Duo at 2.4 GHz using gcc 4.3.2

9.1.3 Link between Walsh transforms and differential characteristics

In 1994, in two independent articles, Chabaud and Vaudenay [CV94], Daemen, Govaerts and Vandewalle [DGV94] discussed the relation between the Walsh transform, or more precisely its square, and the value of differential characteristics. In this section, we illustrate this relation by explicitly describing a Walsh transform based algorithm for computing differential characteristics.

As with the case of linear characteristics, we start by studying S-boxes with a single bit of output. In order to represent these S-boxes, we are going to use polynomials in a multivariate polynomial ring. More precisely, we start from the commutative ring over $\mathbb{Z}$ in n variables $X_0, X_1, \ldots, X_{n-1}$. Then we form an ideal I, generated by the n polynomials $X_0^2 - 1, X_1^2 - 1, \ldots X_{n-1}^2 - 1$. Finally, we form the quotient ring:

$$
\mathbb{K} = \frac{\mathbb{Z}[X_0, X_1, \cdots, X_{n-1}]}{(I)}. \tag{9.5}
$$

In the ring $\mathbb{K}$, we encode an S-box S by the polynomial f_S defined as:

$$
f_S = \sum_{i=0}^{2^n-1} (-1)^{S(i)} M_i. \tag{9.6}
$$

In this expression M_i denotes the following monomial in $\mathbb{K}$:

$$
M_i = \prod_{j=0}^{n-1} X_j^{B_j(i)}, \tag{9.7}
$$

where $B_j(i)$ denotes the j-th bit in the binary representation of i. Note that in this representation, we use the trick of representing S-boxes multiplicatively by 1 and -1 instead of 0 and 1, as we already did in the table T in Section 9.1.2.

In this polynomial representation, we first remark that the multiplication of monomials modulo I is such that $M_i = M_j \cdot M_k$ if and only if $i = j \oplus k$. This implies that the coefficient of M_Δ in f_S^2 is:

$$
\sum_{i \oplus j = \Delta} (-1)^{S(i)} (-1)^{S(j)}. \tag{9.8}
$$

Since each term in the sum is equal to 1 when $S(i) = S(j)$ and to -1 when $S(i) \neq S(j)$, the sum is clearly equal to $\mathcal{D}_\Delta^0(S) - \mathcal{D}_\Delta^1(S)$. Moreover, it is clear from a simple counting argument that $\mathcal{D}_\Delta^0(S) + \mathcal{D}_\Delta^1(S)$ is 2^n. Thus, we easily recover $\mathcal{D}_\Delta^0(S)$ and $\mathcal{D}_\Delta^1(S)$.

As a direct consequence, computing f_S^2 yields the full table of differential characteristics for the single bit output S-box S. Thus, finding a fast multiplication algorithm in $\mathbb{K}$ is sufficient to quickly compute the differential characteristics of S.

What we need to do now, is a simpler analog of the well-known fact that in the polynomial ring $\mathbb{Z}[x]$, fast multiplication can be performed using fast Fourier transform, pointwise multiplication and inverse Fourier transform.

To describe fast multiplication in $\mathbb{K}$, a simple way is to proceed by recursion on the number of variables. Assume that we already know how to multiply polynomials on $n-1$ variables. To multiply f and g, we first write f as $f_0 + X_{n-1}f_1$ and g as $g_0 + X_{n-1}g_1$. With this notation we have:

$$f \cdot g = (f_0 g_0 + f_1 g_1) + X_{n-1}(f_0 g_1 + f_1 g_0). \tag{9.9}$$

Denoting the product by h and writing $h_0 = f_0 g_0 + f_1 g_1$ and $h_1 = f_0 g_1 + f_1 g_0$, we now remark that:

$$\begin{aligned} h_0 + h_1 &= (f_0 + f_1) \cdot (g_0 + g_1) \\ h_0 - h_1 &= (f_0 - f_1) \cdot (g_0 - g_1). \end{aligned} \tag{9.10}$$

Viewing f, g and fg as tables with 2^n entries, where entry i is the coefficient of M_i, and going through the recursion, we find the following relation between the Walsh transforms of f, g and h:

$$W(h) = W(f) \times W(g), \tag{9.11}$$

where $\times$ denotes pointwise multiplication, i.e.,

$$\forall i : \quad W_i(h) = W_i(f) \cdot W_i(g), \tag{9.12}$$

Then h itself is obtained by computing the inverse Walsh transform of the pointwise product.

In order to compute f_S^2, it suffices to perform a single multiplication in $\mathbb{K}$. More precisely, we compute the Walsh transform of f_S, square it pointwise and apply the inverse Walsh transform. Thus, using this algorithm the complexity of computing f_S^2 requires $O(n \cdot 2^n)$ bit operations. As a consequence, we build the complete table of differential characteristics for a S-box S with a single bit output using $O(n \cdot 2^n)$ bit operations, compared to $O(2^{2n})$ for the basic Algorithm 9.1.

9.1.3.1   Differential characteristics for general S-boxes

In order to deal with general S-boxes with more than a single bit of output, we need to generalize the above method, in order to account for the additional output bits. We now describe two slightly different approaches that apply to this generalized case.

9.1.3.1.1   Theoretical approach   The first approach, is to look at the problem from a theoretical point-of-view, replacing the function with several bits of output by another function with a single bit of output. The simplest way to proceed is probably to replace the S-box S by its characteristic function as proposed in [CV94]. Indeed, the characteristic function is a function from $n + t$ bits to a single bit and we can encode it and use the Walsh transform almost as in the case of a single bit output. More precisely, the characteristic function of S is the function χ_S is defined as:

$$\chi_s(y\|x) = \begin{cases} 1, & \text{if } y = S(x) \text{ and} \\ 0, & \text{otherwise.} \end{cases} \tag{9.13}$$

In this definition $y\|x$ is a decomposition of the input to χ_S into a lower part x on n bits and an upper part y on t bits.

Then, we use monomials on $n + t$ variables in an enlarged copy of $\mathbb{K}$ and define:

$$F_S = \sum_{i=0}^{2^{n+t}-1} \chi_S(i) M(i) = \sum_{i=0}^{2^n-1} M(i + 2^n \cdot S(i)). \tag{9.14}$$

Note that, this time, we do not use a multiplicative encoding with 1 and -1. Using a simple computation, we can check that the coefficient of $M(\Delta + 2^n \cdot \delta)$ in F_S^2 is exactly $\mathcal{D}_\Delta^\delta(S)$. Thus we can compute all the differential characteristics using a Walsh transform followed by a pointwise multiplication and finally an inverse Walsh transform. The time complexity is $O((n + t) \cdot 2^{n+t})$. However, the drawback of this approach is that we need to store in main memory a large table with 2^{n+t} entries.

9.1.3.1.2 Practical Variant

To avoid this main drawback, we can use a slightly different approach and rely on disk storage instead of main memory to store the 2^{n+t} entries. The goal here is to reduce the amount of main memory needed to the maximum of 2^n and 2^t. The key ingredient is to alternatively view a large table with 2^{n+t} elements either as 2^t columns of length 2^n or 2^n rows of length 2^t. First, to create the table by columns, we enumerate the 2^t possible[1] linear output masks m and define $S_m(x) = (m|S(x))$ with a single bit of output as in Section 9.1.2. For each value of m, we define f_{S_m}, compute its square and save it to disk as a column containing 2^n elements. Once this is finished, we consider each of the 2^n possible values[2] for Δ and read from disk all the numbers $f_{S_m}^2(\Delta)$ into a table T_Δ with 2^t elements. We now study the relation between this table and the table $\mathcal{T}_\Delta$ of all differential characteristics $\mathcal{D}_\Delta^\delta(S)$, for fixed input Δ. We claim that when the output difference of S is

---

[1] Clearly, for the zero mask, the computation is independent of S and the constant result can be specified within the algorithm. However, to simplify the exposition, we skip this minor improvement.

[2] Once again, we could ignore the case where $\Delta = 0$.

δ, the output difference of S_m is clearly $(m|\delta)$. As a consequence:

$$T_\Delta(m) = \mathcal{D}_\Delta^0(S_m) - \mathcal{D}_\Delta^1(S_m) \tag{9.15}$$

$$= \sum_{(m|\delta)=0} \mathcal{D}_\Delta^\delta(S) - \sum_{(m|\delta)=1} \mathcal{D}_\Delta^\delta(S)$$

$$= \sum_\delta (-1)^{(m|\delta)} \mathcal{D}_\Delta^\delta(S) = \sum_\delta (-1)^{(m|\delta)} \mathcal{T}_\Delta(\delta).$$

Thus, T_Δ is the Walsh transform of the table $\mathcal{T}_\Delta$. This shows that by computing an additional inverse Walsh transform on each table T_Δ we can recover the full table of differential characteristics. The complexity is again $O((n+t) \cdot 2^{n+t})$, but we require less main memory since we use disk space to efficiently transpose the table from columns with fixed m to rows with fixed Δ. In fact, this approach does not even need the distinction between input and output bits, it can be generalized to any Walsh transform computation which overflows the available main memory, by regrouping bits arbitrarily.

When we compare this to the complexity of the basic algorithm, we see that it is very useful when t is small compared to n but much less when t is almost as large as n. The case where t is larger than n is rarely considered; however, in this extreme case, the basic algorithm is faster than the Walsh based technique described here. However, for S-boxes with many algebraic properties, it may happen that even for large t a fast Walsh transform based algorithm exists. It is for example the case of the AES S-box, due to the linear redundancy between its bits described in [FM03]. Another possible application is the computation of generalized differences with more than two inputs. In this case, the cost of the basic algorithm grows to 2^{3n} (with three inputs), while the Walsh based algorithm cost remains at $(n+t)2^{n+t}$.

9.1.4   Truncated differential characteristics

Truncated differential cryptanalysis is a variation of differential cryptanalysis introduced in [Knu94]. The basic idea is to study partially known differences, where the difference is fixed on some bits and left unknown on other bits. For this variation, it is possible to devise a variation of the basic algorithm and build truncated differential characteristics by first enumerating an input and an output masks, which describe the input and output bits where the difference is known. Then, for each pair of masks, one enumerates all possible input pairs and increments the counter corresponding to the truncated input and output difference of the pair. This is shown in pseudo-code as Algorithm 9.5. The time complexity of this algorithm is $O(\max(2^{3n+t}, 2^{2n+2t}))$; it requires a table of 2^{n+t} elements in terms of memory. With this basic approach, computing all possible truncated differential characteristics costs a lot more than computing ordinary differentials. Before giving an improved algorithm, we first need to count the total number of possible truncated differentials, in order to obtain a lower bound for the complexity. The easiest way is to remark that in general a truncated differential is defined by specifying for each input and output bits whether it should be zero, one or unknown. Thus, there are 3^{n+t} different values to compute.

Algorithm 9.5 Algorithm for truncated differential characteristics

---
Require: Input Table S containing 2^n elements on t bits
  Create bidimensional array `Count` of $2^n \times 2^t$ integers
  for `InMask` from 0 to $2^n - 1$ **do**
    for `OutMask` from 0 to $2^t - 1$ **do**
      Set table `Count` to zero
      for x from 0 to $2^n - 1$ **do**
        for y from 0 to $2^n - 1$ **do**
          Increment `Count`$[(x \oplus y)\&\texttt{InMask}, (S[x] \oplus S[x \oplus \Delta])\&\texttt{OutMask}]$
        end for
      end for
      Print table `Count`: `Count`$[\Delta, \delta]/2^n$ is the probability of transition from partial input difference Δ to partial output difference δ
    end for
  end for

Looking beyond the basic approach, another idea is to compute the truncated differentials from the table of ordinary differential. More precisely, we can see that any truncated differential is the sum of all the ordinary differentials that are compatible with this truncated differential. Here the word "compatible" is understood in the natural sense and means that when the truncated differential has a 0 or 1 bit in either the input or output mask, the ordinary differentials we sum have the same bit value. When the truncated differential has an undetermined bit, we sum over all possibilities for this bit. As a consequence, to compute a truncated differential with k bits left undetermined, we need to sum 2^k ordinary differentials. For each choice of k positions for the undetermined bits, there are 2^{n+t-k} possible truncated differentials, obtained by choosing the values of the bits in the other positions. As a consequence, the total cost to compute all the truncated differential with a given set of undetermined bits is 2^{n+t}. Since there are 2^{n+t} possible sets of undetermined bits, the total complexity of this second approach is $2^{2n+2t} = 4^{n+t}$.

In order to compute truncated differentials, let us first consider the simplest case where we limit the possibility of truncating to a single bit. To make things even simpler, let us assume that the only bit which can be truncated is the first input bit of S. In that case, given a fixed output difference δ and a fixed difference on $n - 1$ input bits Δ', we need to build the two differential characteristics $\mathcal{D}_{0\|\Delta'}^{\delta}(S)$ and $\mathcal{D}_{1\|\Delta'}^{\delta}(S)$ to cover the two possible cases where we do not truncate. We also need to compute a differential where the bit is really truncated:

$$\mathcal{D}^{\delta}_{*\|\Delta'}(S) = \mathcal{D}^{\delta}_{0\|\Delta'}(S) + \mathcal{D}^{\delta}_{1\|\Delta'}(S). \tag{9.16}$$

Here, the notation $*$ indicates that a bit of difference is left unspecified.

There are several ways to use Equation (9.16) in order to compute truncated differential. One possibility, remembering that differential characteristics are computing by a Walsh transform, followed by squaring and an inverse Walsh transform, is to embed the computation of the truncated characteristics within a modified inverse Walsh transform algorithm for the final step. Starting with a single unspecified bit in high-order position, we can deal with it in the last iteration of the inverse Walsh transform. Thus, we now concentrate on this last iteration and find out how it can be modified to compute the additional truncated difference. The required modification is very simple, since it suffices in addition to Sum and Diff to store a copy of Sum + Diff, i.e., a copy of the original content of $T[\mathtt{Pos} + j]$ somewhere. Doing a similar reasoning on other bits easily yields an extended variation of the inverse Walsh transform that works on tables of size a power of 3. It is then possible to plug this modified inverse Walsh in the algorithms of Section 9.1.3.1. Note that, for the practical variant presented in Section 9.1.3.1.2 that uses two inverse Walsh transform steps, both need to be modified, one to allow truncated input differences, the other to permit truncated outputs. One difficulty is that right before calling the modified inverse transform, we need to convert tables of size 2^n (resp. 2^t) into tables of size 3^n (resp. 3^t). This is done by sending the entry in position i with binary decomposition $i = \sum_{j=0}^{\ell} b_j 2^j$ to position $\sum_{j=0}^{\ell} b_j 3^j$. Thus, the binary decomposition of i is interpreted in basis 3. All unspecified positions are filled with zeros. A convenient convention is to organize the modified inverse Walsh transform in such a way that for the entry in position i the corresponding truncated difference is obtained by looking at the base 3 decomposition of i. A digit set to 0 means a 0 difference on the corresponding bits, a 1 means 1 and a 2 means $*$ (unknown). From a complexity point-of-view, we first study the runtime of the modified Walsh transform on a table of size 3^{ℓ}. At first, it seems that the complexity is $O(\ell\, 3^{\ell})$. However, taking into account that a large part of the table is initially zero and that it is useless to add/subtract zeros, the complexity can be lowered to $O(3^{\ell})$. To implement this improvement, it is easier to incorporate the binary to base 3 renumbering within the modified inverse Walsh transform itself. With this improvement, the main contribution to the runtime of the complete truncated differential algorithm is the second layer of modified Walsh transform. Indeed, the first layer performs 2^t transforms on tables of size 3^n, while the second layer does 3^n transforms on tables of size 3^t. Thus, we obtain a total runtime of $O(3^{n+t})$ which is optimal, since we need to compute 3^{n+t} values.

For completeness, we give on the book's website two different C programs computing truncated differentials with this algorithm: one implementing the modified inverse Walsh with complexity $O(\ell\, 3^{\ell})$, the other slightly more complicated with complexity $O(3^\ell)$. For the sake of simplicity, in these programs everything is stored in main memory. However, for larger tables, it would be worth doing the transpose from columns to lines on disk as explained in Section 9.1.3.1.2.

9.2   Algebraic normal forms of Boolean functions

For a table S with n input bits and a single bit of output or equivalently for the associated Boolean function f on n bits, another very interesting representation is to write the Boolean function f as a polynomial in n variables over the polynomial ring $\mathbb{F}_2[x_0, x_1, \ldots, x_{n-1}]$. This representation is called the algebraic normal form of f and has many applications in cryptography and coding theory. In particular, for a cryptanalyst, it is often interesting, in the context of algebraic cryptanalysis, to look at the degree of the algebraic norm form of f.

More precisely, the algebraic normal form of f is obtained by writing:

$$f(x_0, \ldots, x_{n-1}) = \bigoplus_{(a_0, \ldots, a_{n-1}) \in \mathbb{F}_2^n} g(a_0, \ldots, a_{n-1}) \prod_i x_i^{a_i} \qquad (9.17)$$

The function g giving the coefficient of the polynomial in Equation (9.17) is called the Moebius transform of the Boolean function f. Since g is also a Boolean function, Moebius transforms can be computed by using operations on bits only. Compared to Walsh transforms this gives a very significant advantage and allows us to represent Boolean functions using a single bit in memory per entry, instead of a full integer. It remains to see whether a fast algorithm can work with this representation.

With a Boolean function f on a single bit entry x_0, it is easy to see that:

$$g(0) = f(0) \quad \text{and} \quad g(1) = f(0) \oplus f(1). \qquad (9.18)$$

Moreover, since the transformation that sends (x, y) to $(x, x \oplus y)$ is its own inverse, we see that for Boolean functions on a single variable, the Moebius transform is an involution on the set of Boolean functions. To generalize to more than one variable, let us consider f a Boolean function on n variables and write:

$$f(x_0, \ldots, x_{n-1}) = f^{(0)}(x_0, \ldots, x_{n-2}) \oplus f^{(1)}(x_0, \ldots, x_{n-2}) \cdot x_{n-1}, \qquad (9.19)$$

where $f^{(0)}$ and $f^{(1)}$ are two Boolean functions on $n-1$ variables. If $g^{(0)}$ and $g^{(1)}$ are the Moebius transforms of respectively $f^{(0)}$ and $f^{(1)}$, they are related to the Moebius transform g of f by the equations:

$$g(x_0, \ldots, x_{n-2}, 0) = g^{(0)}(x_0, \ldots, x_{n-2}) \text{ and} \qquad (9.20)$$
$$g(x_0, \ldots, x_{n-2}, 1) = g^{(1)}(x_0, \ldots, x_{n-2}).$$

Moreover, $f^{(0)}$ and $f^{(1)}$ are related to f by the equations:

$$f^{(0)}(x_0, \ldots, x_{n-2}) = f(x_0, \ldots, x_{n-2}, 0) \text{ and} \qquad (9.21)$$
$$f^{(1)}(x_0, \ldots, x_{n-2}) = f(x_0, \ldots, x_{n-2}, 0) \oplus f(x_0, \ldots, x_{n-2}, 1).$$

From these equations, we can devise a Moebius transform algorithm which is very similar to the Walsh transform Algorithm 9.3. This yields Algorithm 9.6, which is its own inverse. It can be as easy to implement as Program 9.2.

Algorithm 9.6 Moebius transform algorithm

Require: Input Truth table S of Boolean function f, with 2^n entries
  Comment: Variable Sz is the small table size
  Comment: Variable Pos is the small table position
  for i from 0 to $n - 1$ **do**
    Let Sz $\longleftarrow 2^i$, Pos $\longleftarrow 0$
    while (Pos $< 2^n$) **do**
      for j from 0 to Sz $- 1$ **do**
        $S[\text{Pos} + \text{Sz} + j] \longleftarrow S[\text{Pos} + j] \oplus S[\text{Pos} + \text{Sz} + j]$
      end for
      Let Pos $\longleftarrow$ Pos $+ 2 \cdot$ Sz
    end while
  end for
  Output overwritten content of S containing Moebius transform

---

9.3 Goldreich-Levin theorem

Goldreich-Levin theorem [LG89] is a fundamental theorem in the theory of cryptography. In particular, it is essential to the proof by Håstad, Impagliazzo, Levin and Luby [HILL99] that a secure pseudo-random generator can be constructed from any one-way function.

THEOREM 9.1
Let us denote by x a fixed unknown n-bit value and denote by f a fixed n-bit to t-bit one-way function. Suppose there exists an algorithm A that given the value of $f(x)$ allows to predict the value of a scalar product $(R|x)$ with probability $\frac{1}{2} + \epsilon$ over the choice R among n-bit strings, using at most T operations. Then there exists an algorithm B, which given $f(x)$ produces in time at most T' a list of at most $4n^2 \epsilon^{-2}$ values that contain x with at least

Program 9.2 C code for Moebius transform

```
#define TYPE unsigned int

/*In place Moebius transform*/
void Moebius(TYPE *Tab, TYPE size)
{
  int Wsize;
  TYPE i,i0,i1;   TYPE step;

  Wsize=size/(8*sizeof(TYPE));

  /*Moebius transform for high order bits, using word ops*/
  for (step=1;step<Wsize;step<<=1) {
    for (i1=0;i1<Wsize;i1+=2*step) {
      for (i0=0;i0<step;i0++) {
        i=i1+i0;
        Tab[i+step]^=Tab[i];
      }
    }
  }
  /*Moebius transform for low order bits, within words*/
  /* Assumes 8*sizeof(TYPE)=32 */
  for(i=0;i<Wsize;i++) {
    TYPE tmp;
    tmp=Tab[i];
    tmp^=(tmp<<16);
    tmp^=(tmp&0xff00ff)<<8;
    tmp^=(tmp&0xf0f0f0f)<<4;
    tmp^=(tmp&0x33333333)<<2;
    tmp^=(tmp&0x55555555)<<1;
    Tab[i]=tmp;
  }
}
```

1/2. *The running time T' is given by:*

$$T' = \frac{2n^2}{\epsilon^2}\left(T + \log\left(\frac{2n}{\epsilon^2}\right) + 2\right) + \frac{2n}{\epsilon^2}T_f. \tag{9.22}$$

Informally, this theorem means that, for a one-way function f, it is not possible to find an efficient algorithm that can given $f(x)$ predict scalar products of the form $(R|x)$ with a non-negligible advantage. Of course, if we remove the fact that the prediction of $(R|x)$ can be noisy, the theorem becomes a straightforward application of linear algebra methods. Indeed, from n scalar products $(R_i|x)$, we completely recover x as soon as the R_i values are linearly independent over $\mathbb{F}_2$.

Similarly, if we are given a probabilistic algorithm which outputs the value of $(R|x)$ with probability $\frac{1}{2} + \epsilon$ for any **fixed** R, recovering x is extremely easy. Indeed, in that case, by choosing for R a value with a single bit set to 1 and all other set to 0, we can learn the exact value of the corresponding bit of x. It suffices to repeat the probabilistic algorithm to obtain many independent predictions for $(R|x)$. After some time, the correct value can be detected through a majority vote (see Chapter 12).

The difficulty with the proof of Goldreich-Levin theorem is that, in some sense, we need to hide the fact that we are repeating queries. To predict the value of $(R|x)$ for a fixed R, we first choose k random auxiliary strings R'_1, ..., R'_k. Then, we call the prediction algorithm A, 2^k times, giving as input $f(x)$ together with a value $R'(\alpha) = R \oplus \bigoplus_{i=1}^{k} \alpha_i R'_i$, where α takes all possible values in $\{0,1\}^k$. Since:

$$(R|x) = (R'(\alpha)|x) \oplus \bigoplus_{i=1}^{k} \alpha_i(R'_i|x), \tag{9.23}$$

if we guess the k values of $(R'_i|x)$ we deduce 2^k different prediction for $(R|x)$. Note that these predictions are not independent, only pairwise independent, yet this suffices to make the proof go through.

Finally, from an algorithmic point-of-view, we need to compute the difference between the number of predicted 0 values and the number of predicted 1 values. This is precisely obtained by using a Walsh transform on the table of values $(R'(\alpha)|x)$.

9.4   Generalization of the Walsh transform to $\mathbb{F}_p$

The Walsh transform as presented in Section 9.1.2 is specific to the binary field. However, it is natural to consider whether a similar method is applicable to other finite fields. Since the Walsh transform is an efficient way to search for approximate linear relations between the input and output of a function, we first need to adapt the formalization of linear relations to larger finite fields. We only consider the case of prime fields $\mathbb{F}_p$. Indeed, any linear relation that holds with good probability over an extension field implies another good linear relation over the corresponding base field. Such a implied relation can be constructed by applying the trace map to the original relation. Such a generalized Walsh transform can be useful in several applications. A first application is to compute characteristic for the non-Boolean generalization of linear cryptanalysis introduced in [BSV07]. Another, more theoretical application presented in Berbain's PhD thesis [Ber07] concerns the security proof of a generalization of the QUAD family of stream ciphers from [BGP06], based on the evaluation of quadratic polynomials, to non-binary fields. This proof relies on a generalization of the Goldreich-Levin theorem to non-binary fields.

In this section, S is a function from $\mathbb{F}_p^n$ to $\mathbb{F}_p^t$ and all scalar products are computed modulo p. We first note a key difference between $\mathbb{F}_2$ and the general case $\mathbb{F}_p$. In the general case, a linear expression can take more than two values. As a consequence, it is no longer possible to regroup all cases into a single number. Over $\mathbb{F}_2$, this was possible, because for any linear relation, we could split all inputs to f into two sets: the matching entries and the non-matching entries. Since the total number of input is fixed and known, the difference between the cardinality of these two sets sufficed to encode all the information we need. Over $\mathbb{F}_p$, we need to distinguish between p different cases. As a consequence, we change our definition accordingly and define $\mathcal{L}_M^m(S)^{(\alpha)}$ as the cardinality of the set of inputs x such that:

$$(M|x) + (m|S(x)) = \alpha \pmod{p}. \tag{9.24}$$

We know that for all input and output masks:

$$\sum_{\alpha=0}^{p-1} \mathcal{L}_M^m(S)^{(\alpha)} = p^n, \tag{9.25}$$

since there are p^n possible inputs to S. Clearly, we now need $p-1$ different numbers for each pair of masks to extract the complete information about linear relations. Moreover, we can remark than there are relations between some pairs of masks. More precisely, for any non-zero constant λ in $\mathbb{F}_p$ we have:

$$\mathcal{L}_M^m(S)^{(\alpha)} = \mathcal{L}_{\lambda \cdot M}^{\lambda \cdot m}(S)^{(\lambda \alpha)}, \tag{9.26}$$

where $\lambda \cdot m$ denotes the multiplication of each component in the mask m by λ.

Of course, as with the binary field, we first need to address the basic case where S outputs a single element in $\mathbb{F}_p$. In that case, thanks to the multiplicative property above we only need to consider output masks of the form $m = 1$. Indeed, when $m = 0$ the values $\mathcal{L}_M^m(S)^{(\alpha)}$ are independent of S and uninteresting. In this context, in order to compute the linear characteristics of S, we split x into $x_0 \| x'$, M into $M_0 \| M'$ and S into p subfunctions S_{x_0} for each fixed value of the first coordinate. Using the definition, we then find:

$$
\begin{aligned}
\mathcal{L}_M^1(S)^{(\alpha)} &= \#\{x|(M|x) + S(x) = \alpha \pmod{p}\} \tag{9.27}\\
&= \#\cup_{x_0 \in \mathbb{F}_p} \{x'|(M'|x') + S(x_0\|x')\} = \alpha - M_0 x_0\}\\
&= \#\cup_{x_0 \in \mathbb{F}_p} \{x'|(M'|x') + S_{x_0}(x')\} = \alpha - M_0 x_0\}\\
&= \sum_{x_0 \in \mathbb{F}_p} \mathcal{L}_{M'}^1(S_{x_0})^{(\alpha - M_0 x_0)}.
\end{aligned}
$$

From this equations, we could easily derive a Walsh transform that operates on $p \times p^n$ values. However, this would not be optimal. Indeed, we have p^n possible input masks, p values of α for each mask and one copy of Equation (9.25) for each mask. As a consequence, it suffices to compute $(p-1) \cdot p^n$ numbers to obtain the complete characterization of linear approximations of S. In order to devise such an optimal variation, we proceed as in the binary case, keeping only the computation of $\mathcal{L}_M^1(S)^{(\alpha)}$, for non-zero values of α. Of course, this means that whenever $\mathcal{L}_M^1(S)^{(0)}$ is needed, we have to replace it using Equation (9.25). To avoid cumbersome constants, it is easier to compute the renormalized values $\bar{\mathcal{L}}_M^1(S)^{(\alpha)} = \mathcal{L}_M^1(S)^{(\alpha)} - p^{n-1}$, which are related by:

$$
\sum_{\alpha=0}^{p-1} \bar{\mathcal{L}}_M^m(S)^{(\alpha)} = 0. \tag{9.28}
$$

Moreover, since the restricted function S_{x_0} only has $n-1$ input coordinates, we also have:

$$
\bar{\mathcal{L}}_M^1(S)^{(\alpha)} = \sum_{x_0 \in \mathbb{F}_p} \bar{\mathcal{L}}_{M'}^1(S_{x_0})^{(\alpha - M_0 x_0)}. \tag{9.29}
$$

The simplest way to implement the algorithm is to recompute values of the form $\bar{\mathcal{L}}_M^1(S)^{(0)}$ when they are needed, right before the innermost loop of the algorithm. To make the algorithm complete, the only missing step is the initial encoding of the function S in an adequate representation. In the binary case, this initial encoding is a simple change between an additive representation by elements of $\{0,1\}$ to a multiplicative representation $\{-1,1\}$. In the general case, this is more complex, since we need to transform the table S of p^n elements into a larger $(p-1) \times p^n$ array. In fact, this initial encoding can be obtained by specifying the Walsh transform of a constant function with no input, say $S_0() = \alpha_0$. Using the definition, since the function can only be evaluated on the (unique) empty input, we have $\mathcal{L}_{M'}^1(S_0)^{(\alpha_0)} = 1$ and $\mathcal{L}_{M'}^1(S_0)^{(\alpha)} = 0$, for $\alpha \neq \alpha_0$. As a direct consequence, we find:

$$
\bar{\mathcal{L}}_{M'}^1(S_0)^{(\alpha_0)} = \frac{p-1}{p} \quad \text{and} \quad \bar{\mathcal{L}}_{M'}^1(S_0)^{(\alpha)} = -\frac{1}{p}. \tag{9.30}
$$

To avoid denominators, it is easier to multiply these numbers by p. Of course, since all computations are linear, this implies that the results are scaled by the same factor of p. It is interesting to remark that in the binary case, we obtain (up to sign) the usual encoding.

In addition, it is easy to remark that this Walsh transform on $\mathbb{F}_p$ is almost its own inverse. As in the binary case, during inversion we need to add a division by p. Moreover, during inversion, we need an additional change of sign when relating S to its restrictions. More precisely, in Equation (9.29), $\alpha - m_0 x_0$ needs to be replaced by $\alpha + m_0 x_0$. Of course, in the binary case, since $1 = -1 \pmod 2$ this change of sign is not visible. We give a pseudo-code description of the Walsh transform over $\mathbb{F}_p$ and its inverse as Algorithm 9.8, the preliminary encoding is given as Algorithm 9.7.

Algorithm 9.7 Pre-Walsh transform encoding over $\mathbb{F}_p$

Require: Input Table S containing p^n elements of $\mathbb{F}_p$
  Create $(p-1) \times p^n$ table $\mathcal{S}$
  for i from 0 to $p^n - 1$ **do**
    for j from 1 to $p - 1$ **do**
      Let $\mathcal{S}[j][i] \longleftarrow -1$
    end for
    if $S[i] \neq 0$ **then**
      Let $\mathcal{S}[S[i]][i] \longleftarrow p - 1$
    end if
  end for
Require: Output encoded table $\mathcal{S}$

9.4.1 Complexity analysis

Since Algorithm 9.8 is presented in iterative form, its complexity analysis is quite easy. First, when p is fixed, we see that only the outer loops depend on n and that the running time is $O(n \cdot p^n)$, where the implicit constant depends on p. When p also varies, we need to be more careful in our analysis. The contribution of the outer loop is easy, it occurs n times. Put together, the next two loops on Pos and i have a total contribution of p^{n-1} iterations, since Pos advances by steps of $p\,\mathsf{Sz}$. At the deepest level, we have a maximum of three additional levels, with a total contribution bounded by $O(p^3)$. As a consequence, when accounting for both n and p, in terms of arithmetic operations the running time can be written as $O(np^{n+2})$. Asymptotically, this can be improved for large values of p by using fast Fourier transforms, this is described in Section 9.5.2.1.

Algorithm 9.8 Walsh transform algorithm over $\mathbb{F}_p$

Require: Input encoded $(p-1) \times p^n$ table S
Require: Input Boolean value `inverse`
  Comment: Variable `Sz` is the small table size
  Comment: Variable `Pos` is the small table position
  Create temporary $p \times p$ table T.
  for l from 0 to $n-1$ **do**
    Let `Sz` $\longleftarrow p^l$, `Pos` $\longleftarrow 0$
    while (`Pos` $< p^n$) **do**
      for i from 0 to `Sz` -1 **do**
        for k from 0 to $p-1$ **do**
          Let `Sum` $\longleftarrow 0$
          for m from 1 to $p-1$ **do**
            Let $T[m][k] \longleftarrow S[m][\text{Pos} + k\,\text{Sz} + i]$
            Let $S[m][\text{Pos} + k\,\text{Sz} + i] \longleftarrow 0$
            Let `Sum` $\longleftarrow$ `Sum` $+ T[m][k]$
          end for
          Let $T[0][k] \longleftarrow -$`Sum`
        end for
        for k from 0 to $p-1$ **do**
          for m from 1 to $p-1$ **do**
            for j from 0 to $p-1$ **do**
              if `inverse` = `false` **then**
                Let $m' = m - kj \pmod{p}$
              else
                Let $m' = m + kj \pmod{p}$
              end if
              Let $S[m][\text{Pos} + k\,\text{Sz} + i] \longleftarrow S[m][\text{Pos} + k\,\text{Sz} + i] + T[m'][j]$
            end for
          end for
        end for
        if `inverse` = `true` **then**
          for k from 0 to $p-1$ **do**
            for m from 1 to $p-1$ **do**
              Let $S[m][\text{Pos} + k\,\text{Sz} + i] \longleftarrow S[m][\text{Pos} + k\,\text{Sz} + i]/p$
            end for
          end for
        end if
      end for
      Let `Pos` $\longleftarrow$ `Pos` $+ p\,\text{Sz}$
    end while
  end for
  Output overwritten content of S containing (inverse) Walsh transform

---

9.4.2 Generalization of the Moebius transform to $\mathbb{F}_p$

As in the binary case, the algebraic normal form of a function f is obtained by writing:

$$f(x_0, \ldots, x_{n-1}) = \sum_{(a_0,\ldots,a_{n-1})\in\mathbb{F}_p^n} g(a_0, \ldots, a_{n-1}) \prod_i x_i^{a_i} \quad (\text{mod } p) \quad (9.31)$$

There are two essential differences here compared to the binary case. The first difference is that g takes values modulo p instead of binary values. The second one is that the exponent of each variables runs from 0 to $p-1$, instead of simply being 0 or 1. The reason behind the second difference is that for all value x in $\mathbb{F}_p$, $x^p - x$ is always zero. This implies that when giving a polynomial expression for f, any exponent greater than p can be reduced modulo $p - 1$. Note that x^{p-1} should not be replaced by 1, because the two expressions differ when $x = 0$.

Starting again with a univariate function f on a single entry x_0 modulo p, we see that:

$$f(0) = g(0),$$

$$f(1) = \sum_{i=0}^{p-1} g(i),$$

$$\vdots \quad (9.32)$$

$$f(j) = \sum_{i=0}^{p-1} g(i)j^i,$$

$$\vdots$$

Thus, the vector of values of f is obtained by multiplying the vector of values of g by a fixed matrix:

$$H_p = \begin{pmatrix} 1 & 0 & 0 & \cdots & 0 \\ 1 & 1 & 1 & \cdots & 1 \\ 1 & 2 & 4 & \cdots & 2^{p-1} \\ \vdots & \vdots & \vdots & \ddots & \vdots \\ 1 & p-1 & (p-1)^2 & \cdots & (p-1)^{p-1} \end{pmatrix}, \quad (9.33)$$

defined modulo p.

For symmetry, it is better to reorder the rows of H_p. For this, choose a generator h of the multiplicative group $\mathbb{F}_p^*$ and write:

$$f(h^j) = \sum_{i=0}^{p-1} g(i)h^{ij} = (g(0) + g(p-1)) + \sum_{i=1}^{p-2} g(i)h^{ij} \quad (9.34)$$

The sum in the right-hand side of this equation can be recognized as a Fourier transform (see next section) of the vector G of $p - 1$ elements defined as $G(0) = g(0) + g(p - 1)$ and $G(i) = g(i)$ and modulo p the inverse transform is given by:

$$g(0) = f(0),$$

$$g(1) = -\sum_{i=0}^{p-2} f(h^i)h^{-i},$$

$$\vdots$$

$$g(j) = -\sum_{i=0}^{p-2} f(h^i)h^{-ij}, \qquad (9.35)$$

$$\vdots$$

$$g(p-1) = -\sum_{i=0}^{p-2} f(h^i) - f(0).$$

This can be used to construct Algorithm 9.9 for Moebius transforms over $\mathbb{F}_p$.

9.5 Fast Fourier transforms

Fourier transforms and fast algorithms for computing them are very general tools in computer science and they are very closely related to Walsh transforms. In order to define Fourier transforms in the general case, we require the existence of a root of unity ξ of order N. The Fourier transform defined from this root ξ maps vectors of N numbers to vectors of N numbers and the transform of a vector X is $\hat{X}$ given by:

$$\forall i : \quad \hat{X}_i = \lambda_\xi \sum_{j=0}^{N-1} X_j \xi^{ij}, \qquad (9.36)$$

where λ_ξ is a renormalization factor. In practice, λ_ξ is chosen to maximize convenience and its value may depend on the context we are considering.

It is easy to show that up to a constant factor, the Fourier transform based on ξ^{-1} is the inverse of the Fourier transform based on ξ. Indeed, the i-th

---

Algorithm 9.9 Moebius transform algorithm over $\mathbb{F}_p$

Require: Input Table S describing function f on n variables, with p^n entries

  Comment: Variable Sz is the small table size

  Comment: Variable Pos is the small table position

  Create table T of $p-1$ elements

  Choose h, generator of $\mathbb{F}_p^*$.

  for i from 0 to $n-1$ **do**

    Let Sz $\longleftarrow p^i$, Pos $\longleftarrow 0$

    while (Pos $< p^n$) **do**

      for j from 0 to Sz -1 **do**

        Set table T to 0

        for k from 0 to $p-2$ **do**

          for l from 0 to $p-2$ **do**

            Let $T[l] \longleftarrow T[l] + S[\text{Pos} + (h^k)\text{Sz} + j]h^{-kl} \bmod p$

          end for

        end for

        for l from 1 to $p-2$ **do**

          Let $S[\text{Pos} + l\text{Sz} + j] \longleftarrow -T[l]$

        end for

        Let $S[\text{Pos} + (p-1)\text{Sz} + j] \longleftarrow -T[0] - S[\text{Pos} + j]$

      end for

      Let Pos $\longleftarrow$ Pos $+ p \cdot$ Sz

    end while

  end for

  Output overwritten content of S containing Moebius transform

---

component after applying both transforms is:

$$C_i = \lambda_{\xi^{-1}} \lambda_\xi \sum_{j=0}^{N-1} \left(\sum_{k=0}^{N-1} X_k \xi^{jk} \right) \xi^{-ij} \tag{9.37}$$

$$= \lambda_{\xi^{-1}} \lambda_\xi \sum_{j=0}^{N-1} \sum_{k=0}^{N-1} X_k \xi^{j(k-i)} = \lambda_{\xi^{-1}} \lambda_\xi \sum_{k=0}^{N-1} \sum_{j=0}^{N-1} X_k \xi^{j(k-i)}$$

$$= \lambda_{\xi^{-1}} \lambda_\xi \sum_{k=0}^{N-1} X_k \sum_{j=0}^{N-1} \xi^{j(k-i)} = \lambda_{\xi^{-1}} \lambda_\xi N X_i.$$

The last equation in this chain comes for the fact that $\sum_{j=0}^{N-1} \xi^{jt}$ is equal 0 whenever t is non-zero modulo N and equal to N when t is 0 modulo N.

Due to this inversion formula, two choices of the renormalization factors are frequently encountered. In theoretical texts, the usual choice is:

$$\lambda_\xi = \lambda_{\xi^{-1}} = \frac{1}{\sqrt{N}}. \tag{9.38}$$

In computer science and especially when writing programs, it is often more convenient to choose:

$$\lambda_\xi = 1 \quad \text{and} \quad \lambda_{\xi^{-1}} = \frac{1}{N}. \tag{9.39}$$

Indeed, with this choice, we do not need to perform any divisions when computing the direct transform. Moreover, all the divisions by N that occur during the inverse transform are known to be exact. These divisions by N are analogous to the divisions by 2 that occur in the inverse Walsh transform Algorithm 9.3. Of course, if the value of ξ is not an integer, this point is moot. In particular, this is the case when the Fourier transforms are performed over the complex numbers, using for example:

$$\xi = e^{2\pi\sqrt{-1}/N}. \tag{9.40}$$

However, when performing Fourier transform computations over finite fields, choosing $\lambda_\xi = 1$ is a real asset.

From the definition of the Fourier transform, it is very easy to write a simple algorithm with quadratic running time to compute Fourier transform, see Exercise 2. However, this algorithm is not optimal and, as in the case of Walsh transforms, it is possible to compute Fourier transforms much faster, in time $N \log N$. Interestingly, the algorithms involved greatly depend on the value of the order N of the root ξ, but the end result holds for all values of N.

9.5.1   Cooley-Tukey algorithm

The most commonly known algorithm for computing discrete Fourier transforms is due to J. Cooley and J. Tukey [CT65]. This algorithm is based on a mathematical expression that allows to write the discrete Fourier transform of order $N = N_1 N_2$ in terms of smaller transforms of order N_1 or N_2. As above, let ξ be a primitive root of order N and define $\xi_1 = \xi^{N_2}$ and $\xi_2 = \xi^{N_1}$. Clearly, ξ_1 is a primitive root of order N_1 and ξ_2 is a primitive root of order N_2. The key idea is to renumber X and $\hat{X}$, writing:

$$X_{i_1,i_2} = X_{i_1 N_2 + i_2} \text{ with } 0 \leq i_1 < N_1 \text{ and } 0 \leq i_2 < N_2 \quad \text{and} \quad (9.41)$$

$$\hat{X}_{j_1,j_2} = \hat{X}_{j_1 \mid j_2 N_1} \text{ with } 0 \leq j_1 < N_1 \text{ and } 0 \leq j_2 < N_2.$$

Note that two different renumberings are used for X and $\hat{X}$. Using these renumbering and ignoring the renormalization factor λ_ξ we can write:

$$\hat{X}_{j_1,j_2} = \sum_{i=0}^{N-1} X_i \xi^{i(j_1 + j_2 N_1)} \tag{9.42}$$

$$= \sum_{i_2=0}^{N_2-1} \sum_{i_1=0}^{N_1-1} X_{i_1,i_2} \xi^{(i_1 N_2 + i_2)(j_1 + j_2 N_1)}$$

$$= \sum_{i_2=0}^{N_2-1} \sum_{i_1=0}^{N_1-1} X_{i_1,i_2} \xi_1^{i_1 j_1} \xi_2^{i_2 j_2} \xi^{i_2 j_1}$$

$$= \sum_{i_2=0}^{N_2-1} \left(\xi^{i_2 j_1} \sum_{i_1=0}^{N_1-1} X_{i_1,i_2} \xi_1^{i_1 j_1} \right) \xi_2^{i_2 j_2}.$$

Defining an intermediate step $\tilde{X}_{k_1,k_2}$ as:

$$\tilde{X}_{k_1,k_2} = \sum_{i_1=0}^{N_1-1} X_{k_1,i_2} \xi_1^{i_1 k_1}. \tag{9.43}$$

We see that $\tilde{X}$ can be computed by N_2 independent evaluations of discrete Fourier transforms of order N_1. After this first series of Fourier transforms, we obtain:

$$\hat{X}_{j_1,j_2} = \sum_{i_2=0}^{N_2-1} \left(\xi^{i_2 j_1} \tilde{X}_{j_1,i_2} \right) \xi_2^{i_2 j_2}. \tag{9.44}$$

Thus, $\hat{X}$ can be obtained by a second series of N_1 independent Fourier transforms of order N_2 applied to $\tilde{\tilde{X}}$ defined by:

$$\tilde{\tilde{X}}_{k_1,k_2} = \xi^{k_1 k_2} \tilde{X}_{k_1,k_2}. \tag{9.45}$$

This approach of Cooley and Tukey works very nicely when N can be written as a product of small primes. In that case, the above decomposition can be applied recursively and yields a fast discrete Fourier transform algorithm. The typical case of application is $N = 2^n$ and is described in Algorithm 9.10.

Algorithm 9.10 Fast Fourier transform algorithm on $N = 2^n$ values

---

Require: Input Table T containing 2^n entries
Require: Input Boolean value `inverse`
  if `inverse` = `false` **then**
    Let $\xi \longleftarrow e^{2\pi\sqrt{-1}/2^n}$
  else
    Let $\xi \longleftarrow e^{-2\pi\sqrt{-1}/2^n}$
  end if
  for k from 0 to $n - 1$ **do**
    Let Sz $\longleftarrow 2^k$, Pos $\longleftarrow 0$
    while (Pos $< 2^n$) **do**
      for j from 0 to Sz $- 1$ **do**
        Let Sum $\longleftarrow T[\text{Pos} + j] + T[\text{Pos} + \text{Sz} + j]$
        Let Diff $\longleftarrow T[\text{Pos} + j] + \xi T[\text{Pos} + \text{Sz} + j]$
        if `inverse` = `true` **then**
          Let $T[\text{Pos} + j] \longleftarrow$ Sum$/2$
          Let $T[\text{Pos} + \text{Sz} + j] \longleftarrow$ Diff$/2$
        else
          Let $T[\text{Pos} + j] \longleftarrow$ Sum
          Let $T[\text{Pos} + \text{Sz} + j] \longleftarrow$ Diff
        end if
      end for
      Let Pos $\longleftarrow$ Pos $+ 2 \cdot$ Sz
    end while
    Let $\xi \longleftarrow \xi^2$
  end for
  Output overwritten content of T containing (inverse) Fourier Transform

---

9.5.1.1 Multiplication of polynomials

The Fourier transform is often used as an ingredient of fast multiplication algorithms. These fast multiplication algorithms can be applied to integers and to polynomials. The case of polynomials is especially interesting since it sheds a different light on Fourier transform algorithms. To explain the relation between the discrete Fourier transform of order N and polynomial multiplication, we first associate to any vector X of N elements a polynomial P_X defined as:

$$P_X(x) = \sum_{i=0}^{N-1} X_i x^i. \tag{9.46}$$

With this notation, we easily remark that the coefficients of the discrete Fourier transform of X corresponds to the evaluations of the polynomial P_X at roots of unity. More precisely:

$$\hat{X}_i = P_X(\xi^i), \tag{9.47}$$

where as before ξ is a primitive root of unity of order N.

Since the discrete Fourier transform is a specific case of polynomial evaluation, its inverse is a special case of Lagrange interpolation. With this dictionary in mind, we can now study the multiplication of polynomials. Given two polynomials P_X and P_Y of degree less than N and their corresponding vectors X and Y, we know that $\hat{X}_i = P_X(\xi^i)$ and $\hat{Y}_i = P_Y(\xi^i)$. Multiplying these two identities, we find:

$$\hat{X}_i \hat{Y}_i = P_X(\xi^i) \cdot P_Y(\xi^i) = (P_X \times P_Y)(\xi^i). \tag{9.48}$$

We now denote by $\hat{Z}$ the pointwise product of $\hat{X}$ and $\hat{Y}$, i.e., let $\hat{Z}_i = \hat{X}_i \hat{Y}_i$ for all $0 \leq i < N$. The inverse transform of $\hat{Z}$ is a vector Z with associated polynomial P_Z, such that:

$$P_Z(\xi^i) = (P_X \cdot P_Y)(\xi^i), \quad \text{for all } 0 \leq i < N. \tag{9.49}$$

This implies that P_Z is equal to $P_X \cdot P_Y$ modulo the polynomial $x^N - 1$. If, in addition, we take care to choose N larger than the sum of the degrees of P_X and P_Y, then P_Z is exactly the product $P_X \cdot P_Y$.

9.5.1.1.1 Convolution product
In fact, even without the above restriction on the sum of the degrees of P_X and P_Y, the vector Z is still worth considering. Writing down explicitly the rule for polynomial multiplication modulo $x^N - 1$ we find that:

$$Z_i = \sum_{j=0}^{N-1} X_j Y_{(i-j) \bmod N} \tag{9.50}$$

$$= \sum_{j=0}^{i} X_j Y_{i-j} + \sum_{j=i+1}^{N-1} X_j Y_{N+i-j},$$

under the convention that the second sum is empty when $i = N - 1$. This vector Z is called the convolution product of X and Y and denoted by $X \star Y$.

Clearly, from an algorithmic point-of-view, convolution products can be computed very efficiently for $N = 2^n$. It suffices to compute the Fourier transform of X and Y, to perform pointwise multiplication of $\hat{X}$ and $\hat{Y}$ and finally to apply the inverse discrete Fourier transform of the pointwise product.

However, the restriction $N = 2^n$ is not necessary, convolution products can be computed efficiently for all values of N. The trick is to make the following remark, given two vectors X and Y of order N (not a power of two), we pad them into vectors X' and Y' of size N', where N' is the smallest power of two greater than N. The padded vectors X' and Y' are defined in two different ways. The first vector X' is a copy of X with as many zeros as needed in positions from N to $N' - 1$. The second vector Y' is obtained by assembling a complete copy of Y and an initial segment of Y, i.e., in positions from N to $N' - 1$ we find $Y_0, Y_1, \ldots$ up to $Y_{N'-N-1}$. Since N' is a power of two, thanks to the algorithm of Cooley and Tukey, the convolution product of X' and Y' is computed efficiently. Moreover, the first N elements of this product form the convolution product of X and Y, see Exercise 3.

9.5.2 Rader's algorithm

Since the approach of Cooley and Tukey yields a fast Fourier transform algorithm only for highly composite values of the order N, a different approach is also needed, in order to cover all possible values of N. Without loss of generality, we may restrict ourselves to prime values of N, relying on Cooley and Tukey approach to reduce the general case to the prime case. Thus, we now assume that $N = p$ is prime and we let g be a generator of the multiplicative group $\mathbb{F}_p^*$. The computation of $\hat{X}$ can then be decomposed in two cases: $\hat{X}_0$ and $\hat{X}_{g^{-i}}$ with $0 \leq i < p - 1$. Indeed, any non-zero number modulo p can be written as a power of g and we can always choose a negative exponent if we desire to do so. For $\hat{X}_0$, we have:

$$\hat{X}_0 = \sum_{j=0}^{p-1} X_j. \tag{9.51}$$

For $\hat{X}_{g^{-i}}$, we can write:

$$\hat{X}_{g^{-i}} = \sum_{j=0}^{p-1} X_j \zeta^{jg^{-i}} \tag{9.52}$$

$$= X_0 + \sum_{k=0}^{p-2} X_{g^k} \zeta^{g^{k-i}} \text{ using } j = g^k.$$

Computing the above sum simultaneously for all values of i is equivalent to a convolution product of the two sequences X_{g^k} and ξ^{g^k}, as defined in Section 9.5.1.1.1. Thus, thanks to the results of this section, it can be computed using discrete Fourier transforms of order 2^t, where t is obtained by rounding $\log_2(p-1)$ upward. Once this power of 2 has been chosen, the sequences are padded and the convolution can be computed using Cooley and Tukey algorithm for the discrete Fourier transform.

An alternative approach is to avoid the padding to a power of two and to remark that $p-1$ is always composite. As a consequence, it is possible to perform the complete Fourier transforms by alternatively using the approaches of Rader and Cooley-Tukey. However, this makes the complexity analysis much more difficult. Moreover, since the approach of Rader involves the computation of convolutions, it requires at least three applications of the Fourier transform of size $p-1$. Thus, using Rader's approach more than once is unlikely to be efficient. Yet, there is one special case, where it is a good idea. This case occurs when sequences of length $p-1$ fit in memory, while padded sequences of length 2^t do not. In that case, using Rader's approach twice guarantees that all computations are made in main memory and it is faster than swapping the computation to disk.

9.5.2.1 Application to Walsh transform over $\mathbb{F}_p$

We saw in Section 9.4 that the Walsh transform over functions from $\mathbb{F}_p^n$ to $\mathbb{F}_p$ can be computed in $O(np^{n+2})$ arithmetic operations. When p is fixed, this can be rewritten as $O(np^n)$; however, when p is allowed to grow as p^n increases, it is not clear whether $O(np^{n+2})$ is optimal. To answer this question, we now extract the core routine of the Walsh transform over $\mathbb{F}_p$. This core runtine transforms a $p \times p$ array T to another array $\hat{T}$. We give its description in Algorithm 9.11.

Algorithm 9.11 Core transform of extended Walsh over $\mathbb{F}_p$

Initialize $p \times p$ array $\hat{T}$ to 0
for k from 0 to $p-1$ **do**
  for m from 1 to $p-1$ **do**
    for j from 0 to $p-1$ **do**
      Let $\hat{T}[m][k] \longleftarrow \hat{T}[m][k] + T[m-kj][j]$
    end for
  end for
end for
Output $\hat{T}$

Equivalently, we can write the transform as in the following equation:

$$\hat{T}_{m,k} = \sum_{j=0}^{p-1} T_{m-kj,j} \tag{9.53}$$

where the index $m - kj$ is taken modulo p. Using Algorithm 9.11, this transform is computed in $O(p^3)$ arithmetic operations. In order to improve the extended Walsh transform as a whole, we are going to exhibit an asymptotically faster algorithm for this core transform.

For any p-th of unity ξ, let us define:

$$C_j^{(\xi)} = \sum_{m=0}^{p-1} \xi^m T_{m,j} \quad \text{and} \quad \hat{C}_k^{(\xi)} = \sum_{j=0}^{p-1} \xi^{kj} C_j^{(\xi)}. \tag{9.54}$$

We can remark that:

$$\hat{C}_k^{(\xi)} = \sum_{j=0}^{p-1}\sum_{m=0}^{p-1} \xi^{m+kj} T_{m,j} = \sum_{j=0}^{p-1}\sum_{n=0}^{p-1} \xi^n T_{n-kj,j} \tag{9.55}$$

$$= \sum_{n=0}^{p-1} \xi^n \sum_{j=0}^{p-1} T_{n-kj,j} = \sum_{n=0}^{p-1} \xi^n \hat{T}_{n,k}.$$

If we now consider a primitive p-th root of unity ξ, we can recover the array $\hat{T}$ from the collection of vectors $\hat{C}^{(\xi^0)}, \hat{C}^{(\xi^1)}, \dots \hat{C}^{(\xi^{p-1})}$. Indeed, we can see that:

$$p\hat{T}_{n,k} = \sum_{i=0}^{p-1} \xi^{-in} \hat{C}_k^{(\xi^i)}. \tag{9.56}$$

Indeed, in this sum, the coefficient in front of $\hat{T}_{n,k}$ is 1 in each summand; while the coefficient in front of $\hat{T}_{m,k}$ is $\xi^{(m-n)i}$. Thus, for $m \neq n$ the sum of the coefficients is 0.

As a consequence, we can compute the $p \times p$ array $\hat{T}$ in three steps:

1. Compute the p^2 values $C_j^{(\xi^i)}$, where ξ is a primitive p-th root of unity, from the values $T_{m,j}$ using:

$$C_j^{(\xi^i)} = \sum_{m=0}^{p-1} \xi^{im} T_{m,j}.$$

2. Compute the p^2 values $\hat{C}_k^{(\xi^i)}$ from the values $C_j^{(\xi^i)}$ using:

$$\hat{C}_k^{(\xi^i)} = \sum_{j=0}^{p-1} \xi^{kj} C_j^{(\xi^i)}.$$

3. Compute the p^2 values $\hat{T}_{n,k}$ from the values $\hat{C}_k^{(\xi^i)}$ using:

$$\hat{T}_{n,k} = \sum_{i=0}^{p-1} \xi^{-in} \, \hat{C}_k^{(\xi^i)} / p.$$

In each of these three steps, we recognize p copies of a Fourier transform on p elements. In the first step, the transform is performed on index m, in the second step on index j and in the third step on index i. Of course, using Rader's algorithm, all these Fourier transforms can be computed using $p \log(p)$ arithmetic operations. As a consequence, replacing the core Algorithm 9.11 with cost $O(p^3)$ by these fast Fourier transforms with total cost $O(p^2 \log(p))$ yields a fast extended Walsh algorithm with complexity $O(n \log(p) p^{n+1})$ arithmetic operations, where the O constant is independent of both p and n.

Since we are considering the asymptotic complexity, it is also important to express it in terms of bit operations, rather than arithmetic operations. For this, we need to specify where the roots of unity ξ are taken and what the size of involved integers is. It is clear that the integers are smaller than p^n in absolute value, and can, thus, be represented on $n \log(p)$ bits. For the roots of unity, two options are available, we can either take complex roots with sufficient precision to recover the array $\hat{T}$ or p-th root of unity modulo a large enough prime Q such that p divides $Q - 1$. In both cases, assuming that we are using a fast asymptotic multiplication algorithm, the bit complexity is $O(\log(n) \log \log(p)(n \log(p))^2 p^{n+1})$. Thus, the fast Fourier transform can be used to asymptotically improve the extension of the Walsh transform algorithm to $\mathbb{F}_p$. However, in practice, p^n cannot be too large, since the representation $\mathcal{S}$ needs to fit in memory. As a consequence, p and n cannot both be large at the same time and this asymptotic approach is very unlikely to be useful.

9.5.3 Arbitrary finite abelian groups

All the previous sections in this chapter deal with Fourier (or Walsh) transforms in abelian groups of various forms, $\mathbb{Z}/N\mathbb{Z}$ or $\mathbb{F}_p^n$. We now generalize this and show that Fourier transforms can be defined and computed efficiently for arbitrary finite abelian groups. Let $(\mathbb{G}, \times)$ be a finite abelian group. A **character** on $\mathbb{G}$ is a **multiplicative** map χ from $\mathbb{G}$ to the field[3] $\mathbb{C}$ of complex numbers. In this context, multiplicative means that for any pair of group elements (g, h), we have $\chi(gh) = \chi(g)\chi(h)$. In particular, this implies the following properties.

1. If e denotes the neutral element of $\mathbb{G}$, we have $\chi(e) = 1$.

---

[3] Here $\mathbb{C}$ could be replaced by any field containing enough roots of unity. For simplicity, we only describe the case of the complex numbers.

2. For any g in $\mathbb{G}$ and any α in $\mathbb{Z}$: $\chi(g^\alpha) = \chi(g)^\alpha$.

3. For any g in $\mathbb{G}$, we have $g^{|\mathbb{G}|} = e$ and $\chi(g)^{|\mathbb{G}|} = 1$. Thus, χ takes its values in the group $M_{\mathbb{G}}$ of $|\mathbb{G}|$-th roots of unity in $\mathbb{C}$. As usual, $|\mathbb{G}|$ denotes the order (or cardinality) of $\mathbb{G}$. Note that depending on $\mathbb{G}$ the range of characters may be a proper subgroup of $M_{\mathbb{G}}$.

4. The set of all characters on $\mathbb{G}$ forms a group denoted $\hat{\mathbb{G}}$ where multiplication of characters is defined by:

$$\forall g \in \mathbb{G}: \ (\chi_1\chi_2)(g) = \chi_1(g)\chi_2(g). \tag{9.57}$$

5. The neutral element of $\hat{\mathbb{G}}$ is the so-called trivial character $\hat{1}$ which sends any group element to 1 in $\mathbb{C}$.

6. The group $\hat{\mathbb{G}}$ is abelian and finite. Commutativity is clear from the definition. Finiteness is a direct consequence of the fact that both $\mathbb{G}$ and the set $M_{\mathbb{G}}$ of $|\mathbb{G}|$-th roots of unity are finite, thus there is only a finite number of possible mappings from $\mathbb{G}$ to the $|\mathbb{G}|$-th roots of unity.

7. If the group $\mathbb{G}$ can be written as a direct product $\mathbb{G}_1 \times \mathbb{G}_2$, then $\hat{\mathbb{G}} = \hat{\mathbb{G}}_1 \times \hat{\mathbb{G}}_2$.

PROOF In one direction, if χ_1 is a character on $\mathbb{G}_1$ and χ_2 a character on $\mathbb{G}_2$ we can define a character $\chi_1 \times \chi_2$ on $\mathbb{G}$ letting $(\chi_1 \times \chi_2)(g) = \chi_1(g_1)\chi_2(g_2)$ for $g = (g_1, g_2) \in \mathbb{G}_1 \times \mathbb{G}_2$. Thus, $\hat{\mathbb{G}}_1 \times \hat{\mathbb{G}}_2$ is naturally embedded into $\hat{\mathbb{G}}$.

In the other direction, let $\chi \in \hat{\mathbb{G}}$ and define χ_1 and χ_2, respectively, in $\hat{\mathbb{G}}_1$ and $\hat{\mathbb{G}}_2$ by the following formulas:

$$\forall g_1 \in \mathbb{G}_1 : \chi_1(g_1) = \chi((g_1, 1)) \text{ and} \tag{9.58}$$
$$\forall g_2 \in \mathbb{G}_2 : \chi_2(g_2) = \chi((1, g_2)). \tag{9.59}$$

Clearly, χ_1 and χ_2 are multiplicative. Furthermore, $\chi = \chi_1 \times \chi_2$. This concludes the proof.   □

8. There is an isomorphism between $\mathbb{G}$ and $\hat{\mathbb{G}}$.

PROOF Thanks to the previous property and decomposing $\mathbb{G}$ into a product of cyclic groups, it suffices to prove this fact when $\mathbb{G}$ is cyclic. We assume that g_0 denotes a generator of $\mathbb{G}$.

In that case, a character χ is entirely determined by its value at g_0. Indeed, any element g in $\mathbb{G}$ can be written as g_0^α and by multiplicativity, we find $\chi(g) = \chi(g_0)^\alpha$. Since, $\chi(g_0)$ belongs to the set $M_{|G|}$ of $|G|$-th roots of unity, there are $|G|$ possible choices for $\chi(g_0)$. Thus, G and $\hat{G}$ have the same cardinality. It remains to prove that $\hat{G}$ is also cyclic to conclude that they are isomorphic. This is done by choosing a primitive $|G|$-th root of unity μ in $\mathbb{C}$ and by proving that the character χ_μ defined by $\chi_\mu(g_0) = \mu$ is a generator for $\hat{G}$. Indeed, for any χ it is possible to express $\chi(g_0)$ as μ^β, this implies that $\chi = \chi_\mu^\beta$. ☐

9. Furthermore, G is naturally isomorphic to $\hat{\hat{G}}$.

PROOF The existence of an isomorphism is clear from the previous item. Moreover, given an element g of G, we have a natural action of g on $\hat{G}$ given by:

$$g(\chi) = \chi(g).$$

To check that this is indeed the group isomorphism we seek, we need to show that two different elements of G induce different actions on $\hat{G}$. Equivalently, it suffices to prove that for $g \neq e$, there exists $\chi \in \hat{G}$ such that $g(\chi)$. As before, it suffices to give a proof for cyclic group. Using the previous notations, $g = g_0^\alpha$, we see that $g(\chi_\mu) = \mu^\alpha \neq 1$. ☐

10. For $g \neq e$ in G, we have:

$$\sum_{\chi \in \hat{G}} \chi(g) = 0. \tag{9.60}$$

In addition:

$$\sum_{\chi \in \hat{G}} \chi(e) = |\hat{G}| = |G|. \tag{9.61}$$

PROOF Equation (9.61) is clear, since for all χ, we have $\chi(e) = 1$. To prove Equation (9.60), we write G as a product of cyclic groups $G = G_1 \times G_2 \times \cdots \times G_k$ and the sum can be rewritten as:

$$\sum_{\chi_1 \in \hat{G}_1} \cdots \sum_{\chi_k \in \hat{G}_k} \chi_1(g_1) \cdots \chi_k(g_k) = 0.$$

As a consequence, it suffices to prove Equation (9.60) for a single cyclic component with $g_i \neq e_i$. In this case, the sum becomes a sum of powers of some root of unity and equal to 0. ☐

Given all characters on a group $\mathbb{G}$, we can now define the Fourier transform of an arbitrary function f from $\mathbb{G}$ to $\mathbb{C}$ as:

$$\hat{f}(\chi) = \sum_{g \in \mathbb{G}} \chi(g) f(g). \tag{9.62}$$

The Fourier transform is invertible and the inverse is obtained by:

$$f(g) = \frac{1}{|\mathbb{G}|} \sum_{\chi \in \hat{\mathbb{G}}} \chi^{-1}(g) \hat{f}(\chi). \tag{9.63}$$

This is a simple consequence of Equations (9.61) and (9.60).

In the general case, the Fourier transform can be used as a tool to compute the so-called convolution product $f_1 \star f_2$ of two functions f_1 and f_2 from $\mathbb{G}$ to $\mathbb{C}$ defined as:

$$(f_1 \star f_2)(g_0) = \sum_{g \in \mathbb{G}} f_1(g) f_2(g_0 g^{-1}). \tag{9.64}$$

Indeed, in general, we have $\widehat{f_1 \star f_2} = \hat{f}_1 \cdot \hat{f}_2$.

9.5.3.1   Link with the previous cases

To illustrate this general setting, let us compare it to the special cases of the Walsh and Fourier transforms we already addressed. In each case, it suffices to specify how the objects we are considering are encoded into functions from a group $\mathbb{G}$ to $\mathbb{C}$. In the case of the Fourier transform, we are considering vectors of N elements and work with the group $\mathbb{Z}/N\mathbb{Z}$. The link with the general case is obtained by mapping the vector X into a function f_X over $\mathbb{Z}/N\mathbb{Z}$ defined by $f_X(i) = X_i$. Since the group $\mathbb{Z}/N\mathbb{Z}$ has cardinality N, the N-th root of unity $\xi = e^{2\pi\sqrt{-1}/N}$ naturally arises.

For the Walsh transform, we are already considering functions from $\mathbb{F}_p^n$ to $\mathbb{F}_p$ and it suffices to represent the range $\mathbb{F}_p$ using the p-th roots of unity in $\mathbb{C}$. Note that there are several possible representations, one is trivial and maps every element of $\mathbb{F}_p$ to 1. There are also $p-1$ non-trivial representations obtained by sending 1 (an additive generator for $\mathbb{F}_p$) to one of the $p-1$ primitive p-th of unity, say ξ and by sending an element x of $\mathbb{F}_p$ to ξ^x. In truth, the algorithms of Section 9.4 can be viewed as p (or $p-1$) parallel applications of the general case, one for each possible representation.

Exercises

1. Looking back at truncated differential and at Equation (9.16), we would like to compute truncated differentials for an S-box S on a single bit from a table of ordinary differentials. Given a fixed input mask with k unknown bits, show that it suffices to sum 2^k values. Write a program which computes the correct sum.

2^h. Directly following Equation (9.36), write a program to compute Fourier transform. What is the complexity of this program?

3. In Section 9.5.1.1.1, we have seen that to compute the convolution product of two vectors X and Y on N elements, we can pad X with zeros up to the next power of 2, pad Y with a copy of its initial segment and compute the convolution product of the padded copies. Check that both convolution products are indeed equal by using the program resulting from Exercise 3. Prove the equality.

4^h. How many multiplications (of reals) are needed to multiply two complex numbers?

5. Implement the Moebius transform on $\mathbb{F}_3$.

6^h. Study the Walsh transform given as Program 9.3 and its timings. Propose a method to avoid the performance penalty due to cache effects.

Chapter 10

Lattice reduction

10.1 Definitions

Lattices are rich mathematical objects, thus before studying the algorithmic issues involving lattices, it is essential to recall some important definitions and properties concerning lattices.

DEFINITION 10.1 *A* **lattice** *is a discrete additive subgroup of* $\mathbb{R}^n$.

While very short, this definition needs to be studied carefully to better understand the notion. First recall that an additive subgroup of $\mathbb{R}^n$ is a set S of elements such that:

- If $\vec{x} \in S$ then $-\vec{x} \in S$.

- If $\vec{x} \in S$ and $\vec{y} \in S$ then $\vec{x} + \vec{y} \in S$.

As a consequence, the zero vector belongs to S and any linear combination of vectors of S with integer coefficients is a vector of S.

However, a lattice L is more than an additive subgroup, it is a discrete additive subgroup. This additional topological property implies that it is not possible to construct a sequence of non-zero vectors in L with a zero limit. Alternatively, it means that there is a small n-dimensional ball centered at zero that contains no non-zero vector in L. As a consequence, if we are given a norm on $\mathbb{R}^n$, usually the Euclidean Norm, there exists a non-zero vector which minimizes this norm. The norm, or length, of this vector is called the first minimum of the lattice and denoted by $\lambda_1(L)$. Let $\vec{v_1}$ be a vector of L with norm $\lambda_1(L)$, then any vector $\vec{w}$ in L such that the pair $(\vec{v_1}, \vec{w})$ is linearly dependent can be written as $\vec{w} = \alpha \vec{v_1}$ for some integer α. Indeed, assuming that α is not an integer, implies that $\vec{w} - \lceil \alpha \rfloor \vec{v_1}$ is a non-zero vector in L with norm smaller than $\lambda_1(L)$ and leads to a contradiction.

This basic consideration with a single short vector illustrates two very important properties about lattices: the existence of bases and the sequence of successive minima.

A **basis** for a lattice L is a set of linearly independent vectors, $\vec{b_1}$, ..., $\vec{b_r}$ belonging to L such that any vector in L can be written as a linear combination, with integer coefficients, of the vectors $\vec{b_i}$. Any lattice L admits a basis, moreover, the cardinality r is the same for all bases of a given lattice and is called the **rank** of L. It is equal to the dimension of the vector subspace of $\mathbb{R}^n$ spanned by L.

In cryptography, we often encounter, as a special case, lattices that only contain vectors with integer coordinates. Such lattices are called **integer lattices**.

Let L be a lattice of rank r in $\mathbb{R}^n$ and B and B' be two bases of L. Each basis can be represented as a matrix of dimension $r \times n$, where each row of the matrix B (resp. B') contains the coordinates of one vector in the basis B. Since all vectors of B' belong to L, they can be expressed as a linear combination of the vectors of B with integer coefficients. As a consequence, there exists a square integer matrix U of dimension r such that:

$$B' = UB. \tag{10.1}$$

Of course, there also is a matrix U' with integer coefficients such that $B = U'B'$. Putting the two relations together, we find that U and U' are inverses of each other. Since both U and U' are integer matrices, this implies that the determinant of U is equal to 1 or -1. Such a matrix is called **unimodular**.

A direct consequence of the above relationship between two bases of L is that for any basis B, the determinant $\det(B \cdot^\top B)$ is independent of the choice of B. Indeed:

$$\det(B'\cdot^\top B') = \det(UB\cdot^\top B\cdot^\top U) = \det(U)\det(B\cdot^\top B)\det(^\top U) = \det(B\cdot^\top B). \tag{10.2}$$

Since in addition this invariant is a positive number, it is traditional to define the **determinant** of the lattice L as:

$$\det(L) = \sqrt{\det(B \cdot^\top B)}. \tag{10.3}$$

The determinant of L is also called the volume or sometimes co-volume of L. In fact, the matrix $B \cdot^\top B$ itself is also important and is known as the Gram matrix of the basis B.

For every lattice L, we can generalize the notion of first minimum $\lambda_1(L)$ and define the sequence of successive minima. The k-th minimum of the lattice L is defined as the smallest positive real number $\lambda_k(L)$ such that there exists at least one set of k linearly independent vectors of L, with each vector of norm at most $\lambda_k(L)$. Clearly, the sequence of successive minima satisfies:

$$\lambda_1(L) \leq \lambda_2(L) \leq \lambda_3(L) \leq \cdots \tag{10.4}$$

It is important to remark that in this definition, we do not only count the number of vectors, but also ask that they are linearly independent. To understand why, note that in any lattice, there are at least two vectors with minimal length, since when x is one, $-x$ is another. Of course, for some lattices, it is possible to have more than two vectors with minimal length.

From a mathematical point-of-view there exists an interesting relationship between the values of the successive minima and the determinant of the lattice. In particular, a theorem of Minkowski states that:

THEOREM 10.1
For every integer $r > 1$, there exists a constant γ_r, such that for any lattice L of rank r and for all $1 \leq k \leq r$:

$$\left(\prod_{i=1}^{k} \lambda_i(L) \right)^{1/k} \leq \sqrt{\gamma_r} \det(L)^{1/r} \qquad (10.5)$$

For more information about lattices, the reader may refer to [MG02].

10.2   Introductory example: Gauss reduction

Before studying general lattices and their reduction, it useful to start with simple examples in small dimension. In this section, we consider integer lattices in dimension 2, i.e., full rank additive subgroups of $\mathbb{Z}^2$, their descriptions and the related algorithms. Note that, in the special case of dimension 2, lattice reduction can also be expressed in the alternate language of binary quadratic forms; for a detailed exposition, see [BV07].

To describe such a lattice, we are given a basis $(\vec{u}, \vec{v})$ formed of two linearly independent vectors. For example, we might consider the lattice given in Figure 10.1. We see on this figure that an arbitrary basis does not necessarily give a pleasant description of the lattice. In fact, for this example, a much better description is the reduced basis presented in Figure 10.2. In particular, we see that the reduced basis contains much shorter vectors than the initial basis; we also see that the vectors are nearly orthogonal. Computing a reduced basis in dimension 2 can be done by using the Gauss reduction algorithm. This algorithm is a greedy algorithm which iteratively reduces the length of the longest vector in the basis. It continues as long as the reduced vector becomes shorter than the other one, then it stops. We describe it as Algorithm 10.1. A sample run is presented in Figure 10.3. The reader may also notice the similarity of this algorithm with Euclid's GCD Algorithm 2.1. Throughout this chapter, the length of vectors is measured using the Euclidean norm $\| \cdot \|$.

In order to better understand Gauss's algorithm, we are going to check a few simple facts and prove its correctness:

  1. *The algorithm always outputs a basis of the correct lattice.* This can be shown by remarking that the lattice generated by $(\vec{u}, \vec{v})$ is a loop

---

Algorithm 10.1 Gauss's reduction algorithm

Require: Initial lattice basis $(\vec{u}, \vec{v})$
  if $\|\vec{u}\| < \|\vec{v}\|$ **then**
    Exchange $\vec{u}$ and $\vec{v}$
  end if
  repeat
    Find integer λ that minimizes $\|\vec{u} - \lambda\vec{v}\|$ by:
    $\lambda \longleftarrow \left\lfloor (\vec{u}|\vec{v})/\|\vec{v}\|^2 \right\rceil$
    Let $\vec{u} \longleftarrow \vec{u} - \lambda\vec{v}$
    Swap $\vec{u}$ and $\vec{v}$
  until $\|u\| \le \|v\|$
  Output $(\vec{u}, \vec{v})$ as reduced basis

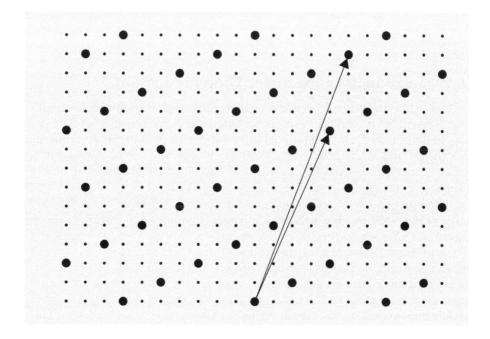

Figure 10.1: A 2-dimensional lattice with a basis

invariant. In other words, the lattice entering any step of the main loop is the same as the lattice exiting this loop. As a consequence, the algorithm as a whole preserves the input lattice. We simply need to check that the bases $(\vec{u}, \vec{v})$ and $(\vec{U}, \vec{V}) = (\vec{u} - \lambda\vec{v}, \vec{v})$ generate identical lattices. Take an element $\vec{x} = a\vec{U} + b\vec{V}$ of the second lattice, i.e., with integer values a and b. Clearly $\vec{x} = a\vec{u} + (b - \lambda a)\vec{v}$ and $\vec{x}$ belongs to the first lattice. Conversely, take $\vec{x} = a\vec{u} + b\vec{v}$ in the first lattice and check

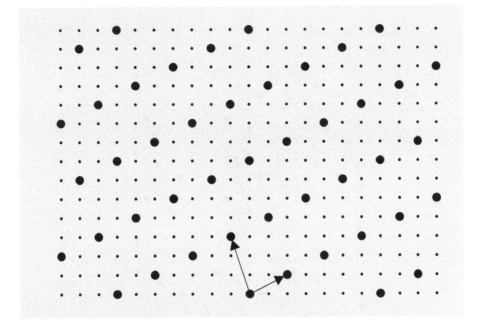

Figure 10.2: A reduced basis of the same lattice

that $\vec{x} = a\vec{U} + (b + \lambda a)\vec{V}$ belongs to the second lattice.

2. *The algorithm always stops.* The norm of the longest vector in the basis decreases from one loop to the next. Since the squared norm is a positive integer which also decreases, the algorithm must stop.

3. *The computation of λ in the algorithm indeed minimizes the norm of $\vec{u} - \lambda \vec{v}$.* Let $\vec{x} = \vec{u} - \lambda \vec{v}$, we have to minimize:

$$\|\vec{x}\| = \|\vec{u}\|^2 - 2\lambda(\vec{u}|\vec{v}) + \lambda^2\|\vec{v}\|^2.$$

This norm is a quadratic function in λ. For real values of λ it is minimized at $(\vec{u}|\vec{v})/\|\vec{v}\|^2$. For integer values of λ the minimum is obtained by rounding this quantity to the nearest integer. The minimum is unique except when the value to be rounded is a half integer, in which case there are two equally good solutions.

4. *The first vector of the reduced basis is the shortest non-zero lattice vector.* First, if the algorithm outputs $(\vec{u}, \vec{v})$ we have $\|\vec{u}\| \leq \|\vec{v}\|$, thanks to the loop exit condition. Then take $\vec{x} = a\vec{u} + b\vec{v}$ any non-zero element of the lattice, i.e., with a and b integers and $(a, b) \neq (0, 0)$. Since the algorithm

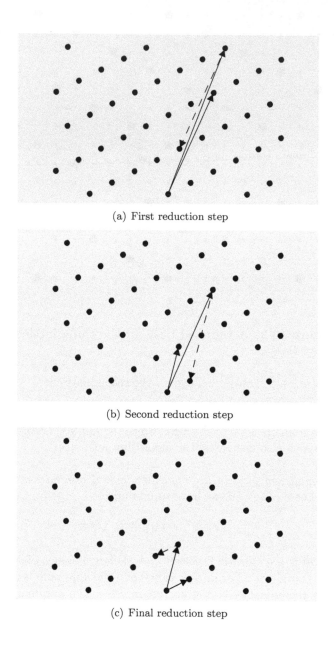

(a) First reduction step

(b) Second reduction step

(c) Final reduction step

Figure 10.3: Applying Gauss's algorithm

ensures that $|(\vec{u}|\vec{v})| \leq \|\vec{u}\|^2/2$ can

$$\|\vec{x}\|^2 = a^2\|\vec{u}\|^2 + 2ab(\vec{u}|\vec{v}) + b^2\|\vec{v}\|^2$$
$$\geq (a^2 - |ab| + b^2)\|\vec{u}\|^2$$

Clearly, $T = a^2 - |ab| + b^2$ is an integer. Moreover, thanks to the symmetry in a and b, we can assume without loss of generality that $b \neq 0$. In that case, we can write:

$$T = |b|^2 \cdot ((|a/b|^2 - |a/b| + 1).$$

From this expression, we conclude that $T > 0$. Indeed $|b|^2 > 0$ and the quadratic expression $x^2 - x + 1$ has negative discriminant and thus $x^2 - x + 1 > 0$ for all real values of x, including $x = |a/b|$. Since T is an integer, we conclude that $T \geq 1$ and finally

$$\|\vec{x}\|^2 \geq \|\vec{u}\|^2.$$

This shows that no non-zero vector in the lattice can be shorter than $\vec{u}$. However, a few other vectors may be as short as $\vec{u}$. Studying our bounds in more details, we see that for all such vectors we have $-1 \leq a \leq 1$ and $-1 \leq b \leq 1$. Clearly $-\vec{u}$ is also shortest vector in any lattice, and whenever $\|\vec{v}\| > \|\vec{u}\|$ it is the only other possibility. When $\|\vec{v}\| = \|\vec{u}\|$, we have two additional solutions, $\vec{v}$ and $-\vec{v}$. If in addition $|(\vec{u}|\vec{v})| = \|\vec{u}\|^2/2$, two more vectors reach the minimum length either $\vec{u} - \vec{v}$ and its opposite or $\vec{u} + \vec{v}$ and its opposite, depending on the sign of $(\vec{u}|\vec{v})$. Lattices illustrating these three possibilities are shown in Figure 10.4. Note that this third option cannot be achieved with integer lattices. Indeed, a typical basis for this is generated by:

$$\vec{u} = \begin{pmatrix} 1 \\ 0 \end{pmatrix} \quad \text{and} \quad \vec{u} = \begin{pmatrix} 1/2 \\ \sqrt{3}/2 \end{pmatrix}.$$

Since $\sqrt{3}$ is irrational, it is clear that this lattice can only be approximated by integer lattices.

We leave as an exercise to the reader to show that no vector in the lattice linearly independent from $\vec{u}$ can be shorter than $\vec{v}$. There are either two vectors $\vec{v}$ and $-\vec{v}$ reaching this minimum or four vectors. With four vectors, the additional pair is formed of $\vec{u} - \vec{v}$ and its opposite or of $\vec{u} + \vec{v}$ and its opposite, depending on the sign of $(\vec{u}|\vec{v})$.

10.2.1 Complexity analysis

In order to give a more precise analysis of Gauss's algorithm it is very useful to introduce a variant called t-Gauss reduction that takes a parameter $t \geq 1$ and is described as Algorithm 10.2.

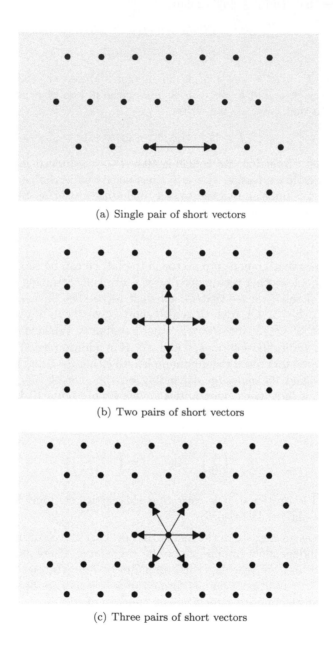

(a) Single pair of short vectors

(b) Two pairs of short vectors

(c) Three pairs of short vectors

Figure 10.4: Typical cases of short vectors in 2-dimensional lattices

---

Algorithm 10.2 t-Gauss reduction algorithm

Require: Initial lattice basis $(\vec{u}, \vec{v})$
Require: Parameter $t \geq 1$
  if $\|\vec{u}\| < \|\vec{v}\|$ **then**
    Exchange $\vec{u}$ and $\vec{v}$
  end if
  repeat
    Find integer λ that minimizes $\|\vec{u} - \lambda\vec{v}\|$ by:
$$\lambda \longleftarrow \left\lfloor (\vec{u}|\vec{v})/\|\vec{v}\|^2 \right\rceil$$
    Let $\vec{u} \longleftarrow \vec{u} - \lambda\vec{v}$
    Swap $\vec{u}$ and $\vec{v}$
  until $\|\vec{u}\| \leq t\|\vec{v}\|$
  Output $(\vec{u}, \vec{v})$ as reduced basis

The only difference between t-Gauss and Gauss's algorithms is the loop exit condition. With t-Gauss, we no longer ask for the new vector to be shorter than previous vectors, we only require it to be not too long compared to the vector $\vec{v}$ that is kept from the previous iteration. With this new condition, the length of the shortest vector decreases by a factor t at every iteration but the last one. This also implies that the length of the longest vector decreases by a factor t at every iteration, except maybe during the first one. Indeed, the longest vector in an iteration is the shortest vector in the next one. As a consequence t-Gauss algorithm for $t > 1$ is a polynomial time algorithm since its number of iterations is smaller than $1 + \log(\max(\|\vec{u}\|, \|\vec{v}\|))/\log(t)$. Moreover, if we denote by $k(t)$ the number of iterations in t-Gauss, we have:

$$k(\sqrt{3}) \leq k(1) \leq k(\sqrt{3}) + 1.$$

This implies that Gauss's algorithm, which is identical to t-Gauss when $t = 1$, also is a polynomial time algorithm. The lower bound is clear. To prove the upper bound, we now let $t = \sqrt{3}$ and study the output basis of $\sqrt{3}$-Gauss. Let α and β denote the two real numbers such that the output basis $(\vec{u}, \vec{v})$ satisfies:

$$\|\vec{u}\| = \alpha\|\vec{v}\| \quad \text{and} \quad (\vec{u}|\vec{v}) = \beta\|\vec{v}\|. \tag{10.6}$$

Several cases are possible:

1. If $\alpha \leq 1$, the basis is already Gauss reduced and thus in this instance, $k(1) = k(\sqrt{3})$.

2. If $\sqrt{3} \geq \alpha > 1$ and $-1/2 < \beta < 1/2$, Gauss algorithm would have an additional iteration, which would simply exchange $\vec{u}$ and $\vec{v}$. In this case, $k(1) = k(\sqrt{3}) + 1$.

3. If $\sqrt{3} \geq \alpha > 1$ and $|\beta| > 1/2$, we know that after the final iteration of $\sqrt{3}$-Gauss, we have:

$$|(\vec{u}|\vec{v})| \leq \|\vec{u}\|^2/2 = \alpha^2/2\|\vec{v}\| \leq 3/2 \cdot \|\vec{v}\|. \qquad (10.7)$$

Thus, the next iteration of the complete Gauss algorithm would replace $\vec{u}$ by either $\vec{u} + \vec{v}$ or $\vec{u} - \vec{v}$, depending on the sign of β. Moreover, due to Equation (10.7), we notice that $|\beta| \leq \alpha^2/2$. This implies that:

$$\|\vec{u} - \vec{v}\|^2 = \|\vec{u}\|^2 - 2(\vec{u}|\vec{v}) + \|\vec{v}\|^2 \qquad (10.8)$$
$$= (\alpha^2 + 1 - 2\beta)\|\vec{v}\|^2 \geq \|\vec{v}\|^2.$$

And similarly that $\|\vec{u} + \vec{v}\|^2 \geq \|\vec{v}\|^2$. Thus, after this additional iteration, Gauss algorithm also stops and $k(1) = k(\sqrt{3}) + 1$.

4. Finally, if $\sqrt{3} \geq \alpha > 1$ and $|\beta| = 1/2$, depending on the rounding convention, one of the two above cases applies. We also have $k(1) = k(\sqrt{3}) + 1$.

Much more is known about Gauss's algorithm and its complexity; useful references are [DFV97] or [Val91].

10.3 Higher dimensions

In order to reduce lattices in higher dimension, it is important to first determine what a reduced lattice is. A first attempt is to generalize the mathematical criteria that says that in dimension 2 a basis of a lattice L is reduced when its first and second vectors realize the first and second minima $\lambda_1(L)$ and $\lambda_2(L)$. Generalizing this, we would ask for a basis that realizes the successive minima of a higher dimensional lattice. However, in large dimension, this is not possible. For example, look at the following 5-dimensional lattice (generated by its rows):

$$L = \begin{pmatrix} 2 & 0 & 0 & 0 & 0 \\ 0 & 2 & 0 & 0 & 0 \\ 0 & 0 & 2 & 0 & 0 \\ 0 & 0 & 0 & 2 & 0 \\ 1 & 1 & 1 & 1 & 1 \end{pmatrix}.$$

The successive minima of L are:

$$\lambda_1 = 2, \lambda_2 = 2, \lambda_3 = 2, \lambda_4 = 2 \text{ and } \lambda_5 = 2.$$

To realize these minima, we may take the following family of vectors:

$$F_L = \begin{pmatrix} 2\,0\,0\,0\,0 \\ 0\,2\,0\,0\,0 \\ 0\,0\,2\,0\,0 \\ 0\,0\,0\,2\,0 \\ 0\,0\,0\,0\,2 \end{pmatrix}.$$

Clearly, this family is not a lattice basis, because it cannot yield the vector $(1, 1, 1, 1, 1)$.

In fact, in large dimension, there is no good unique definition of a reduced lattice. We can try to minimize the basis with respect to a variety of properties; however, the resulting definitions are not equivalent. In 1982, Lenstra, Lenstra and Lovász [LLL82] proposed a polynomial time algorithm able to produce reasonably good basis, using a relaxed notion for reduced basis. This notion, called LLL-reduction is defined as follows.

DEFINITION 10.2 *A basis B of a lattice L is said to be δ-**LLL** reduced for a parameter $1/4 < \delta \leq 1$ if the following conditions are satisfied:*

$$\forall\, i < j : \left| (\vec{b}_j | \vec{b}_i^*) \right| \leq \frac{\|\vec{b}_i^*\|^2}{2}, \tag{10.9}$$

$$\forall\, i : \delta \|\vec{b}_i^*\|^2 \leq \left(\|\vec{b}_{i+1}^*\|^2 + \frac{(\vec{b}_{i+1}|\vec{b}_i^*)^2}{\|b_i^*\|^2} \right), \tag{10.10}$$

where the vectors $\vec{b}_i^*$ result from the Gram-Schmidt orthogonalization of the basis B.

A basis that satisfies the single condition of Equation (10.9) is said to be size-reduced. Moreover, Equation (10.10) is often referred to as Lovász condition.

10.3.1 Gram-Schmidt orthogonalization

Since Gram-Schmidt orthogonalization is already present in the definition of LLL-reduction, before presenting Lenstra, Lenstra and Lovász lattice reduction algorithm, it is essential to recall the details of Gram-Schmidt orthogonalization process. This algorithm takes as input a basis $B = (\vec{b}_1, \vec{b}_2, \cdots, \vec{b}_n)$ of any vector subspace of $\mathbb{R}^m$ (with $m \geq n$) and outputs an orthogonal basis of the same vector subspace $B^* = (\vec{b}_1^*, \vec{b}_2^*, \cdots, \vec{b}_n^*)$. In addition, B^* satisfies the following conditions:

1. $\vec{b}_1^* = \vec{b}_1$,

2. $\vec{b}_i^*$ is the projection of $\vec{b}_i$, orthogonally to the vector subspace generated by the $i - 1$ first vectors of B.

Together with B^* and assuming that vectors are represented by rows, Gram-Schmidt algorithm computes a lower triangular matrix M, with 1s on its diagonal, such that $B = MB^*$. Equivalently, this means that for any index j, we have:

$$\vec{b}_j = \vec{b}_j^* + \sum_{i=1}^{j-1} m_{j,i} \vec{b}_i^*. \qquad (10.11)$$

Gram-Schmidt orthogonalization is given in pseudo-code as Algorithm 10.3. The algorithm is illustrated in dimension 3 in Figure 10.5.

The output of the Gram-Schmidt algorithm also allows to write the Gram matrix and the determinant of a lattice in a convenient form, indeed:

$$B \cdot {}^\top B = M \cdot B^* \cdot {}^\top B^* \cdot {}^\top M. \qquad (10.12)$$

Moreover, $B^* \cdot {}^\top B^*$ is a diagonal matrix with diagonal entries equal to $\|\vec{b}_i^*\|^2$. Since the determinant of the square matrix M is 1, this shows that the determinant of a lattice can be computed as the product of the norms of the vectors in any Gram-Schmidt orthogonal basis.

Algorithm 10.3 Gram-Schmidt algorithm

Require: Initial basis of the vector space $B = (\vec{b}_1, \vec{b}_2, \cdots, \vec{b}_n)$
  Create basis B^* and transformation matrix M
  for i from 1 to n **do**
    Let $\vec{b}_i^* \longleftarrow \vec{b}_i$
    for j from 1 to $i-1$ **do**
      Let $m_{i,j} \longleftarrow \dfrac{(\vec{b}_i | \vec{b}_j^*)}{\|\vec{b}_j^*\|^2}$
      Let $\vec{b}_i^* \longleftarrow \vec{b}_i^* - m_{i,j} \vec{b}_j^*$
    end for
  end for
  Output B^* and M

10.3.2 Lenstra-Lenstra-Lovász algorithm

The algorithm of Lenstra-Lenstra-Lovász, also called the LLL or L^3 algorithm is obtained by combining Gauss reduction in dimension two, together with Gram-Schmidt orthogonalization. The basic principle consists of applying Gauss reduction to a sequence of projected sublattices of dimension 2. More precisely, it considers the lattice generated by the orthogonal projection of two consecutive vectors $\vec{b}_i$ and $\vec{b}_{i+1}$ on the vector subspace generated by the

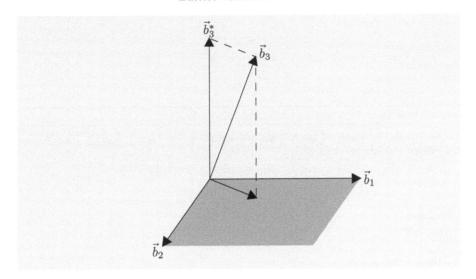

Figure 10.5: Computing $\vec{b}_3^*$ from $\vec{b}_3$

previous vectors $(\vec{b}_1, \cdots, \vec{b}_{i-1})$ and applies one iteration of t-Gauss reduction[1] to this projected lattice. In order to also modify the high dimension lattice, all swap and translation operations are pulled back from the projected space and applied to the vectors $\vec{b}_i$ and $\vec{b}_{i+1}$ themselves. In addition, these vectors may also be changed by adding to them some vector from the sublattice with basis $(\vec{b}_1, \cdots, \vec{b}_{i-1})$. The goal of this operation is to make sure that $\vec{b}_i$ is not too far from the corresponding projected vector $\vec{b}_i^*$. Once again, we illustrate this in dimension 3 in Figure 10.6

Algorithm 10.4 gives a description of the L^3 algorithm, using Algorithm 10.5 as a subroutine for length reduction. This subroutine is called RED within Algorithm 10.4. In this description, all coordinates in B are integers and all numbers in B^* and M are represented as rationals.

To better understand this algorithm, let us detail the conditions of the translation and exchange steps within the L^3 algorithm. The goal of translations is to modify the off-diagonal coefficients in the triangular matrix M and reduce them in absolute value below $1/2$. In other words, these translations are used to make sure that each $\vec{b}_k$ is not too far from the corresponding $\vec{b}_k^*$ and "almost orthogonal" to its predecessors. This is achieved by calling subalgorithm RED to translate a vector $\vec{b}_i$ along $\vec{b}_j$ (with $j < i$). This subalgorithm also modifies M to reflect this translation. The updates performed on M are

---

[1]The use of t-Gauss ensures polynomial time behavior. Note that, the parameter t and the parameter δ of LLL-reduction are related by $\delta = 1/t$.

Algorithm 10.4 LLL algorithm using rationals

Require: Initial basis of an integer lattice $B = (\vec{b}_1, \vec{b}_2, \cdots, \vec{b}_n)$
Require: Parameter $1/4 < \delta \leq 1$
  Compute B^* and M using Gram-Schmidt Algorithm 10.3
  for i from 1 to n **do**
    Let $L_i \longleftarrow \|\vec{b}_i^*\|^2$
  end for
  Erase B^*
  Let $k \longleftarrow 2$
  while $k \leq n$ **do**
    Apply length reduction $\text{RED}(k, k-1)$
    if $(L_k + m_{k,k-1}^2 L_{k-1}) \geq \delta\, L_{k-1}$ **then**
      for l from $k-2$ downto 1 **do**
        Apply length reduction $\text{RED}(k, l)$
      end for
      Increment k
    else
      Let $mm \longleftarrow m_{k,k-1}$
      Let $LL \longleftarrow L_k + mm^2 L_{k-1}$
      Let $m_{k,k-1} \longleftarrow mm L_{k-1}/LL$
      Let $L_k \longleftarrow L_{k-1} L_k / LL$
      Let $L_{k-1} \longleftarrow LL$
      Exchange $\vec{b}_k$ and $\vec{b}_{k-1}$
      for l from 1 to $k-2$ **do**
        Let $\mu \longleftarrow m_{i,k}$
        Let $m_{i,k} \longleftarrow m_{i,k-1} - \mu mm$
        Let $m_{i,k-1} \longleftarrow \mu + m_{k,k-1} m_{i,k}$
      end for
      if $k > 2$ **then**
        Decrement k
      end if
    end if
  end while

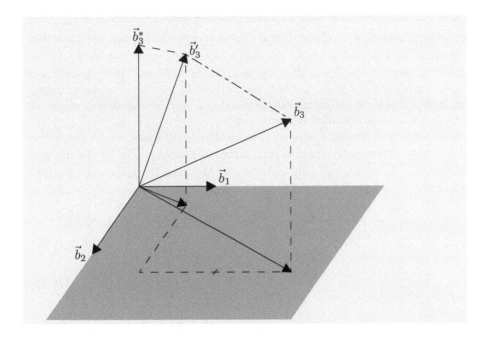

Figure 10.6: An elementary reduction step of L^3 in dimension 3

Algorithm 10.5 Length reduction subalgorithm RED(i,j)

Require: Current lattice basis B and matrix M
Require: Integer i and j, where $\vec{b}_i$ is to be reduced by $\vec{b}_j$
  if $|m_{i,j}| > 1/2$ **then**
    Let $r \longleftarrow \lfloor m_{i,j} \rceil$
    Let $\vec{b}_i \longleftarrow \vec{b}_i - r\vec{b}_j$
    $m_{i,j} \longleftarrow m_{i,j} - r$
    for w from 1 to $j-1$ **do**
      Let $m_{i,w} \longleftarrow m_{i,w} - rm_{j,w}$
    end for
  end if

---

easy to follow and detailed in subalgorithm RED. Note that the orthogonalized vector $\vec{b}_i^*$ remains unchanged during a call to RED.

The exchange steps are used to simultaneously reduce the norm of $\vec{b}_{k-1}^*$ and increase the norm of $\vec{b}_k^*$. Note that B^* itself is not used during the algorithm and that the knowledge of the squared norms $\|\vec{b}_i^*\|^2$ denoted by L_i in the algorithm and of the lower diagonal matrix M suffices. During exchange steps, the values of the norms and the matrix M also need to be updated. Since these updates are more complicated than during translation steps, we now detail the mathematics behind them.

Given a pair of vectors $\vec{b}_i$ and $\vec{b}_{i+1}$, let us denote by $\vec{u}$ and $\vec{v}$ their respective projections orthogonally the subspace spanned by $(\vec{b}_1, \cdots, \vec{b}_{i-1})$. In the special case $i = 1$, we have $\vec{u} = \vec{b}_1$ and $\vec{v} = \vec{b}_2$. We have the following identities before the exchange:

$$\vec{b}_i^* = \vec{u}, \qquad \vec{b}_{i+1}^* = \vec{v} - \frac{(\vec{v}|\vec{u})}{\|\vec{u}\|^2}\vec{u}, \qquad m_{i+1,i} = \frac{(\vec{v}|\vec{u})}{\|\vec{u}\|^2},$$

$$\forall k > i+1: \ m_{k,i} = \frac{(\vec{b}_k|\vec{b}_i^*)}{\|\vec{b}_i^*\|^2}, \qquad m_{k,i+1} = \frac{(\vec{b}_k|\vec{b}_{i+1}^*)}{\|\vec{b}_{i+1}^*\|^2}.$$

$$(10.13)$$

and after the exchange:

$$\hat{b}_i^* = \vec{v}, \qquad \hat{b}_{i+1}^* = \vec{u} - \frac{(\vec{v}|\vec{u})}{\|\vec{v}\|^2}\vec{v}, \qquad \hat{m}_{i+1,i} = \frac{(\vec{v}|\vec{u})}{\|\vec{v}\|^2},$$

$$\forall k > i+1: \ \hat{m}_{k,i} = \frac{(\vec{b}_k|\hat{b}_i^*)}{\|\hat{b}_i^*\|^2}, \qquad \hat{m}_{k,i+1} = \frac{(\vec{b}_k|\hat{b}_{i+1}^*)}{\|\hat{b}_{i+1}^*\|^2}.$$

$$(10.14)$$

Moreover, since the determinant of a lattice is an invariant that does not depend on the chosen basis, we necessarily have:

$$\|\hat{b}_i^*\|^2 \cdot \|\hat{b}_{i+1}^*\|^2 = \|\vec{b}_i^*\|^2 \cdot \|\vec{b}_{i+1}^*\|^2$$

$$(10.15)$$

As a consequence, we derive the following update formulas:

$$\|\hat{b}_i^*\|^2 = \|\vec{b}_{i+1}^*\|^2 + m_{i+1,i}^2\|\vec{b}_i^*\|^2, \quad \|\hat{b}_{i+1}^*\|^2 = \frac{\|\vec{b}_i^*\|^2\|\vec{b}_{i+1}^*\|^2}{\|\hat{b}_i^*\|^2},$$

$$\forall k < i: \quad \hat{m}_{i+1,k} = m_{i,k}, \qquad \qquad \hat{m}_{i+1,i} = \frac{\|\vec{b}_i^*\|^2}{\|\hat{b}_i^*\|^2}m_{i+1,i},$$

$$\hat{m}_{i,k} = m_{i+1,k},$$

$$\forall k > i+1: \hat{m}_{k,i+1} = m_{k,i} - m_{i+1,i}\,m_{k,i+1},$$

$$\hat{m}_{k,i} = m_{k,i+1} + \hat{m}_{i+1,i}\,\hat{m}_{k,i+1}.$$

$$(10.16)$$

Looking at the conditions enforced during L^3 reductions, it is clear that this algorithm yields a LLL-reduced basis of the input lattice. In addition, L^3 is a polynomial time algorithm and LLL-reduced bases are reasonably good.

The L^3 lattice reduction can also, with a minor modification, starting from a family of vectors that generates a lattice, construct a reduced basis for this lattice. This modified algorithm simply removes any zero vectors that appear among the vectors $\vec{b}_i$ during the computations.

Complexity of the L^3 algorithm

In order to bound the complexity of the L^3 algorithm, we are going to rely on the notion of determinant described in Section 10.1. Given a basis $(\vec{b}_1, \cdots, \vec{b}_n)$ of a lattice L, we can construct a family of sublattices $L_1, \ldots, L_n$, where each lattice L_i is generated by an initial segment of the given basis of L: $(\vec{b}_1, \cdots, \vec{b}_i)$. We let d_i denote the determinant $\det(L_i)$ of the i-th sublattice. At each step of the L^3 algorithm, the values of d_i for the current basis may change. More precisely, we can study the possible behavior of these values during translations and during exchanges. During translations, we see that a single vector is changed by addition of multiples of previous vectors. Clearly, this operation does not modify any of the sublattices and only changes the representations of these lattices. During exchanges, the situation is slightly more complicated. Assume that $\vec{b}_k$ and $\vec{b}_{k-1}$ are exchanged, then we see that none of the sublattices L_i for $i < k-1$ are modified. Similarly, none of the sublattices L_i for $i \geq k$ are modified, since up to order the basis of each is preserved. However, the sublattice L_{k-1} is modified. Thus, the determinant d_{k-1} can change. Since this determinant is the product of the norms of the vectors $\vec{b}_i^*$ up to $k-1$ and since $\|\vec{b}_{k_1}^*\|$ decreases by a factor at least δ^{-1}, so does d_{k-1}. More precisely, if $\hat{d}_{k-1}$ denotes the new value after the exchange, we have:

$$\hat{d}_{k-1} \leq \delta \cdot d_{k-1}. \tag{10.17}$$

Thus, during each exchange of the L^3 algorithm, one value of d_i goes down, while the others are unchanged. During all other operations in L^3, all values d_is are left unchanged. As a consequence, the product of all d_is

$$D = \prod_{i=1}^{n} d_i \tag{10.18}$$

is decreasing throughout the L^3 algorithm. Moreover, whenever it changes, it is divided by at least by δ^{-1}. In addition, D^2 is a non-zero integer, thus D must remain greater than 1. Letting D_{Init} denote the initial value of D, we see that the L^3 algorithm performs at most $\log(D_{\text{Init}})/\log(t)$ exchanges. This allows to bound the number of arithmetic operations performed during L^3; however, it does not suffice to compute the running time of L^3. We also need to analyze the size of the numbers that occur throughout the algorithm. In fact, the real difficulty is to check that the denominators of fractions remain small enough. With the original L^3 algorithm, they are bounded by the largest value among the squared determinants d_i^2. As a consequence, it is possible to show that, using ordinary multiplication, the complexity of L^3 reduction for an integer lattice of rank r in dimension n, with all entries in the basis vectors bounded by B is $O(nr^5 \log^3(B))$.

This complexity can be improved by using floating point numbers instead of rationals, within the L^3 algorithm, keeping only the integer coordinates of basis vectors in exact form. A side effect of this change is that we can no longer round numbers to their nearest integer without mistakes and that we need to relax the length reduction condition given by Equation (10.9). It suffices to replace the constant $1/2$ in this algorithm by a slightly larger constant, such as 0.501 or something similar. From a theoretical point-of-view the fastest algorithm is described in [NS05] and has complexity $O(nr^4 \log^2(B))$. In practice, floating point numbers with fixed precision are often used; however, in bad cases the round-off errors prevent the algorithm from terminating. Many heuristics have been proposed to avoid this bad behavior, for example see [SE94], but, despite their good performances in many cases, they sometimes fail.

Properties of LLL-reduced bases

Given a LLL-reduced basis for a lattice L, we first remark that any non-zero vector of L has its norm at least as large as the norm of the shortest vector $\vec{b}_i^*$ in the orthogonal basis. Due to Lovász conditions, we find:

$$\left\| \vec{b}_{i+1}^* \right\|^2 \geq \left\| \vec{b}_i^* \right\|^2 \left(\delta - \frac{1}{4} \right). \tag{10.19}$$

If λ_1 denotes the first minimum of L, it is greater than the norm of the shortest $\vec{b}_i^*$. It follows that:

$$\lambda_1 \geq \left(\delta - \frac{1}{4} \right)^{(n-1)/2} \left\| \vec{b}_1 \right\|. \tag{10.20}$$

As a consequence, the first vector of a LLL-reduced basis is not too large when compared to the first lattice minimum. This relation is often used in the special case $\delta = 3/4$, yielding:

$$\left\| \vec{b}_1 \right\| \leq 2^{(n-1)/2} \lambda_1. \tag{10.21}$$

Another very useful inequality related the size of $\left\| \vec{b}_1 \right\|$ with the determinant of the lattice. Since the determinant is equal to the product of the norms of all $\vec{b}_i^*$ and multiplying together the equations:

$$\left\| \vec{b}_i^* \right\| \geq \left(\delta - \frac{1}{4} \right)^{(i-1)/2} \left\| \vec{b}_1 \right\|, \tag{10.22}$$

we find:

$$\det(L) \geq \left(\delta - \frac{1}{4} \right)^{n(n-1)/4} \left\| \vec{b}_1 \right\|^n. \tag{10.23}$$

Taking the n-th root and letting $\delta = 3/4$, we obtain:

$$\left\| \vec{b}_1 \right\| \leq 2^{(n-1)/4} \det(L)^{1/n}. \tag{10.24}$$

10.4 Shortest vectors and improved lattice reduction

Once we have computed a LLL-reduced basis for a lattice L, we already have obtained some reasonably short vectors. However, we might wish to obtain stronger results. One typical problem is to determine the first minimum $\lambda_1(L)$ of the lattice and find a non-zero vector achieving this minimal length. Note that we cannot use a straightforward brute force approach to find such a shortest vector. Indeed, the set of lattice vectors is infinite and we cannot directly enumerate it. Instead, we first need to analyze the properties the lattice and determine a finite set of candidates that contains the shortest vector. This is the basic idea of enumeration algorithms [Kan83].

10.4.1 Enumeration algorithms for the shortest vector

Assume that we are given a basis B for a lattice L of rank r. Then, for any r-uples α of integers, we can compute a vector $\vec{V}(\alpha)$ in L by the following formula:

$$\vec{V}(\alpha) = \sum_{i=1}^{r} \alpha_i \vec{b}_i, \tag{10.25}$$

where the vectors $\vec{b}_i$ are in the basis B. Can we determine a finite set S_B such that the shortest non-zero vector in L is equal to $\vec{V}(\alpha)$ for some α in S_B? Kannan showed in [Kan83] that such a finite set exists and can be used to construct an algorithm for finding the shortest vector.

As before in this chapter, we denote by B^* the Gram-Schmidt orthogonalized basis corresponding to B and by $\vec{b}_i^*$ the vectors from B^*. Let α by a non-zero r-uple as above such that α_k is the last non-zero component in α, i.e., for any integer $i > k$, we have $\alpha_i = 0$. Then, we have the following identity:

$$\vec{V}(\alpha) = \sum_{i=1}^{k} \alpha_i \vec{b}_i = \sum_{i=1}^{k-1} \beta_i \vec{b}_i^* + \alpha_k \vec{b}_k^*, \tag{10.26}$$

for some $r - 1$-uple β. Since B^* is an orthogonal basis, this implies that the norm of $\vec{V}(\alpha)$ is greater than α_k times the norm of $\vec{b}_k^*$. As a consequence, we know that if α is such that $\vec{V}(\alpha)$ is the shortest non-zero vector in L then for any vector $\vec{v}$ in L we have:

$$|\alpha_k| \leq \frac{\|\vec{v}\|}{\|\vec{b}_k^*\|}. \tag{10.27}$$

As a consequence, we have shown that there are only finitely many possibilities for the last non-zero integer in α. Note that the quality of bound given by Equation (10.27) varies depending on the vector $\vec{v}$ we choose and on the norm of $\vec{b}_k^*$. Using a LLL-reduced basis in this bound is a good idea because it gives two improvements. First, the first vector in the reduced basis is quite small and is a good candidate for $\vec{v}$. Second, thanks to Equation (10.19), the norm of $\vec{b}_k^*$ in a reduced basis is not too large. More precisely, with a L^3 reduced basis, with parameter t, we can derive the following bound for α_k:

$$|\alpha_k|^2 \le \left(\delta - \frac{1}{4}\right)^{k-1} \tag{10.28}$$

It remains to see that the other component in α can similarly be bounded when $\vec{V}(\alpha)$ is the shortest non-zero vector in L. More precisely, substituting Equation (10.11) into Equation (10.26) and letting $\beta_k = \alpha_k$ we may remark that for a L^3 reduced basis, we have:

$$\beta_j = \alpha_j + \sum_{i=j+1}^{k} \beta_i m_{i,j}. \tag{10.29}$$

This implies that:

$$\left|\alpha_j + \sum_{i=j+1}^{k} \beta_i m_{i,j}\right| \le \frac{\left\|\vec{v} - \sum_{i=j+1}^{k} \beta_i \vec{b}_i^*\right\|}{\|\vec{b}_j^*\|} \le \frac{\|\vec{v}\|}{\|\vec{b}_j^*\|}. \tag{10.30}$$

Thus, for a L^3 reduced basis, there are fewer than $2\left(\delta - \frac{1}{4}\right)^{j-1}$ values to consider for α_j.

All in all, starting from a L^3 reduced basis, it is possible to obtain the shortest vector of a lattice of rank r by considering at most $2^{O(r^2)}$ vectors. In addition to Kannan in [Kan83], this has also been studied by Fincke and Pohst in [FP85] and, later, by Schnorr and Euchner in [SE94]. A completely different algorithm for finding short vectors was proposed by Ajtai, Kumar and Sivakumar in [AKS01]. It is asymptotically much more efficient, but does not seem to be practically competitive for the current dimensions that can be addressed with lattice reduction.

Note that a technical difficulty when programming this algorithm is that we need to write down r level of embedded loops. Were r fixed, this would cause no problem, but with varying values of r, we have to either write a recursive routine or remove the recursivity by explicitly remembering a stack of loop counters inside an array. A basic enumeration algorithm using this second approach is listed as Algorithm 10.6. One important practical feature of this algorithm is that whenever a new shortest vector is found, its length is used to bound the coefficients used during the enumeration. Moreover, the coefficients are bounded using the first inequality of Equation (10.30), which is tighter than the second one.

In practice, Schnorr and Euchner propose in [SE94] to speed up the enumeration algorithm by changing the order in which the various possibilities for α are considered. Instead of enumerating the values at each level from the

---

Algorithm 10.6 A basic short vector enumeration algorithm

Require: Input L^3 reduced basis B of rank r, orthogonalized B^* and M
Require: Input round-off error tolerance ϵ

Let array $\alpha_{\text{best}}[1 \cdots r] \longleftarrow (1, 0, 0, \cdots, 0)$
Let $L_{\text{best}} \longleftarrow \|\vec{b}_1\|^2$
Declare array $\alpha[1 \cdots r + 1]$
Let $\alpha[r + 1] \longleftarrow 0$ {Defined to avoid memory overflows}
Let $\alpha[r] \longleftarrow -\lfloor \sqrt{L_{\text{best}}/\|\vec{b}_r^*\|^2} + \epsilon \rfloor$
Declare array $\tilde{L}[1 \cdots r + 1]$
Let $\tilde{L}[r + 1] \longleftarrow 0$
Let $t \longleftarrow r$ {Initialize loop level index to outer loop}
while $t \le r$ **do**
  if $t = 0$ **then**
    Let $\vec{c} \longleftarrow \sum_{i=1}^r \alpha[i]\vec{b}_i$
    Let $L \longleftarrow \|\vec{c}\|^2$ {Recompute exact squared norm}
    if $L < L_{\text{best}}$ **then**
      Let $L_{\text{best}} \longleftarrow L$
      Let $\alpha_{\text{best}} \longleftarrow \alpha$
    end if
    Increment $\alpha[1]$
    Let $t \longleftarrow 1$
  else
    Let $\beta \longleftarrow \alpha[t] + \sum_{i=t+1}^r \alpha[i]m_{i,t}$
    Let $\tilde{L}[t] \longleftarrow \tilde{L}[t + 1] + \beta^2 \|\vec{b}_t^*\|^2$ {Approximate contribution to norm}
    if $\tilde{L}[t] < L_{\text{best}} + \epsilon$ **then**
      if $t > 1$ **then**
        Let $\beta \longleftarrow \sum_{i=t}^r \alpha[i]m_{i,t-1}$
        Let $\alpha[t - 1] \longleftarrow -\lfloor \sqrt{(L_{\text{best}} - \tilde{L}[t])/\|\vec{b}_{t-1}^*\|^2} + \epsilon \rfloor - \beta$
      end if
      Decrement t {Go to next inner loop level}
    else
      Increment $\alpha[t + 1]$
      Increment t {Go to next outer loop level}
    end if
  end if
end while
Output coordinates α_{best} of shortest vector in basis B.

---

smallest to the largest possibility, they propose to use a centered approach. For example, at the first level of the enumeration, i.e., when considering $\alpha[r]$, since the enumeration starts from the end, they enumerate the possibilities in the order 0, 1, −1, 2, −2, ... For inner levels, the enumeration starts from the middle of the range of available option and proceeds outwards, alternating on the left and right sides.

10.4.2   Using shortest vectors to improve lattice reduction

Thanks to the above enumeration algorithms for short vector, it is possible to construct lattice bases which are more strongly reduced than L^3 reduced bases. Moreover, Kannan showed in [Kan83] that the relationship between short vector and strong lattice reduction runs both ways. In one direction, we can use short vectors to improve the quality of lattice basis, in the other direction, the short vector enumeration algorithms run faster when applied to strongly reduced basis. Using this bidirectional relationship, Kannan devised a recursive algorithm that constructs strongly reduced basis and short vector faster than by running the enumeration algorithm directly with a L^3 reduced basis. Recall that for a lattice of rank r, we showed in Section 10.4.1 that starting from a L^3 reduced basis, it suffices to enumerate $2^{O(r^2)}$ vectors to find the shortest lattice vector. With the recursive algorithm of Kannan, the complexity is lowered to $2^{O(r \log r)}$.

This algorithm relies on the notion of Hermite-Korkine-Zolotarev or HKZ reduced basis. A basis B of a lattice L, with orthogonalized basis B^* such that $B = M \cdot B^*$ is HKZ reduced, if and only if, the following properties are satisfied:

1. The basis B is size-reduced, as in Equation (10.9), i.e., all off-diagonal coefficients of M satisfy $|m_{i,j}| \leq 1/2$.

2. The vector $\vec{b_1}$ realizes the first minimum $\lambda_1(L)$.

3. The projection of the vectors $\vec{b_2}, \cdots, \vec{b_r}$ orthogonally to $\vec{b_1}$ form an HKZ reduced basis.

The third condition is recursive; however, it simply means that $\vec{b_2}$ is chosen to minimize $\|\vec{b_2^*}\|$ and so on for the subsequent vectors. It is easy to remark that an HKZ reduced basis is L^3 reduced.

During Kannan's algorithm, we also need a slightly relaxed notion as an intermediate computation step. We say that a basis B is quasi-HKZ reduced, when $\|\vec{b_1}\| \leq 2\|\vec{b_2^*}\|$ and the projection of the sublattice $(\vec{b_2}, \cdots, \vec{b_r})$ orthogonally to $\vec{b_1}$ is HKZ reduced. The key point is that enumerating the shortest vector in a lattice given a quasi-HKZ reduced basis is much more efficient than with an L^3 reduced basis. A sketch of Kannan's algorithm is given as Algorithm 10.7. This algorithm uses L^3 on a generating family as subroutine.

It has been thoroughly analyzed when the underlying L^3 uses exact arithmetic [HS07]. However, in practice, it is usually replaced by floating point heuristics. Due to its high time complexity, Kannan's algorithm for HKZ reduction can only be used for lattices of small enough rank.

Algorithm 10.7 Kannan's HKZ reduction algorithm

Require: Input basis B of a lattice L

    repeat

        Reduce B using L^3

        Use HKZ reduction recursively on projected lattice $(\vec{b}_2, \cdots, \vec{b}_r)$ orthogonaly to $\vec{b}_1$.

    until $\|\vec{b}_1\| > \|\vec{b}_2^*\|$ {Loop output of quasi-HKZ reduced basis}

    Find $\vec{b}_{\min}$ shortest vector in B using Algorithm 10.6.

    Apply L^3 on family $(\vec{b}_{\min}, \vec{b}_1, \cdots \vec{b}_r)$ to obtain a L^3 reduced basis with $\vec{b}_{\min}$ As first vector. Put the output in B.

    Use HKZ reduction recursively on projected lattice $(\vec{b}_2, \cdots, \vec{b}_r)$ orthogonally to $\vec{b}_1$.

    Output HKZ reduced B, including lattice shortest vector.

10.4.2.1 Schnorr's block reduction

To conclude this chapter on lattice reduction, let us mention that for lattices of large rank, where HKZ reduction is not feasible, there exists a family of reduction algorithms introduced by Schnorr in [Sch87] called block Korkine-Zolotarev reduction. The basic idea is to use HKZ reduction on projected sublattices of fixed rank. Without going into the details, let us mention that each algorithm in the family has a polynomial running time. Of course, the degree of this polynomial grows with the size of the considered blocks. Heuristic versions of this algorithm are available in some of the computer packages cited in the preface.

10.5 Dual and orthogonal lattices

When dealing with lattice reduction in cryptography, the notions of dual and orthogonal of a given lattice are frequently encountered. Since these two notions are important and distinct from each other, we describe them in this section.

10.5.1   Dual of a lattice

Let L be a lattice of rank r in a vector space of dimension n. Consider span(L), the vector subspace spanned by L. We first define the dual lattice L^* as a set:

$$L^* = \{\vec{u} \in \text{span}(L) \mid \forall \vec{v} \in L : (\vec{u}|\vec{v}) \in \mathbb{Z}\}.$$

To check that L^* is indeed a lattice, we need to prove that it is a discrete subgroup of $\mathbb{Z}^n$. In fact, in this case, it is even a discrete subgroup of span(L). The subgroup part is left as exercise to the reader. To prove that L^* is a discrete subgroup, we proceed by contradiction. If L^* is not discrete, then there exists a sequence of vectors $\vec{u}_i$ of L^*, whose norms tend to 0 as i tends to infinity. Of course, this limit also holds for any equivalent norm. We now consider $(\vec{v}_1, \cdots, \vec{v}_r)$ a basis of L and the norm $\|\cdot\|_L$ defined on span(L) by:

$$\|\vec{u}\|_L = \sum_{i=1}^{r} (\vec{u}|\vec{v}_i)^2.$$

Now, by definition of L^*, the norm of any vector in the sequence $\vec{u}_i$ is a sum of squares of integers and, thus, a non-negative integer. As a consequence, the sequence $\|\vec{u}_i\|_L$ is a sequence of non-negative integers which tends to zero. Moreover, since none of the $\vec{u}_i$ is the zero vector, none of the norms $\|\vec{u}_i\|$ can be equal to zero and the norms are lower bounded by 1. This contradiction concludes the proof.

When the lattice L has full rank, i.e., when $r = n$ a basis for L^* can simply be obtained by the transpose of the inverse matrix of a basis of L. In the general case, in order to explicitly construct a basis of the dual lattice, the easiest is to start from a Gram-Schmidt orthogonalization of a basis of the original lattice. Let us consider $(\vec{b}_1, \cdots, \vec{b}_r)$ a basis of L and $(\vec{b}_1^*, \cdots, \vec{b}_r^*)$, its Gram-Schmidt orthogonalization.

We first remark that $\vec{\tilde{b}}_r = \vec{b}_r^* / \|\vec{b}_r^*\|^2$ belongs to the dual lattice L^*. Indeed, the scalar product of $\vec{b}_r^*$ with $\vec{b}_i$ is 0 when $i < r$ and $\|\vec{b}_r^*\|^2$ when $i = r$. Thus, decomposing any vector $\vec{v}$ in L on the given basis, we find that the scalar product of $\vec{v}$ with $\vec{\tilde{b}}_r$ is a integer. In truth, this integer is the coefficient of $\vec{b}_r$ in the decomposition of $\vec{v}$. Next, we consider the vector $\vec{\tilde{b}}_{r-1}^* = \vec{b}_{r-1}^* / \|\vec{b}_{r-1}^*\|^2$ and see that, in general, this vector does not belong to L^*. Indeed, its scalar products with all $\vec{b}_i$s are zeros when $i < r - 1$ and its scalar product with $\vec{b}_{r-1}$ is equal to one; however the scalar product with $\vec{b}_r$ is not necessarily an integer. More precisely, using the notation of Section 10.3.1, we have:

$$(\vec{b}_r|\vec{\tilde{b}}_{r-1}^*) = m_{r,r-1}\|\vec{b}_{r-1}^*\|^2. \tag{10.31}$$

It follows that:

$$(\vec{b}_r|\vec{\tilde{b}}_{r-1}^*) = m_{r,r-1}. \tag{10.32}$$

Clearly, adding a multiple of $\vec{b}_r^*$ to $\vec{b}_{r-1}^*$ preserves the scalar product with $\vec{b}_i$ for $i < r$ and modifies the scalar product with $\vec{b}_r$. In particular, if we consider the vector $\vec{b}_{r-1} = \vec{b}_{r-1}^* - m_{r,r-1}\vec{b}_r$, the scalar product with $\vec{b}_r$ becomes 0 and we obtain a second, linearly independent, vector in L^*.

To construct a complete basis for L^*, we first specify its Gram-Schmidt orthogonalized basis:

$$(\vec{\hat{b}}_r^*, \cdots, \vec{\hat{b}}_1^*) = \left(\frac{\vec{b}_r^*}{\|\vec{b}_r^*\|^2}, \cdots, \frac{\vec{b}_1^*}{\|\vec{b}_1^*\|^2} \right). \qquad (10.33)$$

In other words, the basis for L^* has the same orthogonalized basis as the initial basis for L, but the order of vectors is reversed. Then, we specify the Gram-Schmidt coefficients for the dual basis. For simplicity, it is preferable to number the basis vectors of the dual lattice in reversed order and consider a basis $(\vec{\hat{b}}_r, \cdots \vec{\hat{b}}_1)$. The Gram-Schmidt coefficients of this basis are chosen as follows:

$$\frac{(\vec{\hat{b}}_i | \vec{\hat{b}}_j^*)}{\|\vec{\hat{b}}_j^*\|^2} = -m_{i,j}.$$

Equivalently, we simply compute $\vec{\hat{b}}_i$ as $\vec{\hat{b}}_i^* - \sum_{j=i+1}^{r} m_{i,j}\vec{\hat{b}}_j^*$.

We leave as an exercise to the reader to check that by applying the same method to compute a dual of the dual basis gives back the original basis of the lattice L. In addition, the reader can also verify that when the initial basis of L is L^3 reduced, so is the resulting basis of the dual lattice.

10.5.2   Orthogonal of a lattice

Let L be an integer lattice in a vector space of dimension n, with rank $r < n$. We define the orthogonal lattice $L^\perp$ as the following subset of $\mathbb{Z}^n$:

$$L^\perp = \{\vec{u} \in \mathbb{Z}^n \mid \forall \vec{v} \in L : (\vec{u}|\vec{v}) = 0\}.$$

It is clear that $L^\perp$ is a subgroup of $\mathbb{Z}^n$ and thus an integer lattice. Its rank is equal to $n - r$. Moreover, lattice reduction and, in particular, the L^3 algorithm gives an efficient way to compute $L^\perp$ from L. Consider the lattice spanned by the columns of the following matrix:

$$\begin{pmatrix} k \cdot B(L) \\ I \end{pmatrix}$$

where $B(L)$ is the matrix of an L^3 reduced basis of L, k is a large enough constant and I is an $n \times n$ identity matrix. After applying the L^3 algorithm to this lattice, we obtain a matrix of the form:

$$\begin{pmatrix} 0 & k \cdot U \\ B(L^\perp) & V \end{pmatrix}$$

where $B(L^\perp)$ is a L^3 reduced basis of the orthogonal lattice $L^\perp$. This notion of orthogonal lattice is frequently used in Chapter 13. Note that it is usually more efficient to use linear algebra rather than lattice reduction to construct a basis of the vector space spanned by $B(L^\perp)$. Moreover, by using a linear algebra method that does not perform divisions, i.e., similar to Hermite normal form computations, we can in fact obtain a basis $B(L^\perp)$. This is essential for the efficiency of the algorithms presented in Chapter 13.

Exercises

1. Let A and N be two coprime integers. Consider the lattice L generated by the rows of:
$$\begin{pmatrix} A & 1 \\ N & 0 \end{pmatrix}.$$

   - Let (u, v) be any vector in this lattice with $v \neq 0$. What can you say about u/v?

   - What is the determinant of the lattice L?

   - If (u, v) is the smallest vector in L, what can you say about u/v?

   - Find a similar result using the continued fraction expansion of A/N.

2^{h}. Prove that the second vector in a Gauss reduced basis for a 2-dimensional lattice achieves the second minimum of this lattice.

3. Consider the n dimensional lattice L generated by the identity matrix.

   - What are the minima of this lattice?

   - Write a program that generates random looking bases of L.

   - How can you easily check that a matrix M is a basis for L?

   - Experiment with an existing implementation of lattice reduction algorithms and try to reduce random looking bases of L. When does the lattice reduction algorithm output the identity matrix (or a permuted copy)?

4^{h}. Show that any lattice reduction algorithm can be modified to keep track of the unimodular transform that relates the input and output matrices.

5. Reduce, by hand, with the L^3 algorithm the lattice generated by:
$$\begin{pmatrix} 1 & 2 & 3 \\ 4 & 5 & 6 \\ 7 & 8 & 10 \end{pmatrix}.$$

Chapter 11

Polynomial systems and Gröbner base computations

Thus, if we could show that solving a certain system requires at least as much work as solving a system of simultaneous equations in a large number of unknowns, of a complex type, then we would have a lower bound of sorts for the work characteristic.

Claude Shannon

As shown by the above quote, written by Claude Shannon in his famous paper [Sha49], multivariate polynomial systems of equations naturally arise when writing down cryptanalytic problems. To illustrate this fact, let us consider two simple examples, one coming from public key cryptography and the other from secret key cryptography.

Factoring Writing down factoring as a multivariate system of polynomial equations is a very simple process. Assume that $N = PQ$ is an RSA number. Writing P and Q in binary, we can define two sets p and q of unknowns over $\mathbb{F}_2$ according to the following correspondence:

$$P = \sum_i p_i 2^i, \quad \text{and} \quad Q = \sum_i q_i 2^i.$$

Replacing P and Q in the equation $N = PQ$ yields a multivariate polynomial equation over the integers. Finding the factorization of N corresponds to finding a $\{0, 1\}$ non-trivial solution to this equation.

In fact, it is possible to go further and rewrite this equation as a system of equations over $\mathbb{F}_2$. In order to get a reasonably simple expression, it is a good idea to add extra variables to keep track of the computation. To give a simple example, assume that we would like to factor 35 into two numbers of 3 bits each: $P = 4p_2 + 2p_1 + p_0$ and $Q = 4q_2 + 2q_1 + q_0$. This easily yields an equation:

$$16p_2q_2 + 8(p_1q_2 + p_2q_1) + 4(p_0q_2 + p_1q_1 + p_2q_0) + 2(p_0q_1 + p_1q_0) + p_0q_0 = 35. \tag{11.1}$$

To replace this equation with a system of equations over $\mathbb{F}_2$, we define a partial sum $S = p_0Q + 2p_1Q$ and see that $N = S + 4p_2Q$. The partial sum S can be represented as $16s_4 + 8s_3 + 4s_2 + 2s_1 + s_0$. Adding two new sets of variables c and c' to deal with the carries in both additions, we finally obtain:

$$
\begin{aligned}
s_0 &= p_0q_0, \\
s_1 &= p_0q_1 \oplus p_1q_0 & c_1 &= p_0p_1q_0q_1, \\
s_2 &= c_1 \oplus p_0q_2 \oplus p_1q_1, & c_2 &= c_1p_0q_2 \oplus c_1p_1q_1 \oplus p_0p_1q_1q_2, \\
s_3 &= c_2 \oplus p_1q_2, & c_3 &= c_2p_1q_2, \\
s_4 &= c_3;
\end{aligned}
$$

$$
\begin{aligned}
1 &= s_0, & 1 &= s_1, \\
0 &= s_2 \oplus p_2q_0, & c'_2 &= s_2p_2q_0, \\
0 &= s_3 \oplus c'_2 \oplus p_2q_1, & c'_3 &= s_3c'2 \oplus s_3p_2q_1 \oplus c'2p_2q_1, \\
0 &= s_4 \oplus c'_3 \oplus p_2q_2, & 1 &= s_4c'3 \oplus s_4p_2q_2 \oplus c'3p_2q_3.
\end{aligned}
$$

$$(11.2)$$

This is a polynomial system of multivariate equations of degree 4. This can be generalized to express the factorization of any RSA number N as a system of $O(\log(N)^2)$ equations of degree 4 over $\mathbb{F}_2$.

Cryptanalysis of DES   We saw in Chapter 5 the bitslice description of the DES algorithm. In this description, the S-boxes are given as logical formulas or equivalently as polynomial expressions over $\mathbb{F}_2$. Since the other components of the DES algorithm are linear, the DES can in fact be written as a sequence of polynomial evaluations. Thus, by substituting the plaintext and ciphertext values he knows into these evaluations, a cryptanalyst may describe the key recovery problem of DES as a polynomial system of equations. In order to limit the total degree of the system and preserve its sparsity, it is convenient to introduce additional variables that represent the intermediate values encountered in this Feistel scheme.

From these two simple examples, we see that multivariate polynomial systems of equations can be used very easily to describe many cryptanalytic tasks. As a consequence, it is important to study whether such multivariate systems of equations are easy to solve and to understand the key parameters that govern the complexity of solving such systems.

11.1   General framework

Throughout the rest of this chapter, $\mathbb{K}$ denotes an arbitrary field and $\bar{\mathbb{K}}$ its algebraic closure. As a final goal in this chapter, we would like to formalize the resolution of a system of m algebraic equations in n unknowns:

$$\left\{ \begin{array}{c} f_1(x_1, \cdots, x_n) = 0 \\ \vdots \\ f_m(x_1, \cdots, x_n) = 0 \end{array} \right\} \tag{11.3}$$

More precisely, we would like to determine if such a system has got solutions and if so, whether there are finitely or infinitely many. Finally, we would like to compute some solutions and, when the number of solutions is finite, possibly list all solutions. We already know that, even when considering univariate polynomials, the notion of roots of polynomials is a delicate one. Indeed, over the real field $\mathbb{R}$, similar polynomials may have distinct behaviors, for example, $x^2 - 1$ has two roots, while $x^2 + 1$ has none. Thankfully, this first difficulty vanishes when the polynomials are considered over the algebraic closure of the real field, i.e., the field $\mathbb{C}$ of complex numbers. There, both polynomials have two roots. For this reason, we systematically consider roots of an algebraic system as in Equation (11.3) as defined over the algebraic closure $\bar{\mathbb{K}}$.

Looking at our system of equations, intuition says that we should expect a finite number of solutions when the number of equations is equal to the number of unknowns, $m = n$. With more unknowns than equations, there will be degrees of freedom left and we expect an infinite number of solutions. With more equations than unknowns, we expect that the equations are incompatible and that no solutions exist. Of course, we already know from the simple case of linear systems of equations that this is not always true. For example, the following system of 2 linear equations in three unknowns has no solution:

$$\left\{ \begin{array}{c} x_1 + x_2 + x_3 = 0 \\ x_1 + x_2 + x_3 = 1 \end{array} \right\}. \tag{11.4}$$

To also give a counter-example for the other case, let us consider the following system of 3 linear equations in two unknowns:

$$\left\{ \begin{array}{rcr} x_1 + x_2 & = & 0 \\ x_1 & = & 1 \\ x_2 & = & -1 \end{array} \right\}, \tag{11.5}$$

this system clearly has a solution. In the case of linear systems, the algorithms from Chapter 3 can solve all these issues. With general systems, the situation is more complicated and we need to introduce some additional mathematical notions. Comparing the systems given by Equations (11.3) and (11.4), we first see a difference in the way of writing the equations. With linear systems, it is convenient to have a linear combination of the unknowns on the left-hand side of each equation and a constant on the right-hand side. With algebraic systems, it is often considered preferable to shift the constants to the left-hand sides and incorporate them in the polynomials $f_1, \ldots, f_m$.

11.2   Bivariate systems of equations

Since the issue of solving general systems of equations is quite complicated and since we already know how to solve univariate polynomial equations, we start by considering the reasonably simple case of bivariate systems of equations before turning to the general case.

First, if we are given a single polynomial $f(x, y)$ in two unknowns, we can prove that this polynomial has a infinite number of roots over the algebraic closure $\bar{\mathbb{K}}$. For this purpose, let us write:

$$f(x, y) = \sum_{i=0}^{d_x} f^{(i)}(y) x^i. \tag{11.6}$$

Furthermore, let $g(y)$ denote the greatest common divisor of $(f^{(1)}, \cdots, f^{(d_x)})$, omitting $f^{(0)}$. This polynomial g has a finite number of roots, possibly zero. For each root ν of g, we see that $f(x, \nu)$ is a constant polynomial and thus does not have any root, unless $f^{(0)}(\nu) = 0$. For any value ν with $g(\nu) \neq 0$, $f(x, \nu)$ is a non-constant polynomial and its number of roots (with multiplicities) is equal to its degree. Since $\bar{\mathbb{K}}$ is always an infinite field, f has necessarily infinitely many solutions: at least one for all possible values of y but the finitely many roots of g.

After considering a single polynomial, the next step is to look at a system of two algebraic equations, $f_1(x, y) = f_2(x, y) = 0$. Proceeding as above, we can write:

$$f_1(x, y) = \sum_{i=0}^{d_{1,x}} f_1^{(i)}(y) x^i \quad \text{and} \quad f_2(x, y) = \sum_{i=0}^{d_{2,x}} f_2^{(i)}(y) x^i, \tag{11.7}$$

thus viewing f_1 and f_2 as polynomials in $\mathbb{K}[y][x]$. With this view, it is natural to compute the GCD of f_1 and f_2 using Euclid's algorithm for univariate polynomial in x. Note that we need to be careful because $\mathbb{K}[y]$ is a ring and not a field. To avoid this complication, one option is to allow rational fractions from $\mathbb{K}(y)$ as coefficients of polynomial in x during the GCD computations. At the end of the computation, we obtain a polynomial in x with coefficients in $\mathbb{K}(y)$. Two cases are possible, either this polynomial is of degree zero in x or it is of higher degree. When the degree is positive, we learn that f_1 and f_2 are both multiples of another polynomial in x and y. Thus, we are back to the case of a single polynomial. When the degree in x is 0, the GCD with coefficient in $\mathbb{K}(y)$ is 1. Thus, at first, one may expect that the system has no roots. However, at that point, we need to clear the denominators in the relation and obtain:

$$\mu(y) f_1(x, y) + \nu(y) f_2(x, y) = g(y). \tag{11.8}$$

Since, at this point, g no longer contains the variable x, we say that we have eliminated x from the algebraic system of equations.

An example of this approach is given in Figure 11.1. This example is quite simple; however, in general, due to the need to clear denominators, using GCD computations for the purpose of elimination is somewhat cumbersome. Thankfully, there exists a different approach which always yields the correct result and simply requires to compute the determinant of a square matrix whose coefficients are univariate polynomials. This approach is the computation of resultants and is addressed in the next section.

11.2.1 Resultants of univariate polynomials

To understand how resultants work, we start with univariate polynomials. Let $g_1(x)$ and $g_2(x)$ be two monic polynomials in $\mathbb{K}[x]$, we define the resultant $\mathrm{Res}(g_1, g_2)$ as:

$$\mathrm{Res}(g_1, g_2) = \prod_{\substack{\alpha \text{ root in } \bar{\mathbb{K}} \text{ of } g_1 \\ \text{(with multiplicity)}}} g_2(\alpha). \tag{11.9}$$

Splitting g_2 into a product of linear factor over $\bar{\mathbb{K}}$ we see that:

$$\mathrm{Res}(g_1, g_2) = \prod_{\substack{\alpha \text{ root in } \bar{\mathbb{K}} \text{ of } g_1 \\ \text{(with multiplicity)}}} \prod_{\substack{\beta \text{ root in } \bar{\mathbb{K}} \text{ of } g_2 \\ \text{(with multiplicity)}}} (\alpha - \beta). \tag{11.10}$$

As a consequence, exchanging the two products and replacing $\beta - \alpha$ by $\alpha - \beta$, we find that:

$$\mathrm{Res}(g_1, g_2) = (-1)^{\deg(g_1)\deg(g_2)}\mathrm{Res}(g_2, g_1). \tag{11.11}$$

Moreover, $\mathrm{Res}(g_1, g_2)$ is 0 if and only if g_1 and g_2 share at least one common root.

A very important fact is that computing the resultant of g_1 and g_2 does not require to compute the roots of either polynomial. Instead, it suffices to construct a simple matrix formed from the coefficients of both polynomials and to compute its determinant using classical linear algebra techniques. If d_1 is the degree of g_1 and d_2 the degree of g_2, we construct a square matrix of dimension $d_1 + d_2$. Writing:

$$g_1(x) = \sum_{i=0}^{d_1} g_1^{(i)} x^i \quad \text{and} \quad g_2(x) = \sum_{i=0}^{d_2} g_2^{(i)} x^i, \tag{11.12}$$

As a sample system of bivariate equations, let us define:

$$f_1(x,y) = y \cdot x^3 + (2y^2 + y) \cdot x^2 + (y^3 + y^2 + 1) \cdot x + (y+1) \text{ and}$$
$$f_2(x,y) = y \cdot x^4 + 2y^2 \cdot x^3 + (y^3 + 1) \cdot x^2 + y \cdot x.$$

Dividing f_1 and f_2 by y, we obtain unitary polynomials g_1 and g_2 in x whose coefficients are rational fractions in y. Using Euclid's GCD Algorithm 2.1 produces the following sequence of polynomials of decreasing degree in x:

$$g_3(x,y) = (g_2 - (x-1)g_1)/(y+1) = x^2 + y \cdot x + \frac{1}{y},$$
$$g_4(x,y) = g_1 - (x+y+1) \cdot g_3 = 0.$$

As a consequence, f_1 and f_2 are both multiples of g_3. It is even possible to clear the denominator and write:

$$f_1(x,y) = (x+y+1) \cdot (y\,x^2 + y^2\,x + 1) \text{ and}$$
$$f_2(x,y) = (x^2 + y\,x) \cdot (y\,x^2 + y^2\,x + 1).$$

Thus, solutions of the original system are either roots of $y\,x^2 + y^2\,x + 1$ or solutions of the lower degree system given by:

$$h_1(x,y) = x+y+1 \text{ and } h_2(x,y) = x^2 + y\,x.$$

Applying Euclid's algorithm once more, we find that:

$$h_3(x,y) = h_2 - (x-1)h_1 = y+1.$$

Thus, any solution to this system of equation has $y = -1$. Substituting this value back into h_1 and h_2, we find that in addition $x = 0$. Since $(x,y) = (0,-1)$ is not a root of $y\,x^2 + y^2\,x + 1$, this illustrates the fact that in the general case, the set of roots of algebraic systems of equations can be decomposed into a union of simpler sets. The reader interested in the theory behind these decompositions should look up the notion of primary decomposition in a textbook about algebraic geometry, such as [Eis95].

For further characterization of the roots of $y\,x^2 + y^2\,x + 1$, it is useful to remark that this polynomial is symmetric in x and y. As a consequence, when (x_0, y_0) is a root, so is (y_0, x_0). This polynomial can be rewritten as $sp+1$, where $s = x+y$ and $p = xy$ are the elementary symmetric polynomials in two variables. Note that there is a bijection between sets $(x_0, y_0), (y_0, x_0)$ and pairs (s_0, p_0). In the forward direction, one computes the sum and product. In the backward direction, one solves the equation $X^2 - s_0 X + p_0$ whose roots are x_0 and y_0.

Figure 11.1: A sample algebraic system in two unknowns

we construct the matrix:

$$M_S(g_1, g_2) = \begin{pmatrix} g_1^{(d_1)} & g_1^{(d_1-1)} & g_1^{(d_1-2)} & \cdots & g_1^{(0)} & 0 & 0 & \cdots & 0 \\ 0 & g_1^{(d_1)} & g_1^{(d_1-1)} & \cdots & g_1^{(1)} & g_1^{(0)} & 0 & \cdots & 0 \\ \vdots & \ddots & \ddots & \cdots & \ddots & \ddots & \ddots & \cdots & \vdots \\ 0 & 0 & 0 & \cdots & \cdots & \cdots & g_1^{(i)} & \cdots & g_1^{(0)} \\ g_2^{(d_2)} & g_2^{(d_2-1)} & g_2^{(d_2-2)} & \cdots & g_2^{(0)} & 0 & 0 & \cdots & 0 \\ 0 & g_2^{(d_2)} & g_2^{(d_2-1)} & \cdots & g_2^{(1)} & g_2^{(0)} & 0 & \cdots & 0 \\ \vdots & \ddots & \ddots & \cdots & \ddots & \ddots & \ddots & \cdots & \vdots \\ 0 & 0 & 0 & \cdots & \cdots & \cdots & g_2^{(j)} & \cdots & g_2^{(0)} \end{pmatrix} \qquad (11.13)$$

This matrix, called the Sylvester matrix of g_1 and g_2 can be interpreted in the following way: each column corresponds to some power of x, 1 for the first column, x for the second and so on, up to $x^{d_1+d_2-1}$ for the last column. Each row corresponds to a polynomial, the first d_2 rows are given by $g_1, xg_1, \ldots,$ $x^{d_2-1}g_1$, the next d_1 rows are given by $g_2, xg_2, \ldots, x^{d_1-1}g_2$. The resultant of g_1 and g_2 is obtained as the determinant of the Sylvester matrix:

$$\text{Res}(g_1, g_2) = \det(M_s(g_1, g_2)). \qquad (11.14)$$

The proof of this important result is out of the scope of this book and can, for example, be found in [Lan05, Chapter IV]. Note that using the matrix of Sylvester, the resultant is also defined for non-monic polynomials. In this case, we have:

$$\text{Res}(g_1, g_2) = \left(g_1^{(d_1)}\right)^{d_2} \left(g_2^{(d_2)}\right)^{d_1} \prod_{\substack{\alpha \text{ root in } \bar{\mathbb{K}} \text{ of } g_1 \\ \text{(with multiplicity)}}} \prod_{\substack{\beta \text{ root in } \bar{\mathbb{K}} \text{ of } g_2 \\ \text{(with multiplicity)}}} (\alpha - \beta). \qquad (11.15)$$

Prime factors of the resultant When g_1 and g_2 are monic polynomials with integer coefficients, each prime p that divides the resultant is such that g_1 and g_2 have at least a common root on the algebraic closure of $\mathbb{F}_p$. Note, however, that when a larger power of a prime, say p^k, divides the resultant, it does not imply that the polynomials have a common root modulo p^k.

11.2.2 Application of resultants to bivariate systems

When g_1 and g_2 are bivariate polynomials in x and y, we can also use resultants to solve the corresponding system of equations. To do this, we need, as when considering GCDs, to view these polynomials as polynomials in x whose coefficients are polynomials in y, i.e., polynomials in $\mathbb{K}[y][x]$. With this view, Sylvester's matrix, given by Equation (11.13), is filled with univariate polynomials in y and its determinant is another polynomial in y. In order to specify which variable is eliminated from the system of equations during the computation of the resultant, we say that we have computed the resultant of g_1 and g_2 in the variable x. This resultant is a polynomial in y. Three cases arise:

- If the resultant is the zero polynomial, the two polynomials g_1 and g_2 have a common factor. Using the example of the polynomials f_1 and f_2 from Figure 11.1, we may check that their Sylvester's matrix (in x) is:

$$\begin{pmatrix} y & 2y^2+y & y^3+y^2+1 & y+1 & 0 & 0 & 0 \\ 0 & y & 2y^2+y & y^3+y^2+1 & y+1 & 0 & 0 \\ 0 & 0 & y & 2y^2+y & y^3+y^2+1 & y+1 & 0 \\ 0 & 0 & 0 & y & 2y^2+y & y^3+y^2+1 & y+1 \\ y & 2y^2 & y^3+1 & y & 0 & 0 & 0 \\ 0 & y & 2y^2 & y^3+1 & y & 0 & 0 \\ 0 & 0 & y & 2y^2 & y^3+1 & y & 0 \end{pmatrix}$$

Moreover, the determinant of this matrix is zero. Indeed, multiplying $M_S(f_1, f_2)$ on the left by the vector $(0, -1, -y, 0, 0, 1, y+1)$ yields the null vector.

- If the resultant is a non-zero constant polynomial, we learn that the system of equations is incompatible over the field of complex numbers. However, assuming that f_1 and f_2 are monic, then modulo prime divisors of the resultant, they have a common factor. If f_1 and f_2 are not monic, the conclusion only holds for prime factors of the resultant which divide none of the leading coefficients.

- If the resultant is a polynomial of higher degree, say $R(y)$, then the y coordinate of any solutions of the system of equations given by f_1 and f_2 is a root of R. However, the converse is not true.

11.2.2.1 Using resultants with more variables

In theory, using resultants with more variables allows to solve arbitrary systems of equations. However, with n polynomials in n unknowns, we see that the computation quickly becomes difficult. A first problem, we have too many possible choices, we can choose to eliminate any of the n variables for any of the $n(n-1)/2$ pairs of polynomials. Moreover, even if we decide to eliminate a specific variable, it is unclear whether we should compute the resultant of all pairs of polynomials or only keep a fraction of them. With the next variable to eliminate, we have even more options and do not know how to proceed. A second problem is that the degrees of the polynomials we construct during computations of resultants grow too quickly.

As a consequence, resultants can only be used to solve systems of polynomial equations with a reasonably small number of unknowns and a different approach is needed to address more general cases.

11.3 Definitions: Multivariate ideals, monomial orderings and Gröbner bases

We now turn to the general case of multivariate systems of equations as in system of equations (11.3). Clearly, any root of this system is also a root of all polynomials in the ideal I generated by $f_1, \ldots, f_m$. Recall that this ideal is given by:

$$I = (f_1, \cdots, f_m) = \left\{ \sum_{i=1}^{m} h_i f_i \mid \forall (h_1, \cdots, h_m) \in \mathbb{K}[x_1, \cdots, x_n]^m \right\}. \quad (11.16)$$

Given such an ideal I, we define the affine algebraic variety of I, as the set $\mathcal{V}(I)$ of the common roots of all polynomials in I over the algebraic closure $\bar{\mathbb{K}}$:

$$\mathcal{V}(I) = \{(\alpha_1, \cdots, \alpha_n) \mid \forall f \in I : f(\alpha_1, \cdots, \alpha_n) = 0\}. \quad (11.17)$$

Clearly, $\mathcal{V}(I)$ does not depend on the exact choice of generating polynomials $f_1, \ldots, f_m$, but only on the ideal I. This is going to be a key property for solving algebraic systems. In fact, the goal of the present chapter can informally be rephrased into: Given a set of generating polynomials for an ideal I, find a new set of generating polynomials better suited to determine $\mathcal{V}(I)$ and its properties.

Conversely, we can ask whether I is uniquely determined by $\mathcal{V}(I)$. More precisely, given a set of points V, let us define:

$$\mathcal{I}(V) = \{f \in \mathbb{K}[x_1, \cdots, x_n] \mid \forall (\alpha_1, \cdots, \alpha_n) \in V : f(\alpha_1, \cdots, \alpha_n) = 0\}. \quad (11.18)$$

We see that $\mathcal{I}(V)$ is an ideal and that $\mathcal{I}(\mathcal{V}(I)) \subset I$. However, equality does not hold in general. To give an example with a single variable, we see that $\mathcal{I}(\mathcal{V}((x^2))) = (x)$. Indeed, the only (double) root of x^2 is 0 and it is also a root of x. The generalization of this example relies on the definition of the radical of an ideal. Given an ideal I, we define its radical $\sqrt{I}$ as:

$$\sqrt{I} = \left\{ f \in \mathbb{K}[x_1, \cdots, x_n] \mid \exists t \in \mathbb{N}; \ f^t \in I \right\}. \quad (11.19)$$

A famous theorem of Hilbert, called the Nullstellensatz theorem relates $\mathcal{I}(\mathcal{V}(I))$ and $\sqrt{I}$:

THEOREM 11.1
If $\mathbb{K}$ is an algebraically closed field and I an ideal of $\mathbb{K}[x_1, \cdots, x_n]$, then:

$$\mathcal{I}(\mathcal{V}(I)) = \sqrt{I}.$$

PROOF See [CLO07, Chapter 4]. ⬚

11.3.1   A simple example: Monomial ideals

We recall that a monomial in a polynomial ring $\mathbb{K}[x_1, \cdots, x_n]$ is a polynomial of the form $\lambda x_1^{\alpha_1} \cdots x_n^{\alpha_n}$. For conciseness, we denote it by $\lambda X^{\vec{\alpha}}$, where $\vec{\alpha} = (\alpha_1, \cdots, \alpha_n)$. We say that an ideal I is a monomial ideal if I can be generated by a family of monomials, i.e., $I = (X^{\alpha^{(1)}}, \cdots, X^{\alpha^{(m)}})$. Note that since $\mathbb{K}$ is a field, the constants appearing in the monomials can be removed after multiplication by their respective inverses.

Monomial ideals are easy to manipulate. First, a polynomial f belongs to a monomial ideal $I = (X^{\alpha^{(1)}}, \cdots, X^{\alpha^{(m)}})$, if and only if all the monomials that occur with a non-zero coefficient in f also belong to I. Moreover, a monomial $\lambda X^{\vec{\beta}}$, with $\lambda \neq 0$ is in I, if and only if, at least one monomial $X^{\alpha^{(i)}}$ divides $X^{\vec{\beta}}$. Furthermore, testing divisibility for monomials is very simple, indeed, $X^{\vec{\alpha}}$ divides $X^{\vec{\beta}}$ if and only if, for all $1 \leq i \leq n$, we have $\alpha_i \leq \beta_i$.

11.3.2   General case: Gröbner bases

Since the case of monomial ideals is very simple, it is useful to link the general case with this special case. For this purpose, we introduce order relations between monomials. We call a monomial ordering, any total order relation $\succeq$ on the set of monomials of $\mathbb{K}[x_1, \cdots, x_n]$ which satisfies the following properties:

- **Compatibility** For any triple of monomials $(X^{\vec{\alpha}}, X^{\vec{\beta}}, X^{\vec{\gamma}})$:

$$X^{\vec{\alpha}} \succeq X^{\vec{\beta}} \text{ implies } X^{\vec{\alpha}} X^{\vec{\gamma}} \succeq X^{\vec{\beta}} X^{\vec{\gamma}}.$$

- **Well-ordering** Any set S of monomials contains a smallest element with respect to $\succeq$, this element is denoted $\min_{\succeq}(S)$.

By abuse of notation, when $X^{\vec{\alpha}} \succeq X^{\vec{\beta}}$, we may also write $\vec{\alpha} \succeq \vec{\beta}$. With this convention, the compatibility property becomes:

- **Compatibility** For any triple $(\vec{\alpha}, \vec{\beta}, \vec{\gamma})$:

$$\vec{\alpha} \succeq \vec{\beta} \text{ implies } \vec{\alpha} + \vec{\gamma} \succeq \vec{\beta} + \vec{\gamma}.$$

To better understand the well-ordering property, let us introduce two order relations on monomials:

- The lexicographic ordering $\succeq_{\text{lex}}$ is defined by: $\vec{\alpha} \succ_{\text{lex}} \vec{\beta}$, if and only if there exists an index $1 \leq i_0 \leq n$ such that:

  - For all $i < i_0$, $\alpha_i = \beta_i$.

     – And $\alpha_{i_0} > \beta_{i_0}$.

- The reverse lexicographic[1] order relation $\succeq_{\text{revlex}}$ is defined by: $\vec{\alpha} \succ_{\text{revlex}} \vec{\beta}$, if and only if there exists an index $1 \leq i_0 \leq n$ such that:

     – For all $i > i_0$, $\alpha_i = \beta_i$.

     – And $\alpha_{i_0} < \beta_{i_0}$.

We leave as an exercise to the reader to check that both of the above order relations satisfy the compatibility property and that the lexicographic ordering also satisfies the well-ordering property. However, the reverse lexicographic order relation is not well ordered. Indeed, with a single variable, we see that $x^i \succ_{\text{revlex}} x^j$, if and only if $j > i$. As the consequence, the infinite set S of monomials defined by:

$$S = \{x^i \mid i \in \mathbb{N}\} \tag{11.20}$$

has no smallest element.

     Thus, the reverse lexicographic order is not a proper monomial ordering. On the contrary, the lexicographic ordering is admissible and is one of the most frequent monomial orderings used with Gröbner bases.

     Several other monomial orderings are also used in this context. To introduce two of them, let us first define the total degree of a monomial $X^{\vec{\alpha}}$, as the sum of the exponents, $\deg(X^{\vec{\alpha}}) = \sum_{i=1}^{n} \alpha_i$. It is called the total degree by opposition to the partial degree in each of the variables. When clear by context, the total degree of a monomial is simply called its degree. With this notion of degree, we can define two additional monomial orderings:

- The graded lexicographic ordering $\succeq_{\text{deglex}}$ is defined by: $\vec{\alpha} \succ_{\text{deglex}} \vec{\beta}$ if and only if:

     – Either $\deg(\vec{\alpha}) > \deg(\vec{\beta})$

     – Or $\deg(\vec{\alpha}) = \deg(\vec{\beta})$ and $\vec{\alpha} \succ_{\text{lex}} \vec{\beta}$.

- The graded reverse lexicographic ordering, often known as "grevlex" $\succeq_{\text{grevlex}}$ is defined by: $\vec{\alpha} \succ_{\text{grevlex}} \vec{\beta}$ if and only if there exists an index $1 \leq i_0 \leq n$ such that:

     – Either $\deg(\vec{\alpha}) > \deg(\vec{\beta})$

     – Or $\deg(\vec{\alpha}) = \deg(\vec{\beta})$ and $\vec{\alpha} \succ_{\text{revlex}} \vec{\beta}$.

---

[1]As explained below, this is **not** a monomial ordering.

Both orderings satisfy the compatibility and well-ordering properties (left as exercise). In the sequel, the expression **graded ordering** or **total degree ordering** is used to collectively refer to these two orders.

Note that the three monomials ordering we have now defined are, in truth, families of order parametrized by the order of the variables themselves. Since n variables can be arranged in $n!$ ways, there are $n!$ monomial orderings in each family. There exists more monomial orderings that can be used with Gröbner bases. We do not describe these orderings and refer the reader to a textbox on Gröbner bases such as [CLO07].

Given a monomial ordering $\succeq$ and a polynomial f, we define the head monomial of f as the largest monomial that appear with non-zero coefficient in f. We denote this head monomial of f by $in_{\succeq}(f)$. When clear by context, we omit $\succeq$ from the notation and simply write $in(f)$. Using head monomials, we can associate to any ideal I a monomial ideal $in(I)$ generated by the head monomials of all polynomials in I. When I already is a monomial ideal, we see that $in(I) = I$. In the general case, $in(I)$ can be used to define Gröbner bases:

DEFINITION 11.1   *A family of polynomials f_1, f_2, ..., f_m is a Gröbner basis for I, if and only if:*

$$I = (f_1, \ldots, f_m) \text{ and} \tag{11.21}$$
$$in(I) = (in(f_1), \cdots, in(f_m)). \tag{11.22}$$

An important fact is that any ideal admits a Gröbner basis (see [CLO07, Chapter 2]). Moreover, a Gröbner basis for an ideal I can be used to construct a Euclidean division rule for I. More precisely:

THEOREM 11.2
If $(f_1, \cdots, f_m)$ is a Gröbner basis for the ideal I it generates, then any polynomial g of $\mathbb{K}[x_1, \cdots, x_n]$ can be written as:

$$g = \sum_{i=1}^{n} g_i f_i + r, \tag{11.23}$$

where none of the monomials with non-zero coefficient in r belong to $in(I)$. This remainder r is unique and does not depend on the choice of the Gröbner basis but only on the ideal I and on the chosen monomial ordering. In particular, if $g \in I$ then the remainder is always 0.

The remainder r is called the **normal form** *of g with respect to the ideal I (and monomial ordering $\succeq$).*

PROOF   See [CLO07, Chapter 2].   □

As a direct consequence, Gröbner bases are often used to determine whether a polynomial belongs to a given ideal or not. The algorithmic version of the computation of a normal form by Euclidean division as described above is given as Algorithm 11.1.

Algorithm 11.1 Computation of normal form

Require: Input polynomial g and Gröbner basis $(f_1, \ldots, f_m)$ of ideal I

  Let $r \longleftarrow 0$

  Let $M \longleftarrow HM(g)$ {HM stands for Head monomial (with respect to monomial order) with convention $HM(0) = 0$}

  while $M \neq 0$ **do**

    for i from 1 to m **do**

      if $HM(f_i)$ divides m **then**

        Let $\mu \longleftarrow m/HM(f_i)$

        Let $\lambda \longleftarrow HC(g)/HC(f_i)$ {HM stands for Head coefficient}

        Let $g \longleftarrow g - \lambda \mu f_i$

        Break from FOR loop

      end if

    end for

    if $M = HM(g)$ **then**

      Let $r \longleftarrow r + HC(g) M$ {M has not been reduced, transfer to r}

      Let $g \longleftarrow g - HC(g) M$

    end if

    Let $M \longleftarrow HM(g)$

  end while

  Output r (Normal form of g)

11.3.3 Computing roots with Gröbner bases

Coming back to our initial motivation of finding roots of systems of algebraic equations, we now consider this problem in the light of Gröbner bases. In fact, Gröbner bases can be used to address a slightly more general problem, the problem of unknowns elimination, based on the following theorem:

THEOREM 11.3

For any ideal I in the polynomial ring $\mathbb{K}[x_1, \cdots, x_n]$ and any integer $t \in [1, n]$, let $I^{(t)}$ denotes the set of polynomials in $I \cap \mathbb{K}[x_t, \cdots, x_n]$. Then, $I^{(t)}$ is an ideal in $\mathbb{K}[x_t, \cdots, x_n]$. Moreover, if $G = (f_1, \cdots, f_m)$ is a Gröbner basis for I with respect to the lexicographic order $\succeq_{lex}$, then the subset $G^{(t)}$ of polynomials in G that belongs to $I^{(t)}$ is a Gröbner basis for $I^{(t)}$.

PROOF   See [CLO07, Chapter 3].   ▯

Note that this theorem only holds for the lexicographic ordering. As a consequence, the applications we now describe require a Gröbner basis with respect to this specific ordering. If we are given a Gröbner basis for a different ordering and want to perform elimination, we first need to convert it to lexicographic ordering (see Section 11.6.4).

To find roots of an ideal using a Gröbner basis for the lexicographic ordering, we start by eliminating all unknowns but one. Equivalently, we consider the elimination ideal $I^{(n)}$. Since this ideal is univariate, we know how to find all possible for x_n. It remains to lift such a partial solution in x_n to complete solution(s) of the original system. We now distinguish several cases.

11.3.3.1   The case of incompatible systems

The simplest case occurs when the initial system of algebraic equations is incompatible, i.e., when no solution for the system exists over the algebraic closure $\bar{\mathbb{K}}$, the Gröbner basis for I contains a constant polynomial, say $f_1 = 1$ after normalization. Note that in this special case, the elimination ideal $I^{(n)}$ is also spanned by 1 and there are no partial solutions in x_n either.

With incompatible system, the good news is that the Gröbner basis contains 1, regardless of the monomial ordering that is used. Thus, the knowledge of a Gröbner basis for I under any monomial ordering is enough to check that no solution exists.

11.3.3.2   The case of dimension 0

When a system algebraic equations has a finite number of solutions over $\bar{\mathbb{K}}$ (at least one), it is said to be of dimension 0. In this case, a Gröbner basis G for I as in Theorem 11.3 has a very special form. Namely, for each unknown x_i, there is a polynomial in G whose head monomial is a power of x_i. In addition, if we are considering the lexicographic ordering, this polynomial only involves the unknowns x_i, ..., x_n. For simplicity, we let f_i denote the polynomial of the Gröbner basis with a power of x_i as head monomial and that only contains x_i to x_n. We might expect that I only contains these polynomials; however, this is not true in general.

In any case, the roots of the system can be determined using a simple backtracking algorithm. First, one finds the roots of f_n, which is a univariate polynomial in x_n. Then plugging each possible value for x_n in f_{n-1}, one computes the corresponding values of x_{n-1} by solving another univariate polynomial. And so on, until the values of all unknowns back to x_1 are computed. In addition, if there are extra polynomials in the Gröbner basis, we should check that they are satisfied by the candidate solutions. Note that the total number of solutions can be quite large; counting multiple roots with their multiplicities, the number of candidates before the final checking is equal to the product of the degree of the head monomials in the polynomials f_1 to f_n.

Thus, except in the good case where many of the head monomials have degree 1, the total number of roots is at least exponential in n. As a consequence, computing all these roots can be very costly. However, if we only ask for one solution of the algebraic system or otherwise restrict the acceptable roots, this process can become quite efficient.

Coming back to the problem of extra polynomials in addition to $f_1, \ldots,$ f_n as defined above, this, in fact, covers two different things. First, there is a general fact that Gröbner basis can contain too many polynomials, this can be addressed by using the notion of reduced Gröbner basis (see Definition 11.2 later). Second, somewhat surprisingly, even a reduced Gröbner basis may contain extraneous polynomials, in addition to the polynomials f_1 to f_n mentioned above.

11.3.3.3   The case of higher dimension

When the system of algebraic equations has an infinite number of solutions over $\bar{\mathbb{K}}$, the system is under-determined and there are some (at least one) degrees of freedom left. The dimension of the corresponding ideal is precisely the number of degrees of freedom that are left. This can be determined from the Gröbner basis of Theorem 11.3. We do not go into more details for this case, because most cryptanalytic applications consider algebraic systems of dimension 0. However, it remains possible in this case to find solutions of the algebraic system.

11.3.4   Homogeneous versus affine algebraic systems

A very frequent special case which is encountered when working with Gröbner bases is the case of homogeneous system of equations. To address this case, we first need to define the notion of homogeneous polynomials. Given a multivariate polynomial F, we say that F is homogeneous, if and only if all the monomials that occur in F with a non-zero coefficient have the same total degree. For example, $F_1(x, y) = x^2 + xy + y^2$ is a homogeneous polynomial. An algebraic system of equations is said to be homogeneous, if and only if, all the polynomials in the system are homogeneous. The definition of homogeneous ideals is more complicated. This stems from the fact that all ideals, including homogeneous ideals, contain non-homogeneous polynomials. Going back to our example, the ideal generated by the polynomial $F_1(x, y)$ contains $(x + 1)F_1(x, y)$ which is not homogeneous. To get rid of this difficulty, we say that an ideal is homogeneous, if and only if, there exists a family of homogeneous generators for this ideal.

Homogeneous ideals satisfy many additional properties. They are often considered when working with Gröbner bases because they are better behaved and easier to analyze. In particular, given a homogeneous ideal I, it is very useful to consider its subset I_d containing only homogeneous polynomials of degree d and the zero polynomial. This subset I_d is a finite dimensional vector space which naturally occurs in Gröbner basis computations.

In this book, for lack of space, we do not consider homogeneous systems but deal with ordinary systems, usually known as affine systems of equations. This choice is more natural for readers who do not already know Gröbner basis; however, it induces many complications when going into the details. To balance this, we sometimes remain imprecise in the following discussion.

To shed some light on the difference between homogeneous and affine systems in Gröbner basis computations, let us quickly mention the notion of "degree fall." This notion refers to the fact that the sum of two homogeneous polynomials of degree d is either a polynomial of degree d or 0. On the contrary, with affine polynomials, this is no longer true, the sum of two polynomials of degree d may be 0, a polynomial of degree d or a non-zero polynomial of degree $< d$. In this last case, we say that a degree fall occurs during the sum. While computing Gröbner basis, degree falls may occur during the process and they greatly complicate matters. One consequence is that in the affine case, in order to obtain vector spaces corresponding to I_d for homogeneous ideals, we need to consider the subset $I_{\leq d}$ of all polynomials in the ideal of degree at most d.

11.4 Buchberger algorithm

The first algorithm for computing Gröbner bases was proposed by Buchberger. To introduce this algorithm, let us consider a simple example of an ideal with 2 generators in 2 unknowns, namely the ideal I generated by $f_1 = x^2y + 1$ and $f_2 = xy^2 - 2$. We first remark that the basis (f_1, f_2) for I is not a Gröbner basis (for any monomial ordering). Indeed, the ideal $in(I)$ is not generated by the head monomials x^2y and xy^2 of f_1 and f_2. In particular, considering $f_3 = yf_1 - xf_2 = y + 2x$, we see that its head monomial, which is either x or y depending on the considered ordering, does not belong to the ideal (x^2y, xy^2).

The method we used to construct f_3 is the key ingredient of Buchberger's algorithm. This polynomial f_3 is constructing by subtracting a multiple of f_1 and a multiple of f_2 in order to guarantee that the two multiples share the same head monomial and that the contribution of this head monomial vanishes in f_3. Clearly, the resulting polynomial f_3 is unique, up to multiplication by a constant. It is called a **principal syzygy** or sometimes a S-polynomial for f_1 and f_2. The principal syzygy of f_1 and f_2 is denoted by $S(f_1, f_2)$. In the sequel, we often write syzygy instead of principal syzygy. Note however that in general, syzygy has a wider meaning.

The first application of syzygies is an algorithm, also due to Buchberger, for testing whether a basis G for an ideal I is a Gröbner basis for I. This algorithm takes all pairs of polynomials from G, constructs their principal syzygies and computes the remainder of each syzygy modulo G using the normal form Algorithm 11.1. A theorem of Buchberger states that G is a Gröbner basis for I, if and only if the normal form obtained for each principal syzygy with respect to G is 0. For a proof, refer to [CLO07, Chapter 2].

A simple version of Buchberger's algorithm can be derived from this decision algorithm. It suffices to add to the initial basis G any non-zero remainder that appears and to repeat the process until the decision algorithm is happy with the current basis. Since each of the computed remainders clearly belongs to the ideal I, the final basis generates the correct ideal and it is a Gröbner basis. This simple version is given as Algorithm 11.2. Note that it needs to compute the GCD of two monomials M_1 and M_2, this is easily done by keeping for each unknown the minimum of the two exponents in M_1 and M_2.

Algorithm 11.2 Basic version of Buchberger's algorithm

Require: Input list of polynomials G {Throughout the algorithm $|G|$ denotes the current size of list G}
  Let $i \longleftarrow 2$
  while $i \leq |G|$ **do**
    Let $g_1 \longleftarrow G[i]$
    Let $M_1 \longleftarrow HM(g_1)$
    Let $\lambda_1 \longleftarrow HC(g_1)$
    for j from 1 to $i - 1$ **do**
      Let $g_2 \longleftarrow G[j]$
      Let $M_2 \longleftarrow HM(g_2)$
      Let $\lambda_2 \longleftarrow HC(g_2)$
      Let $M \longleftarrow GCD(M_1, M_2)$
      Let $h \longleftarrow \lambda_2(M_2/M)g_1 - \lambda_1(M_1/M)g_2$
      Let r be the normal form of h with respect to G {Algorithm 11.1}
      if $r \neq 0$ **then**
        Append r to G {This increments $|G|$}
      end if
    end for
    Let $i \longleftarrow i + 1$
  end while
  Output Gröbner basis G

Buchberger's algorithm can be improved from this basic version by adding rules to avoid the computation of any principal syzygy, for which we can predict a reduction to zero; see [CLO07, Chapter 2].

We now continue our example, using a monomial ordering such that $y \succ x$. Thus y is the head monomial of f_3. We see that the principal syzygy of f_1

and f_3 is $S(f_1, f_3) = f_1 - x^2 f_3 = -2x^3 + 1$, since x^3 is not a multiple of the previously encountered head monomial, the remainder of this syzygy is $f_4 = -2x^3 + 1$. Similarly, the syzygy of f_2 and f_3 is $S(f_2, f_3) = f_2 - xyf_3 = -2yx^2 - 2$, adding $2f_1$, we see that the corresponding remainder is 0. Next, we turn to $S(f_1, f_4) = 2xf_1 + yf_4 = 2x + y$, whose remainder is 0 since $S(f_1, f_4) = f_3$. We also see that $S(f_2, f_4) = 2x^2 f_2 + y^2 f_4 = -4x^2 + y^2 = (-2x + y)f_3$ has remainder 0 and that $S(f_3, f_4) = x^2 f_3 + f_4 = yx^2 + 1 = f_1$ has remainder 0. We conclude that (f_1, f_2, f_3, f_4) is a Gröbner basis for I.

However, let us remark that this Gröbner basis is too large. Indeed, $f_1 = x^2 f_3 + f_4$ and $f_2 = (xy - 2x^2)f_3 - 2f_4$, thus f_1 and f_2 can safely be removed and (f_3, f_4) also is a Gröbner basis for I. This also shows that Gröbner bases are not uniquely determined when an ideal and monomial ordering are fixed. Should unicity be required, the notion of **reduced Gröbner basis** can be used.

DEFINITION 11.2   *A family of polynomials $F = (f_1, f_2, \cdots, f_m)$ is a reduced Gröbner basis for I, if and only if:*

1. *F is a Gröbner basis for I*

2. *For all f_i in F, no monomial appearing in f_i is a head monomial in the subfamily $F - \{f_i\}$.*

3. *The coefficient of the head monomial in each polynomial f_i is 1.*

It is easy to obtain a reduced Gröbner basis from a Gröbner basis with a simple reduction procedure given as Algorithm 11.3. Moreover, any non-zero ideal I has a unique reduced Gröbner basis; see [CLO07, Chapter 2].

11.5   Macaulay's matrices

In Section 11.2, we saw that in the bivariate case, polynomial systems can be computed using resultants, which rely on the computation of determinants of Sylvester's matrices. In the general case, this can be somehow generalized by using Macaulay's matrices.

Macaulay's matrices are usually defined to represent homogeneous ideals (see Section 11.3.4). Following our convention in this chapter, we adapt the definition to the affine case. Given an ideal I, a Macaulay's matrix of degree d for I is a matrix that describes a basis for the vector space $I_{\leq d}$ introduced in Section 11.3.4).

Given a Gröbner basis for I, it is easy to construct a Macaulay's matrix $M_{\leq d}(I)$ of $I_{\leq d}$. The columns of this matrix are labelled by the monomials

---

Algorithm 11.3 Reduction of a Gröbner basis

Require: Input Gröbner basis G as list of polynomials {Throughout the algorithm $|G|$ denotes the current size of list G}

Let $i \longleftarrow 1$

while $i \le |G|$ **do**

    Let $g \longleftarrow G[i]$

    Let r be the normal form of g with respect to $G - G[i]$ {Algorithm 11.1}

    if $r = 0$ **then**

        Append $G[i]$ from G {This decrements $|G|$ and renumber polynomials}

    else

        Let $\lambda \longleftarrow HC(r)$

        Let $r \longleftarrow r/\lambda$

        Replace $G[i]$ by r

        Let $i \longleftarrow i + 1$

    end if

end while

Output reduced Gröbner basis G

---

of degree $\le d$. With n variables in degree at most d, there are $\binom{n+d}{d}$ such monomials. The rows of the matrix are labelled by simple multiples of the polynomials in the Gröbner. For example, if f is a polynomial of degree d' in the Gröbner basis, for each monomial m of degree at most $d - d'$ there is a row labelled mf. In particular, the matrix of Sylvester can be interpreted in that way. Starting from f_1 and f_2 of respective degree d_1 and d_2 in x, the matrix contains $d_1 + d_2$ columns labelled from x^0 to $x^{d_1+d_2-1}$ and $d_1 + d_2$ rows labelled $x^0 f_1$ to $x^{d_2-1} f_1$ and $x^0 f_2$ to $x^{d_1-1} f_2$. The only noticeable difference is that for Sylvester's matrices, the highest degree monomials are traditionally the leftmost columns, while for Macaulay's matrices they are usually the rightmost ones. Redefining the order of rows and columns in Sylvester's matrices would, depending on the degrees d_1 and d_2, affect the sign of resultants.

Conversely, Lazard showed in [Laz83] that by performing linear algebra on Macaulay's matrices of the ideal I, up to high enough a degree, it is possible to obtain a Gröbner basis for I. This property is, in particular, used for the algorithms presented in the next section.

11.6 Faugère's algorithms

When computing Gröbner bases with Buchberger's algorithm, the practical limits are quickly encountered. As a consequence, many applications, especially in cryptanalysis need to use more advanced algorithms. Among those, two algorithms by Faugère, called F_4 and F_5 (see [Fau99, Fau02]) are often considered. The papers presenting these two algorithms focus on different aspects of Gröbner basis computations. On the one hand, F_4 considers how to improve the computation of reduced syzygies that occurs in Buchberger's algorithm, by using (sparse) linear algebra algorithms on Macaulay's matrices. On the other hand, F_5 focuses on the issue of avoiding reduction to zero during the computation. Indeed, it is well known that a large fraction of the syzygies considered during Gröbner basis computations yields the zero polynomial after reduction modulo the current basis of the ideal.

The basic idea of the F_4 algorithm is to use Macaulay's matrices in order to represent polynomial ideals, compute syzygies and speed up the computation required by Buchberger's algorithm. The basic idea of the F_5 algorithm is to organize the computation at the abstract algebraic level in order to predict and remove many of the reductions to zero. For optimal performance of Gröbner basis computations, it is best to use both the F_4 and F_5 ideas at the same time.

11.6.1 The F_4 approach

As said above, the F_4 approach heavily relies on the use of Macaulay's matrices. Thus, it is useful to explicitly specify the correspondence between polynomial ideals and matrices. If we restrict ourselves to polynomials of total degree at most d within an ideal, we need to represent a vector space of finite dimension.

Let us start by considering a vector space V_d be a vector space generated by an arbitrary (finite) set S_d of polynomials of degree at most d. We can easily describe V_d with a matrix if we label the columns of the matrix by monomials and fill each row with a polynomial by placing in the column corresponding to a monomial the coefficient of this monomial in the polynomials. In particular, if we multiply this matrix by the vector formed of all monomials in the same order, we obtain a vector whose coordinates are the polynomials of S_d. To make sure that this encoding is meaningful in the context of Gröbner basis, the mapping between monomials and columns should conform to the monomial ordering for which the Gröbner basis is computed. Moreover, for the sake of compatibility with the linear algebra algorithms, it is preferable to consider that this mapping is in decreasing order starting from the leftmost column. With this convention, the head monomial of a polynomial corresponds to the leftmost non-zero entry in the corresponding row.

In the case of ideals, we encounter complications because a family of polynomials which generates an ideal does not suffice to generate the full vector space. Temporarily forgetting about the degree limit, we can construct an infinite generating family simply by putting together all multiples of the original polynomials. Putting back the degree limit, two very different cases arise. With homogeneous ideals, it suffices to consider the multiples of the original polynomials up to the degree limit. With affine ideals, because of degree falls, this is no longer true and we might need additional polynomials.

To illustrate this by a basic example, let us consider the ideal generated by $f_1(x, y) = x$ and $f_2(x, y) = x^2 + y$. Up to degree 1, there is a single multiple to consider: f_1 itself. However, the ideal generated by f_1 and f_2 contains the polynomial $f_2 - f_1^2 = y$, which is of degree one.

Thus, we see that for fixed degree d, we are faced with two challenges. First, we need to make sure that the matrix we construct truly covers all polynomials of degree up to d. Second, we should represent this matrix in a convenient form which allows us to quickly test whether a given polynomial belongs to the vector subspace spanned by this matrix.

The second issue is quite easy to address, it suffices to transform the matrix into row echelon form as described in Section 3.3.3 of Chapter 3. However, in the context of Gröbner basis computations, we need to slightly change the implementation of the linear algebra algorithm, in order to mix some dense and some sparse rows within the same matrix. Indeed, on the one hand, in some rows we simply need to store multiples of polynomials of lower degree and it is preferable to store them in factored form. On the other hand, in some rows we need to store polynomials which have been computed as syzygies and it is often easier to store them in dense form. Note that in tuned implementations, it can be very useful to use various techniques, described in [Fau99], to also compress these polynomials.

The first issue is more complicated. Indeed, due to degree fall, when computing the row echelon form of Macaulay's matrix up to degree d, new polynomials of degree smaller than d may appear due to a degree fall. In that case, we should not only add this polynomial to the matrix but also all its multiples. Due to this possibility, it is not enough to compute the multiples of the initial polynomials up to degree d and to perform linear algebra in order to make sure that Macaulay's matrix is complete. Instead, the F_4 approach couples this linear algebra approach with the notion of syzygy as in Buchberger's algorithm.

Still, the basic idea of constructing all multiples of the given polynomial up to some degree d, putting them in a matrix and reducing this matrix in row echelon form is a very useful thought experiment. Note that, if d is high enough, this yields a Gröbner basis. The main problem is that, short of taking an unreasonably high degree, we do not know beforehand how high d should be. Moreover, the degree of monomials that needs to be considered with such a direct approach is usually larger than the maximum degree that occurs during more advanced Gröbner basis algorithm. In addition, except for systems of equations with specific properties, in order to find the right value of d, we essentially need to compute a Gröbner basis of the system. Yet, we give for further reference an algorithmic version of this basic idea as Algorithm 11.4. Note that this algorithm is essentially what has been called linearization/relinearization in [KS99].

Algorithm 11.4 A basic linear algebra based Gröbner basis algorithm

---

Require: Number of unknowns n
Require: Input list of polynomials G
Require: Maximal degree D to consider
  Let $N \longleftarrow \binom{n+d}{d}$ {Number of monomials, see Exercise 6}
  Create array A (with N columns)
  for i from 1 to $|G|$ **do**
    for all M monomial of degree at most $D - \deg G[i]$ **do**
      Let $h \longleftarrow M \cdot G[i]$
      Add a new row in A encoding h
    end for
    Compute row echelon form for A
    Let m denote the number of non-zero rows in A
    Create empty list H
    for i from m downto 1 **do**
      Transform i-th row of A into polynomial h {This loop considers head monomials in increasing order.}
      if $HM(h)$ not divisible by any head monomial of polynomial in H **then**
        Add h to H
      end if
    end for
  end for
  Output H

11.6.2   The F_5 approach

With Buchberger's algorithm and also when using the F_4 approach, it has been remarked that a large fraction of the polynomials that are constructed reduces to zero. As a consequence, the algorithms spend a large fraction of their running time computing and reducing polynomials which are then discarded.

To analyze this problem, let us consider the thought experiment Algorithm 11.4. Assume that f_1 and f_2 are two polynomials of degree d_1 and d_2 in the initial system. If we construct the matrix containing all multiples up to degree d_1+d_2, the identity $f_2f_1 - f_1f_2 = 0$ is going to induce a reduction to zero during the linear algebra. To see why, it is useful to label the rows we are inserting in the matrix. Given an initial polynomial f_k and a monomial m, we label the rows containing a description of the polynomial mf_k by the label mF_k. Once this is done, we can easily label a linear combination of rows, in the following way: given a polynomial $g = \sum_{i=1}^{N} \alpha_i m_i$ then gF_k denotes the linear combination of rows: $\sum_{i=1}^{N} \alpha_i(m_i F_k)$. With this convention, any linear combination of rows can be described as a sum of products of the form $g_i F_i$. We now remark that the above identity implies that the linear combination $f_2F_1 - f_1F_2$ is equal to 0. This is a generic reduction to zero which occurs for an arbitrary system of equations.

The goal of the F_5 approach is precisely to avoid all such generic reductions to 0. In the context of Algorithm 11.4, let us describe all possible generic reductions to 0. Assume that the initial system contains polynomials from f_1 to f_k, labeled F_1 to F_k and denote by I_ℓ the ideal generated by the initial sequence of polynomial from f_1 to f_ℓ. Let g be an arbitrary polynomial in I_ℓ, which can be written as $\sum_{i=1}^{\ell} g_i f_i$. Assuming that $\ell < k$, we have an identity $gf_{\ell+1} - f_{\ell+1}g = 0$. From this identity, we easily see that the linear combination

$$gF_{\ell+1} - \sum_{i=1}^{\ell}(f_{\ell+1}g_i)F_i$$

is equal to zero. To avoid the corresponding reduction to zero, we should make sure that at least one monomial involved in this combination is not included in Macaulay's matrix. Moreover, if we remove a single row for this reduction to zero, we are not changing the vector space spanned by the matrix. Thus, the row echelon form we obtain is unaffected[2]. In [Fau02], Faugère proposed to remove the row $mF_{\ell+1}$ corresponding to the product of the head monomial m of g by the polynomial $f_{\ell+1}$. Of course, this should be done for all possible head monomials in I_ℓ. Faugère also proved in [Fau02] that for homogeneous ideals, this approach removes all generic reduction to 0.

Up to now, we have described Faugère's idea in the context of Algorithm 11.4. However, we would like to use it with practical Gröbner basis algorithms. In

---

[2] More precisely, the only change to the row echelon form is the removal of a row of zeros.

order to do this, we need to be able to label all the polynomials that occur during these computations. The problem is that, with these algorithm, we are no longer inserting multiples of the initial polynomials but more complicated combinations obtained from syzygies. Of course, we could always label the linear combination as above with a label of the form $\sum_{i=1}^{k} g_i F_i$. However, this would be cumbersome to manipulate. The good news is that to remove generic reduction to zero, it suffices to keep the "head term" of this complicated label. More precisely, if g_ℓ is the first non-zero polynomial in the complete label and m_ℓ is its head monomial, then we simply keep $m_\ell F_\ell$ as simplified label. These labels can be easily tracked during the computations of syzygies. Indeed, let us consider a syzygy between f and g of the form $\alpha_f m_f f - \alpha_g m_g g$ and assume that f and g are respectively labelled $m_\ell F_\ell$ and $m_{\ell'} F_{\ell'}$ with $\ell' > \ell$. Then the syzygy is simply labelled by $(m_f m_\ell) F_\ell$. If $\ell = \ell'$, we should keep the largest of the two monomials m_ℓ and $m_{\ell'}$ in the label. Note that we cannot have both $\ell = \ell'$ and $m_\ell = m_{\ell'}$ because it would correspond to inserting twice the same row in Macaulay's matrix and lead to a generic reduction to zero.

11.6.3 The specific case of $\mathbb{F}_2$

A specific case which arises quite often in cryptography is the computation of Gröbner basis over the finite field $\mathbb{F}_2$. Moreover, in this case, we usually want to find solutions of the system over the finite field itself and not over its algebraic closure. In order to make this information available to the Gröbner basis, a traditional method is to add the so-called "field equations." For an unknown x, the field equation over $\mathbb{F}_2$ simply is $x^2 + x = 0$. It is clear that over $\mathbb{F}_2$ both 0 and 1 satisfy this equation. Moreover, any value strictly belonging to an extension field of $\mathbb{F}_2$ does not satisfy it. Thus, adding field equations let us focus on the ground field solutions. Note that this can be generalized to $\mathbb{F}_p$ by considering the equations $x^p - x = 0$. However, when p becomes even moderately large, the degree of the field equations is too big to be useful in Gröbner basis computations. With $\mathbb{F}_2$, the field equations are only quadratic and very helpful to speed up the computations of Gröbner bases.

However, this is not the only optimization which can be used over $\mathbb{F}_2$. We can also improve the linear algebra as in Chapter 3 using low-level optimization tricks. In fact, similar tricks can be used to add the field equations implicitly, redefining multiplication to reflect the property that $x^2 = x$. This allows to represent the monomials in a more compact way. However, a bad side effect is that the syzygies between a polynomial and an implicit field equation should be addressed in a specific fashion.

For illustration purposes, a simple implementation of Gröbner basis computation over $\mathbb{F}_2$ simultaneously using the F_4 and F_5 approaches is available on the book's website.

11.6.4 Choosing and changing monomial ordering for Gröbner bases

In theory, all the algorithms that we have seen for computing Gröbner bases can be applied with an arbitrary monomial ordering. However, in practice, it has often been remarked that the choice of monomial ordering greatly impacts the running time of the computation. As a general rule, total degree orderings are to preferred. More precisely, the ordering grevlex is usually the best possible choice. Depending on the specific problem at hand, one can also fine-tune the order of the individual variables within the monomial ordering.

This becomes a problem when the application we have in mind requires a specific monomial ordering. For example, we have seen that to solve a algebraic system of equations, the lexicographic ordering should be preferred. The good news is that, once we have computed a Gröbner basis of an ideal for some monomial ordering, it is easier to derive from this a Gröbner basis for another ordering than to recompute this second basis from scratch. Moreover, specific algorithms have been proposed to address this problem. They are called ordering change algorithms for Gröbner bases. In this section, we briefly discuss these algorithms.

When considering ordering change, two cases need to be distinguished. For ideals of dimension 0, i.e., ideals with finitely many solutions, there is a method due to Faugère, Gianni, Lazard and Mora (see [FGLM93]), which is called FGLM. This method is based on linear algebra and very efficient. We sketch its main idea below in a simplified version. Thanks to this method, the best approach to compute Gröbner bases in dimension 0 is usually to start by computing a basis for the grevlex ordering and then to convert this basis to the desired ordering. For ideals of larger dimension, there exists a technique called the Gröbner Walk [CKM97] which can be used for conversion. However, in practice, it is much less efficient than FGLM. Thus, problems which involve ideals of positive dimension are often harder to solve than similar problems in dimension 0. The details of the Gröbner Walk algorithm are out of the scope of this book; however, it is useful to know that this algorithm is often available in Gröbner basis computation packages.

11.6.4.1 A simplified variant of FGLM

Assume that we are given a reduced Gröbner basis G for some ideal I and some monomial ordering, usually grevlex. We know that to solve the algebraic system, it is useful to convert this to obtain a Gröbner basis G' for the lexicographic ordering. Moreover, if I has dimension 0, there exists in G' a univariate polynomial $f(x_1)$ involving only the smallest variable x_1 in the lexicographic ordering.

To illustrate FGLM, we now present a simplified version of this algorithm, whose goal is to efficiently recover $f(x_1)$ from G. In order to do this, we first define the set $B(G)$ of all monomials which are not divisible by a head monomial of G, with respect to the monomial ordering used for computing G.

A very important fact is that, since I has dimension 0, $B(G)$ is a finite set. Let us also consider the set $V(G)$ of all polynomials after reduction modulo G. This set is a finite dimensional vector space and $B(G)$ is a basis for $V(G)$.

In the vector space $V(G)$, we can consider the action of the multiplication by x_1. More precisely, starting from a polynomial P in $V(G)$, we multiply it by x_1 and perform reduction modulo G. The result $x_1 P \pmod{G}$ is a polynomial in $V(G)$. Since, $V(G)$ is finite dimensional, any large enough family of polynomials in $V(G)$ is linearly dependent. In particular, the sequence of monomials $1, x_1, \ldots x_1^N$ reduced modulo G is linearly dependent when N is large enough. Moreover, such a dependence relation can be translated into a polynomial f such that $f(x_1)$ is equal to 0 modulo G. Thus, $f(x_1)$ belongs to I. If f is minimal, i.e., if N is minimal, this polynomial is the first polynomial of G', that we were looking for.

From an algorithmic point-of-view, we need to consider two issues. First, we need a fast method for multiplication by x_1 in order to produce the sequence x_1^i. Second, we need to recover f from the expression of the monomials. The first issue can be addressed by remarking that to multiply a polynomial by x_1, it suffices to multiply each of its terms by x_1 and to sum the results. For monomials, the multiplication by x_1 is very easy and we only need to study the reduction modulo G. For the monomial m, two cases arise: either $x_1 m$ is in $B(G)$ and there is nothing to do, or $x_1 m$ is divisible by the head monomial of a polynomial g in G. In that case, we need to perform a full reduction modulo G. An important improvement on this basic method is proposed in [FGLM93] to replace this full reduction by a simple subtraction of a multiple of g, followed by the re-use of already computed monomial multiplications. We do not fully describe this improvement which requires to simultaneously compute the multiplication of monomials by each possible variable x_i.

The second issue is a simple matter of using linear algebra algorithms to find a linear dependency between the reduced expressions for the monomials x_1^i. The rest of the polynomials in G' are found by computing extra linear dependencies in other families of monomials.

11.7 Algebraic attacks on multivariate cryptography

To illustrate the use of Gröbner basis computations in cryptography, we are going to consider attacks against a specific public key cryptosystem based on multivariate cryptography and called HFE (Hidden Field Equations). This section presents a description of HFE together with an account of Gröbner-based attacks against this cryptosystem. Note that these attacks do not work on all the variations of the HFE system as proposed in [Pat96], they target the basic HFE scheme. For more information about multivariate public key cryptography, the reader may refer to [BBD07, pages 193–242].

11.7.1   The HFE cryptosystem

The HFE scheme proposed by Patarin in [Pat96] is a multivariate cryptographic scheme based on the hardness of solving quadratic systems of multivariate equations over finite fields. Its binary version uses quadratic systems over $\mathbb{F}_2$. The rationale behind this scheme is that solving quadratic systems over $\mathbb{F}_2$ is an NP-complete problem.

In order to describe the HFE cryptosystem, we are going to alternate between two different representations of the polynomial ring $\mathbb{F}_{2^n}[X]$. In what we call the secret view, we consider the finite field with its complete algebraic structure and take a univariate polynomial ring over this finite field. In the public view, we strip a large amount of structure and consider the multivariate polynomial ring $\mathbb{F}_2[x_0, \cdots, x_{n-1}]$. To link the two representations, we choose a basis for $\mathbb{F}_{2^n}$, for example a polynomial basis given by 1, α, ..., α^{n-1}, replace X by $\sum_{i=0}^{n-1} x_i \alpha^i$ and take the coordinates of the result. Using this link, any polynomial F in X is translated into n polynomials f_0, ..., f_{n-1} (with f_i denoting the α^i-coordinate). Moreover, if the degree of the monomials appearing in F is well chosen, it is possible to make sure that the coordinates f_i are quadratic. Indeed, it suffices to remark that any monomial X^{2^i} is obtained by iterating the Frobenius map i times and has a linear expression in the x_is. As a consequence, any monomial $X^{2^i+2^j}$ with $i \neq j$ leads to a quadratic expression.

The key idea behind HFE is that in the secret view, evaluating F at a point X is easy and solving the equation $Y = F(X)$ for a given value of Y is also easy. In the public view, evaluating n quadratic polynomials remains easy, but solving a system of n quadratic equations is hard in general. Thus, F is a candidate trapdoor one-way function, the knowledge of the secret view being the trapdoor. Note that to make sure that F remains easy to invert, we need to put an upper bound D on the degree of the monomials X^{2^i} and $X^{2^i+2^j}$ appearing in F. In order to make the description of HFE complete, a single ingredient is missing: how do we guarantee that the secret view is hidden and not revealed by the public view? The answer proposed in [Pat96] is astonishingly simple: perform two (invertible) linear changes, one on the input variables and one on the resulting system of equations. Despite the simplicity of this ingredient, no known simple attack can recover the secret view from the public view of an HFE system.

To make the description of HFE more precise, we first fix the two parameters n and D and choose a random polynomial F in $\mathbb{F}_{2^n}[X]$ of degree at most D that only contains that constant monomial and monomials of the form X^{2^i} and $X^{2^i+2^j}$. We also choose two random invertible $n \times n$ matrices over $\mathbb{F}_2$ and produce the public system of quadratic functions by considering the composition $T \circ f \circ S$. In other words, we proceed as follows:

- Let $x_0, \ldots, x_{n-1}$ be the input variables and first compute a linear change of variables by writing:

$$
\begin{pmatrix} x'_0 \\ x'_1 \\ \vdots \\ x'_{n-1} \end{pmatrix} = S \cdot \begin{pmatrix} x_0 \\ x_1 \\ \vdots \\ x_{n-1} \end{pmatrix}. \tag{11.24}
$$

- Let $X = \sum_{i=0}^{n-1} x'_i \alpha^i$, consider the polynomial $F(X)$ and take its coordinates writing:

$$
F(X) = \sum_{i=0}^{n-1} f'_i(x'_0, \cdots, x'_{n-1}) \alpha^i. \tag{11.25}
$$

- Compute the second change of coordinates, letting:

$$
\begin{pmatrix} f_0 \\ f_1 \\ \vdots \\ f_{n-1} \end{pmatrix} = T \cdot \begin{pmatrix} f'_0 \\ f'_1 \\ \vdots \\ f'_{n-1} \end{pmatrix}. \tag{11.26}
$$

After computing the polynomials f_i, they are published as the public key of the HFE scheme. Since these polynomials are quadratic in n variables over $\mathbb{F}_2$, they can be described quite compactly. Indeed, there are $n(n-1)/2$ quadratic monomials $x_i x_j$, n linear monomials x_i and a single constant monomial. All in all, the n polynomials can be encoded using about $n^3/2$ bits. The HFE public operation simply consists of evaluating the n public polynomials on a n-uple of bits. For encryption purposes, it is important to make sure that each ciphertext has a unique decryption. Since this is not guaranteed for an arbitrary function F, it is necessary in this case to add redundancy in the encrypted n-uple, in order to allow unique decryption. We do not describe this mechanism here.

A important remark about HFE is that inverting F, with knowledge of the secret view, requires time polynomial in n and D. As a consequence, to express this time in terms of a single parameter, it is natural to assume that there exists a parameter γ such that $D = O(n^\gamma)$. Letting t denote $\log_2(D)$, this implies that $t = O(\log n)$. Note that t is an upper bound on the values i and j that may appear in the monomials X^{2^i} and $X^{2^i+2^j}$. In practice, a value of D around 128 is frequently considered.

11.7.2 Experimental Gröbner basis attack

Clearly, in order to break HFE, it suffices to find a method that allows inversion of an algebraic system of equations based on the polynomials f_i

and on a target n-uple of bits. In general, for large values of n, currently available algorithms completely fail to compute a Gröbner basis for n random quadratic polynomials over $\mathbb{F}_2$. In particular, for $n = 80$ which is one of the parameters proposed as a challenge by Patarin, random quadratic systems are out of range of Gröbner algorithm.

However, quadratic systems based on the HFE trapdoor are not random. In particular, this was illustrated by Faugère who was able to break an HFE challenge for $n = 80$, see [FJ03]. The experimental fact is that, when faced with HFE systems, the Gröbner basis algorithm only involves the computation of polynomials and syzygies of surprisingly low degree. This degree is related to the degree D of the secret polynomial, for example for $17 \leq D \leq 128$ no polynomial of degree higher than 4 appears. Moreover, the resulting Gröbner basis contains a very large number of linear polynomials, almost n. As a consequence, once the Gröbner basis is obtained for the grevlex ordering, there is no need to perform a change of ordering, it suffices to compute all the solutions of the linear system and test each of them against the original system of equations.

This experimental result was published, together with a partial heuristic explanation based on monomial count in [FJ03].

11.7.3   Theoretical explanation

The reason for the experimental behavior of HFE systems can be analyzed in a much more precise fashion. The key argument is that the degree of the polynomials and syzygies that appear in a Gröbner basis computation is essentially unaffected by the two sorts of linear changes that are used when constructing the HFE secret key.

As a consequence, it suffices to analyze the highest degree that can be reached for a Gröbner basis computation, when the linear transforms are omitted. Thanks to the use of normal bases, it is even possible to study an even simpler system of equations over the extension field $\mathbb{F}_{2^n}$. The idea is to represent an element X in $\mathbb{F}_{2^n}$ in a redundant manner as a n-uple $(X, X^2, X^4, \cdots, X^{2^{n-1}})$ containing X and all its copies by the Frobenius map, i.e., by squaring. There is a linear change, with coefficients in $\mathbb{F}_{2^n}$ between this n-uple and the coordinates of X in any fixed basis for $\mathbb{F}_{2^n}$ over $\mathbb{F}_2$. For simplicity, we give a name to each unknown in the above n-uple and let X_i denote X^{2^i}. With this representation of X, the polynomial F can be written as a quadratic polynomial $G(X_0, X_1, \cdots, X_{n-1})$. The good news is that, when looking more precisely at the construction of G from F, this quadratic polynomial only involves a subset of the unknowns $(X_0, \cdots, X_t)$ where t is the parameter $t = \log_2 D$ that we introduced earlier. Moreover, applying the Frobenius maps i times to the equation $G(X_0, X_1, \cdots, X_{n-1}) = Y$, we obtain new equations:

$$G(X_i, X_{i+1}, \cdots, X_{i+n-1}) = Y^{2^i}, \tag{11.27}$$

where the indices are reduced modulo n in case of overflow.

Of course, we also have extra quadratic equations that translate the Frobenius relations, i.e., the equations $X_{i+1} = X_i^2$. Putting all this together, if we take t copies of G by applying the Frobenius map up to $t-1$ times, we have a total of $3t-1$ quadratic equations in $2t$ unknowns. Since the number of equations in this system is 1.5 times the number of unknowns, it is possible to use generic complexity analysis results on Gröbner bases computation to show that the running time is bounded by $t^{O(t)}$; see [GJS06]. Note that our specific use of this complexity result involves a heuristic argument. However, it can be made rigorous if we only aim at constructing a distinguisher between HFE public keys and random quadratic systems with equivalent parameters.

To summarize, the relative weakness of HFE schemes to Gröbner techniques comes from the fact that the system hides by linear changes a system that can be obtained from the secret view and that involves much fewer equations.

11.7.4   Direct sparse approach on Macaulay's matrix

To conclude this chapter, it is important to mention that it is, sometimes, possible to consider a very direct approach to Gröbner basis computations for cryptographic applications, similar to Algorithm 11.4. To give a single example, consider once again the case of HFE systems, we know that a Gröbner basis computation only involves polynomial up to some predictable degree and that the resulting Gröbner basis for the corresponding ideal contains many linear polynomials.

Then, an option is to consider a matrix similar to Macaulay's matrix defined in Section 11.5 and to search for linear polynomials in the vector space spanned by this matrix. To construct this matrix, we proceed exactly as in Section 11.5 or Algorithm 11.4. Note that, since we a not starting from a Gröbner basis for our ideal, we obtain an incomplete copy of the Macaulay matrix of the same degree. If we consider high enough a degree, we may hope, thanks to degree falls, to find linear polynomials in the vector space spanned by this matrix. Ignoring this difficulty for now, we may see that linear polynomials in the space spanned by Macaulay's matrix can be found by looking for kernel elements of a truncated Macaulay's matrix from which the columns labelled by the constant and linear monomials have been removed. Clearly, such a kernel element defines a linear combination of rows that is non-zero on all monomials of degree 2 or more. This yields either the zero polynomial, the unit polynomial or a non-trivial linear polynomial. Note the unit polynomial can only occur if the algebraic system of equations generates the unit ideal, in that case, the system of equations does not have any solutions.

Since we are looking for kernel elements in a matrix, we may use the algorithms of Chapter 3. Moreover, looking more precisely at our variation of Macaulay's matrix, we may remark that it is a sparse matrix. Indeed, each row in the matrix represents a multiple of a polynomial of low degree and, thus, it cannot contain too many non-zero coefficients. As a consequence, we can apply sparse linear algebra algorithms to find kernel elements. In fact, we can do better than that. The key argument is to remember that, with sparse linear algebra algorithm, the main requirement is to be able to efficiently multiply vectors on the left by the matrix we are considering and also by its transpose. Since Macaulay's matrix is obtained by putting together several shifted copies of the matrix encoding the initial polynomials, we can implement multiplication of a vector by the complete matrix by executing several multiplications of well-chosen subvectors by the original matrix and by reassembling the results into a single vector. This yields a faster matrix by vector multiplication than a straightforward sparse approach. Indeed, this approach is compatible with bitslice operations and also cache-friendly.

11.8   On the complexity of Gröbner bases computation

The complexity of a specific Gröbner basis computation depends on several parameters, the underlying field, the number of unknowns, the number of equations and their degrees. However, the most important parameter is the maximum degree of the polynomials that occurs during the computation. This maximum degree may greatly vary even for similarly looking systems of equations. In fact, this is the parameter which explains why HFE systems can be solved, while random systems with similar specifications cannot.

When using F_5 with the grevlex technique on homogeneous system, it is possible by using sophisticated mathematical techniques to compute an upper bound on a technical parameter called the degree of semi-regularity of the system.

Without entering the technical details, let us simply mention that the degree of semi-regularity provides a heuristic bound on the maximum degree of polynomials occuring in the Gröbner basis computation.

A good reference for this, written in French, is the Ph.D. thesis of Bardet [Bar04]. Let us quote a few very useful theorems from [Bar04]. We start by two theorems which hold for computing Gröbner bases over large finite fields.

THEOREM 11.4 Bardet 4.1.2
For a fixed value of k and n tending to infinity, the degree of semi-regularity for a system of $n + k$ equations of degree D in n unknowns is asymptotically upper bounded by:

$$n\frac{D-1}{2} - \alpha_k\sqrt{n\frac{D^2-1}{6}} + o(\sqrt{n}), \qquad (11.28)$$

for some constant α_k.

THEOREM 11.5 Bardet 4.4.1

For a fixed value of $\alpha > 1$ and n tending to infinity, the degree of semi-regularity for a system of αn quadratic equations in n unknowns is asymptotically upper bounded by:

$$(\alpha - 1/2 - \sqrt{\beta})\, n + \frac{-a_1}{2\,\beta^{1/6}}\, n^{1/3} - \left(2 - \frac{2\alpha - 1}{4\beta^{1/2}}\right) + o(n^{-1/3}), \qquad (11.29)$$

where $\beta = \alpha(\alpha - 1)$ and $a_1 \approx -2.33811$ is a constant (the largest root of Airy's Ai function).

For algebraic systems over $\mathbb{F}_2$, the results are slightly different. We have:

THEOREM 11.6 Bardet 4.4.3

For a fixed value of $\alpha > 1$ and n tending to infinity, the degree of semi-regularity reached for a system of αn quadratic equations in n unknowns over $\mathbb{F}_2$ is asymptotically upper bounded by:

$$\left(-\alpha + \frac{1}{2} + \frac{1}{2}\sqrt{2\alpha^2 - 10\alpha - 1 + 2(\alpha + 2)\sqrt{\alpha(\alpha + 2)}}\,\right) n + O(n^{1/3}). \quad (11.30)$$

In this theorem, the field equations are implicitly counted and should not be counted when determining the value of α. For large values of α, the bound given by this theorem becomes close to the bound obtained by using Theorem 11.5 for $(\alpha + 1)n$ equations. However, when α is small the specific $\mathbb{F}_2$ bound is much tighter.

Exercises

1. Consider the monomial orderings described in this chapter and show that they indeed satisfy the well-ordering and compatibility properties. Show that the reverse lexicographic order relation is not a monomial ordering.

2. Compute the resultant of x and $x + 25$. Also compute the resultant of x^2 and $x^2 + 5$. Which of these two pairs of equations has a common root modulo 25? Conclude that resultants only give information about roots modulo prime, not prime powers.

3^h. Let f be a polynomial of degree d. Recall that the reciprocal polynomial of f is $x^d f(1/x)$. Given two polynomials f and g, what is the relation between the resultant of f and g and the resultant of their reciprocal polynomials?

4. Consider the polynomials $xy + 1$ and $x^2 y + x + y$ and compute their resultant, eliminating x. What are the roots of the resultant? Show that they cannot be completed into solutions of the bivariate system.

5^h. Construct an example of a bivariate ideal of dimension 0 whose Gröbner contains more than 2 polynomials.

6^h. What is the number of monomials of degree d in n unknowns? The number of monomials of degree at most d? Write a program that assigns a number to each monomial of degree at most d and explicitly computes the bijection between a monomial and its number.

7. Let $f(x)$ be a polynomial of degree d. Compute the resultant of f and $ax + b$. This expression is especially useful for the algorithms of Chapter 15.

This chapter can be a source for numerous implementations projects. A good start is to implement the basic linear algebra approach given as Algorithm 11.4. This can be completed by considering algorithmic techniques to reduce the amount of memory: compressing the polynomial representations or using iterative algorithms. An alternative approach is to add F_5 criterias to this algorithm to construct smaller matrices with very few reductions to zero during the row echelon form computation.

Part III

Applications

Chapter 12

Attacks on stream ciphers

Stream ciphers are widely encountered in applications where resources are limited and speed is critical. They come in two flavors: keystream generators and self-synchronizing stream ciphers. However, keystream generators are much more frequent and we only consider this kind of stream ciphers in this chapter.

Before studying keystream generators, let us first recall Shannon's one time pad. In its binary version, each message is viewed as a sequence of bits and encrypted by bitwise xoring each bit of message with a corresponding bit of key. Despite its extreme simplicity, this encryption algorithm provably ensures the confidentiality of encrypted messages. The only information that can be learned by an eavesdropper is the length of the message. However, the security proof of the one time pad requires a perfect key: each bit of key should be generated at random from a uniform distribution and should be independent from other key bits. In particular, it is well know that re-using the key material with a one time pad yields a severely broken system; it is known as the parallel message attack. As a consequence, using the one time pad in practice is very challenging and rarely done.

Even with a perfect key, the security of the one time pad holds against a very limited class of adversaries: passive eavesdroppers. If active attacks are allowed, the one time pad is no longer secure. There are two main directions that allow active adversaries to attack one time pad systems. The first approach is to remark that encrypted strings are malleable and thus do not protect the integrity of transmitted messages. The second approach is to remark that from a plaintext/ciphertext pair, the encryption key can be directly recovered. As a consequence, if the attacker can trick the receiver into decrypting a fake message, he can obtain the key and decrypt the real message later on.

Keystream generators can be seen as a practical way of using the one time pad. Instead of using an extremely long key, the two parties of the encrypted communication possess a common short secret key and use the keystream generator to generate a large random looking string. This idea of random looking strings can be formalized into polynomial time indistinguishability from random, called IND$ in Chapter 1.

Note that encryption schemes based on keystream generators inherit the weaknesses of the one time pad. The same keystream should never be re-

used to avoid the parallel message attack. Moreover, a secure integrity check
mechanism should be used together with the keystream generator is order to
protect the integrity of messages. For example, following the Encrypt-then-
MAC approach presented in Chapter 1, it suffices to add a MAC tag of the
encrypted message to avoid the malleability issue.

Another feature of practical keystream generators is that they usually accept on auxiliary input, the initial value or IV that allows the users to generate
different keystreams for different messages without requiring too much has-
sle to manage these keystreams and to avoid re-using key material between
messages.

12.1 LFSR-based keystream generators

Linear Feedback Shift Registers (LFSR) are often used as a basis for pseudo-
random generators. One important reason is that these generators have a long
period, in particular, when the feedback polynomial is primitive (see Chapter 2, Section 2.5.2), the period of an LFSR with n cells is $2^n - 1$. However,
while directly using an LFSR as pseudo-random source in a video game or
in scientific computing is a good idea, where cryptographic applications are
concerned, this is not acceptable. Indeed, after observing $2n$ output bits, an
adversary can completely reconstruct the state and feedback polynomial of
the LFSR being observed, using either a simple linear algebra approach or
the more specific Berlekamp-Massey algorithm presented in Chapter 2.

However, the long period property, together with the good statistical properties of output sequences are highly desirable and it remains tempting to use
LFSRs as core components of keystream generators. Several basic constructions are frequently used. They all aim at using the good properties of LFSRs
while hiding the linearity.

Throughout the rest of this chapter, we are mostly going to study the
security of one particular type of LFSR-based stream ciphers: **filtered generators**. However, several kinds of LFSR-based generators exist; let us now
review some of these constructions, starting with the filtered generator.

The filtered generator   The filtered generator tries to hide the linearity of
a core LFSR by using a complicated non-linear output function on a few bits.
At each step, the output function takes as input t bits from the inner state
of the LFSR. These bits are usually neither consecutive, nor evenly spaced
within the register.

This output function produces a single bit, which is given as the LFSR
output. The main difficulty when using the filtered generator is to choose an
adequate function. It should not be too simple, otherwise the keystream gen-

erator becomes weak; it should not be too complicated, otherwise it becomes the bottleneck of the generator.

The function f is usually described either as a table of values or as a polynomial. Note that, using the techniques of Section 9.2, f can always be expressed as a multivariate polynomial over $\mathbb{F}_2$.

Non-linear output from multiple generators A variant of the filtered generator is to run several LFSRs in parallel and to output at each clock a non-linear function of bits coming from these generators. A particularly simple version is the Geffe generator, it is constructed from three LFSRs, one serves as a control LFSR and the other two as output LFSRs. When the control LFSR produces a 0, the output of the keystream comes from the first of the output LFSRs. When it produces a 1, the output comes from the second generator. With each clock signal, all three LFSRs advance.

The Geffe generator can be generalized to the selection from one generator among many, using a control value. Geffe's generators are highly vulnerable to correlation attacks.

Another variation on this idea is the shrinking generator. Here we only have two LFSRs. The first one serves as a control LFSR, the second one as an output generator. When the control LFSR produces a '0', no keystream output is generated. When the control LFSR produces a '1', the output bit of the output generator is added to the keystream. Once again, with each clock signal, both LFSRs advance. Implementing the shrinking generator requires special care to hide the irregular rhythm of the output production. Indeed, if this rhythm can be precisely measured, then the state of both LFSRs can easily be recovered through a basic side-channel attack.

Keystream generators with memory In order to obtain more complex keystream generators, it is also possible to add memory that preserves some non-linear information from one clock step to the next.

A typical example is the summation generator. From a mathematical point-of-view, this generator views the output of each of its n internal LFSRs as the binary representation of a large integer, with low bits coming first. Then, it sums these n numbers and outputs the binary representation of the sum.

In practice, this is realized by using a small binary adder and a carry register. Initially, the carry is initialized to 0. The carry register can store a small integer in the range $[0, n-1]$ and the binary adder computes the sum of the n LFSR output and of the previous carry. The result of this sum S is a number in the range $[0, 2n-1]$. Its parity is the output of the keystream generator, the high order bits of S, i.e., the value $\lfloor S/2 \rfloor$ is recycled as the next value of the carry register to allow carry propagation.

Clock controlled generators Another way of combining LFSRs in a non-linear fashion is to use the output of a control LFSR or of a simple keystream

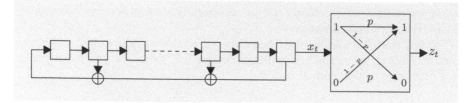

Figure 12.1: Noisy LFSR (Binary Symmetric Channel) model

generator to clock the output generator. For example, with a control LFSR and two output LFSRs, we obtain the alternating step generator. As each step in time, the output of the keystream is the XOR of the two output LFSRs. However, at each step, only one of the two LFSRs states is advanced. When the control bit is a '0', the first output LFSR advances, when the control bit is a '1', the second one does.

12.2 Correlation attacks

12.2.1 Noisy LFSR model

A very convenient way of modeling LFSR based keystream generators is to consider that the keystream is the output of a regular LFSR masked by some noise. This approach is called the noisy LFSR model or the binary symmetric channel model. In this model, if x_t denotes the output of the LFSR at time t, the keystream z_t is constructed as follows:

$$z_t = \begin{cases} x_t & \text{with probability } p, \\ x_t \oplus 1 & \text{with probability } 1 - p. \end{cases} \qquad (12.1)$$

This is often represented as in Figure 12.1.

In this model, when the probability p is equal to $1/2$, z_t is purely random and carries no information about x_t. When p is equal to 0 or 1, z_t is simply a (possibly inverted) copy of x_t. Due to symmetry of the construction, we may safely assume that $p > 1/2$. Otherwise, if suffices to flip z_t in order to replace p by $1 - p$. Under this assumption, it is convenient to write $p = (1 + \epsilon)/2$ or equivalently to let $\epsilon = 2p - 1$. This value ϵ is called the bias of the binary symmetric channel.

In order to see why this model is relevant in the case of the filtered generator, we can use the techniques of Chapter 9. Assume that the bit of keystream z_t is obtained from k bits of output as the LFSR as $f(x_t, x_{t-\delta_1}, \cdots, x_{t-\delta_{k-1}})$. Then f is a function from $\{0,1\}^k$ to $\{0,1\}$. Using a Walsh transform based approach, we compute the best possible linear approximation L_f of f, this linear function coincide with the output of f for some fraction p of the 2^k possible inputs. Let $y_t = L_f(x_t, x_{t-\delta_1}, \cdots, x_{t-\delta_{k-1}})$ and remark that when the k-uple on the right-hand side belongs to the set of input where f and L_f agree, then $y_t = z_t$. Moreover, since y_t is a linear combination of outputs of the same LFSR, with shifted initial values, y_t itself can be produced by the LFSR when it is initialized with the XOR of all these initial values. Assuming for simplicity that the k-uples $(x_t, x_{t-\delta_1}, \cdots, x_{t-\delta_{k-1}})$ are random and independent, it should be clear that z_t is the image of y_t under a binary symmetric channel with probability p and bias $\epsilon - 2p - 1$.

Of course, this independence hypothesis is clearly false. However, k-uples which are generated by LFSRs are well balanced and despite the dependence between successive k-uples, attacks based on the binary symmetric channel do work well in practice. Correlation attacks precisely work within this framework, forget about the exact output function f and simply focus on the bias of the best possible linear approximation.

12.2.2 Maximum likelihood decoding

Given a noisy LFSR with bias ϵ, a natural question is to ponder whether the real output of the LFSR can be recovered from N bits of noisy output. This question arises both in coding theory and in cryptography. It is very useful to first solve the problem with unlimited computing power. Clearly, if the LFSR is fully specified, then its feedback polynomial is already known and recovering its exact output is equivalent to finding its initial state.

With unlimited computing power, it is possible to compute the LFSR output on N bits for all initial states. Comparing all these sequences to the noisy output z, we can measure the Hamming distance of each candidate to the noisy output, i.e., the number of bits where the sequences differ. For the correct guess, the average Hamming distance is $(1 - p)N$. For incorrect guesses, assuming that they behave essentially like random strings, we expect an average distance of $N/2$. The average distance for the correct guess simply comes from the way the noise is generated. However, the exact behavior of incorrect guesses is slightly harder to explicit. In fact, the difference between the observed stream and an incorrect guess is the XOR of three terms, the LFSR output for the guessed initial state, the LFSR output for the correct initial state and the noise. By linearity, the two LFSR outputs can be grouped into a single LFSR output whose initial state is obtained by xoring the two initial states. As a consequence, the difference is a noisy copy of this LFSR output. Thus, it is almost balanced, i.e., it contains about half 0s and half 1s. This implies that the average distance is $N/2$ as expected. However, since there are exponentially many incorrect guesses, a few of these bad guesses may be much closer to the observed strings.

One important question with maximum likelihood decoding is to determine the parameters for which the correct guess is likely to correspond to the smallest Hamming distance to the noisy output. Indeed, when this happens, it is possible to determine the correct guess by keeping the candidate corresponding to the minimal Hamming distance. Note that in some contexts, a weaker condition may suffice. For example, if it is possible to test whether a likely candidate is indeed correct, then we can be satisfied if the maximum likelihood decoding produces a short list of candidates that contains the correct one.

In practical cryptanalysis, this exhaustive approach can be used to cryptanalyze keystream generators made of several independent LFSRs together with a non-linear output rule. Indeed, in this context, each LFSR has a small internal state compared to the complete cryptosystem and an exhaustive search for the initial state of a single LFSR is much more efficient than a global exhaustive key search. As a consequence, if a correlation of bias ϵ can be found between the output of the keystream generator and one of the internal LFSRs with a state of length n, then the initial state of this register can be recovered in time $N2^n$ using $N \approx 1/\epsilon^2$ bits of keystream. In particular, this can be applied to Geffe's generators. This attack initially proposed in [Sie84, Sie85] is called a correlation attack.

12.2.2.1 Necessary amount of keystream for correlation attacks

In order to determine the amount of keystream required for correlation attacks, we need to determine how close to the observed string a wrong guess can be. Assume that we are observing N bits of keystream, that the bias for the correct guess is ϵ and that we are testing C incorrect candidates. We need determine the probability of having an incorrect candidate closer to the observed stream than the correct one. If we are interested by the setting where seemingly valid candidates can be further tested, it also useful to determine the average number of incorrect candidates closer than the correct one.

On average, the correct candidate agrees with the observed keystream in $(1+\epsilon)N/2$ positions, but the exact number can vary around this average. Similarly, on average, each incorrect candidate agrees with the observed keystream in $N/2$ positions, but the exact number can vary. Thus, to answer our questions, we need to understand these variations around the average.

We first consider a simpler case offering the choice between two candidates, a correct and an incorrect one. In this case, the problem is equivalent to distinguishing between a perfect random generator[1] and a slightly biased generator which outputs 0 with probability $(1 + \epsilon)/2$. For each of the two generators, we first compute the probability that it outputs T zeros and $N - T$ ones after N measurement, taking into account the number of possible arrangements of the zeros, this probability can be written as:

$$P_T(p) = \binom{N}{T} p^T (1-p)^{N-T}.$$

---

[1]That is, a generator which outputs each successive bit independently, with probability $1/2$ of giving a 0 and probability $1/2$ of giving a 1.

λ	$1 - \frac{1}{2}\mathtt{erfc}\left(\lambda/\sqrt{2}\right)$
0	1/2
0.1	0.5398
0.5	0.6915
1	0.8413
1.5	0.9332
2	0.9772
2.5	0.9938
3	0.9987

Table 12.1: Typical probabilities with binomial distributions

When we observe T zeros, it seems natural to compare $P_T(1/2)$ and $P_T((1+\epsilon)/2)$ and to keep the candidate corresponding to the largest probability. In fact, this is the best that can be done. When ϵ is small, $P_T(1/2)$ is the larger probability for $T < T_0$, where $T_0 = (1 + \epsilon/2) \cdot N/2$. To determine the probability of success of this strategy, we need to compute sums of $P_T(p)$ for T in interval of the form $[0, T_0]$ and $[T_0, N]$. The problem is that there are no closed formulas for these sums. However, when N is large, there is a standard way of approximating the formulas. Without going into the details, the idea is to first view T as a random variable obtained by summing N values 0 or 1 (with probabilities p and $1 - p$). By definition, T is said to follow a binomial distribution. Then, the binomial distribution is approximated by a Gaussian (or normal) distribution with average $E = (1 - p)N$ and variance $\sigma^2 = p(1-p)N$. For a normal distribution, the probability to measure a value smaller than $E + \lambda\sigma$ is:

$$Q(E + \lambda\sigma) = \frac{1}{\sqrt{2\pi}} \int_{-\infty}^{\lambda} e^{-u^2/2} du. \tag{12.2}$$

Moreover, with most computer algebra systems, this probability can be computed using the complementary error function, usually denoted by $\mathtt{erfc}$, according to the relation:

$$Q(E + \lambda\sigma) = 1 - \frac{1}{2}\mathtt{erfc}\left(\lambda/\sqrt{2}\right). \tag{12.3}$$

We give some typical values in Table 12.1.

With this tool, computing the probability of successfully distinguishing the correct and incorrect candidates becomes easy. Write the number of experiments N as $(2\lambda/\epsilon)^2$, i.e., let

$$\lambda = \frac{1}{2} \cdot \sqrt{N\epsilon^2} \tag{12.4}$$

and define a threshold $T_0 = (1 + \epsilon/2)N/2$. For the incorrect candidate, the average value of T is $T_i = N/2$ and $\sigma_i = \sqrt{N}/2 = \lambda/\epsilon$. As a consequence, $T_0 -$

$T_i = \lambda^2/\epsilon$ is $\lambda\sigma_i$. For the correct candidate, we have $T_c = (1+\epsilon)N/2$ and $\sigma_c = \sqrt{N/(1-\epsilon^2)}/2$. When ϵ is small, we can conveniently ignore the difference between σ_c and σ_i and approximate $T_0 - T_c$ by $-\lambda\sigma_i$. As a consequence, if we predict that the observed distribution is the correct one when $T \geq T_0$ and the incorrect one otherwise, the probability of success is obtained by looking up λ in Table 12.1.

The more general case with C incorrect candidates and a correct one can be viewed in several different ways which yield the same threshold T_0. One option is to regroup the candidates in two groups, a bad one that occurs with probability $C/(C+1)$ and a good one that occurs with probability $1/C$. Using the previous notations, the probability of observing T zeros is $P_T(1/2)$ in the bad case and $P_T((1+\epsilon)/2)$ in the good case. Using this option, we can apply Neyman-Pearson lemma which says that an optimal distinguisher should predict that the good case occured, if and only if, $P_T((1+\epsilon)/2) > C \cdot P_T(1/2)$.

12.2.3 Fast correlation attacks

While correlation attacks are very effective against stream ciphers which involve small LFSRs, they quickly become impractical as the size of the target registers grows. With large LFSRs, we would like a different algorithm to exploit the correlation without having to pay with exponential complexity in terms of the LFSR's length. Fast correlation attacks proposed by Meier and Staffelbach (see [MS89]) offer this possibility.

Note that there exists a much wider variety of correlation attacks than can be presented here. For example, conditional correlation attacks use the fact that some specific configurations of output bits gives the value of some linear combination of LFSR bits much more precisely than in the average case. To use this fact, we need parity checks that only use bit positions where these specific configurations occur. For more information, see [LMV05]. Let us also mention iterative correlation attacks [MG90], multipass correlation attacks [ZF06] and vectorial fast correlation attacks [GH05].

12.2.3.1 Binary symmetric channel with repetitions

In order to illustrate one essential idea needed to devise such a correlation attack, we start with an easy case. This easy case can be formulated as a learning problem. In this problem, we try to learn a secret sequence S of length n. To this effect, we are given access to several independent noisy copies of S obtained through a binary symmetric channel with bias ϵ. How many copies are needed to recover S?

We start by solving the case of a single bit s. After obtaining one measurement of s through the binary symmetric channel, we can predict s with probability $p = (1 + \epsilon)/2$. With many measurements, the best we can do is to use a majority rule in order to determine our prediction. If we see more 0s than 1s, we predict a 0 and similarly in the opposite case. In the unlikely case where there are as many 0s and 1s, we output a random prediction. Following the results of Section 12.2.2.1, the probability of correctly predicting 0 is thus:

$$P = \begin{cases} \sum_{t=(N+1)/2}^{N} P_t & \text{for odd } N \text{ and} \\ \sum_{t=N/2+1}^{N} P_t + \frac{1}{2} \cdot P_{N/2} & \text{for even } N. \end{cases} \tag{12.5}$$

Due to the symmetry between the cases $s = 0$ and $s = 1$, P is also the overall probability of correct prediction. Once again, this probability can be computed using the complementary error function. The only difference with the previous case comes from the fact that since we have two opposite biases, we now compute $\lambda = \sqrt{N\epsilon^2}$, i.e., rewrite Equation (12.4) without a factor $1/2$. In the reverse direction, if we specify the desired probability of success, this fixes λ and we need to make $N = (\lambda/\epsilon)^2$ measurement.

This is usually summarized by saying that a correlation attack with bias ϵ requires $O(\epsilon^{-2})$ measurements.

12.2.3.2 A basic attack on LFSRs

With an LFSR-based keystream generator, unless the cryptosystem as a whole presents some unexpected weakness, it is unlikely that the cryptanalysis has access to many noisy copies of the same LFSR output. As a consequence, we cannot directly use the above method. Instead, we need to exploit the redundancy within the LFSR output in order to create such repeated copies of individual bits. Note that we have encountered a similar problem in Chapter 9, Section 9.3 with Goldreich-Levin theorem. However, with Goldreich-Levin theorem, we could freely choose the noisy scalar products we wanted to learn. Here, we no longer have this freedom and we need to work with noisy values which are fixed by the LFSR specifications.

Remember that in order to know the complete state of an LFSR it suffices to learn the exact values of n consecutive bits. Since n is a relatively small number, we can first focus on the recovery of a single bit. The key idea is the construction of parity check equations for the LFSR. A parity check equation is simply a systematic linear equality of the form:

$$x_{i_1} \oplus x_{i_2} \oplus \cdots \oplus x_{i_t} = 0. \tag{12.6}$$

Here, systematic means that the equation is true regardless of the initial value of the LFSR.

Each such parity check yields a biased prediction for x_{i_1}:

$$x_{i_1} = z_{i_2} \oplus \cdots \oplus z_{i_t}. \tag{12.7}$$

Since each value $z_{i_2}, \ldots, z_{i_t}$ is an approximation for $x_{i_2}, \ldots, x_{i_t}$ with bias ϵ and since all these approximations involve independent random choices in the binary symmetric channel model, we can show that we obtain an approximation for x_{i_1} with bias ϵ^{t-1}. This can be proved by repeated application of the following lemma.

LEMMA 12.1
Given two independent approximations $\hat{x}$ and $\hat{y}$ of x and y with respective biases ϵ_x and ϵ_y, then $\hat{x} \oplus \hat{y}$ is an approximation of $x \oplus y$ with bias $\epsilon_x \epsilon_y$.

PROOF   We see that $\hat{x} \oplus \hat{y}$ is a correct approximation of $x \oplus y$, if either $\hat{x}$ and $\hat{y}$ are either both correct or both incorrect. The first event where both are correct occurs, due to independence, with probability $1/4 \cdot (1+\epsilon_x)(1+\epsilon_y)$. The second event occurs with probability $1/4 \cdot (1-\epsilon_x)(1-\epsilon_y)$. Summing the two probabilities, we find that the total probability of success is:

$$\frac{1}{4} \cdot (2 + 2\epsilon_x \epsilon_y) = \frac{1}{2} \cdot (1 + \epsilon_x \epsilon_y). \tag{12.8}$$

Thus the bias of the combined approximation is $\epsilon_x \epsilon_y$.   ☐

By repeatedly applying this lemma, we obtain the bias of any sum of independent approximations simply by multiplying the elementary biases. This can be applied to parity check involving t bits as above, called a parity check of weight t, to obtain an approximation of one of the involved bits with bias ϵ^{t-1}. As a consequence, if we obtain a large number of parity checks all involving the same bit x_{i_1}, we can predict this bit with high confidence thanks to a majority vote between the individual predictions. Using the results of Section 12.2.3.1, we see that this number is of the order of ϵ^{-2t+2}.

To evaluate the performance of this basic attack, it is important to take several parameters into account. The first parameter is, of course, the amount of time needed to evaluate all the parity checks and perform the majority vote. However, the amount of keystream required for the attack is also very important. The other two important parameters are the initial bias ϵ and the length n of the register. In order to understand the relationship between these parameters, we now study the conditions of existence of enough parity checks to perform the basic attack. For this analysis, we do not consider algorithms for generating the parity checks, merely their existence.

Due to the linearity of the LFSR, for each output bit, there exists fixed constants, which only depend on the feedback polynomial such that:

$$x_j = \sum_{i=0}^{n-1} c_i^{(j)} x_i. \tag{12.9}$$

This allows us to express each output bit as a linear function of the bits contained in the initial value of the LFSR. It is convenient to group the coefficients $c_i^{(j)}$ in a vector $c^{(j)}$ and view the initial value of the LFSR as a vector $\vec{X}$.

With these notations, we can rewrite:

$$x_j = (c^{(\vec{j})}|\vec{X}).\tag{12.10}$$

To construct parity checks of weight t, we look for sets of t vectors $\vec{c_{j_1}}, \ldots,$ $\vec{c_{j_t}}$ such that:

$$\bigoplus_{i=1}^{t} c^{(\vec{j_i})} = 0.\tag{12.11}$$

Indeed, by linearity of the scalar product, this implies:

$$\bigoplus_{i=1}^{t} x_{j_i} = (\bigoplus_{i=1}^{t} c^{(\vec{j_i})}|\vec{X}) = 0.\tag{12.12}$$

To count the number of parity checks of weight t, all involving some fixed position, say $j_1 = 0$, we need to choose the remaining positions j_2 to j_t and check that the sum of the n-bit coefficient vectors is 0. Assuming that we have access to a noisy output of the LFSR of length N, each of the above positions is bounded by N, and there are $\prod_{i=1}^{t-1}(N-i)/(t-1)!$ unordered choices for the $t-1$ positions. We expect that a fraction 2^{-n} of these choices sum to zero. As a consequence, the average number of available parity checks for j_1 is:

$$\binom{N-1}{t-1} \cdot 2^{-n} = \frac{(N-1)!}{2^n(t-1)!(N-t)!}.\tag{12.13}$$

With an elementary bias ϵ, we compute the probability of correct prediction by letting:

$$\lambda = \sqrt{\frac{(N-1)!}{2^n(t-1)!(N-t)!}\epsilon^{2t-2}}\tag{12.14}$$

and by looking up λ in Table 12.1.

Note that, given a set of parity checks allowing recovery of x_{j_1}, it is easy to modify it to recover x_{j_1+1} by shifting all indices by 1. Similarly, we can obtain $x_{j_1+2}, \ldots, x_{j_1+n-1}$. Of course, since the indices are shifted, the highest required position is no longer N but can go up to $N+n-1$. However, since n is normally very small when compared to N, this issue can be neglected.

12.2.4 Algorithmic aspects of fast correlation attacks

The advantage of fast correlation attacks compared to basic correlation attacks is that we no longer need to exhaustively search the $2^n - 1$ possible initialization values. Instead, we need to find and evaluate all the necessary parity checks. We address these algorithm issues in this section.

12.2.4.1 Computing parity checks

The first algorithmic problem we encounter is to find a large enough number of parity checks for fast correlation attacks. This can be done in many ways. However, a common pre-requisite is to express each output bit of a given LFSR has a linear combination of the input bits, i.e., following the notation of Section 12.2.3.2, to compute the vectors $c^{(j)}$. A very simple way to do this is to initialize n copies of the LFSR, with the same feedback register, with n initial values containing $n-1$ zeros and a single 1. In the copy numbered i, the only non-zero initial bit is in position 1. By linearity, the n bits of output obtained at time j form the vector $\vec{c}^{(j)}$. An easy way to speed up this process is to use a bitslicing approach as in Chapter 5 and apply the LFSR recursion to n-bit words, thus running the n copies in parallel. This is described in pseudo-code as Algorithm 12.1.

Algorithm 12.1 Computing formal expression of LFSR outputs

Require: Input coefficients of LFSR $(\alpha_0, \ldots, \alpha_{n-1})$
Require: Bound on stream length N
  Allocate array A of N words on n bits
  Initialize $A[0 \cdots n - 1]$ to 0
  for i from 0 to $n - 1$ **do**
    Set i-th bit of $A[i]$ to 1
  end for
  for i from n to N **do**
    Let $A[i] = \sum_{j=0}^{n-1} \alpha_j A[i - n + j]$
  end for
  Output array A {$A[i]$ contains the expression of the i-th bit of LFSR output as a linear combination of bits 0 to $n - 1$.}

Once the n-bit vectors of coefficients are known, we need to form parity checks of weight t, by summing together one fixed vector, corresponding with a position we want to predict and $(t-1)$ other vectors corresponding to other positions.

Brute force. The first idea that comes to mind to solve this problem is simply to exhaustively try all possible set of positions and keep the sets which sum to zero. The running time is $\binom{N-1}{t-1}$ and can be closely approximated by $N^{t-1}/(t-1)!$, by ignoring the difference between N and $N - i$ for small values of i.

Birthday approach. The next idea is to remark that each parity check can be written as an equality between the vectors corresponding to two half parity checks in two smaller lists. Indeed, it is possible to equate the expressions of two sums of $t/2$ positions each. This method has been the method of choice for constructing parity checks for a long time. Note that each parity check can be split into two parts in many different ways and that we need to apply some care in order to avoid generating the same parity check many times, one for each different split.

Quadrisection approach. Going further, it is also possible to use the algorithms presented in Chapter 8 to compute parity checks. This does not change the running time compared to the birthday approach; however, it greatly reduces the required amount of memory. Indeed, each parity check can be viewed as a sum of 4 parts, each part containing (almost) $t/4$ values. Thus, as in [CJM02], we obtain a four set sum problem over $\mathbb{F}_{2^n}$ which can be directly addressed using the algorithms of Chapter 8. As with the birthday approach, each parity check can be split into four parts in many ways and we need to be careful to avoid generating the same parity check many times.

Rho approach. A natural question is to wonder whether we can get rid of memory altogether when generating parity checks. After all, a parity check is nothing but a collision in a function that maps $t/2$ different positions to the sum of their expression in terms of the LFSR initial value. Since collisions in functions can be found by cycle finding algorithms, it may appear that such an approach can be used to find parity checks. However, it is not known how to do this, unless we are ready to increase the value of t beyond the value determined in Section 12.2.3.2. Since the bias of parity checks decreases quickly when t increases, cycle finding algorithms are usually not used to construct parity checks.

If we are ready to increase t, we can, as in Section 7.5.1, proceed by using a function defined in the following way. First, any subset of $t/2$ positions can be mapped to a n-bit string describing the corresponding linear combination of bits of the initial value. This n-bit string can be viewed as an n-bit integer and this number can in turn be interpreted to construct a new set of $t/2$ positions. Any cycle finding algorithm can obtain a collision for this function. However, the collision may occur in two different places, either when mapping sets to bitstrings or when mapping bitstrings to sets. The first type of collision yields a parity check. Unfortunately, the second type of collision is meaningless. The problem is that when t is too small, there are much more bitstring values than sets of $t/2$ positions. As a consequence, an overwhelming fraction of the collisions is of the useless type. If we increase t, roughly doubling it, we can make sure that the mapping from bitstrings to sets is injective and thus that all collisions are of the useful type. Note that doubling t makes the rest of the fast correlation attack very inefficient. Thus, it is best to avoid this Rho based approach.

Discrete logarithm based approach. For parity checks involving $t = 3$ positions, Kuhn and Penzhorn proposed in [PK95] to compute the discrete logarithm of $1 + \alpha^i$ in basis α in the finite field $\mathbb{F}_{2^n}$, for i smaller than N, the maximal amount of accessible keystream. Here, α is a root of the feedback polynomial of the LFSR under attack. If the discrete logarithm j is also smaller than N, we learn than $1 + \alpha^i + \alpha^j = 0$, which can be converted into a parity check $x_0 \oplus x_i \oplus x_j = 0$. An extension of this idea was recently proposed in [DLC07].

12.2.4.2 Improving the basic attack

The basic attack described in Section 12.2.3.2 presents two main difficulties:

- When the length n of the LFSR increases, the number of parity checks decreases and we need to increase t and/or N to implement the attack. This is due to the 2^{-n} factor in Equation (12.13).

- When t increases, the number of necessary parity checks increases exponentially with t and evaluating these parity checks on the observed keystream takes a long time.

It is possible to reduce the impact of these two difficulties by using an approach similar to the method used for the Goldreich-Levin theorem, in Section 9.3. We isolate a subset S_0 of n_0 bits in the initial state of the LFSR. These bits are going to be addressed separately using an exhaustive search. As a consequence, when counting and constructing parity checks, we only cancel the $n - n_0$ other bits. For each acceptable parity check, we also keep the linear expression specifying the contribution of the bits in S_0. To count these modified parity checks, we replace the factor 2^{-n} by $2^{-(n-n_0)}$ in Equation (12.13). This increases the number of available parity checks and allows to reduce t and/or N. Note, however, that since the n_0 bits are fixed and cannot be shifted, we cannot use shifted parity checks this time and need to compute them independently $n - n_0$ times, once for each of the bits that need to be recovered.

Without using the Walsh transform, recovering each of the $n - n_0$ remaining bits would require to evaluate and count the number of predictions 2^{n_0} times. Instead, we start by regrouping the parity checks that share the same linear expression on S_0. Then, within each group we partially evaluate these parity checks and keep the difference between the number of 0 and 1 values. This difference is stored in an array of 2^{n_0} elements in the position given by the binary encoding of the linear expression of the group. After applying a Walsh transform to this array, position i contains the difference between the numbers of 0 and 1 values for the corresponding guess of the bits of S_0. As in the analysis of Section 12.2.2.1, we have $2^{n_0} - 1$ bad candidates and a single good candidate. With well-chosen parameters, the position corresponding to the good candidate contains the largest value in absolute value. Learning this position gives the value of the n_0 bits in S_0. Moreover, the sign of this largest value determines the value of the bit for which the parity checks were constructed (see [CJM02] for more details).

12.3 Algebraic attacks

In addition to the binary symmetric channel model, the filtered generator is also well-suited to a modelization by a system of algebraic equations. Indeed, if in the basic expression:

$$z_t = f(x_t, x_{t-\delta_1}, \cdots, x_{t-\delta_{k-1}}), \tag{12.15}$$

we replace each bit of x by its linear expression in terms of the initial values $x_1, \ldots x_n$ of the LFSR resulting from Algorithm 12.1, we obtain a transformed equation:

$$z_t = f_t(x_1, x_2, \cdots, x_n). \tag{12.16}$$

Assuming a known plaintext attack, the cryptanalyst knows the value of z_t and wants to recover the initial state. At this point, two important remarks are needed. First, we can collect a large number of equations, one for each bit of known plaintext. Second, the degree of each polynomial f_t is the same as the degree of f, since f_t results from a linear change of variables.

As a consequence, the filtered generator can be vulnerable to algebraic attacks when its parameters are badly chosen. It is especially important not to have a low degree for f. For example, let us detail what happens when f can be expressed as a quadratic polynomial. In that case, each equation $f_t = z_t$ can be rewritten as:

$$z_t = \sum_{i=1}^{n} \alpha_i^{(t)} x_i + \sum_{i=1}^{n-1} \sum_{j=i+1}^{n} \beta_{i,j}^{(t)} x_i x_j. \tag{12.17}$$

Using (re)linearization techniques, we can define $y_{i,j}$ as a new unknown with value equal to $x_i x_j$ and substitute $y_{i,j}$ in the above. This yields a linear equation in $n(n+1)/2$ unknowns. Collecting $n(n+1)/2$ (or a few more) such equations, it suffices to solve a moderately large linear system of equations to recover the initial state of the LFSR. Equivalently, we can use Algorithm 11.4 on this set of algebraic equations using a degree limit of 2.

12.3.1 Predicting an annihilator polynomial

Alternatively, one can also look for linear combinations of the z_t such that the contribution of all terms $y_{i,j}$ vanishes. Unless we are especially unlucky, there are linear terms x_i left and we obtain linear equations in the n initial unknowns. At first, this seems essentially equally costly as applying Algorithm 11.4. However, it was shown recently by Rønjom and Helleseth in [RH07] that this alternative can be made extremely efficient. In order to see why, let us briefly recall Wiedemann's algorithm from Chapter 3. In this algorithm, there are two steps, the first one determines the minimal polynomial of some linear sequence and the second one uses this minimal polynomial to invert a linear system. In the case of an algebraic attack against filtered LFSR, a multiple of the minimal polynomial can be predicted in advance and the first step of Wiedemann's algorithm can be skipped. Tweaking in a modified second step, it is then possible to directly derive linear equations in the n initial unknowns, without performing linear algebra on a large system of equations.

To understand the method of [RH07], it is best to view the current LFSR state as a element of $\mathbb{F}_{2^n}$ and the advance step of the LFSR as a multiplication by some element α in $\mathbb{F}_{2^n}$. If the initial state is X_0, the state at time t is $X_t = \alpha^t X_0$. With this notation, it is also possible to write the output bit x_t as a trace of βX_t for some constant β in $\mathbb{F}_{2^n}$. We recall that the trace of X in $\mathbb{F}_{2^n}$ is defined as:

$$\mathrm{Tr}(X) = \sum_{i=0}^{n-1} X^{2^i}. \tag{12.18}$$

Let us now focus on an arbitrary quadratic term $x_t x_{t+\delta}$. We can write:

$$
\begin{aligned}
x_t x_{t+\delta} &= \mathrm{Tr}(\beta X_0 \alpha^t) \cdot \mathrm{Tr}(\beta X_0 \alpha^{t+\delta}) \\
&= \left(\sum_{i=0}^{n-1} (\beta X_0 \alpha^t)^{2^i} \right) \cdot \left(\sum_{i=0}^{n-1} (\beta X_0 \alpha^{t+\delta})^{2^i} \right) \\
&= \sum_{i=0}^{n-1} \sum_{j=0}^{n-1} (\beta X_0)^{2^i + 2^j} \alpha^{2^j \delta} \cdot (\alpha^{2^i + 2^j})^t.
\end{aligned} \tag{12.19}
$$

Thus, each quadratic term is a linear combination with complicated but fixed coefficients of powers of t. This powers are of the form $\alpha^{\ell t}$, where ℓ has at most two non-zero bits in its binary decomposition. We can regroup the values α^ℓ by considering the difference $\Delta = i - j$ modulo n and form polynomials:

$$P_\Delta(X) = \prod_{i=0}^{n-1} X - \alpha^{2^i + 2^{(i+\Delta) \bmod n}}. \tag{12.20}$$

This polynomial is invariant under the action of Frobenius, thus, it has coefficients in $\mathbb{F}_2$. Note that $P_0(X)$ regroups the values α^{2^i}, thus it is the feedback polynomial of the considerered LFSR.

Following [RH07] and proceeding as with Wiedemann's algorithm, we see that P_Δ annihilates all powers of values $\alpha^{2^i+2^{i+\Delta}}$ in Equation (12.19). As a consequence, the product of all P_Δ for $\Delta \neq 0$ annihilates all non-linear terms of Equation (12.19). Unless we are extremely unlucky, the linear terms are not canceled and it suffices to solve a system of n linear equations in n unknowns to recover the initial state X_0.

For monomials of higher degree d, it is possible to proceed similarly, by considering products of term $X - \alpha^\ell$ for values of ℓ with d non-zero bits in their binary decomposition.

12.4 Extension to some non-linear shift registers

In order to avoid many of the weaknesses of linear feedback shift registers, non-linear feedback shift registers were recently considered for cryptographic uses. Non-linear shift registers (NLFSR) are very similar to LFSR, the only difference is that the new bit entering the register at each step is computed using a non-linear function instead of a linear one. One difficulty with these registers is to make sure that the period is long enough for all initial values. To overcome this difficulty, several approaches are possible. One option is to use registers of small size, to make sure through exhaustive search that all these registers have a long enough periods and to combine several into the keystream generator. Another option is to consider registers of a specific form that allow the period to be computed.

At first, it seems that algebraic and correlation attacks cannot apply to non-linear shift registers. Indeed, concerning algebraic attacks, we see that the iteration of the low degree feedback function quickly produces equations of very high degree, which cannot be solved by the techniques of Chapter 11. Concerning correlation attacks, getting biased prediction of inner bits of the NLFSR remains easy. However, without linearity, we can no longer construct parity checks.

Yet, in this section, we show that the correlation attacks presented in this chapter can be directly applied to a specific subclass of non-linear shift register based keystream generators. This subclass is formed of keystream generators based on a non-linear shift register whose output is a linear combination of the registers cells.

To make things more precise, we can view such a keystream generator as built around an inner sequence x_i defined from n initial bits x_0 to x_{n-1}, together with a recursion formula:

$$x_{n+i} = F(x_i, x_{i+1}, \ldots, x_{i+n-1}), \tag{12.21}$$

for some efficiently computable non-linear function F, for example a low-degree function. The output z_i at time i is obtained by applying a fixed linear function L to the inner sequence. Assuming, for simplicity of exposition, that x_i is always used when computing z_i, we write:

$$z_i = x_i \oplus L(x_{i+1}, \ldots, x_{i+n-1}). \tag{12.22}$$

Note that, depending on the choice of L, this construction can be trivially broken. For example, if $L = 0$, z_i simply is a copy of x_i. In this case, it suffices to collect n consecutive output bits to break the construction. Indeed, using Equation (12.21), the rest of the sequence can then be predicted.

We now show that for any choice of L, this NLFSR based generator can be vulnerable to correlation and algebraic attacks. The key argument is that for each inner bit of the NLFSR, we have two different equations, the non-linear Equation (12.21) and a linear equation which is obtained by rewriting Equation (12.22) as:

$$x_i = z_i \oplus L(x_{i+1}, \ldots, x_{i+n-1}). \tag{12.23}$$

Moreover, once the keystream z has been observed, we can substitute Equation (12.23) into itself for various values of i. As a consequence, each inner bit x_i can be expressed as a linear equation $L_i(x_0, \ldots, x_{n-1})$ of the initial bits. The constant term in L_i is obtained by linearly combining bits of the output stream z. Substituting each bit of x by the corresponding linear expression in Equation (12.21) produces a low degree algebraic equation in the unknown $x_0, \ldots, x_{n-1}$. Since there are many such equations, we can use the methodology of algebraic attacks.

Similarly, if we take the best linear approximation by the inputs of F and its output, then after replacing in this linear expression each value of x by its linear expression, we obtained a noisy linear equation in the bits of initial state. Of course, we can apply the correlation methodology to the resulting set of noisy linear equations.

12.5   The cube attack

A third class of attack against the filtered generator was recently proposed by Dinur and Shamir [DS09]. It is called the cube attack, it is a new way to use a well-known identity on multivariate polynomials over $\mathbb{F}_2$ and transform it into an attack against some secret-key ciphers. In particular, this can be applied to the filtered generator.

To explain the underlying identity, let us start with a univariate polynomial F over $\mathbb{F}_2$. When x is in $\mathbb{F}_2$, we know that $x^2 = x$. As a consequence, the polynomial f obtained by keeping the remainder of the Euclidean division of

F by f induces the same function as F on $\mathbb{F}_2$. Moreover, f is of degree at most one, i.e., $f(x) = ax \oplus b$. We now see that:

$$f(0) \oplus f(1) = a. \tag{12.24}$$

In fact, by summing on the two points 0 and 1, we are taking a derivative of f and we can recover the coefficient of the term x in f. Note that this is based on the same basic equation as differential cryptanalysis, but views things from a different angle.

With more unknowns, we proceed similarly. Let $F(x_1, \cdots, x_k)$ be a polynomial in k unknowns. After accounting for the identities $x_i^2 = x_i$ for each unknown, we obtain a polynomial f, whose partial degree in each unknown is at most one. Factoring x_1 out where possible, we can write:

$$f(x_1, \cdots, x_k) = a(x_2, \cdots, x_k) \cdot x_1 + b(x_2, \cdots, x_k).$$

As a consequence, the function $f(0, x_2, \cdots, x_k) \oplus f(1, x_2, \cdots, x_k)$ is equal to $a(x_2, \cdots, x_k)$.

The key idea of the cube attack is to repeat this process until we reach easy-to-solve linear equations. For example, to repeat the approach on x_1, x_2 and x_3, we first write:

$$
\begin{aligned}
f(x_1, \cdots, x_k) &= a_7(x_4, \cdots, x_k) \cdot x_1 x_2 x_3 \oplus a_6(x_4, \cdots, x_k) \cdot x_1 x_2 \oplus \\
&= a_5(x_4, \cdots, x_k) \cdot x_1 x_3 \oplus a_4(x_4, \cdots, x_k) \cdot x_1 \oplus \\
&= a_3(x_4, \cdots, x_k) \cdot x_2 x_3 \oplus a_2(x_4, \cdots, x_k) \cdot x_2 \oplus \\
&= a_1(x_4, \cdots, x_k) \cdot x_3 \oplus a_0(x_4, \cdots, x_k).
\end{aligned}
$$

Then, we remark that:

$$
\begin{aligned}
a_7(x_4, \cdots, x_k) &= f(0,0,0,x_4, \cdots, x_k) \oplus f(0,0,1,x_4, \cdots, x_k) \oplus \\
&= f(0,1,0,x_4, \cdots, x_k) \oplus f(0,1,1,x_4, \cdots, x_k) \oplus \\
&= f(1,0,0,x_4, \cdots, x_k) \oplus f(1,0,1,x_4, \cdots, x_k) \oplus \\
&= f(1,1,0,x_4, \cdots, x_k) \oplus f(1,1,1,x_4, \cdots, x_k).
\end{aligned}
$$

From this example, the reader can easily see that when we repeat the process with t unknowns, we need to decompose f into 2^t parts and to sum over 2^t points. The first t coordinates of these points cover all the possible values that can be achieved by setting each coordinate to 0 or 1. These points are the vertices of a hypercube in the t-dimensional space.

It is shown in [DS09] that for any polynomial f and any set of unknowns S, we can write:

$$f(x_1, \cdots, x_k) = \prod_{i \in S} x_i f_S(x_1, \cdots, x_k) + q(x_1, \cdots, x_k),$$

where each monomial in q misses at least one unknown from S and f_s is a polynomial of the variables not in S. When summing on a hypercube induced by S, we obtain a value for the polynomial f_S. In [DS09], f_S is called the super-polynomial of S in f. Here, we instead called it the derived function of f at S. The basic idea of cube attacks is very similar to a generalization of differential cryptanalysis called higher order differential cryptanalysis [Knu94, Lai94, MSK98]. However, it emphasizes a different attack scenario.

12.5.1 Basic scenario for the cube method

In order to use the cube attack, we need to be able to sum many evaluations of the same polynomial along some hypercube. As a consequence, we need a scenario where the attacker controls at least part of the unknowns entering the polynomial. Since this process of summing produces values for a lower degree polynomial derived from the first, it is natural to use this lower degree polynomial to extract information about other unknowns. For this reason, in [DS09], it is proposed to consider systems involving polynomials where some unknowns are controlled by the attackers and some unknowns contain secret key material. To simplify the extraction of information, the easiest is to choose the hypercube we are summing on, in such a way that the derived polynomial is linear in the secret unknowns. If this can be done, we obtain a linear equation in the secret key. Repeating this with several hypercubes, we obtain a linear system of equations, from which the secret key can hopefully be recovered.

Since the cube attack is very recent, its scope of application is not yet known. However, this attack seems to have a lot of potential. One very surprising application is already given in [DS09]. This application allows to break cryptosystems, which satisfy some specific criteria, in a black box way, without even knowing the full details about these cryptosystems. The idea is, in an initial phase, to choose hypercubes at random and look at the derived function that we obtain, when running test copies of the system for which the attacker controls the secret key. If the derived function is, or seems to be, constant, the hypercube contains too many unknowns. If the derived function is complicated, the hypercube contains too few unknows. If the derived function is a simple linear function of the secret key, then we have achieved the right balance and found one linear equation to be used later in an attack phase. Note that there are several ways to test the linearity of the derived function. One option is to check that if we add any of the secret key unknowns to the hypercube variables, the resulting derived function becomes a constant.

Among the possible applications of the cube attacks, it is proposed in [DS09] to consider an extremely large filtered LFSR (on 10,000 unknowns) which mixes IV and secret key variables in a complex way. The only real constraint is that the degree of the output in terms of the variables is at most 16. The cube attack successfully breaks this incredibly large filtered LFSR using 2^{20} different IV values and a single bit of output for each. Note that this filtered LFSR is completely of range for more conventional correlation or algebraic attacks.

12.6 Time memory data tradeoffs

In Chapter 7, we describe some time-memory tradeoffs against block ciphers. Since stream ciphers behave quite differently from block ciphers, it is interesting to revisit these attacks in this different context. Let us start by listing the relevant differences between the two kinds of ciphers. First, instead of combining a key with a block of plaintext to get a block of ciphertext as block ciphers do, stream ciphers generate a pseudo-random sequence from an inner state, usually obtained by mixing together a key and an initialization value. During the generation of the pseudo-random sequence, the inner state takes many different consecutive values, where each value is obtained from the previous one through an update function. At the same time, the output sequence is obtained by applying an output function to each inner state. In order to cryptanalyze the stream cipher, it suffices to recover one value of the inner state at any point in time. Indeed, this suffices to predict the sequel of the pseudo-random stream. Moreover, most of the time, the update function is in fact reversible and recovering a single inner state allows to find the complete sequence of inner states. From this point, depending on the precise cipher being considered, it might also be possible to recover the encryption key. Thus, with a stream cipher, the goal of an attacker is slightly different; instead of having to directly find the key, it suffices to recover one of the many inner states taken during the encryption process.

From this remark, it is already possible to devise a simple attack that works better than exhaustive search. To do this, we start by defining a function F, which takes as input an inner state and outputs a prefix of the corresponding pseudo-random stream. For the size of the prefix, we choose either the size of the inner or something slightly larger. Clearly, from some output of the stream, we can deduce the image by F of the corresponding inner states of the stream. As a consequence, if we know how to invert F we can recover the inner state and break the cipher. Moreover, if we know how to invert F only for a small fraction of the states, this already yields an attack. Indeed, by observing the output of the stream cipher, we cannot only compute F for the initial inner state, but also for most of the subsequent inner states taken during the encryption process. In fact, we obtain F for all of these inner states except the few final ones, for which we do not have enough keystream available. To simplify the analysis of the attack, one generally assumes that each inner state gives a new independent chance of inverting F. Of course, successive states are not really independent, but from the cryptanalyst point-of-view, the simplified modelization is a good rule of thumb. In this model, this basic attack, independently discovered by Babbage and Golic, works as follows. First, during a precomputation step, we compute F on some fraction α of possible inner states and store these values in a sorted table. Then to recover an unknown state, we obtain the corresponding pseudo-random sequence. From this sequence, we deduce N different values of F for N of the inner states seen during the construction of the pseudo-random sequence (whose length is a little above N). If $\alpha N \approx 1$, we expect that one among the F values can be found in our table and we deduce the corresponding inner state.

This simple attack illustrates the fact that with stream ciphers, we have many available targets instead of a single one and this helps the cryptanalyst. This was further extended by Biryukov and Shamir in [BS00] into time/memory/data tradeoff attacks. By playing with three parameters instead of two, it is possible to obtain a better compromise. For example, assuming that the space of inner states contains N values, it is possible to recover one state in time $N^{2/3}$ using output sequences of length $N^{1/3}$ and memory of the order of $N^{1/3}$. This is clearly more efficient than Hellman's usual compromise with time and memory $N^{2/3}$. For more details, we refer the reader to [BS00].

Exercises

1^{h}. Assume that the same one-time pad key is used twice to encrypt two texts in English, M and M'. Show how both texts can be recovered from the encrypted strings.

2. Work out a scenario where the malleability of messages encrypted by the one-time pad yields a major security hole.

3^{h}. Consider a filtered LFSR where the input bits to the non-linear function f are consecutive. Assume that f maps k bits to one. Show that the best linear correlation between input and output can be improved by considering several consecutive output bits. Try this approach for randomly selected functions.

4. Consider a Geffe generator based on three LFSRs: a control LFSR C on n bits and two outputs LFSR A and B on m bits each. How much does exhaustive search on this generator cost? Given the initial value of C, how can you efficiently recover the initial values A and B? Conclude that n should not be too small.

5^{h}. Continuing with the Geffe generator, assume that n is too large for exhaustive search. Construct a correlation attack and recover A and B. How much keystream is needed? What happens if A and B use the same feedback polynomial? Once A and B are known, how do you reconstruct C?

6. Recompute the necessary amount of keystream N required for a correlation attack on a filtered LFSR with bias ϵ if we want the correct candidate to appear among the list of L candidates generating sequences closest to the output stream.

7^{h}. Write a program for computing parity checks. Try the birthday based and the quadrisection approach. Let n denote the number of bits taken in account in the parity checks. What values of n can you achieve with each approach?

8. Following Section 12.3.1, consider a LFSR on n bits and the output sequence $x \oplus y \oplus xy$, where x and y are two consecutive bits, then construct a linear combination of output bits where all quadratic terms vanish.

9^{h}. This exercise considers non-linear shift registers on n bits. Show that there are 2^n possible shift registers of this sort. Study the possible structure of the oriented graph whose vertices are the inner states and whose arrows are obtained by advancing the NLFSR. Which NLFSR would you keep for cryptographic purposes? Write a program to enumerate the possibilities and select the good ones.

Finally, here are possible implementation projects on this topic:

i. Write a complete set of programs for fast correlation attacks. Compare the performance of using the Walsh transform or not during the evaluation of parity checks. Look up and implement some advanced correlation attack from the literature (conditional, iterative, ... correlation attacks).

ii. Generalize the method of Section 12.3.1 to polynomials of arbitrary degree. Write a program that computes, given a feedback polynomial and output function for a LFSR, a linear combination of output bits such that the non-linear terms vanish. Show that for some badly chosen output functions the method can work with short output sequences. Construct such bad examples.

iii. Write a working implementation of the cube attack as presented in [DS09]. Compare this to other attacks against the filtered generator.

Chapter 13

Lattice-based cryptanalysis

Lattice-based cryptanalysis uses lattice reduction algorithms in order to discover short linear relations between integer vectors that give a useful insight into various cryptographic systems. They can roughly be classified into two main classes: direct attacks, where the cryptanalytic problem at hand can directly be expressed as a lattice reduction problem and Coppersmith's based attacks, which rely on several algorithms which can recover small roots of polynomial systems of equations using lattice reduction.

Due to the complexity of lattice reduction algorithms, a frequent way of analyzing lattice-based attacks consists of assuming the existence of a lattice reduction oracle that solves some lattice problem such as the shortest vector problem (see Chapter 10). Of course, no such oracle exists; however, it is a very convenient heuristic based on the good practical performance of lattice reduction algorithms. In some cases, the analysis can be refined to reflect the proved properties of available lattice reduction algorithm. This can be necessary either to obtain a proved attack or when faced with lattice reduction problems that are out of reach of current lattice reduction programs.

13.1 Direct attacks using lattice reduction

13.1.1 Dependence relations with small coefficients

One very important family of lattice reduction attacks makes use of lattices to find linear relations with relatively small coefficients between numbers or vectors. These relations may be considered over the integers or in some modular ring. Before addressing this problem, it is nice to look at its combinatorial properties and determine for which instances we should expect small linear relations.

13.1.1.1 Combinatorial properties

When searching short linear dependency between elements of a family of numbers or vectors, two cases occur. We can either be looking for an "abnormally" small relation whose existence is guaranteed by the specific properties of the family and may, for example, reflect the existence of a trapdoor in a cryptographic scheme, or we may simply desire an ordinary relation whose existence is generically guaranteed. The goal of this section is to analyze the expected size of such generic relations.

LEMMA 13.1

Let $\vec{v}_1, \ldots, \vec{v}_n$ be a family of vectors with integer coefficients in t coordinates with $t < n$. Let M denote an upper bound for the absolute values of all coefficients of the various $\vec{v}_i$s. Then there exists a non-zero integer relation

$$\sum_{i=1}^{n} \lambda_i \vec{v}_i = \vec{0}, \tag{13.1}$$

such that $\max |\lambda_i| \leq B$, where B is given by

$$\log_2 B = t \cdot \frac{\log_2 M + \log_2 n + 1}{n - t} \tag{13.2}$$

PROOF   Consider all possible linear combinations

$$\sum_{i=1}^{n} \mu_i \vec{v}_i \tag{13.3}$$

with $0 \leq \mu_i < B$. There are B^n such relations and the resulting vectors have all their coordinates upper bounded by nBM. Since there are less than $(2nBM)^t$ such vectors, two distinct relations have to compute the same value, as soon as $(2BM)^t \leq B^n$, which amounts to the relation given in Equation (13.2). This collision gives

$$\sum_{i=1}^{n} \mu_i \vec{v}_i = \sum_{i=1}^{n} \mu_i' \vec{v}_i, \tag{13.4}$$

with $0 \leq \mu_i < B$ and $0 \leq \mu_i' < B$. After regrouping both sides by subtraction, we find a non-zero relation as per Equation (13.1).

Note that in this proof, we do not invoke the birthday paradox, because the various linear combinations are related and highly dependent.   ☐

Of course, for a specific instance of the problem, the shortest dependence relation (say w.r.t. the Euclidean length) can be much shorter than the generic bound stated in the above lemma. A similar lemma exists in the case of modular relations:

LEMMA 13.2

Let $\vec{v}_1, \ldots, \vec{v}_n$ be a family of vectors with integer coefficients in t coordinates with $t < n$. Let N be an integer. Then there exists a non-zero integer relation *modulo N:*

$$\sum_{i=1}^{n} \lambda_i \vec{v}_i = \vec{0} \quad (\mathrm{mod}\ N), \tag{13.5}$$

such that $\max |\lambda_i| \le \min(B, N/2)$, *where* B *is given by*

$$B = N^{t/n}. \tag{13.6}$$

PROOF As before, consider all possible linear combinations

$$\sum_{i=1}^{n} \mu_i \vec{v}_i \quad (\mathrm{mod}\ N) \tag{13.7}$$

with $0 \le \mu_i < B$. There are B^n such relations and that the resulting vectors are expressed modulo N. Thus, there are less than N^t possible vectors and two distinct relations have to compute the same value, as soon as $N^t \le B^n$, which is equivalent to Equation (13.2). This collision gives

$$\sum_{i=1}^{n} \mu_i \vec{v}_i = \sum_{i=1}^{n} \mu'_i \vec{v}_i \quad (\mathrm{mod}\ N), \tag{13.8}$$

with $0 \le \mu_i < B$ and $0 \le \mu'_i < B$. Once again, this yields a non-zero relation. Moreover, using reduction modulo N, this relation can be expressed with coefficients in the range $[-N/2, N/2]$. □

13.1.1.2 Lattice reduction based search for short relations

Given a family of integer vectors on t coordinates $\vec{v}_1, \ldots, \vec{v}_n$ as above, we now form the integer lattice generated by the columns of the following matrix:

$$L(\vec{v}) = \begin{pmatrix} K\vec{v}_1 & K\vec{v}_2 & \cdots & K\vec{v}_n \\ 1 & 0 & \cdots & 0 \\ 0 & 1 & \cdots & 0 \\ \vdots & \vdots & \ddots & \vdots \\ 0 & 0 & \cdots & 1 \end{pmatrix}, \tag{13.9}$$

where K is a constant that is determined in the sequel.

Any linear relation as per Equation (13.1) can be mapped into a lattice point:

$$^{\top}(0 \cdots 0 \lambda_1 \lambda_2 \cdots \lambda_n),$$

whose first t coordinates are equal to 0. Conversely, any lattice point starting with t zeros corresponds to a linear relation. As a consequence, the main restriction when choosing K is that this constant should be large enough to ensure that the first vector in a reduced basis of the lattice has zero components in its upper part consisting of the first t coordinates that correspond to the contributions of the $\vec{v}_i$s. Indeed, when the contribution of the first t coordinates is not all zero, then the Euclidean norm of the lattice vector is at least K.

To make the choice of K more precise, it is important to specify the lattice reduction process that is used. With a lattice reduction oracle, it suffices to choose a value K larger than the norm of the expected linear relation. Using the expected size of generic relation as in the previous section, it suffices to take:

$$K = \left\lceil \sqrt{n}(2M)^{\frac{1}{n-t}} \right\rceil. \tag{13.10}$$

When using the L^3 lattice algorithm, K should be multiplied by a safety coefficient $2^{n/2}$, to account for the fact that the first vector of the reduced basis may be larger than the shortest lattice vector by a factor up to $2^{n/2}$.

With this choice for K, the lattice reduction process outputs short vectors whose upper part is guaranteed to be zero and these vectors clearly correspond to linear relations with small coefficients. As pointed out above, the coefficients of the linear relation appear as coordinates of rank $t+1, \cdots, t+n$ of the output vector.

13.1.1.3 Generalization to approximate relations

In some cases, instead of searching for exact relations between the vectors, we may look for approximate relations, where $\sum_{i=1}^{n} \lambda_i \vec{v}_i$ is no longer $\vec{0}$ but a vector of small norm. A typical application of approximate relations occurs when the values of the vectors $\vec{v}_i$s correspond to approximations of vectors with real-valued coordinates.

In the case of approximate relations, one option is to choose $K = 1$. With this choice output vectors are short but there is no special reason to obtain an upper part formed of zeros. This clearly corresponds to approximate dependencies with small coefficients.

Other choices for K are possible and the choice of K allows to control the balance between the quality of the approximate relation and the size of the coefficients. When K is larger, the linear relation is more precise but the coefficients are larger.

13.1.1.4 Modular relations

Another possible extension is to consider linear relations which are no longer exact but, instead, hold modulo some integer N. In other words, we need to modify the lattice that is considered to deal with modular relations. Assuming that N is large enough, the answer is very simple and consists of adding to the lattice basis a few columns that ensure modular reduction. This is done by considering the lattice generated by the columns of the following matrix:

$$\begin{pmatrix} K\vec{v}_1 & K\vec{v}_2 & \cdots & K\vec{v}_n & KNI \\ 1 & 0 & \cdots & 0 & 0 \\ 0 & 1 & \cdots & 0 & 0 \\ \vdots & \vdots & \ddots & \vdots & 0 \\ 0 & 0 & \cdots & 1 & 0 \end{pmatrix}$$

where I is a $t \times t$ identity matrix. Thanks to the additional columns, a modular relation $\sum_{i=1}^{n} \lambda_i \vec{i} = \vec{0}$ (mod N) corresponds to a lattice point given by the vector:

$$^\top(0 \ \cdots \ 0 \ \lambda_1 \ \lambda_2 \ \cdots \ \lambda_n).$$

Of course, this is not sufficient, we also need to make sure that such a vector corresponding to a modular linear relation may appear as the result of lattice reduction. To discuss this question, it is important to remark that the above lattice includes short vectors which are not related to the existence of any linear relation. These vectors can be obtained by multiplying any of the first n vectors in the initial basis by a factor N and then by reducing upper part mod N. This process yields vectors whose coordinates are all zero, with the exception of a single coordinate with value N. Applying the above construction to all $\vec{v}_i$s, we get a family of n vectors of norm N that are mutually orthogonal. Experiments show that, if N is too small, this family appears in sequence as the first output vectors of a reduced basis, and thus completely masks any useful information about linear relations.

Clearly, assuming a lattice reduction oracle, we need to check that the short vector that corresponds to the linear relation coming from Lemma 13.2 is shorter than these systematic vectors of norm N. Since the short vector associated with the linear relation has norm at most $\sqrt{n}B$, we need to check that $\sqrt{n}N^{t/n} < N$ or equivalently that:

$$N > n^{\frac{n}{2(n-t)}}. \tag{13.11}$$

If the reduction is done using L^3, an additional factor of $2^{n/2}$ is required to guarantee that the first vector of the reduced basis corresponds to a genuine linear relation.

An interesting special case is the binary case where $N = 2$. In this case, any linear relation with four or more non-zero coefficients yields a vector with norm greater than the systematic vectors of norm N. Thus, we cannot hope that a lattice reduction approach finds a linear relation with more than 3 non-zero coefficients. Moreover, relations with 3 coefficients or fewer can easily be found using exhaustive search or birthday based algorithms. For this reason no known lattice reduction algorithms are used for attacking binary problems, such as finding short codewords in linear codes.

13.1.2   Some applications of short dependence relations

13.1.2.1   Knapsack problems

Solving knapsack problems is an interesting special case of finding linear relations between given numbers. Historically, breaking knapsack problems was one major application of lattice based cryptanalysis in the 80s and 90s. In addition to this historical role, solving a knapsack problem requires more involved techniques than the general case, because the expected relations have all their coefficients in $\{0, 1\}$. In cryptographic scenarios, we know that such a relation exists between the given elements of the knapsack $a_1, \ldots, a_n$ and the target sum $s = \sum_{i=1}^{n} \epsilon_i a_i$. Moreover we know that the Euclidean norm of the vector that encodes this relation is $\sqrt{\alpha n}$, where α is the proportion of ones in the relations. Depending on the considered cryptosystem, α may or may not be known to the cryptanalyst but, in many practical examples it is a part of the cryptographic system itself. Furthermore α is an important parameter when trying to analyze the performances of lattice-based attacks against knapsack problems. When each coefficient of ϵ is chosen uniformly at random, the number of ones is close to $n/2$ with high probability and we can assume that $\alpha = 1/2$.

Another important parameter with knapsack problems is their density d defined as:

$$d = \frac{n}{\log_2(\max_i a_i)}. \tag{13.12}$$

Since this parameter is the ratio between the number of elements in the knapsack and the number of bits in each element, it controls the expected size of "parasitical" linear relations of short norms between the a_i, with coefficients not restricted to $\{0, 1\}$.

It was shown in [LO85], that, when the density is low, parasitical vectors are large and, thus, the shortest vector gives the solution to any random instance of the knapsack problem. If we use the lattice from Equation (13.9), for vectors on one coordinate only, and if we assume that shortest lattice-vectors can be efficiently computed (even if this is not totally accurate), then low density means $d < 0.6463$. Later in [CJL$^+$92], this condition was improved to $d < 0.9408$. In order to reach that bound, either of the two following lattices (generated by columns) can be used:

$$
\begin{pmatrix}
Ka_1 & Ka_2 & \cdots & Ka_n & -Ks \\
n+1 & -1 & \cdots & -1 & -1 \\
-1 & n+1 & \cdots & -1 & -1 \\
\vdots & \vdots & \ddots & \vdots & \vdots \\
-1 & -1 & \cdots & n+1 & -1 \\
-1 & -1 & \cdots & -1 & n+1
\end{pmatrix}
\begin{pmatrix}
Ka_1 & Ka_2 & \cdots & Ka_n & Ks \\
1 & 0 & \cdots & 0 & 1/2 \\
0 & 1 & \cdots & 0 & 1/2 \\
\vdots & \vdots & \ddots & \vdots & \vdots \\
0 & 0 & \cdots & 1 & 1/2
\end{pmatrix}.
$$

Before we close this section, let us warn the reader on the meaning of the low-density attacks. The inequality $d < 0.9408$, provides a *provable* guarantee that, from a shortest vector for a lattice computed from the problem one can, with high probability, solve a random instance of the original knapsack problem. Note, however, that in some cases, even lattice with larger density can sometimes be solved using the approach. Moreover, the critical density given here is for knapsack problems where the solution is balanced, with about half zeros and half ones. For knapsacks with an unbalanced solution, i.e., when the parameter α given above differs from $1/2$, the critical density can be much higher.

13.1.2.2 Polynomial relations

Finding the minimal polynomial of a real algebraic number x of degree d corresponds to searching a linear dependency between 1, x, x^2, ..., x^d. Since we are working with integer lattices, we choose a large integer K and we try to find an approximate relation between the closest integers to K, Kx, Kx^2, ..., Kx^d. More precisely, we reduce the following lattice, given by columns:

$$\begin{pmatrix} K & \lfloor Kx \rceil & \lfloor Kx^2 \rceil & \cdots & \lfloor Kx^d \rceil \\ 1 & 0 & 0 & \cdots & 0 \\ 0 & 1 & 0 & \cdots & 0 \\ 0 & 0 & 1 & \cdots & 0 \\ \vdots & \vdots & \vdots & \ddots & \vdots \\ 0 & 0 & 0 & \cdots & 1 \end{pmatrix}$$

The first vector of the reduced lattice can be written as:

$$^T(\epsilon \; a_0 \; a_1 \; \cdots \; a_d).$$

Since we wish to interpret a_0, ..., a_d as the coefficients of the minimal polynomial of x, i.e., we want to conclude that $a_0 + a_1 x + a_2 x^2 + \cdots + a_d x^d = 0$. The most important parameters here are K and d. If d is smaller than the degree of the minimal polynomial of x then this technique cannot succeed. Likewise, if K is too small, then it cannot succeed either. To see this, assume for example that x is between 0 and 1 and apply Lemma (13.1): this yields a linear combination of the elements on the first row of the above matrix with coefficients bounded above by B, where B satisfies:

$$\log B = \frac{\log K + \log d + 1}{n - 1}$$

If K is small, this relation is much more likely to appear as an output to lattice reduction algorithms than the one corresponding to the minimal polynomial. In fact, a reasonable choice is to take $K \geq (\max |a_i|)^{2d}$. In other words, K should be much larger than the expected size of the coefficients of the minimal polynomial. If d is not exactly known, for example if we only know an upper bound on the degree of the minimal polynomial of x, then the following trick can be applied: take the first two or three vectors appearing in the output reduced lattice, transform them into polynomials and compute their GCD. If K was large enough the minimal polynomial of x is usually obtained.

It is very important to know that the heuristic procedure we just described can give positive results, i.e., it can find a minimal polynomial, but cannot give a negative result. Moreover, if the method succeeds, we have a candidate for the minimal polynomial; we can then check this candidate either up to arbitrary large precision or formally if the situation permits.

13.1.2.3 NTRU lattices

Another kind of lattices has been an important source of progress for practical lattice reduction, lattice obtained from the NTRU cryptosystem. Without entering into the details of this cryptosystem, let us simply state that the resulting lattices have some very specific properties. In fact, they are very similar to knapsack lattices, except that we are considering modular knapsacks on vectors. Depending on the exact version of the NTRU system, these knapsack instances contain 0, 1 and also -1 values; moreover, the exact number of 1 and -1 values is usually known. The modulo used in these knapsacks is very small, for example, 64 is a typical value. In addition, there is some extra structure coming from the fact that the vectors in this modular knapsack are rotated copies of each other. Another difference with pure knapsack systems is that the sum of vectors we are searching does not exactly sum to zero but instead to a vector with a fixed proportion of 1 and -1 values.

More precisely, starting from a vector with n coordinates, we are considering a lattice spanned by the columns of a $2n \times 2n$ matrix of the following form:

$$\begin{pmatrix} a_1 & a_2 & \cdots & a_n & q & 0 & \cdots & 0 \\ a_2 & a_3 & \cdots & a_1 & 0 & q & \cdots & 0 \\ \vdots & \vdots & \ddots & \vdots & \vdots & \vdots & \ddots & \vdots \\ a_n & a_1 & \cdots & a_{n-1} & 0 & 0 & \cdots & q \\ 1 & 0 & \cdots & 0 & 0 & 0 & \cdots & 0 \\ 0 & 1 & \cdots & 0 & 0 & 0 & \cdots & 0 \\ \vdots & \vdots & \ddots & \vdots & \vdots & \vdots & \ddots & \vdots \\ 0 & 0 & \cdots & 1 & 0 & 0 & \cdots & 0 \end{pmatrix}$$

This lattice has many interesting properties:

- Starting from any vector in the lattice and rotating the coordinates of the first and of the second half by the same amount, we obtain another lattice vector.

- The short vectors corresponding to the NTRU secret key contain a small number of 1 and -1 values and have a very short norm.

- The initial lattice basis given above already contains short vectors, of norm q, corresponding to the reduction modulo q rule.

Despite these extra properties, fully reducing the NTRU lattice in large dimension in order to recover the secret key is a real challenge. This shows that even when working with small numbers, lattice reduction is not an easy task. A state-of-the-art of lattice reduction methods applied, in particular, to NTRU lattices is given in [GN08].

It is also worth mentioning that combined attacks mixing lattice reduction and birthday paradox methods can be devised for NTRU (see [HG07]).

13.1.2.4 Cryptanalysis of Damgård's hash function

In [Dam90], among other options, Damgård proposed to base a hash function on a knapsack compression function using 256 (non-modular) numbers a_i of size 120 bits. His idea was to divide the message to be hashed into blocks of 128 bits, and to apply the following process:

- Start with a fixed initial value on 128 bits. Appending the first 128-bit block of the message, one gets a block B of 256 bits.

- (Compression phase.) Compute the knapsack transform of these 256 bits, i.e., starting from zero, add up all a_is whose index corresponds to the position of a one bit of B. The resulting number can be encoded using 128 bits.

- Append the next block to get 256 bits and iterate the compression phase.

In order to find a collision for this hash function, it is clearly enough to find two different 128-bit blocks that, when appended to the initial value, yield the same hash value. This clearly corresponds to finding a collision in a knapsack transform based on 128 numbers of 120 bits. In the sequel, we study how collisions in such a knapsack transform can be found using lattice reduction, and we show that it is feasible to build collisions for Damgård's hash function. A completely different kind of attack against this construction has already appeared in the work of P. Camion and J. Patarin ([CP91]). Still, it has never been implemented, and besides, it could only find collisions for the compression function rather than for the hash function itself. In contrast to this approach, our attack runs on a computer and actually outputs collision for the size of the parameters suggested by Damgård.

Unfortunately, our attack cannot be proven, even in the lattice oracle setting described in Section 13.1.2.1. Nevertheless, for a slightly weaker notion of a collision, which we call pseudo-collision, a correct mathematical analysis can be carried through. A **pseudo-collision** for Damgård's hash function consists of two messages whose hash values coincide except for the 8 leading bits. The practical significance of pseudo-collisions is obvious since pseudo-collisions have a non-negligible chance of being actual collisions.

13.1.2.4.1 The basic strategy In this section, we associate a lattice to any given knapsack-based compression-function in such a way that collisions correspond to short vectors. Before describing the reduction, we make our definitions and notations a bit more precise: we fix a sequence of integers, $\mathbf{a} = a_1, \ldots, a_n$. The knapsack-compression function $S_{\mathbf{a}}$, which we simply denote by S, takes as input any vector x in $\{0, 1\}^n$ and computes

$$S(x) = \sum_{i=1}^{n} a_i x_i$$

A *collision* for this function consists of two values x and x' such that $S(x) = S(x')$.

In order to search collisions, we reduce the lattice given by the columns of the following matrix:

$$B = \begin{pmatrix} Ka_1 & Ka_2 & \cdots & Ka_n \\ 1 & 0 & \cdots & 0 \\ 0 & 1 & \cdots & 0 \\ \vdots & \vdots & \ddots & \vdots \\ 0 & 0 & \cdots & 1 \end{pmatrix}.$$

Note that this lattice is exactly the lattice used in the original Lagarias-Odlyzko attack for solving knapsack problems (see [LO85]). Let us consider the possible output of lattice reduction. Since K is large, it is clear that the first coordinate of a short vector is 0. As for the other coordinates, we expect them to be all 0, 1 or -1. Indeed, if this happens we clearly get a collision: from an element of the lattice

$$e = {}^{\top}(0 \ \epsilon_1 \ \epsilon_2 \ \cdots \ \epsilon_n).$$

with all coordinates 0, 1 or -1, we find that

$$\sum_{i=1}^{n} \epsilon_i a_i = 0$$

and thus obtain a collision:

$$\sum_{\epsilon_i=1} a_i = \sum_{\epsilon_i=-1} a_i.$$

Analysis of the attack With the attack as was stated above, finding a collision for Damgård's hash function can be done in practice, using L^3 to compute collisions for a knapsack compression function based on 128 numbers with 120 bits each. This was described in [JG94].

Surprisingly, despite these practical results, a theoretical analysis shows that asymptotically, the attack as presented does not work. However, it is possible to slightly modify the attack and obtain an exponential attack with complexity around $2^{n/1000}$. This explains why the attack works for the values of n suggested in [Dam90].

13.2 Coppersmith's small roots attacks

13.2.1 Univariate modular polynomials

In general, the problem of finding a root of a univariate polynomial $f(x)$ modulo an integer N of unknown factorization is a difficult problem. Indeed, it is well known that finding a root of $x^2 - r^2 \pmod{N}$ for a random value r, usually yields a (partial) factorization of N; see Section 2.3.3.1. On the contrary, when the factorization of N is known, it is easy to find roots of f modulo each prime factor and paste them together with the Chinese remainder theorem, using a Hensel lifting step to deal with multiple factors of N. On the other hand, when f has roots over the integers, it is easy to find them and use them as roots modulo N. For example, finding a root of $x^2 - 4$ modulo a large integer N is easy. This example can be generalized to any polynomial f with small enough coefficients. Write $f(x) = \sum_{i=0}^{d} f^{(i)} x^i$ and denote by $|f|$ the polynomial defined by $|f|(x) = \sum_{i=0}^{d} |f^{(i)}| x^i$ whose coefficients are the absolute values of the coefficients of f. If B is a positive integer such that $|f|(B) < N$, then for all integers x such that $-B \le x \le B$, we have $-N < f(x) < N$. As a consequence, any root x of f modulo N that lies the interval $[-B, B]$ is also a root of f over the ring of integers. As a consequence, it can be easily found.

The first small root algorithm of Coppersmith extends this idea beyond its initial range. It plays on the two sides of the inequality $|f|(B) < N$ in order to allow larger values for B. On the right-hand side, N is replaced by some power of N; on the left-hand side, f is replaced by another polynomial F with smaller coefficients. One basic remark behind Coppersmith's construction is that if we denote by $F_{i,j,k}$ the polynomial $F_{i,j,k}(x) = x^i f(x)^j N^k$, then any root r of f modulo N is also a root of $F_{i,j,k}$ modulo N^{j+k}. Let us choose a parameter t and look at polynomials of the form $F_{i,j,t-j}$; they all share the root r modulo N^t. In addition, the degree of $F_{i,j,t-j}$ is $i + j \cdot \deg(f)$. It is convenient to choose a bound D on the allowed degree and consider a set of polynomials $F_{i,j,t-j}$ of degree at most D.

13.2.1.1 Howgrave-Graham's variation

In fact, it is easier to first describe a variation of Coppersmith's first small root algorithm, as presented in [HG97]. In this version, we use the fact that since all polynomials, in the set of polynomials $F_{i,j,t-j}$ of bounded degree, have the common root r modulo N^t, any linear combination of these polynomials also has the same root modulo N^t. Our goal is now to find such a linear combination F that maximizes the possible bound B satisfying the constraint $|F|(B) < N^t$. Invoking our early remarks, we can then find the root r under the condition $|r| < B$. Once again writing, $F(x) = \sum_{i=0}^{D} F^{(i)} x^i$, we can see that $|F|(B)$ is the $\|\cdot\|_1$ of the row vector:

$$\vec{V}_F = (F^{(0)}, BF^{(1)}, B^2 F^{(2)}, \cdots, B^D F^{(D)}). \qquad (13.13)$$

Thus, the best possible polynomial that can be obtained from a family of polynomials $F_{i,j,t-j}$ corresponds to the shortest non-zero vector in norm $\|\cdot\|_1$ in the lattice generated by the row vectors $\vec{V}_{F_{i,j,t-j}}$.

Finding this best possible polynomial is not possible for two reasons: first available lattice reduction algorithms work on the Euclidean norm and not the $\|\cdot\|_1$ norm; second, even considering the Euclidean norm, we do not generally obtain the shortest non-zero vector in the lattice. However, the good news is that the shortest vector that can be found using existing lattice reduction algorithms suffices to obtain a polynomial F with $|F|(B) < N^t$ for a large value of B. We now analyze this lattice in order to establish the relationship between the bound B, the parameters t and D, the size of the coefficients of f and of the modulus N.

13.2.1.1.1 Properties of the lattice of Howgrave-Graham Throughout the analysis, we assume that the bound B is a fixed parameter. We assess the largest value of B that can be obtained at the very end of the analysis.

We start from a set of integer pairs S and we construct a lattice $L_S(f)$ generated by the family of vectors $V_{F_{i,j,t-j}}$ defined as in Equation (13.13) from the polynomials $F_{i,j,t-j}$, for $(i,j) \in S$. When choosing S, it is useful to make sure that this generating family is in fact a basis of $L_S(f)$, thus avoiding trivial dependencies between the vectors. For example, if S contains $(0,2)$ and $(0,1)$, $(1,1)$, $\ldots (d,1)$ such a trivial dependency occurs. Indeed, assuming that $t = 2$, these pairs correspond to the polynomials, f^2, Nf, Nxf, $\ldots$, $Nx^d f$ which are related by:

$$N \cdot f^2 = \sum_{i=0}^{d} f^{(i)}(Nx^i f). \qquad (13.14)$$

Thus, it is a good strategy to keep the parameter i below the degree d of the starting polynomial f. With this in mind, a natural choice for the set S is the direct product $[0 \ldots d - 1] \times [0 \ldots t]$. With this choice, we have $d(t + 1)$ polynomials of degree at most $D = d(t + 1) - 1$ and the generating family of L_S is a square matrix of dimension $D + 1 = d(t + 1)$. The next step to obtain information about the lattice is to compute the determinant of the square matrix. To do this, remark that since the lattice is generated by using the encoding described in Equation (13.13) we can factor B out of the second column of the matrix, B^2 out of the third and so on. All in all, this implies that $B^{D(D+1)/2}$ can be be factored out of the determinant. Similarly, we can factor N^t out of each polynomial $F_{i,0,t}$, N^{t-1} out of $F_{i,1,t-1}$ and so on. The total contribution to the determinant is $N^{dt(t+1)/2} = N^{(D+1)t/2}$. After this, there remains a triangular matrix whose diagonal entries are powers of the highest degree coefficient of f.

At that point, we need to mention that multiplying f by the modular inverse of its high degree coefficient, we can always assume that f is unitary. The only way this could fail is by revealing of factor of N; in that case, we can work modulo each of the revealed factors[1]. When f is unitary, the determinant of the lattice L_S as above is $N^{(D+1)t/2}B^{D(D+1)/2}$. Using Equation (10.24) from Section 10.3.2, we see that we can thus find a short vector $\vec{V}$ in L_S of Euclidean norm satisfying:

$$\|\vec{V}\| \leq 2^{D/4}N^{t/2}B^{D/2}. \tag{13.15}$$

Since $\|\vec{V}\|_1 \leq \sqrt{D+1}\|\vec{V}\|$ we associate to $\vec{V}$ a polynomial F such that:

$$|F|(B) \leq \sqrt{D+1} \cdot 2^{D/4}N^{t/2}B^{D/2}. \tag{13.16}$$

In order to make sure that $|F|(B) < N^t$ we need to choose t and D that satisfy:

$$\sqrt{D+1}(\sqrt{2} \cdot B)^{D/2} < N^{t/2}. \tag{13.17}$$

Ignoring the $\sqrt{D+1}$ factor and taking logarithms, the largest value of B that can be achieved satisfies:

$$(\sqrt{2} \cdot B)^D \approx N^t. \tag{13.18}$$

Letting t grow, we find that asymptotically, we can achieve $B \approx N^{1/d}/\sqrt{2}$.

It is nice to note that in this context of Coppersmith's attack, the $2^{D/4}$ approximation factor only changes the bound B by a factor $\sqrt{2}$ compared to an ideal lattice reduction oracle. Since a constant factor on B can be gained by exhaustively trying a small number of bits of the root we seek, we see that where Coppersmith's attack is concerned, the L^3 algorithm is essentially as satisfactory as a lattice reduction oracle.

13.2.1.2 Coppersmith's original method

Starting from the same set of polynomials $F_{i,j,t-j}$ of degree at most D, Coppermith proceeds differently. He remarks that, since r is a common root of all these polynomials, the vector:

$$W(r) = (1, r, r^2, \dots, r^D)$$

is orthogonal to the vectors representing each of these polynomials. As in Chapter 11, we map vectors to polynomials and vice versa using a monomial ordering. Moreover, with univariate monomials, it is clear that monomials are simply sorted by degree.

Thus, instead of looking for a short vector in the lattice generated by the vectors corresponding to the polynomials $F_{i,j,t-j}$, Coppersmith's method

---

[1]In fact, in many cases, if the factorization is revealed, the attacker succeeds and can abort his attack.

looks for the vector $W(r)$ which belongs to the orthogonal of this lattice. As with Howgrave-Graham's variation, ad-hoc multipliers should appear in each coordinate to guarantee that we look for a short vector. To choose the multipliers, recalling that $|r| < B$ it suffices to remark that:

$$W'(r) = \frac{1}{\sqrt{D+1}} \cdot (1, r/B, (r/B)^2, \ldots, (r/B)^D)$$

is a short vector of norm less than 1, since all of its $D + 1$ coordinates are smaller than $1/\sqrt{D+1}$ in absolute value.

More precisely, Coppersmith let $\delta = 1/\sqrt{D+1}$ and considers the lattice L generated by the rows of the following matrix:

$$\begin{pmatrix}
\delta & 0 & 0 & \cdots & 0 & [x^0]F_{0,0,t} & [x^0]F_{1,0,t} & \cdots & [x^0]F_{d-1,t,0} \\
0 & \delta \cdot B^{-1} & 0 & \cdots & 0 & [x^1]F_{0,0,t} & [x^1]F_{1,0,t} & \cdots & [x^1]F_{d-1,t,0} \\
0 & 0 & \delta \cdot B^{-2} & \cdots & 0 & [x^2]F_{0,0,t} & [x^2]F_{1,0,t} & \cdots & [x^2]F_{d-1,t,0} \\
\vdots & \vdots & \vdots & \ddots & \vdots & \vdots & \vdots & \ddots & \vdots \\
0 & 0 & 0 & \cdots & \delta \cdot B^{-D} & [x^D]F_{0,0,t} & [x^D]F_{1,0,t} & \cdots & [x^D]F_{d-1,t,0} \\
0 & 0 & 0 & \cdots & 0 & N^t & 0 & \cdots & 0 \\
0 & 0 & 0 & \cdots & 0 & 0 & N^t & \cdots & 0 \\
0 & 0 & 0 & \cdots & 0 & 0 & 0 & \cdots & N^t
\end{pmatrix},$$

where $[x^i]F$ denotes the coefficient of x^i in F. Clearly $W'(r)$ is a short vector in the sublattice of L with zeros everywhere on the right coordinates. In fact, the matrix given in [Cop96b] is obtained by factoring out extra powers of N in the right-hand part of the matrix. Indeed, it is possible to factor N^{t-j} out the coefficients of $F_{i,j,t-j}$.

One important feature of Coppersmith's attack is that instead of looking at the first vector of the L^3 reduced basis in order to find $W'(r)$, it focuses on the last vector of the lattice. If the norm of the Gram-Schmidt orthogonalized of this last vector is large enough, then it cannot appear in the decomposition of $W'(r)$. As a consequence, $W'(r)$ is orthogonal to this orthogonalized vector. In other words, the orthogonalized vector can be transformed into a polynomial with r as a root over $\mathbb{Z}$.

In [HG97], it is shown that the original approach of Coppersmith and Howgrave-Graham's method are related through lattice duality. Following Section 10.5, this relationship explains why we need to consider the first vector with one lattice and the last one with the other.

13.2.2   Bivariate polynomials

Another difficult problem is to find integral solutions of polynomial equations in two unknowns. For example, if we could find the roots of $xy - N$, we could clearly factor N. The second small root algorithm of Coppersmith precisely addresses this case. Assume that we are given an irreducible polynomial

$f(x, y)$ with integer (or rational) coefficients and an integral root (x_0, y_0) such that $|x_0| \leq B_x$ and $|y_0| \leq B_y$ for some bounds B_x and B_y. We would like to determine a method that recovers (x_0, y_0) assuming that B_x and B_y are small enough. Note that this notion of "smallness" can no longer be related to the modulus; instead, we need to relate B_x and B_y with the size of coefficients occuring in f. More precisely, writing $f(x, y) = \sum_{i=0}^{d} \sum_{j=0}^{d} f_{i,j} x^i y^j$, the relevant parameter is $M(f) = \max_{i,j} B_x^i B_y^j |f_{i,j}|$. Let $F_{i,j}$ denote the polynomial $x^i y^j f(x, y)$. Of course, (x_0, y_0) is a common root of all polynomials in this family. Let us choose a set S of pairs of non-negative integers. We say that a polynomial $g(x, y)$ has its support in S if and only if all monomials $x^i y^j$ that appear with non-zero coefficients in g are such that $(i, j) \in S$. When g has its support in S, we can write:

$$g = \sum_{(i,j) \in S} g^{(i,j)} x^i y^j, \tag{13.19}$$

in this case, we let $|g|$ denote the polynomial:

$$|g| = \sum_{(i,j) \in S} |g^{(i,j)}| x^i y^j, \tag{13.20}$$

We can now encode g into a vector, following a graded monomial ordering:

$$\vec{V}_g^{(S)} = (g^{(0,0)}, B_x g^{(1,0)}, B_y g^{(0,1)}, \cdots, B_x^i B_y^j g^{(i,j)}, \cdots). \tag{13.21}$$

In fact, the order of coefficients of g in this vector is unessential, as long as we make the same choice for all polynomials that we encode. However, following a monomial ordering makes things easier to write down. When S is clear from the context, we omit it and write $\vec{V}_g$.

In parallel, based on the same order, we may encode the (unknown) root (x_0, y_0) into an (unknown) vector:

$$\vec{W}_{(x_0, y_0)}^{(S)} = \frac{1}{\sqrt{|S|}} \cdot (1, x_0/B_x, y_0/B_y, \cdots, (x_0/B_x)^i (y_0/B_y)^j, \cdots). \tag{13.22}$$

When g has its support in S and when (x_0, y_0) is a root of g, we see that:

$$(\vec{V}_g^{(S)} | \sqrt{|S|} \cdot \vec{W}_{(x_0, y_0)}^{(S)}) = \sum_{(i,j) \in S} g^{(i,j)} B_x^i B_y^j (x_0/B_x)^i (y_0/B_y)^j$$

$$= \sum_{(i,j) \in S} g^{(i,j)} x_0^i y_0^j = g(x_0, y_0) = 0. \tag{13.23}$$

Thus, the vector $\vec{W}_{(x_0, y_0)}^{(S)}$ is orthogonal to $\vec{V}_g^{(S)}$.

Now, given a set S, we construct a lattice $L_S(f)$ as follows:

- Consider all polynomials $F_{i,j}$ with support in S and construct the corresponding vector $V_{F_{i,j}}^S$.

- Let $L_S(f)$ be the lattice spanned by the above vectors.

If (x_0, y_0) is a root of f, it is also a root of each $F_{i,j}$, thus $W_{(x_0,y_0)}^S$ is orthogonal to the lattice $L_S(f)$. In addition, if $|x_0| \leq B_x$ and $|y_0| \leq B_y$, we see that $\|W_{(x_0,y_0)}^S\| \leq 1$. As a consequence, $W_{(x_0,y_0)}^S$ is a short vector in the orthogonal of the lattice $L_S(f)$.

Of course, (x_0, y_0) can be recovered from $W_{(x_0,y_0)}^S$, thus a natural approach to solve the equation $f(x, y) = 0$ is to use lattice reduction to find a short vector in the orthogonal lattice. However, as with Coppersmith's first algorithm, to provably obtain the solution (x_0, y_0), we need to make sure that no other short vector can hide the solution. To avoid this difficulty, Coppersmith instead proves that, under some conditions, $W_{(x_0,y_0)}^S$ belongs to the sublattice obtained by removing the final vector in a L^3 reduced basis. As a consequence, the Gram-Schmidt vector corresponding to this final basis element is orthogonal to $W_{(x_0,y_0)}^S$; thus, it encodes a bivariate polynomial $h(x, y)$ that vanishes at (x_0, y_0). Since f is irreducible, it suffices to prove that h is not a multiple of f in order to make sure that the system of equation $f(x, y) = 0$, $h(x, y) = 0$ has a finite number of solution over the complex field $\mathbb{C}$ and thus to guarantee that (x_0, y_0) can be recovered from the knowledge of f and h.

In [Cop96a], Coppersmith studied the case of polynomials of degree d in each variable. In that case, (x_0, y_0) can be recovered as long as $B_x B_y < M(f)^{2/(3d)}$. He also considered the case of bivariate polynomials of total degree d. In this other case, the corresponding bound is $B_x B_y < M(f)^{1/d}$. The method can be generalized to many different shapes of polynomials, for a survey, refer to [BM05].

In [Cor04] and [Cor07], Coron proposed a variation of Coppersmith's algorithm for bivariate polynomials using a modular approach similar to Howgrave-Graham's method.

13.2.2.1   Coppersmith's algorithm with more variables

Coppersmith's second algorithm can be extended to polynomials with more variables. It suffices to choose an appropriate set of monomials S. With this adaptation, we easily obtain a second polynomial h from the initial polynomial f. However, using the results of Chapter 11, we know that with more than two unknowns, two polynomials do not suffice to determine a zero-dimensional ideal. So, the key question in this context is to obtain extra polynomials with the same root. A simple heuristic method consists of taking several vectors in Coppersmith's reduced lattice instead of one and to construct a polynomial from each of these vectors. The same heuristic idea also works with modular polynomials with more than a single variable. The problem with this method is that it does not guarantee algebraic independence of polynomials. However, many cryptographic attacks, such as the one presented in Section 13.2.4 are based on this heuristic method and work extremely well in practice.

In some cases, it is possible to replace this heuristic method by a different approach which mixes lattice reduction and Gröbner basis techniques [BJ07].

13.2.3 Extension to rational roots

A natural question concerning Coppersmith's algorithm is to ask whether it can also discover "small" rational roots. Of course, we need to define the meaning of the word "small" in this context. The easiest is to define the size of a fraction p/q in irreducible form as $\max(|p|, |q|)$. With this definition of size, finding a small rational root of a polynomial $f(x_1, \dots, x_k)$ is equivalent to finding a small integer root of $F(X_1, Y_1, X_2, Y_2, \dots, X_k, Y_k)$, where F is derived from f by independently homogenizing each of the k variables in f. The algorithms of Coppersmith can then easily be adapted. We now detail the case of Coppersmith's first algorithm, using Howgrave-Graham's method, to show how this affects the bound on the size of the root. Let f be a univariate polynomial of degree d, with a rational root $r = x_0/y_0$ modulo N satisfying $|x_0| < B$ and $|y_0| < B$. After homogenization, we obtain an homogeneous polynomial in two variables, x and y, namely $y^d f(x/y)$. Given a fixed degree bound D, we can define the family of polynomials $F_{i,j,k}(x, y) = y^{D-i} x^i f(x/y)^j N^k$. Each polynomial in the family is homogeneous of degree D in x and y. Then, we construct a lattice as in Section 13.2.1.1 but instead of using the encoding of Equation (13.13) we instead use:

$$\vec{V_F} = (F^{(0)}, F^{(1)}, F^{(2)}, \cdots, F^{(D)}). \tag{13.24}$$

The only difference with Equation (13.13) is that we omit the factor B used to balance the size of different power of x when searching integer roots. Indeed, with a homogeneous polynomial and the same bound on x and y, this is no longer needed.

With this change, the determinant of the lattice adapted from Howgrave-Graham's is $N^{(D+1)t/2}$. We can guarantee that the shortest vector in the adapted lattice corresponds to a polynomial with the exact rational root r if the sum of the absolute values of its coefficients are smaller than N^t/B^D. Thanks to the L^3 bound, the sum of coefficients can be bounded by $\sqrt{D+1} \cdot 2^{D/4} \cdot N^{t/2}$. Ignoring the factor $\sqrt{D+1}$, we find that the largest value of B we can obtain satisfies:

$$(\sqrt[4]{2} \cdot B)^D \approx N^{t/2}. \tag{13.25}$$

Asymptotically, we can achieve:

$$B \approx \sqrt{N^{1/d}/\sqrt{2}}.$$

In other words, in this case, the bound on the numerator and denominator x_0 and y_0 of the rational root is the square root of the original bound for integer roots.

13.2.4   Security of RSA with small decryption exponent

One typical application of Coppersmith's method in cryptography is the Boneh-Durfee attack for recovering the secret key of an RSA instance, when the decryption exponent is small. Let $N = pq$ be an RSA modulus and $\phi(N) = (p-1) \cdot (q-1)$. We know that the encryption exponent e and the decryption exponent d are related by the following equation:

$$ed = 1 + k\phi(N). \tag{13.26}$$

In the sequel, we assume that d is normalized in the interval $[-\phi(N)/2, \phi(N)/2]$ and we say that d is α-small when $|d| \leq \phi(N)^\alpha$. We also assume that e is in the interval $[0, \phi(N)]$. Clearly, in this setting, if d is α-small then so is k. Moreover, since $\phi(N) = N - (p+q) + 1$, it is possible to write $\phi(N) = N - z$, with a value of z below a small multiple of $\sqrt{N}$. Thus, Equation (13.26) can be rewritten as:

$$ed - kN - kz - 1 = 0. \tag{13.27}$$

In this equation, the term kz is smaller than the terms ed and kN. This remark was used by Wiener in [Wie90]. It means that the equation can be rewritten as $e/N \approx k/d$. As a consequence, if d and k are small enough, the fraction k/d naturally arises as an approximation of e/N using the techniques presented in Section 2.2.2. Using a continued fraction algorithms, Wiener showed in [Wie90] that this allows recovery of the decryption exponent d under the condition $d < N^{1/4}$.

In [BD99], in order to increase the bound to N^α with $\alpha > 1/4$, Boneh and Durfee rewrite Equation (13.27) as a bivariate modular equation $kN + kz + 1 = 0 \pmod{e}$. In this equation, we search a solution with z of the order of $N^{1/2}$ and k of the order of N^α. Using the heuristic extension of Coppersmith univariate modular algorithm to the bivariate modular case adapted to the shape of this polynomial, they first obtain a new bound $\alpha \approx 0.285$. With a specific improvement, involving rectangular matrices, they derive an improved bound $\alpha \approx 0.292$.

Exercises

1. Consider the floating point number:

$$x = -8.4431161058379455039313851 7.$$

Show that x is a close approximation of a real root of a polynomial of degree 3, with coefficient bounded by 20 (in absolute value).

2[h]. Let p be a large prime such that the polynomial $f_1(x) = x^3 + 2$ has a root X_0 modulo p. Find a polynomial $f_2(X)$ of degree 2 with coefficients of the order of $p^{1/3}$ and such that $f_2(X_0) = 0 \pmod{N}$. What can you say about the resultant of f_1 and f_2.

3[h]. Generalize the above method for a polynomial f_1 of degree d with small coefficients.

4[h]. Show that the polynomials f_1 and f_2 from the two previous exercises can serve as a basis for a number field sieve computation modulo p (see Chapter 15). Compare with the bound given in Section 15.4.3.

5. Assume that a positive integer x has been encrypted under 3 different RSA keys, N_1, N_2 and N_3 (x is smaller than these three values to allow decryption), under the public exponent 3. Show that x can be recovered from the three encrypted values c_1, c_2 and c_3.

6[h]. Consider an RSA number $N = pq$, with $\sqrt{N}/2 \le p \le \sqrt{N}$. Assume that a fraction of the high order bits of p are known. In that case, we can write $p = p_0 + x$.

- Give an upper bound on x.

- Show that the high order bits of q are also known.

- When can Coppersmith's method be used to solve $N = (p_0 + x) \cdot (q_0 + y)$?

7. Consider Equation (13.26). Check that by using the methods for approximating fractions given in Chapter 2, it is possible to recover the RSA secret key as long as it remains below $N^{1/4}$.

8. Consider a variation of Equation (13.26), namely:

$$ed = r + k\phi(N).$$

Assuming that d and r are both small, under which condition can these numbers be recovered?

Chapter 14

Elliptic curves and pairings

Elliptic curves are an essential tool in today's cryptography. They are often used to construct cryptosystems but there are also some cryptanalytic applications where the use of elliptic curves is necessary. In this chapter, we give a self-contained introduction to elliptic curves, together with a description of the Weil pairing. This pairing was initially introduced in cryptography as a cryptanalytic tool to attack the discrete logarithm problem on some special elliptic curves. It is interesting to note that the Weil pairing[1] has now become an essential tool for constructing new cryptosystems.

The constructions and proofs given in this chapter are often ad-hoc shortcuts which hide deeper and nicer theory. In particular, we only consider elliptic curves over finite fields of characteristic $p \geq 5$. In order to learn more about elliptic curves and their cryptographic applications, interested readers should refer to more specific textbooks such as [CF05, JN08, Sil86].

14.1 Introduction to elliptic curves

Over a finite field $\mathbb{F}_q = \mathbb{F}_{p^n}$, with $p \geq 5$, an elliptic curve is described by a Weierstrass equation:

$$y^2 = x^3 + ax + b, \tag{14.1}$$

where a and b are elements of $\mathbb{F}_q$. For this equation, it is essential to be able to define the tangent line to the curve at every point on the curve. If it is not the case, we say that the curve is singular and not really an elliptic curve. The tangent line at a point P of coordinates (x_P, y_P) is given by the equation:

$$2y_P(y - y_P) = (3x_P^2 + a)(x - x_P). \tag{14.2}$$

It is well defined unless $2y_P = 3x_P^2 + a = 0$. Since we also have $y_P^2 = x_P^3 + ax_P + b$, the tangent is well defined unless both $3x_P^2 + a = 0$ and

---

[1] Together with a few cousins which can be computed faster, such as the Tate, Ate or Eta pairings.

417

$x_P^3 + ax_P + b = 0$. Put together, these two conditions imply:

$$(6ax_P^2 - 9bx_P + 4a^2) \cdot (3x_P^2 + a) + (-18ax_P + 27b) \cdot (x_P^3 + ax_P + b) = 0. \quad (14.3)$$

The reader can easily check that the above quantity is equal to $\Delta(a,b) = 4a^3 + 27b^3$, it is called the discriminant of the curve. The above discussion shows that the curve is non-singular (or smooth) if and only if $\Delta(a,b) \neq 0$.

Given an elliptic curve E in Weierstrass equation, we can clearly define the set of points on this curve:

$$E(\mathbb{F}_q) = \left\{ P = (x_P, y_P) \mid (x_P, y_P) \in \mathbb{F}_q^2, y_P^2 = x_P^3 + ax_P + b \right\} \bigcup \{O\}. \quad (14.4)$$

The additional point O is called the point at infinity on the elliptic curve. Similarly, for any extension of the finite field, we can define $E(\mathbb{F}_{q^e})$, by taking points with coordinates in the extension field.

14.1.1   The group structure of elliptic curves

The fundamental property of elliptic curves is that we can add a group structure to these sets of points. This group structure is denoted additively and the point at infinity is the zero element in the group. This group structure can be introduced in several ways. Here, we give an algebraic construction in order to define other mathematical objects that are needed for the introduction of the Weil pairing.

14.1.1.1   Divisors

The first step of the construction is to define the divisor group of the elliptic curve. This divisor group **Div** is the group of maps from the curve to the set of integers $\mathbb{Z}$ which are equal to zero except on a finite[2] set of points. Clearly, if we define the sum of two maps D_1 and D_2 as the map $D_1 + D_2$ whose value at a point P is $D_1(P) + D_2(P)$, we obtain an additive group. The zero element is the zero mapping and the opposite of a map D is $-D$, with value $-D(P)$ at P. Furthermore, this group is abelian, since addition of integers is commutative. To represent these maps in the divisor group, it is traditional to write them as formal sums $\sum D(P)(P)$. For example, the map D with value 3 at P, -3 at O and equal to zero elsewhere is represented as $3(P) - 3(O)$. From now on, an element from the divisor group will simply be called a divisor.

Given any divisor D, we define its degree as the (finite) sum of the values of D at all points. For example, $\deg(2(P) + 2(Q) - 3(O)) = 1$. Since $\deg(D_1 + D_2) = \deg(D_1) + \deg(D_2)$, deg is a group morphism from the divisor group **Div** to $\mathbb{Z}$. As a consequence, its kernel is a subgroup, called the subgroup of degree 0 divisors and denoted by **Div$_0$**.

---

[2]Of course, since $E(\mathbb{F}_q)$ is already a finite set, this restriction is irrelevant in our case.

14.1.1.2 Functions

The second step of the construction is to define functions on the elliptic curve. We start from an arbitrary polynomial $F(x, y)$. Clearly, for any point $P = (x_P, y_P)$ other than O on the elliptic curve $E(\mathbb{F}_q)$, we can define $F(P) = F(x_P, y_P)$. Note that $F(P)$ is an element of $\mathbb{F}_q$. If $F(P) = 0$, we say that P is a zero of F. Unless F is multiple of $y^2 - x^3 - ax - b$, the set of zeros of F is finite. In fact, adding any multiple of $y^2 - x^3 - ax - b$ to a polynomial F preserves the values $F(P)$ and the set of zeros. As a consequence, if we so wish, we can replace y^2 by $x^3 + ax + b$ in F and write $F(x, y)$ as $F_0(x) + yF_1(x)$. When P is a zero of F, we can define its order or multiplicity denoted by $\mathbf{ord}_P(F)$. It is the largest integer o such that some multiple FH of F, with $H(P) \neq 0$ can be written as a product (modulo $y^2 - x^3 - ax - b$) of o polynomials, each with value 0 at P. When P is not a zero of F we say that $\mathbf{ord}_P(F) = 0$. With this definition, we have the nice property that P is a zero of a product GH if and only if it is a zero of either G or H, in addition:

$$\mathbf{ord}_P(GH) = \mathbf{ord}_P(G) + \mathbf{ord}_P(H). \tag{14.5}$$

To make this notion of order of a function at P more precise, let us analyze in detail the zeros of $F(x, y) = F_0(x) + yF_1(x)$. Let $P = (x_P, y_P)$ be a zero of F. If $y_P = 0$, it is always possible to factor y out of F (after potentially multiplying by a polynomial $H(x, y)$ with $H(P) \neq 0$). If $F_0(x) = 0$ this is clear. Otherwise, since x_P is a root of $x^3 + ax + b$, $x - x_P$ divides $x^3 + ax + b$, e.g., there exists a polynomial $H(x)$ such that $x^3 + ax + b = H(x)(x - x_P)$. Moreover, x_P is a root of $F_0(x)$ and $(x - x_P)$ divides $F_0(x)$. As a consequence, $x^3 + ax + b$ divides $H(x)F_0(x)$. Thus, replacing $x^3 + ax + b$ by y^2 we can factor y out of $H(x)F(x, y)$.

Similarly, if $y_P \neq 0$, it is always possible to factor $x - x_P$ out of F after potential multiplication by H. First, remark that $x - x_P$ divides either both $F_0(x)$ and $F_1(x)$ or none of them. If $x - x_P$ divides both, the conclusion follows. If $x - x_P$ divides none, then letting $H(x, y) = F_0(x) - yF_1(x)$, we see that $H(P) \neq 0$. Moreover, $H(x, y) \cdot F(x, y) = F_0(x)^2 - y^2 F_1(x)^2$ can be written as a polynomial in the single unknown x by replacing y^2 by $x^3 + ax + b$. Since x_P is a root of this polynomial, we can factor $x - x_P$.

The two polynomials y or $x - x_P$ that occur in the above discussion are called **uniformizers** at P. With this notion in mind, the order of F at P can be equivalently defined as the maximal power of the uniformizer at P that can be factored out of F. Using uniformizers, we can more easily analyze the zeros of $F(x, y) = F_0(x) + yF_1(x)$ and their orders. When $F_1(x) = 0$, this reduces to finding the points P such that $F_0(x_P) = 0$ and their multiplicity. Clearly, we need to factor F_0 as a univariate polynomial. If x_P is a root of F_0 with multiplicity o, two cases arise. In the first case, $x_P^3 + ax_P + b \neq 0$ and we find two zeros of F: $P = (x_P, y_P)$ and $P' = (x_P, -y_P)$, each of these two zeros has order o. Indeed, in that case $x - x_P$ is a uniformizer both at P and at P'. In the second case, $x_P^3 + ax_P + b = 0$ and we find a single zero

$P = (x_P, 0)$, with order $2o$. To understand why the order is doubled, let us again write $x^3 + ax + b = H(x)(x - x_P)$. Thus, modulo the curve equation, y^{2o} divides $H(x)^o F_0(x)$. Since y is the uniformizer at P in this case, the conclusion follows.

Continuing the analysis, we now turn to the case $F_0(x) = 0$ and are left with $yF_1(x)$. Clearly, we already know how to find the zeros of $F_1(x)$ and simply need to look at the zeros of y. In fact, y has three zeros, one for each root of $x^3 + ax + b$. Note that, since the curve has a non-zero discriminant, the three roots are distinct. Moreover, since y is a uniformizer at these three points, the order of y at each point is 1.

Finally, let us look at $F(x, y) = F_0(x) + yF_1(x)$ when neither F_0 nor F_1 is zero. Factoring out the greatest common divisor of F_0 and F_1, we see that it suffices to deal with the case where F_0 and F_1 are coprime. Assume that $P = (x_P, y_P)$ is a root of F. If $y_P = 0$, then $\mathbf{ord}_P(F) = 1$ indeed we know that P is a root of order at least 2 of F_0. Thus, if we had $\mathbf{ord}_P(F) > 1$, we could write $yF_1(x) = F_0(x) - F(x, y)$ with $F_1(P) \neq 0$ and conclude that $\mathbf{ord}_P(y) > 1$, this would contradict the above analysis which says that $\mathbf{ord}_P(y) = 1$. Finally, if P is a root of F with $y_P \neq 0$, we see that neither $F_0(x_P)$ nor $F_1(x_P)$ can be 0, otherwise both would be and F_0 and F_1 would not be coprime. Thus, multiplying F by $H(x, y) = F_0(x) - yF_1(x)$, we find that the order of P at F is equal to the multiplicity of $x - x_P$ in the univariate polynomial $F_0(x)^2 - (x^3 + ax + b)F_1(x)^2$.

When looking at the zeros of a polynomial F, it is important to remember that the zeros are not necessarily points with coordinates in $\mathbb{F}_q$. In many cases, the zeros have their coordinates in some extension field. If we construct all the zeros of a non-zero polynomial $F(x, y) = F_0(x) + yF_1(x)$ including zeros over extension fields, it is possible to prove that the sum of their orders is $\max(2 \deg(F_0), 3 + 2 \deg(F_1))$, using the usual convention that $\deg(0) = -\infty$. It is useful to define the order of F at the point at infinity O as $\mathbf{ord}_O(F) = -\max(2 \deg(F_0), 3 + 2 \deg(F_1))$. With this convention, we now define the divisor of a non-zero polynomial F as:

$$\mathbf{div}(F) = \sum_P \mathbf{ord}_P(F)(P), \qquad (14.6)$$

where the sum includes points over extension fields and the point at infinity. We see that $\mathbf{div}(F)$ is a divisor of degree 0.

The above definitions can easily be extended to quotients F/G letting:

$$\mathbf{ord}_P(F/G) = \mathbf{ord}_P(F) - \mathbf{ord}_P(G) \text{ and}$$
$$\mathbf{div}(F/G) = \mathbf{div}(F) - \mathbf{div}(G). \qquad (14.7)$$

When $\mathbf{ord}_P(F/G) > 0$ we say that P is a zero of F/G and when $\mathbf{ord}_P(F/G) < 0$ we say that P is a pole of F/G. Any fraction F/G is called a function on the elliptic curve. Note that functions are determined modulo the curve equation $y^2 - x^3 - ax - b$ and can thus be represented by several possible fractions. Any divisor which can be written as the divisor of a function is called a **principal divisor**.

14.1.1.3 Principal divisors and the group structure

Looking at the set of principal divisors, we may easily see that it is a subgroup of the group $\mathbf{Div}_0$ of degree 0 divisors. Indeed, each principal divisor has degree 0; the empty[3] divisor is principal and equal to $\mathbf{div}(1)$ since the constant function 1 has neither zeros nor poles; the opposite of a principal divisor is principal since $\mathbf{div}(F/G) = -\mathbf{div}(G/F)$; and the sum of two principal divisors is principal, $\mathbf{div}(F_1/G_1 \cdot F_2/G_2) = \mathbf{div}(F_1/G_1) + \mathbf{div}(F_2/G_2)$. It is thus possible to form a quotient group by considering degree 0 divisors modulo principal divisors. It is usual to say that two divisors which differ by a principal divisor are **linearly equivalent**. We are now going to see that this quotient group and this notion of linear equivalence can be used to give a group structure to the elliptic curve itself.

The key ingredient is to show that any degree zero divisor D can be written as $D = (P) - (O) + \mathbf{div}(f)$ for some point P on the elliptic curve and some function f. In fact, this can be done in a completely effective and computationally efficient manner. Clearly, since any divisor is a finite sum of points, it suffices to show how to compute the sum and the difference of $D_1 = (P_1) - (O)$ and $D_2 = (P_2) - (O)$ in the above form. To compute the sum, we write $P_1 = (x_{P_1}, y_{P_1})$, $P_2 = (x_{P_2}, y_{P_2})$ and we consider the following cases:

1. If $P_1 = O$, then D_1 is the empty divisor and $D_1 + D_2 = D_2$ already is of the correct form. If $P_2 = O$ this remark also applies.

2. If $x_{P_1} = x_{P_2}$ and $y_{P_1} = -y_{P_2}$ then:

$$\mathbf{div}(x - x_{P_1}) = (P_1) + (P_2) - 2(O) = D_1 + D_2, \qquad (14.8)$$

   thus $D_1 + D_2$ is linearly equivalent to the empty divisor $(O) - (O)$. Note that this also covers the special case where $P_1 = P_2$ and $y_{P_1} = 0$.

3. If $P_1 = P_2$ and $y_{P_1} \neq 0$ then let $\mathcal{L}(x, y)$ be the equation of the tangent to the elliptic curve at P_1. More precisely:

$$\mathcal{L}(x, y) = (y - y_{P_1}) - \lambda(x - x_{P_1}),$$

   with $\lambda = \frac{3x_{P_1}^2 + a}{2y_{P_1}}$.

   From our general analysis, we know that the sum of the orders of the zeros of $\mathcal{L}(x, y)$ is 3 and thus that:

$$\mathbf{div}(\mathcal{L}(x, y)) = 2(P_1) + (P_3) - 3(O), \qquad (14.9)$$

---

[3]Remember that the empty divisor is the group neutral element.

for some point P_3. Moreover, the coordinates of P_3 are in the same field as the coordinates of P_1. We can explicitly write:

$$x_{P_3} = \lambda^2 - x_{P_1} - x_{P_2} \quad \text{and} \tag{14.10}$$
$$y_{P_3} = y_{P_1} + \lambda(x_{P_3} - x_{P_1}). \tag{14.11}$$

This shows that $D_1 + D_2$ is linearly equivalent to $(O) - (P_3)$. However, we are not done because this divisor does not have the expected form. We now write:

$$\mathbf{div}(x - x_{P_3}) = (P_3) + (P_4) - 2(O), \tag{14.12}$$

with $P_4 = (x_{P_3}, -y_{P_3})$. As a consequence,

$$D_1 + D_2 = (P_4) - (O) + \mathbf{div}\left(\frac{\mathcal{L}(x,y)}{x - x_{P_3}}\right). \tag{14.13}$$

4. If $x_{P_1} \neq x_{P_2}$, let $\mathcal{L}(x,y)$ be the equation of the line passing though P_1 and P_2. More precisely:

$$\mathcal{L}(x,y) = (y - y_{P_1}) - \lambda(x - x_{P_1}),$$

with $\lambda = \frac{y_{P_2} - y_{P_1}}{x_{P_2} - x_{P_1}}$. The very same reasoning as above apply and once again:

$$D_1 + D_2 = (P_4) - (O) + \mathbf{div}\left(\frac{\mathcal{L}(x,y)}{x - x_{P_3}}\right),$$

where P_3 is the third point on the line of equation $\mathcal{L}(x,y)$ with coordinates as in Equations 14.10 and 14.11, and where P_4 is its symmetric with coordinates $(x_{P_3}, -y_{P_3})$.

To compute differences, it suffices to use the above formula that transform $(O) - (P_3)$ into a linearly equivalent divisor $(P_4) - (O)$ repeatedly. As a consequence, iterating these formulas we are able to reduce any divisor D to a linearly equivalent divisor $(P) - (O)$. In fact, we even recover an expression for a function f such that:

$$D = (P) - (O) + \mathbf{div}(f). \tag{14.14}$$

In order to turn the elliptic curve itself into a group, it now suffices to interpret any point P as the divisor $(P) - (O)$ and vice versa. To make sure that this interpretation is unequivocal, we need to check that two different points P_1 and P_2 yield two different group elements. Thus, we need to verify that $D = (P_1) - (P_2)$ is not a principal divisor when $P_1 \neq P_2$; see Exercise 3 or refer to [Was03, Lemma 11.3]. Using this interpretation, we obtain the following properties for the group structure:

- The point at infinity O is the neutral element in the group. Indeed, $(O) - (O)$ is the empty divisor.

- The opposite of a point $P = (x_P, y_P)$ is its symmetric $-P = (x_P, -y_P)$. Indeed, $(P) - (O) + (-P) - (O) = \mathbf{div}(x - x_P)$ is principal.

- The sum of P_1 and P_2 is the symmetric of the third point of intersection P_3 of the line through P_1 and P_2 and the elliptic curve. When $P_1 = P_2$, we take the tangent at P_1 as the line through P_1 and P_2.

14.1.2 Double and add method on elliptic curves

With the above addition law, it is possible to compute xP for any point P on the elliptic curve E and any integer x. When x is zero, the result is the point at infinity O. When $x < 0$, we replace the computation of xP by the computation of $(-x)P'$ where P' is the symmetric, i.e., opposite, of P. By definition, xP with $x > 0$ is obtained by adding together x copies of the point P. However, for large values of x this is not efficient. To speed up the computation of xP, it suffices to generalize the square and multiply algorithms used for exponentiation of modular integers to the case of elliptic curve. Adapted to this specific case, the basic idea is simply to remark that the sequence $P, 2P, 4P, \ldots 2^t P$ can be efficiently computed by writing each term as the sum of the previous term in the sequence with itself, i.e., as the double of the previous term. Moreover, from the binary decomposition of x, we can rewrite xP as the sum of the terms $2^i P$ such that 2^i appear in the decomposition of x. Thus, xP can be computed using $\log_2(x)$ doublings and at most $\log_2(x)$ additions. We leave the adaptation of Algorithms 2.9 and 2.10 as an exercise to the reader.

14.1.3 Number of points on elliptic curves

In order to use an elliptic curve over a finite field as the basis for a discrete logarithm based cryptographic problem, it is useful to know the cardinality of this elliptic curve. This is not mandatory and there are systems that work with groups of unknown cardinality. However, the most frequent cryptosystems over elliptic curves require knowledge of the cardinality. In addition, it is useful to describe the group structure, i.e., to find an isomorphic group, written as a product of cyclic groups of the form $\mathbb{Z}/r\mathbb{Z}$. With elliptic curves over a finite field $\mathbb{F}_p$, a lot of information is known about the cardinality and the group structure. In particular, we have the following theorem:

THEOREM 14.1
Let E be an elliptic curve over $\mathbb{F}_p$ and let C_E be the number of points of E with both coordinates in $\mathbb{F}_p$, including the point at infinity O. Then we have:

$$p + 1 - 2\sqrt{p} \leq C_E \leq p + 1 + 2\sqrt{p}. \tag{14.15}$$

Moreover, there exist two positive integers r_1 and r_2 such that:

1. *r_2 divides r_1.*

2. *The number of points C_E is equal to $r_1 r_2$.*

3. *The group structure of the elliptic curve E is $\mathbb{Z}/r_1\mathbb{Z} \times \mathbb{Z}/r_2\mathbb{Z}$.*

PROOF   See [Sil86, Chapter III], Corollary 6.4 and [Sil86, Chapter V], Theorem 1.1. ▯

We do not develop point counting algorithms here and refer the reader to [CF05, Part IV]. For the Schoof-Elkies-Atkin point counting method, the noisy Chinese remainder reconstruction algorithm discussed in Section 8.4.1 can be used to speed up the final phase.

14.2   The Weil pairing

14.2.1   Weil's reciprocity law

Since we have already defined divisors and functions, we are almost ready to go forward and define pairings. Before that, we need to learn how to evaluate a function on a divisor. Let $f = F/G$ be a function on an elliptic curve E and let D be an arbitrary divisor. We define the support of D to be the finite set of points of E that appear with non-zero coefficients in D. When the support D contains none of the zeros or poles of f, writing once again $D = \sum D(P)(P)$ we can define:

$$f(D) = \prod f(P)^{D(P)}. \qquad (14.16)$$

Note that we have already defined $f(P)$ as $F(P)/G(P)$ for all points P not equal to the point at infinity (O). If the support of D does not contain O, nor any zero or pole of f, $f(D)$ is a well-defined and non-zero value. We shall determine, later on, the precise finite field where this value is defined. For now, let us simply say that it clearly belongs to some extension of $\mathbb{F}_p$. To define $f(D)$ in all cases, we need to define $f(O)$ when O is neither a zero nor a pole of f. In fact, this can be done by looking only at high order terms. More precisely, if we define the degree of a monomial $y^{d_y} x^{d_x}$ in the function field as $3d_y + 2d_x$ and the degree of a polynomial F as the maximum of the degrees of the monomials in F, looking back at Section 14.1.1.2 we see that the order of f at O is then equal to the degree of F. Thus, saying that O is neither a zero nor a pole of f means that the degrees of F and G are equal. In that case, we define $f(O)$ as the quotient of the coefficients in front of the highest degree monomials in F and G.

With this definition of $f(D)$, we easily remark that when D is of degree 0, $f(D)$ is equal to $(\lambda f)(D)$ for any constant λ. This is extremely interesting because if we only consider f up to a multiplicative constant, we are in fact looking at an object which is determined by the divisor $\mathbf{div}(f)$. This connection is emphasized by the following theorem:

THEOREM 14.2 Weil's reciprocity
Let f and g be two functions in the function field of an elliptic curve E. If $\mathbf{div}(f)$ and $\mathbf{div}(g)$ have disjoint support, or equivalently if the zeros and poles of f and g do not intersect, then:

$$f(\mathbf{div}(g)) = g(\mathbf{div}(f)). \tag{14.17}$$

We give here an elementary proof using resultants; for a shorter proof, the reader may refer to [BSS05, pages 212–213].

PROOF Since f and g need only be defined up to multiplicative constants, we may assume the polynomials at the numerators and denominators of f and g are normalized, i.e., that the coefficients in front of the highest degree monomials are everywhere 1s.

Under this hypothesis, all evaluations at O yield 1. As a consequence, we can consider a simplified variation of Weil's reciprocity that ignores the point at infinity. For this variation, let us take two polynomials on the curve $F(x, y) = F_0(x) + yF_1(x)$ and $G(x, y) = G_0(x) + yG_1(x)$, with no common zeros. We know that $\mathbf{div}(F)$ can be written as $D_F - \mathbf{ord}_O(F)(O)$, where D_F describes the zeros of F. Similarly, $\mathbf{div}(G)$ can be written as $D_G - \mathbf{ord}_O(G)(O)$. We now claim that:

$$F(D_G) = \pm G(D_F). \tag{14.18}$$

Moreover, the $\pm$ sign in the above equation is uniquely determined by the degrees of F and G.

Weil's reciprocity law easily follows from Equation (14.18). Indeed, thanks to the normalization we chose, if we write $f = F/H$ and $g = G/K$ then we see that:

$$f(\mathbf{div}(g)) = \frac{F(D_G)H(D_G)}{F(D_K)H(D_K)} = g(\mathbf{div}(f)), \tag{14.19}$$

since all evaluations at O are equal to 1. The only tricky part is to verify that the signs really cancel out. We leave the verification to the reader. Anyway, since we only intend to define the Weil pairing on ℓ-torsion points where ℓ is an odd prime, this verification is not even required here.

In order to prove Equation (14.18), we need to consider three elementary possibilities for F and G. For F, the options are $F(x, y) = F_0(x)$, $F(x, y) = y$ and $F(x, y) = F_0(x) + yF_1(x)$ with F_0 and F_1 coprime. Indeed, any other case can be obtained by multiplication of these elementary cases and of course Equation (14.18) is preserved by multiplication, since clearly for $F = F^{(1)}F^{(2)}$, if Weil's reciprocity is satisfied by $(F^{(1)}, G)$ and $(F^{(2)}, G)$ we have:

$$F(D_G) = F^{(1)}(D_G) \cdot F^{(2)}(D_G) = G(D_{F^{(1)}})G(D_{F^{(2)}}) = G(D_F), \quad (14.20)$$

since the divisor of a product is the sum of the divisors of each term.

Combining all possibilities for F and G, there is a total of 9 cases to consider. Taking into account the symmetry betwen F and G, this is reduced to 6 cases. In addition, since F and G have no common zero, we can ignore[4] the case $F(x, y) = G(x, y) = y$. As a consequence, there are five remaining cases to consider. It is important to remember that all highest degree coefficients are 1s. This allows us to replace some expressions by resultants as defined in Chapter 11. This makes these expressions easier to manipulate and greatly simplifies the proof. The five cases are:

1. When $F(x, y) = F_0(x)$ and $G(x, y) = G_0(x)$, for each root α of F_0 in the algebraic closure of the field of definition of E, there are either two corresponding points (α, y_α) and $(\alpha, -y_\alpha)$ or a point $(\alpha, 0)$. In the first case, the order at each point is equal to the multiplicity of α in F_0. In the second case, the order is equal to twice the multiplicity. Thus:

$$G(D_F) = \prod_{\substack{\alpha \text{ root of } F_0 \\ \text{(with multiplicity)}}} G_0(\alpha)^2 = \mathbf{Res}(F_0, G_0^2) = \mathbf{Res}(F_0, G_0)^2.$$

   Clearly, by symmetry, we also have:

$$F(D_G) = \mathbf{Res}(G_0, F_0)^2.$$

   Moreover, since $\mathbf{Res}(F_0, G_0) = \pm\mathbf{Res}(G_0, F_0)$, both expressions are equal.

2. When $F(x, y) = F_0(x)$ and $G(x, y) = y$, once again we can look at the roots α of F_0. If there is a corresponding point of the form $(\alpha, 0)$, then F and G have this point as a common zero and $D_F(G) = D_G(F) = 0$. Otherwise, we see that:

$$G(D_F) = \prod_{\substack{\alpha \text{ root of } F_0 \\ \text{(with multiplicity)}}} -y_\alpha^2$$

$$= -\prod_\alpha \alpha^3 + a\alpha + b = -\mathbf{Res}(F_0, x^3 + ax + b)$$

---

[4]Anyway, it is clear in that case that $D_F(G) = D_G(F) = 0$.

Letting $(\beta_1, 0)$, $(\beta_2, 0)$ and $(\beta_3, 0)$ denote the three distinct points with y coordinate 0, i.e., letting β_1, β_2 and β_3 be the roots of $x^3 + ax + b$, we also have:

$$F(D_G) = F_0(\beta_1)F_0(\beta_2)F_0(\beta_3) = \mathbf{Res}(x^3 + ax + b, F_0).$$

As a consequence, we conclude that $D_F(G) = \pm D_G(F)$.

3. When $F(x, y) = F_0(x)$ and $G(x, y) = G_0(x) + yG_1(x)$, with G_0 and G_1 coprime, once more, we start by looking at the roots α of F_0. If a point of the form $(\alpha, \pm y_\alpha)$ is a zero of G, then F and G have this point as a common zero and $D_F(G) = D_G(F) = 0$. Otherwise, we first compute $G(D_F)$ as:

$$G(D_F) = \prod_{\substack{\alpha \text{ root of } F_0 \\ \text{(with multiplicity)}}} (G_0(\alpha) + y_\alpha G_1(\alpha))(G_0(\alpha) - y_\alpha G_1(\alpha))$$

$$= \mathbf{Res}(F_0, \mathcal{G}),$$

where $\mathcal{G}$ denotes the polynomial $\mathcal{G}(x) = G_0(x)^2 - (x^3 + ax + b)G_1(x)^2$. Indeed, we can group the zeros of F in pairs (α, y_α) and $(\alpha, -y_\alpha)$. Note that this also holds when $y_\alpha = 0$, since in that case the order of the point is twice the multiplicity of α in F_0.

To evaluate the second term $F(D_G)$, we remark that each zero (β, y_β) of G with order e corresponds to a root β of $\mathcal{G}$ of multiplicity e. Also note that from β, we can obtain y_β as $-G_0(\beta)/G_1(\beta)$. Since F does not depend on y, we do not need this expression of y_β for this case, but it will be useful for the next cases. Thanks to this remark, we can write:

$$F(D_G) = \prod_{\substack{\beta \text{ root of } \mathcal{G} \\ \text{(with multiplicity)}}} F_0(\beta) = \mathbf{Res}(\mathcal{G}, F_0).$$

Clearly, the two expressions we obtain for $G(D_F)$ and $F(D_G)$ are equal, up to sign. This concludes the third case.

4. When $F(x, y) = y$ and $G(x, y) = G_0(x) + yG_1(x)$, with G_0 and G_1 coprime, let us denote $\mathcal{G}(x) = G_0^2(x) - (x^3 + ax + b)G_1^2(x)$ as above, we

find:

$$F(D_G) = \prod_{\substack{\beta \text{ root of } \mathcal{G} \\ \text{(with multiplicity)}}} y_\beta = \prod_\beta \frac{-G_0(\beta)}{G_1(\beta)} = \frac{\mathbf{Res}(\mathcal{G}, -G_0)}{\mathbf{Res}(\mathcal{G}, G_1)}$$

$$= \frac{\mathbf{Res}(-(x^3 + ax + b)G_1^2(x), -G_0)}{\mathbf{Res}(G_0^2(x), G_1)}$$

$$= \mathbf{Res}((x^3 + ax + b), -G_0) \cdot \frac{\mathbf{Res}(-G_1^2(x), -G_0)}{\mathbf{Res}(G_0^2(x), G_1)}$$

$$= \pm\mathbf{Res}((x^3 + ax + b), G_0).$$

For the other expression, recalling that the zeros of y correspond to the roots of $x^3 + ax + b$, we write:

$$G(D_F) = \prod_{\alpha \text{ root of } x^3 + ax + b} G_0(\alpha)$$

$$= \mathbf{Res}((x^3 + ax + b), G_0).$$

This concludes the fourth case.

5. Finally, we need to address the most complicated case where $F(x, y) = F_0(x) + yF_1(x)$ and $G(x, y) = G_0(x) + yG_1(x)$, with F_0, F_1 coprime and G_0, G_1 coprime. For this case, it is useful to define $\mathcal{G}$ as above, $\mathcal{F}$ in the same way and to write the rational fraction F_1/G_1 in irreducible form as f_1/g_1. We then write:

$$G(D_F) = \prod_{\substack{(\alpha, y_\alpha) \text{ zero of } F \\ \text{(with order)}}} (G_0(\alpha) + y_\alpha G_1(\alpha))$$

$$= \prod_{(\alpha, y_\alpha)} G_0(\alpha) - F_0(\alpha)G_1(\alpha)/F_1(\alpha)$$

$$= \frac{\mathbf{Res}(\mathcal{F}, \Delta)}{\mathbf{Res}(\mathcal{F}, f_1)}.$$

where $\Delta = f_1 G_0 - F_0 g_1$.

By symmetry:

$$F(D_G) = \frac{\mathbf{Res}(\mathcal{G}, -\Delta)}{\mathbf{Res}(\mathcal{G}, g_1)}.$$

We can now prove that:

$$\mathbf{Res}(\mathcal{F}, \Delta) = \pm\mathbf{Res}(\Delta, \mathcal{F}) = \pm\mathbf{Res}(\Delta, F_0(x)^2 + (x^3 + ax + b)F_1(x)^2)$$

$$= \pm\mathbf{Res}(\Delta, F_0(x)^2 + (x^3 + ax + b)G_1(x)^2 \left(\frac{f_1}{g_1}\right)^2)$$

$$= \pm\mathbf{Res}(\mathcal{G}, \Delta) \cdot \left(\frac{\mathbf{Res}(\Delta, f_1)}{\mathbf{Res}(\Delta, g_1)}\right)^2.$$

We also have:

$$\mathbf{Res}(\mathcal{F}, f_1) = \pm\mathbf{Res}(f_1, \mathcal{F}) = \pm\mathbf{Res}(f_1, F_0(x)^2 + (x^3 + ax + b)F_1(x)^2)$$

$$= \pm\mathbf{Res}(f_1, F_0(x)^2 + (x^3 + ax + b)G_1(x)^2 \left(\frac{f_1}{g_1}\right)^2)$$

$$= \pm\left(\frac{\mathbf{Res}(f_1, g_1 F_0)}{\mathbf{Res}(f_1, g_1)}\right)^2 = \pm\left(\frac{\mathbf{Res}(f_1, \Delta)}{\mathbf{Res}(f_1, g_1)}\right)^2.$$

And by symmetry:

$$\mathbf{Res}(\mathcal{G}, g_1) = \pm\left(\frac{\mathbf{Res}(g_1, \Delta)}{\mathbf{Res}(g_1, f_1)}\right)^2.$$

Putting everything together concludes the final case.

$\square$

14.2.2   The Weil pairing on ℓ-torsion points

Using Weil's reciprocity, it is now possible to define the Weil pairing of two ℓ-torsion points P and Q on an elliptic curve E. Remember that a point P is said to be an ℓ-torsion point when $\ell P = O$ on the elliptic curve. In that case, $\ell(P) - \ell(O)$ is a principal divisor and we can express it as $\mathbf{div}(f_P)$. More generally, for any divisor D_P which sums to P, ℓD_P is principal. To define $e_\ell(P, Q)$ the Weil pairing of P and Q, we choose two arbitrary divisors D_P and D_Q with respective sums P and Q and with distinct supports. Then, we define the two functions f_P and f_Q such that $\mathbf{div}(f_P) = \ell D_P$ and $\mathbf{div}(f_Q) = \ell D_Q$. Finally, we define:

$$e_\ell(P, Q) = \frac{f_P(D_Q)}{f_Q(D_P)}. \tag{14.21}$$

THEOREM 14.3
The Weil pairing is a well-defined function of P and Q, i.e., it is independent of the choice of D_P and D_Q. It satisfies the following properties:

1. *For all ℓ-torsion points P and Q, $e_\ell(P, Q)$ is an ℓ-th root of unity, i.e., $e_\ell(P, Q)^\ell = 1$.*

2. For all ℓ-torsion points P and Q: $e_\ell(P, Q) = e_\ell(Q, P)^{-1}$.

3. In particular, for all ℓ-torsion point P: $e_\ell(P, P) = 1$.

4. For all ℓ-torsion points P, P' and Q: $e_\ell(P+P', Q) = e_\ell(P, Q) \cdot e_\ell(P', Q)$.

5. For all ℓ-torsion points P, Q and Q': $e_\ell(P, Q+Q') = e_\ell(P, Q) \cdot e_\ell(P, Q')$.

6. For all ℓ-torsion points P, Q and all integers a, b: $e_\ell(aP, bQ) = e_\ell(P, Q)^{ab}$.

7. The pairing e_ℓ is non-degenerate, i.e., there exists ℓ-torsion points P and Q, such that $e_\ell(P, Q) \neq 1$.

PROOF　　To show that e_ℓ is well defined, let us see that for two different choices of divisor summing to P, say D_P and D'_P we obtain the same value for $e_\ell(P, Q)$. Since, D_P and D'_P both sum to P, their difference is principal and we can write $D'_P = D_P + \mathbf{div}(h)$ for some function h. Thus:

$$f_Q(D'_P) = f_Q(D_P) \cdot f_Q(\mathbf{div}(h)).$$

If we let f_P and f'_P be the functions corresponding to ℓD_P and $\ell D'_P$, we see that $f'_P = h^\ell f_P$. Evaluating at D_Q, we find:

$$
\begin{aligned}
f'_P(D_Q) &= f_P(D_Q) \cdot h(D_Q)^\ell \\
&= f_P(D_Q) \cdot h(\ell D_Q) \\
&= f_P(D_Q) \cdot h(\mathbf{div}(f_Q)).
\end{aligned}
$$

Thanks to Weil's reciprocity, we know that $h(\mathbf{div}(f_Q)) = f_Q(\mathbf{div}(h))$. Thus, after dividing the numerator by the denominator, we find equality for our two expressions of $e_\ell(P, Q)$. Clearly, the same reasoning also applies for D_Q and Weil pairing does not depend on the chosen divisors, only on the points P and Q. A frequently encountered choice is to take:

$$D_P = (P) - (O) \text{ and}$$
$$D_Q = (Q + X) - (X) \text{ for an arbitrary point } X \neq P, -Q. \quad (14.22)$$

In order to verify that $e_\ell(P, Q)$ is an ℓ-th root of unity, let us look at the value of the numerator raised to the power ℓ:

$$f_P(D_Q)^\ell = f_P(\ell D_Q) = f_P(\mathbf{div}(f_Q)).$$

Thanks to Weil's reciprocity, we see that this is equal to the denominator $f_Q(D_P)^\ell$, as a consequence $e_\ell(P, Q)^\ell = 1$.

To check that $e_\ell(P+P', Q) = e_\ell(P, Q)e_\ell(P', Q)$, we first remark that $D_P + D'_P$ is a possible choice for the divisor $D_{P+P'}$. With this choice, we clearly have:

$$
\begin{aligned}
f_Q(D_{P+P'}) &= f_Q(D_P)f_Q(D_{P'}) \text{ and} \\
f_{P+P'} &= f_P f_{P'}.
\end{aligned}
\quad (14.23)
$$

Thus the property follows. The same argument also implies $e_\ell(P, Q + Q') = e_\ell(P, Q) e_\ell(P, Q')$.

The bilinearity $e_\ell(aP, bQ) = e_\ell(P, Q)^{ab}$ follows by induction for positive values of a and b. For $a = 0$, it suffices to check that $e_\ell(O, Q) = 1$. Similarly, for $b = 0$. For negative a and b, we first remark that $e_\ell(-P, Q) = e_\ell(P, -Q) = e_\ell(P_Q)^{-1}$ and conclude by induction.

For the proof of non-degeneracy, we refer the reader to [Sil86, Proposition 8.1]. □

In [Mil04], it is shown that by choosing normalized functions f_P and f_Q with highest degree coefficients equal to one, such that $\mathbf{div}(f_P) = \ell(P) - \ell(O)$ and $\mathbf{div}(f_Q) = \ell(Q) - \ell(O)$, then the computation of the Weil pairing $e_\ell(P, Q)$ can be simplified to $f_P(Q)/f_Q(P)$.

14.2.2.1 Miller's algorithm for the Weil pairing

In [Mil04], Miller describes a very useful algorithm generally known as Miller's algorithm for computing pairing or more precisely to evaluate a function f_P at a point Q. Note that using this algorithm is essential for constructive application of pairings. Trying to compute f_P itself rather than evaluating it on the fly is doomed because, even in factored form, storing this function is very costly.

Miller's algorithm considers intermediate functions, $f_P^{(i)}$ specified by:

$$\mathbf{div}(f_P^{(i)}) = (i)(P) - (iP) - (i-1)(P).$$

Note that the divisor on the right-hand side is principal. Remark that $\mathbf{div}(f_P^{(0)})$ and $\mathbf{div}(f_P^{(1)})$ are both empty. Thus, we can choose $f_P^{(0)} = f_P^{(1)} = 1$. Moreover, $f_P^{(\ell)} = f_P$.

We now remark that it is easy to compute $f_P^{(i+j)}$ from $f_P^{(i)}$ and $f_P^{(j)}$. Indeed:

$$\mathbf{div}(f_P^{(i+j)}) = \mathbf{div}(f_P^{(i)}) + \mathbf{div}(f_P^{(j)}) + (iP) + (jP) - ((i+j)P) - (O).$$

Moreover, following Section 14.1.1.3, we know that there exists a linear polynomial $\mathcal{L}(x, y)$ such that:

$$\mathbf{div}(\mathcal{L}(x, y)) = (iP) + (jP) + (-(i+j)P) - 3(O).$$

Moreover, if x_0 is the x coordinate of $(i+j)P$, we have:

$$\mathbf{div}(x - x_0) = ((i+j)P) + (-(i+j)P) - 2(O).$$

It follows that:

$$\mathbf{div}(f_P^{(i+j)}) = \mathbf{div}(f_P^{(i)}) + \mathbf{div}(f_P^{(j)}) + \mathbf{div}(\mathcal{L}(x, y)) - \mathbf{div}(x - x_0). \quad (14.24)$$

As a consequence, we can choose:

$$f_P^{(i+j)} = f_P^{(i)} \cdot f_P^{(j)} \cdot \frac{\mathcal{L}(x,y)}{x - x_0}. \tag{14.25}$$

Miller's method can be incorporated into any efficient algorithm for computing ℓP, such as the double and add method, see Algorithm 14.1. To optimize the efficiency of this algorithm, the reader should refer to [CF05].

Algorithm 14.1 Miller's algorithm with double and add

Require: Input integer $\ell > 0$, points P and Q of ℓ-torsion
Require: Finite field $\mathbb{F}_q$
  Write ℓ in binary $n = \sum_{i=0}^{k-1} \ell_i 2^i$
  Let $R \longleftarrow P$
  Let $y \longleftarrow 1$
  for i from $k-1$ down to 0 **do**
    Let $\mathcal{L}$ be the tangent line at R
    Let $R \longleftarrow 2R$
    Let $y \longleftarrow y^2 \cdot \mathcal{L}(Q)/(x_Q - x_R)$ in $\mathbb{F}_q$
    if $n_i = 1$ **then**
      Let $\mathcal{L}$ be the line through P and R
      Let $R \longleftarrow R + P$
      Let $y \longleftarrow y \cdot \mathcal{L}(Q)/(x_Q - x_R)$ in $\mathbb{F}_q$
    end if
  end for
  Output y {Value of $f_P(Q)$}

14.3   The elliptic curve factoring method

Another interesting use of elliptic curves in cryptography is the elliptic curve factoring method (ECM) which can be used to detect small enough factors of large integers. This method is closely related to Pollard's $p - 1$ factoring algorithm. Thus, for simplicity, we start by describing this method.

14.3.1   Pollard's $p - 1$ factoring

Pollard's $p - 1$ factoring algorithm can efficiently find factors of a special form in composite numbers. Its name comes from the fact that a factor p can be efficiently found with this method when $p - 1$ is a product of small enough primes. Here, small enough means that all prime factors of $p-1$ are smaller than some bound B specified in advance. Alternatively, we say that $p-1$ is B-smooth. Given p and B, let us define:

$$P(p, B) = \prod_{q \text{ prime}, \ q \leq B} q^{\left\lceil \frac{\log(p)}{\log(q)} \right\rceil}. \tag{14.26}$$

This product is constructed in a way that guarantees that $p-1$ is B-smooth if and only if $p-1$ divides $P(p, B)$.

As a consequence, when $p-1$ is B-smooth, for any invertible element x modulo p, we have $x^{P(p,B)} = 1 \pmod{p}$. Thus, if a composite N has a factor p with a B-smooth value of $p-1$, we see that p divides the GCD of N and $(x^{P(p,B)} - 1) \pmod{N}$ for any invertible x modulo N. Moreover, this GCD is not N itself, except when all factors of N satisfy the $p-1$ property. This last case is quite unlikely. Moreover, from a practical point-of-view, it is better to work with a sequence of exponents $P(p, B')$ for values of B' that increase up to B. This practical adaptation makes the exceptional bad case even rarer. In pseudo-code, this is described as Algorithm 14.2. In the code, since p is not known in advance, we upper bound $P(p, B)$ by $P(N, B)$.

Algorithm 14.2 Pollard's $p-1$ factoring algorithm

Require: Input composite N and smoothness bound B
  Choose random element x modulo N
  if $\text{GCD}(x, N) \neq 1$ **then**
    Print '$\text{GCD}(x, N)$ is a factor' and Exit
  end if
  for All primes q from 2 to B **do**
    Let $Q \longleftarrow q^{\left\lceil \frac{\log(N)}{\log(q)} \right\rceil}$
    Let $x \longleftarrow x^Q \bmod N$
    if $\text{GCD}(x - 1, N) \neq 1$ **then**
      Print '$\text{GCD}(x - 1, N)$ is a factor' and Exit
    end if
  end for
  Print 'No factor found'

14.3.2 Elliptic curve factoring

To generalize Pollard's $p-1$ to elliptic curves, it is useful to study the behavior of the group law of an elliptic curve E when considered modulo a composite integer N. Assume that we are given points modulo N on the elliptic curve and that we try to apply the usual additions formulas to these points. We would like to look at the behavior that these formulas induced modulo a factor p of N. Since the formulas for addition are algebraic expressions, it is clear that when we apply an addition formula modulo N, we are in fact applying the same formula modulo p. As a consequence, adding two points modulo N usually induces an addition of the same point modulo p. Indeed, in most cases, we need to use the same formula modulo each factor of N. However, there are some exceptional cases where the point addition would require different formulas modulo the prime factors. Typically, if neither of the points P and Q to be added is the point at infinity, this means that modulo a factor q the two points P and Q are neither equal nor opposite and modulo another factor p they are equal or opposite, i.e., $x_P = x_Q \pmod{p}$. When the points are considered modulo N, it is clear that $x_P \neq x_Q \pmod{N}$, thus we use the formula for unrelated points and start by computing $\lambda = (y_Q - y_P)/(x_Q - x_P)$. However, since p divides $x_Q - x_P$, it is not possible to invert this value modulo N and λ is not defined. The good news is that when trying to compute the inverse of $x_Q - x_P$ modulo N using the extended Euclidean algorithm, we are going to detect the problem and to discover the factor p of N. Similarly, when trying to double a point P with $y_P = 0 \pmod{p}$, the doubling rule tries to invert y and discovers p.

From this preliminary discussion, we see that most of the time, it is possible to add points and double points by computing the classical addition formulas modulo N. Moreover, when this fails, it yields a factor of N. In truth, it is even possible to define an addition law on the curve modulo N that works in all cases. However, this is not required for ECM since the goal is to factor N. As a consequence, a failure of the addition formula becomes our goal. Putting the cases of failure together, we see that addition/doubling fails when the result is the point at infinity modulo some factor of N but not modulo N. Thus, the question becomes: "How can we reach the point at infinity modulo some factor p of N using point addition?"

Since an elliptic curve modulo p is a group, we know that multiplying any point on the curve by the cardinality of the group yields the point at infinity. Thus, in a nutshell the ECM algorithm works by choosing an elliptic curve modulo N, a point P on the curve and by multiplying P by a multiple of the cardinality of the curve when taken modulo a factor p of N. At first, this idea may seem unworkable. A first obstruction is that given an elliptic curve E modulo N, finding a point P on E seems to be a difficult problem, indeed if we fix the x coordinate, we need to compute the square root of $x^3 + ax + b$, a hard problem modulo a composite. Similarly, if we fix y, we need to solve a polynomial equation of degree 3 modulo N, which we do not know how to achieve in general. A second obstruction is that there is no known algorithm to compute the cardinality of E modulo N or modulo a factor of N.

To remove the first obstruction, the key argument is to remark that we are free to choose any convenient curve E. Thus, in practice, instead of choosing E first, we may decide to first choose a point P and then to take a curve E going through P. An even simpler alternative is to choose curves of the form $y^2 = x^3 + ax + 1$ and remark that the point $P = (0, 1)$ belongs to all these curves. Moreover, since the parameter a can still be freely chosen, there are enough possible choices of curve that remain.

The second obstruction says that we do not know how to determine a good multiplier k to ensure that kP goes to infinity modulo a factor p of N. Instead, we are going to choose multipliers that maximizes the probability of going to infinity. The core remark is that when the number of points k on E modulo p is a product of primes smaller than some bound B, then we can choose for k as in Pollard's $p-1$. More precisely, we can take for k a value $P(\cdot, B)$, as we did in Algorithm 14.2. The missing parameter in this expression needs to be replaced by an upper bound on the number of points of an elliptic curve modulo the smallest prime factor of N (see complexity analysis below).

14.3.2.1 Complexity analysis

In order to minimize the average running time of the ECM algorithm, we assume that we are given a (tight) upper bound on p, say P. From this bound, we can derive an upper bound $\tilde{P} = P + \lceil 2\sqrt{P} \rceil + 1$ on the number of points of an elliptic curve modulo p. We need to find the best value for the smoothness bound B. The cost of the ECM is the product of two contributions, the number of curves to try times the individual cost of each trial. On each curve, we compute a multiplication by $k = P(\tilde{P}, B)$ for a cost of $O(B \log(\tilde{P}))$ arithmetic operations. Moreover, the expected number of curve to try is the inverse of the probability that the number of points on the curve modulo p is B-smooth.

Plugging in the smoothness probabilities given in Chapter 15 and ignoring the logarithmic factors, we can optimize the complexity by balancing the individual cost of each trial and the number of trials. Thus, we need to achieve:

$$B \approx e^{\frac{\log \tilde{P}}{\log B} \cdot \log\left(\frac{\log \tilde{P}}{\log B}\right)}.$$

Choosing B of the form:

$$B = e^{\alpha \sqrt{\log \tilde{P} \log \log \tilde{P}}},$$

we can achieve the balance when $\alpha = 1/\sqrt{2}$. The expected runtime is $B^{2+o(1)}$. Moreover, we can forget the distinction between $\tilde{P}$ and p without affecting the expression of the complexity. For this reason, the complexity of ECM is usually given as:

$$e^{\sqrt{2 \log p \log \log p}},$$

where p is the smallest factor of N.

The practical efficiency of ECM can be improved by adding a second phase in the algorithm, often called ECM Step 2. This phase allows the algorithm to succeed when the cardinality of the chosen curve has one prime factor larger than the bound B, but not too large. This is discussed in details in [Mon92].

Exercises

1^h. Implement elliptic curve operations over a small finite field, say $\mathbb{F}_{65521}$. Do experiments with the group operation, check that it is indeed a group, compute the order of points for several curves. Determine the group structure of each curve. It is also interesting to consider a singular curve, i.e., a curve with $4a^3 + 27b^2 = 0$. Can you use the addition formulas in this case? Try to find out whether there is a group and, if so, what its structure is.

2. Write a program to work with small functions of the form $F_0(x) + yF_1(y)$ on an elliptic curve over a small finite field. This should compute the zeros of a function and be able to evaluate a function at a point. Experiment Weil's reciprocity with your program.

3. Assuming that a divisor $(P_1) - (P_2)$ is principal, show that you can write it as:

$$(P_1) - (P_2) = \mathbf{div}\frac{F_1(x) + yF_2(x)}{G(x)}.$$

Show that this expression can be simplified until G is a polynomial of degree at most 1. Prove that in that case, $F_2 = 0$ and conclude.

4^h. Program a simple implementation of Miller's algorithm to compute $f_P(Q)$ for a well-chosen elliptic curve and some well-chosen points. Try to write explicitly f_P in expanded form as $F_0(x) + yF_1(x)$ on a computer algebra system. For which size is this approach successful?

5. Find out how to compute the order of a point on an elliptic curve using a birthday algorithm. Can you do it without using a large amount of memory?

6^h. Assume that you are given an elliptic curve together with a pairing $e_\ell(\cdot, \cdot)$ from $\mathbb{G}_1 \times \mathbb{G}_2$ to the group of ℓ-th roots of unity in a finite field $\mathbb{F}_{p^k}$. Each of the groups $\mathbb{G}_1$ and $\mathbb{G}_2$ is a subgroup of ℓ-torsion points containing exactly ℓ element. Let Q be a point in $\mathbb{G}_2$ and assume that there exists an efficient algorithm e^{-1} that given a ℓ-th root of unity ξ outputs a point P in $\mathbb{G}_1$ such that $e_\ell(P, Q) = \xi$. Further assume that $\ell - 1$ is a product of small primes, show that there exists an efficient algorithm that makes use of e_ℓ and e_ℓ^{-1} to compute discrete logarithms in $\mathbb{G}_1$. How would you proceed for general values of ℓ?

7. Let $N = pq$ be an RSA number and E be an elliptic curve. Consider E modulo N. Show using the Chinese remainder theorem that this forms a group and determine the full structure of this group. There are some "exceptional" points on this curve, what are they?

8^{h}. Now consider the curve E modulo p^2 where p is a prime.

- Determine the structure of the curve in the general case. Show that the exceptional points (as in the previous exercise) form a group. Determine the order of this group. Write the group law on the exceptional points. Can you simplify the expression of this group law?

- In the special case where the cardinality of the curve is p, what are the possible group structures. Given a point P, not the point at infinity, on the curve modulo p, show that there are p different points modulo p^2 whose reduction modulo p is equal to P. Let P_1 and P_2 be two such lifts of P, what can you say about pP_1 and pP_2?

- In the same special case, let P and Q be two points on the elliptic curve modulo E. We would like to solve the discrete logarithm problem $Q = \lambda P$. Let P_1 be a lift of P and Q_1 a lift of Q, what can you say about the relation between pP_1 and pQ_1? Under which condition can you break the discrete logarithm problem?

- Sometimes the above condition is not satisfied. How can you proceed to make the method work despite that?

Chapter 15

Index calculus algorithms

15.1 Introduction to index calculus

Index calculus is a rich area of research in number theory and cryptography. This technique is common to a large family of algorithms that can be used for many applications. From a number theoretic point-of-view, these algorithms allow to factor large integers, to compute discrete logarithms in finite fields, to compute the class numbers of number field and even apply to discrete logarithms on the jacobians of some algebraic curves. From a cryptographic perspective, the above applications are directly interesting but in addition, index calculus techniques have recently been used to undermine the security of several cryptographic primitives in some specific contexts. With these cryptographic primitives which, at the present time, include the plain RSA e-th root extraction problem [JNT07] and the static Diffie-Hellman problem, an adversary learns enough to solve the problem on his own after asking a large number of questions about well-chosen instances of the problem. The total cost of index calculus in these cryptographic applications is less than the cost of the corresponding index calculus algorithm against the underlying number theoretic problem: factoring or computation of discrete logarithms.

To tackle this large scope, we are going to dive into index calculus algorithms carefully. Indeed, some of these algorithms require much more mathematical machinery than others. Thus, to better focus on the key ideas, we mostly study a typical but simple example about the computation of discrete logarithms in finite fields with a well-chosen structure.

Before that, let us have a quick bird's-eye view of the general principle of index calculus algorithms. This principle puts two basic ideas into play. First, in order to learn information about a mathematical structure A, we enlarge this structure in two different but compatible ways. More precisely, we try to construct a commutative diagram:

$$
\begin{array}{ccc}
 & \mathcal{A} & \\
{}^{\psi_1}\swarrow & & \searrow^{\psi_2} \\
\mathcal{A}_1 & & \mathcal{A}_2 \\
{}_{\phi_1}\searrow & & \swarrow_{\phi_2} \\
 & A &
\end{array}
\tag{15.1}
$$

With such a diagram, any element x in $\mathcal{A}$ can be sent into A along two different paths, either going through $\mathcal{A}_1$ or through $\mathcal{A}_2$. In both cases, we reach the same value, i.e., $\phi_1(\psi_1(x)) = \phi_2(\psi_2(x))$.

The second idea is to choose $\mathcal{A}_1$ and $\mathcal{A}_2$ in a way that allows a non-negligible fraction of elements of either set to be written as a product of mathematical objects belonging to reasonably small sets. The values which can be written in this way are usually called **smooth**. And the sets used to write the products are traditionally called the **smoothness bases** for $\mathcal{A}_1$ and $\mathcal{A}_2$. Combining the two ideas means that if both $\psi_1(x)$ and $\psi_2(x)$ can be written as products over their respective smoothness bases, then we get an "equality" between these two products. Note that for some index calculus algorithms, some functions among the ψ_i or ϕ_i are either the identity function or some other natural map and the above diagram can be simplified, by removing some of the arrows or sets that are involved.

In this setting, index calculus algorithms create and collect many "equalities." Once these equalities are constructed, they need to be combined in specific ways, in order to solve the problem at hand. The way of combining the equalities greatly vary from one problem to the next, as we show in the sequel. However, linear algebra always plays a role in this process.

In addition to the basic structure described above, index calculus algorithms also have their asymptotic complexities in common. Indeed, they all have complexities expressed as

$$L_N(\alpha, c) = \exp\left((c + o(1))\log(N)^\alpha \log\log(N)^{1-\alpha}\right), \qquad (15.2)$$

where N is the number that characterizes the problem being considered and where $0 < \alpha < 1$ and $c > 0$ are two constants. The notation $L_N(\alpha)$ is frequently used as a shorthand when c is left unspecified. Moreover, when N is clear from the context, we simply write $L(\alpha)$. These strange looking functions arise from the probability for values to be smooth. Two values of α are widely encountered in index calculus algorithms, they usually have complexity either $L(1/2)$ of $L(1/3)$. The reader should be aware that a change from $\alpha = 1/2$ to $\alpha = 1/3$ means much more in terms of complexity than a similar change for the value of the second parameter c. Indeed, complexity of the type $L(\alpha)$ becomes polynomial in $\log(N)$ when α tends to zero and exponential in $\log(N)$ when α tends to one. Between these two extreme values of α, the complexity is called subexponential. Thus, α controls the change from polynomial to exponential, and it clearly has much more impact than c which, in the polynomial case, merely controls the degree of polynomial. Note that $\log(N)$ is the bitsize of the value N which characterizes the problem and thus the relevant parameter to use to define polynomial or exponential complexity.

15.2 A simple finite field example

15.2.1 Overview

To describe in details the principle of index calculus, without having to tackle too many mathematical obstructions, we start by considering the problem of computing discrete logarithms in a finite field $\mathbb{F}_{p^n}$, where the relative values of p and n are chosen to make things as simple as possible. More precisely, we let Q denote p^n and consider the case where:

$$p = L_Q(1/3, \theta) \quad \text{and} \tag{15.3}$$

$$n = \frac{\log(Q)}{\log(p)} = \frac{1}{\theta} \cdot \left(\frac{\log(Q)}{\log \log(Q)} \right)^{2/3},$$

where θ is a constant to be determined later on. We let $d = \lceil \sqrt{n} \rceil$ and choose two polynomials of degree d: f_1 and f_2. These two polynomials are used to implicitly define the finite field $\mathbb{F}_Q$. The idea is to fix two polynomial relations modulo p between two variables x and y:

$$y = f_1(x) \quad \text{and} \quad x = f_2(y). \tag{15.4}$$

For these two relations to hold, we need to have $x = f_2(f_1 x)$. Thus, x needs to be a root of $f_2(f_1(x)) - x$. Assume that this polynomial, whose degree is $d^2 \geq n$ has an irreducible factor of degree n and denote this factor by I_x. Then I_x can be used to define the finite field $\mathbb{F}_Q$. Let α denote the image of x in the finite field using the canonical projection, i.e., α is the class of the polynomial x modulo I_x (and of course modulo p). Let $\beta = f_1(\alpha)$, then in the finite field, we also have $\alpha = f_2(\beta)$. Indeed, by construction, $f_2(\beta) = f_2(f_1(\alpha))$ is equal to α plus a multiple of $I_x(\alpha)$. And, by definition of the finite field, $I_x(\alpha) = 0$ in $\mathbb{F}_Q$.

The element β of the finite field has a characteristic polynomial I_y of degree n. Indeed, the degree is at most n, since β belongs to $\mathbb{F}_{p^n}$, and if the degree is less than n, then β belongs to a proper subfield and so does $\alpha = f_2(\beta)$. Since by definition α generates $\mathbb{F}_Q$, this is not possible. Thanks to the relation $\beta = f_1(f_2(\beta))$, we see that I_y is a divisor of degree n of the polynomial $f_1(f_2(y)) - y$ modulo p.

As a consequence of the existence of I_x and I_y, we have two different representations of the finite field $\mathbb{F}_Q$, one with generator α and the other with generator β, together with explicit low degree isomorphisms between the two representations, given by $\beta = f_1(\alpha)$ and $\alpha = f_2(\beta)$. Putting these two representations together, we obtain the following commutative diagram:

$$
\begin{array}{ccc}
 & \mathbb{F}_p[x,y] & \\
\nearrow{\scriptstyle y\to f_1(x)} & & \searrow{\scriptstyle x\to f_2(y)} \\
\mathbb{F}_p[x] & & \mathbb{F}_p[y] \\
\searrow{\scriptstyle x\to\alpha} & & \nearrow{\scriptstyle y\to\beta} \\
 & \mathbb{F}_Q &
\end{array}
\qquad (15.5)
$$

Using this commutative diagram, we now take polynomials of the form $xy + ax + by + c$ for arbitrary values of a, b and c in $\mathbb{F}_p$. Each polynomial of this form is transformed into a univariate polynomial in x of degree $d + 1$ when we substitute $f_1(x)$ for y and into a univariate polynomial in y of degree $d + 1$ when we substitute $f_2(y)$ for x. Thus, we obtain the following equality in the finite field $\mathbb{F}_q$:

$$
\alpha f_1(\alpha) + a\alpha + bf_1(\alpha) + c = \beta f_2(\beta) + af_2(\beta) + b\beta + c. \qquad (15.6)
$$

We only keep a fraction of these equalities. More precisely, we focus on the equations where both sides factor into linear polynomials. To obtain these equations, at least three different approaches can be considered. The first method is to simply factor $L(x) = xf_1(x) + ax + bf_1(x) + c$ and $R(y) = yf_2(y) + af_2(y) + by + c$ and to check that both only contain factors of degree 1. Recalling the algorithm for factoring polynomials over a finite, this can be slightly improved by checking that $x^p \equiv x \pmod{L(x)R(x)}$, in $\mathbb{F}_p[x]$. The second method is to use a sieving algorithm as in Chapter 4. The third method is to use Bernstein's algorithm for avoiding sieving, also presented in Chapter 4. For implementations of the methods, please refer to the book's website.

Each equation left after selection can be rewritten as a multiplicative equality between $d+1$ linear polynomials $\alpha - u$ for constants u in $\mathbb{F}_p$ and $d+1$ linear polynomials $\beta - v$ for constants v in $\mathbb{F}_p$. Given any multiplicative generator γ of $\mathbb{F}_Q$, the multiplicative equalities can be transformed into additive relations between the logarithms in base γ. As a consequence, each equation becomes:

$$
\sum_{i=1}^{d+1} \log_\gamma(\alpha - u_i) \equiv \sum_{i=1}^{d+1} \log_\gamma(\beta - v_i) \pmod{Q - 1}, \qquad (15.7)
$$

where the modulus is the order of γ. Clearly, the total number of unknowns, i.e., of discrete logarithm values, to determine is $2p$. Thus, given $2p$ equations, we hope to be able to obtain the logarithm values. However, there is a simple obstruction that prevents this method from working directly. This obstruction stems from the fact that our system of equation possesses a parasitical solution. Indeed, each equation has $d + 1$ unknowns on each side. As a consequence, setting every unknown to 1 gives a solution to the system of equations. To remove this parasitical solution, it suffices to find a single equation of a different type. This can be done by considering the polynomial

$x + a$ for a value of a such that $f_2(y) + a$ splits into linear factors. This yields an equation with a single unknown on the left-hand side and d unknowns on the right-hand side. Once this is done, there are no more evident parasitical solutions. Moreover, in practice, with a few extra equations thrown in as a precaution, the system of equations has a kernel of dimension 1. As a consequence, up to an arbitrary multiplicative factor, the system has a unique solution. In fact, to each different choice of the logarithm's base γ corresponds a different choice of a multiplicative constant. Note that trying to solve the linear system modulo $Q - 1$ is often a bad idea. The problem is that $Q - 1$ may contain small factors, which can lead to trouble when the linear algebra is done using an iterative algorithm. Instead, we should, if possible, factor $Q - 1$ and work modulo each factor, solving the above system for large factors and using a baby step, giant step or another birthday paradox based algorithm for the small factors. Note that we already know that $p - 1$ is a factor of $Q - 1$. When $Q - 1$ is hard to factor, we can still obtain the logarithm by using an iterative algorithm to solve the linear system modulo composite factors of $Q - 1$ after eliminating the small factors of $Q - 1$.

15.2.1.1 Individual logarithms

After solving the linear system, we obtain the discrete logarithms of all the values $\alpha - u$ or $\beta - v$ in the finite field. However, the work is not complete. Indeed, we would like to be able to compute the discrete logarithms of arbitrary values in $\mathbb{F}_Q$. The key to this is to use a descent algorithm, where the discrete logarithm of an arbitrary polynomial is expressed in terms of discrete logarithms of polynomials with smaller degrees. This idea is used iteratively, until we reach the point where everything is expressed as a function of discrete logarithms of linear polynomials, which are already known. To express the logarithm of a polynomial $q(x)$ of degree d_q on the left-hand side in terms of polynomials of lower degree, we consider bivariate polynomials $T(x, y)$ of degree t, separately in each variable, such that $q(x)$ divides the left-hand side projection $T(x, f_1(x))$. Constructing these polynomials can be done using linear algebra modulo p. More precisely, we build a matrix whose columns are indexed by monomial $x^{d_x} y^{d_y}$ with $0 \le d_x \le t$ and $0 \le d_y \le t$. Each of the $(t + 1)^2$ columns gives the representation of the projection $x^{d_x} f_1(x)^{d_y}$ (mod $q(x)$) of the corresponding monomial. In the first row, we find the constant coefficients of these polynomials, in the second row the x coefficient and so on. Since $q(x)$ has degree d_q, there are d_q rows in the matrix. Now, an element of the kernel of the matrix can be interpreted as a polynomial in x and y whose projection is equal to 0 modulo q. Reciprocally, each polynomial $T(x, y)$ of degree at most t in x and y, whose projection is a multiple of q, belongs to the kernel of the matrix. As a consequence, to enumerate the polynomials $T(x, y)$ it suffices to compute a basis of the kernel and to look at all the possible linear combinations of the kernel elements.

Polynomials $q(y)$ on the right-hand side are treated in the same way, reversing the roles of x and y. Note that during the descent algorithm, we alternatively encounter both cases.

At each step, we find a relation between a polynomial q and several polynomials of lower degree. As a consequence, the descent produces a tree of computation. The root of the tree is the original values whose discrete logarithm is desired. The leaf of the tree are the linear polynomials. To make sure that the descent is feasible, we need to ensure that each step of the descent can be performed quickly enough and that the tree of descent has a reasonable number of nodes. To satisfy this condition, we need to carefully balance the speed of the descent. If we try to descend too quickly, then each individual step is too costly. If we descend too slowing, the size of the tree becomes too large and the complete computation cannot be done. The complexity analysis below shows how to balance the speed of the descent adequately.

15.2.1.2 Complexity analysis

To perform the complexity analysis, we need to know the probability for a polynomial of degree Δ to split into irreducible factors of degree δ. In fact, in the early steps of the computation, we only need the case where $\delta = 1$. With this case, the analysis is quite simple, the total number of (monic) polynomials of degree Δ is p^Δ, while the number of polynomials that split is approximately $p^\Delta/\Delta!$. Thus, the probability that a given polynomial splits is $1/\Delta!$. For the analysis, it is convenient to write that the logarithm in base p of the probability is close to $-\Delta \log_p(\Delta)$. We discuss smoothness probabilities in more details in Section 15.5.

For arbitrary δ, the result generalizes into the following theorem from [PGF98].

THEOREM 15.1
Let P be a random polynomial of degree Δ over a finite field $\mathbb{F}_p$. Let $pr(\Delta, \delta)$ be the probability that P factors into a product of irreducible polynomials of degree at most δ.
Then:

$$- \log_p(pr(\Delta, \delta)) \approx -(\Delta/\delta) \log_p(\Delta/\delta)$$

For completeness, we prove the lower bound part of this theorem in Section 15.5. Indeed, such a lower bound suffices for the complexity analysis.

In our analysis, as in many index calculus algorithms, we make an essential heuristic hypothesis. When we generate the equations, we simultaneously consider two related polynomials $L(x) = xf_1(x) + ax + bf_1(x) + c$ and $R(y) = yf_2(y) + af_2(y) + by + c$, and ask for both to be smooth. When the two polynomials are related in this fashion, it is not known how to analyze the probability for both to factor in the right way. Instead, we make the heuristic assumption that $L(x)$ and $R(y)$ behave as random independent polynomials in this respect. We need a similar hypothesis during the descent phase.

With this hypothesis, the probability that both $L(x)$ and $R(y)$ split is $((d + 1)!)^{-2}$ since both have degree $d + 1$. Taking the logarithm, we find $2(d + 1) \log(d + 1)$. Another way to state the heuristic is to say that $L(x)R(x)$ behaves as a random polynomial with respect to the splitting property. This yields a probability $1/(2d+2)!$ and taking logarithm we find $2(d+1) \log(2(d + 1))$. We can check asymptotically that both expressions are equivalent and can be written as $2d \log(d)(1 + o(1))$.

Complexity of finding and solving the linear system of equations.
Here, there are two important parameters to take into account:

$$p = L_Q(1/3, \theta) \quad \text{and} \quad d = \frac{1}{\sqrt{\theta}} \cdot \left(\frac{\log(Q)}{\log \log(Q)} \right)^{1/3}. \tag{15.8}$$

We have a total of $2p$ unknowns and thus need to generate a little more than $2p$ equations. The linear algebra step essentially costs $(2p)^2$ operations. While generating the equations, we have three degrees of freedom a, b and c. Each triple (a, b, c) yields an equation with probability $(1/d!)^2$. We can hope to generate enough equations when:

$$\frac{p^3}{d!^2} \geq 2p \quad \text{or} \quad \frac{p^2}{2} \geq d!^2. \tag{15.9}$$

Note that when $d!^2$ is close to the upper bound $p^2/2$ the generation of the equations step costs roughly p^3 operations, while the linear algebra costs $4p^2$. In the opposite direction, when $d!^2 < p$ the linear algebra step costs more than the generation of the equations. The two steps are balanced when $4p \approx d!^2$. Let us analyze this middle case. Taking the logarithm of the condition we find:

$$\log(p) \approx 2d \log(d) \quad \text{i.e.}$$
$$\theta \log(Q)^{1/3} \log \log(Q)^{2/3} \approx \frac{2}{3\sqrt{\theta}} \log(Q)^{1/3} \log \log Q^{2/3}, \tag{15.10}$$

since $\log(d) \approx \log(Q)/3$. Thus, the balance condition becomes:

$$\theta = \frac{2}{3\sqrt{\theta}} \quad \text{or} \quad \theta = \left(\frac{2}{3} \right)^{\frac{2}{3}} = \sqrt[3]{4/9}. \tag{15.11}$$

Since in this middle case the complexity of generating and solving the system of equations is approximately p^2, it can be written as:

$$L_Q(1/3, 2\theta) = L_Q\left(1/3, \sqrt[3]{32/9} \right).$$

When p is larger than this value, the complexity is dominated by the linear algebra and as a result costs more than in the balance case. When p is smaller and still conforms to Equation (15.9), the complexity slowly decreases, at the limit, we find $p^2/2 = d!^2$ which corresponds to:

$$2\theta = \frac{2}{3\sqrt{\theta}} \text{ or } \theta = \left(\frac{1}{3}\right)^{\frac{2}{3}} = \sqrt[3]{1/9}. \tag{15.12}$$

In this case, the complexity is dominated by the generation of the equations which needs to consider p^3 candidate triples. Thus, the total cost is $L_Q(1/3, 3/9^{1/3}) = L_Q(1/3, 3^{1/3})$.

Complexity of individual discrete logarithms To determine the complexity of the individual discrete logarithms, we need a detailed analysis of the descent parameters. Assume that we are given a polynomial[1] $q(x)$. At most the degree d_q of q is $n-1$ since q represents an element of the finite field. We use a descent step to express q in terms of polynomials of smaller degree. Remember that this is done by searching for a bivariate polynomial $T(x, y)$ of degree t, in each variable, such that $q(x)|T(x, f_1(x))$. Without the divisibility condition, there are $(t+1)^2 - 1$ degrees of freedom that determine[2] T. The divisibility condition adds d_q linear constraints. Thus, the total number of polynomials we can consider is:

$$N_T = p^{(t+1)^2 - d_q - 1}. \tag{15.13}$$

We want both $T(x, f_1(x))$ and $T(f_2(y), y)$ to split into small enough factors. These two polynomials have degree td, their probability to simultaneously split into factors of degree t' has a logarithm approximately equal to $-2(td/t') \log(td/t')$. For the descent to be effective, each step should not cost too much. It is tempting to try making the probability comparable to the probability encountered while generating the main set of equations. Making this choice, we now need to make sure that we can really find an equation during this step of the descent. This means that we need to check that the number of polynomials N_T is larger than the inverse of the probability, i.e.:

$$(t+1)^2 - d_q - 1 \geq 2(td/t') \log_p(td/t') = 2d \log_p(d) \approx \frac{2}{3\theta^{3/2}}. \tag{15.14}$$

Thus, $t+1$ can be chosen as the integer obtained by rounding up the square root of d_q plus some constant $\mu = 1 + \frac{2}{3\theta^{3/2}}$. Clearly, with this choice of parameters for the descent, each individual step does not cost too much, since its cost is comparable to the cost of computing a single equation during the initial construction of the system of equation. Moreover, the tree of descent has a small height. Indeed, we start from $d_q^1 = n$ and at step i we go down from $d_q^{(i)}$

---

[1] Or $q(y)$, both cases are completely symmetric.
[2] We subtract 1 from $(t+1)^2$ because multiplying T by a constant does not change the degrees of its factors. This remark removes one degree of freedom.

to $d_q^{(i_1)} = \lceil \sqrt{d_q^{(i)} + \mu} \rceil$. This means that $\log(d_q)$ roughly follows a geometric progression of ratio $1/2$ and that we have a height bounded by $O(\log\log(n))$. As a consequence, the descent tree contains a polynomial number of nodes and does not impact the constants in the subexponential complexity.

Since $\theta \geq 3^{-1/3}$ according to Section 15.2.1.2, we have $\mu < 1.39$, and the above analysis allows us to descent until we reach $d_q = 2$. Indeed, for $d_q = 3$ we choose $t + 1 = \lceil \sqrt{3 + \mu} \rceil = 3$ or $t = 2$ and we split the polynomials into factors of degree 2. Since our factor basis only contains polynomials of degree 1, we need to analyze separately an extra step to descent from degree 2 to degree 1. In this final step of descent, we have $d_q = 2$. The first option is to consider polynomials $T(x, y) = xy + ax + by + c$ for triples (a, b, c) of coefficients in $\mathbb{F}_p$. Since we have two linear conditions to make sure that the degree 2 polynomial divides either $T(x, f_1(x))$ or $T(f_2(y), y)$, there are p possible choices for T. After substitution, the degree on each side is $d + 1$ and it can be reduced to $d - 1$ on one side by removing the known factor of degree 2. The probability of having two polynomials that split into degree one is:

$$\frac{1}{(d-1)!(d+1)!}.$$

Asymptotically, this is $L_Q(1/3, -\frac{2}{3\sqrt{\theta}})$. We can hope to find a relation if this probability is at least $1/p$, i.e., when $\theta \geq (2/3)^{2/3}$. This corresponds either to the balanced case of Section 15.2.1.2 or to large prime p.

When p is a smaller prime, we need to consider a second option and to look at unbalanced polynomials[3], $T(x, y) = x^2 y + axy + bx^2 + cx + dy + e$. Removing the degrees of freedom due to the two linear conditions, we are left with p^3 possible polynomials. The probability of splitting correctly is:

$$\frac{1}{(2d-1)!(d+1)!}.$$

Asymptotically, this is $L_Q(1/3, -\frac{1}{\sqrt{\theta}})$. In this case, there is enough freedom to find a relation if this probability is at least $1/p^3$, i.e., when $\theta \geq (1/3)^{2/3}$. As a consequence, we can asymptotically address all the cases of Section 15.2.1.2.

However, the reader should note that for practical computations, this option of a smaller prime should be handled with care. The difficulty stems from the fact that for moderate values of d, such as, for example, $d = 7$, there is a considerable gap between $d!^2$ and $(2d)!$. Thus for some practical choices of parameters, we may encounter discrete logarithm computations where the initial system of linear equations can be constructed and solved reasonably quickly, while the final step of the descent from degree 2 to degree 1 remains unpractical. For example, when choosing $p = 65537$ and $n = 49$, i.e., $d = 7$

---

[3]We assume here that the degree 2 polynomial we are trying to express is on the x side. Otherwise, we need to reverse the roles of x an y.

we run into this exact problem. It is possible to perform the early steps reasonably easily on today's computers and obtain the discrete logarithms of basis elements; moreover, the early steps of the descent down to polynomials of degree 2 do not present any special difficulty. However, the last descent step from polynomials of degree 2 to linear polynomials does not work, because we do not have enough degrees of freedom and we cannot conclude the discrete logarithms computation.

15.2.2 A toy example

In this section, we illustrate the above method by computing discrete logarithm in $\mathbb{F}_{101^4}$. Let us define $p = 101$ and $n = 4$ together with the two polynomials:

$$f_1(x) = x^2 + 1 \quad \text{and} \quad f_2(y) = y^2 + 2. \tag{15.15}$$

Then the polynomial $f_2(f_1(x)) - x = x^4 + 2x^2 - x + 3$ is irreducible modulo p and so is the polynomial $f_1(f_2(y)) - y = y^4 + 4y^2 - y + 5$. Thus, the two relations $y = f_1(x)$ and $x = f_2(y)$ implicitly define the finite field $\mathbb{F}_{101^4}$.

With this definition for f_1 and f_2, we now search for smooth multiplicative relations involving the two finite field representation. Before considering polynomials of the form $xy + ax + by + c$, we first look at polynomials $x + a$, $y + a$ and $x + ay + b$. The advantage of these polynomials is that, since they give relations between polynomials in x and y of lower degree, we expect a better probability of smoothness. In addition, having equations of this type in our system removes the all '1' parasitical solution from the linear system we are constructing.

In our example, we obtain 51 equations using polynomials of the form $x + a$ and also 51 equations from polynomials $y + a$. This can be easily interpreted by remarking that $x + a$ is always expressed as a linear term on the x side and that on the y side we are looking at the factorization of $y^2 + 2 + a$. This polynomial splits into linear factors if and only if $-(a + 2)$ is either 0 or a quadratic residue. Since in $\mathbb{F}_{101}$ there are 50 quadratic residues and 50. In addition, we only need another hundred equations and they are obtained by considering polynomials $x + ay + b$ with $a \neq 0$. In fact, it suffices to consider values of a in $[1 \cdots 5]$ and test all values of b to obtain enough equations. Typical equations are:

$$x + 7 = (y + 30) \cdot (y + 71) \pmod{101}, \tag{15.16}$$
$$(x + 44) \cdot (x + 57) = y + 83 \pmod{101},$$
$$(x + 6) \cdot (x + 96) = (y + 49) \cdot (y + 53) \pmod{101} \text{ [from } x + y + 70],$$
$$3(x + 64) \cdot (x + 71) = (y + 4) \cdot (y + 100) \pmod{101} \text{ [from } x + 3y + 95].$$

It is interesting to remark that the fourth among Equations (15.16) involves a finite field constant 3 in addition to the linear polynomials $x + u$ or $y + v$. Clearly, in general, we may expect arbitrary constants to appear, since from

$x + ay + b$ we generate an equation which has a as coefficient of the highest degree monomial on the x side. A natural question to consider is the impact of these constants on the discrete logarithm computations. Indeed, when writing the linear equations we get as additional unknowns the logarithms of the $p-1$ non-zero constants. At first, this seems to imply that we need to collect $3p$ equations instead of $2p$. However, there are simple ways of sidestepping this issue. One possible idea is to simply avoid equations from polynomials of the form $x + ay + b$ since these equations are the only ones that involve constants. In fact, this is not a good idea, because these polynomials have a better chance of yielding equations than the more general form $xy + ax + by + c$, thus using them speeds up the computations.

To get rid of the extra p unknowns coming from the constants, it is better to remark that since $\mathbb{F}_p^*$ is a multiplicative subgroup of $\mathbb{F}_{p^n}^*$, its order $p-1$ divides the order $p^n - 1$ of the full group. Thus, thanks to the Pohlig-Hellman method, see Section 6.4.1, we are going to deal separately with the discrete logarithms modulo $p-1$ and modulo $q = (p^n - 1)/(p-1)$. Since $p-1$ is small, we can use Pollard's rho algorithm to determine this part of the discrete logarithm and thus we do not require any equation at all modulo $p-1$. Thus, we are only looking at our equations modulo q. At that point, we should remember that to focus on the discrete logarithm modulo q using Pohlig-Hellman method, we need to raise everything to the power $p-1$ in order to get rid of this factor. This means that in our equations, we do not need the logarithms of the constants in $\mathbb{F}_p^*$ but the logarithms of the non-zero constants raised to the power $p-1$. Of course, thanks to Fermat's little theorem, all these powers are equal to 1. As a consequence, since the logarithm of 1 is zero, we can completely forget the constants when setting up the linear system for discrete logarithm modulo q.

The equations we need are summarized in three tables. Table 15.1 describes all equations coming from $x + a$, Table 15.2 describes all equations coming from $y + a$ and Table 15.3 contains enough equations of the form $x + ay + b$ to get a complete system.

15.3 Generalization to finite fields with small enough characteristic

The method described in Section 15.2 can in fact be generalized to all values of the characteristic smaller than $L(1/3)$. More precisely, discrete logarithms in finite fields $\mathbb{F}_Q = \mathbb{F}_{p^n}$, with p smaller than $L_Q\left(1/3, 3^{-1/3}\right)$ can be computed using such a generalization. In the range of p from $L_Q(1/3, 3^{-1/3})$ to $L_Q(1/3, \Theta)$, for some value of Θ we can use the algorithm of Section 15.2 itself. Beyond that point, we need a variation of the number field sieve described

$(2, 81, 20)$	$(3, 55, 46)$	$(4, 87, 14)$	$(7, 30, 71)$	$(11, 54, 47)$
$(12, 17, 84)$	$(14, 40, 61)$	$(15, 65, 36)$	$(17, 48, 53)$	$(18, 9, 92)$
$(19, 22, 79)$	$(20, 33, 68)$	$(21, 52, 49)$	$(22, 73, 28)$	$(23, 51, 50)$
$(28, 77, 24)$	$(29, 26, 75)$	$(31, 13, 88)$	$(34, 41, 60)$	$(35, 8, 93)$
$(41, 19, 82)$	$(43, 64, 37)$	$(45, 85, 16)$	$(47, 31, 70)$	$(50, 7, 94)$
$(52, 59, 42)$	$(54, 67, 34)$	$(56, 89, 12)$	$(62, 21, 80)$	$(63, 95, 6)$
$(66, 29, 72)$	$(68, 58, 43)$	$(69, 63, 38)$	$(74, 96, 5)$	$(75, 78, 23)$
$(76, 86, 15)$	$(77, 74, 27)$	$(78, 83, 18)$	$(79, 11, 90)$	$(80, 25, 76)$
$(82, 57, 44)$	$(83, 4, 97)$	$(85, 69, 32)$	$(86, 66, 35)$	$(90, 3, 98)$
$(93, 62, 39)$	$(94, 45, 56)$	$(95, 99, 2)$	$(98, 100, 1)$	$(99, 0, 0)$
$(100, 91, 10)$				

Table 15.1: Equations $(x + u) = (y + v_1) \cdot (y + v_2)$ **as triples** (u, v_1, v_2)

$(91, 10, 0)$	$(81, 20, 3)$	$(55, 46, 4)$	$(14, 87, 5)$	$(30, 71, 8)$
$(54, 47, 12)$	$(84, 17, 13)$	$(61, 40, 15)$	$(36, 65, 16)$	$(53, 48, 18)$
$(9, 92, 19)$	$(22, 79, 20)$	$(68, 33, 21)$	$(49, 52, 22)$	$(73, 28, 23)$
$(50, 51, 24)$	$(24, 77, 29)$	$(75, 26, 30)$	$(13, 88, 32)$	$(41, 60, 35)$
$(93, 8, 36)$	$(19, 82, 42)$	$(64, 37, 44)$	$(85, 16, 46)$	$(70, 31, 48)$
$(94, 7, 51)$	$(59, 42, 53)$	$(34, 67, 55)$	$(89, 12, 57)$	$(80, 21, 63)$
$(6, 95, 64)$	$(29, 72, 67)$	$(43, 58, 69)$	$(63, 38, 70)$	$(5, 96, 75)$
$(23, 78, 76)$	$(15, 86, 77)$	$(74, 27, 78)$	$(83, 18, 79)$	$(90, 11, 80)$
$(76, 25, 81)$	$(44, 57, 83)$	$(4, 97, 84)$	$(32, 69, 86)$	$(66, 35, 87)$
$(3, 98, 91)$	$(62, 39, 94)$	$(45, 56, 95)$	$(99, 2, 96)$	$(100, 1, 99)$
$(0, 0, 100)$				

Table 15.2: Equations $(x + u_1) \cdot (x + u_2) = (y + v)$ **as triples** (u_1, u_2, v)

$(1,4)\ \rightarrow (27,75,25,77)$	$(1,10)\rightarrow (10,92,43,59)$	$(1,38)\ \rightarrow (30,72,45,57)$
$(1,44)\rightarrow (94,8,89,13)$	$(1,50)\rightarrow (56,46,74,28)$	$(1,51)\ \rightarrow (28,74,66,36)$
$(1,52)\rightarrow (36,66,78,24)$	$(1,53)\rightarrow (78,24,33,69)$	$(1,54)\ \rightarrow (33,69,62,40)$
$(1,55)\rightarrow (40,62,26,76)$	$(1,58)\rightarrow (95,7,55,47)$	$(1,61)\ \rightarrow (19,83,16,86)$
$(1,69)\rightarrow (12,90,96,6)$	$(1,70)\rightarrow (6,96,49,53)$	$(1,74)\ \rightarrow (52,50,51,51)$
$(1,75)\rightarrow (51,51,61,41)$	$(1,79)\rightarrow (31,71,5,97)$	$(1,80)\ \rightarrow (5,97,65,37)$
$(1,88)\rightarrow (98,4,34,68)$	$(1,91)\rightarrow (11,91,15,87)$	$(1,94)\ \rightarrow (99,3,42,60)$
$(1,95)\rightarrow (42,60,73,29)$	$(1,96)\rightarrow (29,73,84,18)$	$(1,97)\ \rightarrow (18,84,2,100)$
$(1,98)\rightarrow (100,2,23,79)$	$(1,99)\rightarrow (23,79,0,1)$	$(2,4)\ \ \rightarrow (72,80,47,56)$
$(2,8)\ \ \rightarrow (44,7,72,31)$	$(2,18)\rightarrow (73,79,54,49)$	$(2,21)\ \rightarrow (57,95,34,69)$
$(2,24)\rightarrow (37,14,52,51)$	$(2,29)\rightarrow (60,92,78,25)$	$(2,36)\ \rightarrow (76,76,9,94)$
$(2,44)\rightarrow (96,56,65,38)$	$(2,46)\rightarrow (21,30,17,86)$	$(2,48)\ \rightarrow (62,90,71,32)$
$(2,51)\rightarrow (88,64,95,8)$	$(2,63)\rightarrow (97,55,22,81)$	$(2,64)\ \rightarrow (59,93,96,7)$
$(2,70)\rightarrow (11,40,64,39)$	$(2,75)\rightarrow (33,18,97,6)$	$(2,76)\ \rightarrow (85,67,79,24)$
$(2,77)\rightarrow (13,38,87,16)$	$(2,78)\rightarrow (98,54,28,75)$	$(2,80)\ \rightarrow (43,8,12,91)$
$(2,84)\rightarrow (3,48,5,98)$	$(2,86)\rightarrow (26,25,70,33)$	$(2,87)\ \rightarrow (71,81,36,67)$
$(2,91)\rightarrow (61,91,99,4)$	$(2,95)\rightarrow (58,94,57,46)$	$(2,96)\ \rightarrow (100,52,3,100)$
$(2,99)\rightarrow (0,51,0,2)$	$(3,5)\ \ \rightarrow (61,74,76,28)$	$(3,6)\ \ \rightarrow (82,53,26,78)$
$(3,8)\ \ \rightarrow (21,13,39,65)$	$(3,12)\rightarrow (70,65,60,44)$	$(3,18)\ \rightarrow (96,39,71,33)$
$(3,22)\rightarrow (84,51,36,68)$	$(3,24)\rightarrow (69,66,83,21)$	$(3,27)\ \rightarrow (45,90,45,59)$
$(3,29)\rightarrow (20,14,94,10)$	$(3,45)\rightarrow (41,94,95,9)$	$(3,46)\ \rightarrow (38,97,90,14)$
$(3,53)\rightarrow (16,18,37,67)$	$(3,54)\rightarrow (30,4,25,79)$	$(3,56)\ \rightarrow (17,17,63,41)$
$(3,59)\rightarrow (27,7,96,8)$	$(3,63)\rightarrow (58,77,87,17)$	$(3,67)\ \rightarrow (55,80,49,55)$
$(3,71)\rightarrow (63,72,97,7)$	$(3,82)\rightarrow (22,12,66,38)$	$(3,85)\ \rightarrow (95,40,82,22)$
$(3,90)\rightarrow (54,81,69,35)$	$(3,95)\rightarrow (64,71,100,4)$	$(3,96)\ \rightarrow (33,1,61,43)$
$(3,97)\rightarrow (28,6,74,30)$	$(3,98)\rightarrow (0,34,85,19)$	$(3,100)\rightarrow (93,42,80,24)$
$(4,1)\ \ \rightarrow (22,54,1,3)$	$(4,2)\ \ \rightarrow (71,5,2,2)$	$(4,6)\ \ \rightarrow (87,90,22,83)$
$(4,11)\rightarrow (39,37,73,32)$	$(4,15)\rightarrow (38,38,49,56)$	$(4,16)\ \rightarrow (43,33,19,86)$
$(4,19)\rightarrow (48,28,38,67)$	$(4,21)\rightarrow (45,31,50,55)$	$(4,24)\ \rightarrow (23,53,35,70)$
$(4,32)\rightarrow (56,20,26,79)$	$(4,35)\rightarrow (84,93,15,90)$	$(4,38)\ \rightarrow (12,64,43,62)$
$(4,39)\rightarrow (52,24,10,95)$	$(4,45)\rightarrow (26,50,21,84)$	$(4,51)\ \rightarrow (68,8,72,33)$
$(4,58)\rightarrow (79,98,69,36)$	$(4,60)\rightarrow (6,70,91,14)$	$(4,67)\ \rightarrow (85,92,97,8)$
$(4,73)\rightarrow (32,44,65,40)$	$(4,79)\rightarrow (99,78,80,25)$	$(4,80)\ \rightarrow (41,35,17,88)$
$(4,83)\rightarrow (74,2,92,13)$	$(4,86)\rightarrow (19,57,59,46)$	$(4,94)\ \rightarrow (75,1,5,100)$
$(4,97)\rightarrow (0,76,64,41)$	$(4,99)\rightarrow (60,16,4,0)$	$(5,3)\ \ \rightarrow (9,72,25,81)$
$(5,4)\ \ \rightarrow (39,42,2,3)$	$(5,9)\ \ \rightarrow (63,18,77,29)$	$(5,10)\ \rightarrow (71,10,79,27)$
$(5,12)\rightarrow (85,97,66,40)$	$(5,15)\rightarrow (36,45,12,94)$	$(5,26)\ \rightarrow (56,25,37,69)$

Table 15.3: Equations $a(x+u_1)\cdot(x+u_2) = (y+v_1)\cdot(y+v_2)$ from $x+ay+b$ represented by $(a,b) \rightarrow (u_1,u_2,v_1,v_2)$

in [JLSV06].

The main difference between the basic algorithm of Section 15.2 and the variation we are now presenting is the choice of the smoothness bases. Instead of considering only linear polynomials, we now consider polynomials of degree D. When p remains expressed as $L_Q(1/3, \theta)$ as the size of the finite field grows, we choose for D a fixed constant. In that case, the analysis is very similar to the case $D = 1$.

However, when p is smaller than that, i.e., when:

$$\log(p) = o(\log(Q)^{1/3} \log\log(Q)^{2/3}),$$

we should make a careful choice for D. In fact, we write:

$$D = \frac{\log(L_Q(1/3, \theta_D))}{\log(p)}. \tag{15.17}$$

Note that thanks to the $o(1)$ in Equation (15.2) defining the L notation, we can round D to the nearest integer without changing Equation (15.2).

In addition to controlling the size of the smoothness bases, the parameter D also changes the algorithm of Section 15.2 in other places. First, it modifies the set of polynomials we consider. Instead of looking at polynomials $xy + ax + by + c$, we generalize to polynomials of the form $a(x) + b(x)y$ where a and b are coprime univariate polynomials of degree at most D in x. Indeed, if a and b are not coprime, we can factor out the common factor and obtain another equation of the same form. Thus, when a and b are not coprime, we obtain duplicates of existing equation. With such polynomials a and b, we now see that when substituting y by $f_1(x)$ we obtain a polynomial of degree (at most) $D + d_1$, where $d_1 = \deg(f_1)$. When substituting x by $f_2(y)$, we obtain a polynomial of degree Dd_2, where $d_2 = \deg(f_2)$. Clearly, we should try to minimize the sum of the two degrees, while keeping $d_1 d_2 \approx n$. Asymptotically, the optimal choice is to have:

$$d_1 = \sqrt{nD} \quad \text{and} \quad d_2 = \sqrt{n/D}. \tag{15.18}$$

With this choice, we have the degree on each side is approximately d_1. Since $n = \log(Q)/\log(p)$, we see that:

$$d_1 = \frac{\log(L_Q(2/3, \sqrt{\theta_D}))}{\log(p)} \quad \text{and} \quad d_2 = \frac{1}{\sqrt{\theta_D}} \left(\frac{\log(Q)}{\log\log(Q)} \right)^{1/3}. \tag{15.19}$$

As usual the logarithm of the probability of splitting into irreducible polynomials of degree at most D on both sides is approximately:

$$-(2d_1/D) \log(d_1/D) = L_Q \left(1/3, -\frac{2}{3\sqrt{\theta_D}} \right). \tag{15.20}$$

In order to make sure that we obtain enough equations, we remark that we have $L_Q(1/3, \theta_D)$ polynomials in the smoothness bases and that the size of the search space $a(x) + b(x)y$ is $L_Q(1/3, 2\theta_D)$. As a consequence, we need :

$$\theta_D = \frac{2}{3\sqrt{\theta_D}}. \tag{15.21}$$

This condition implies $\theta_D = (2/3)^{2/3}$. Note that making this specific choice also balances the runtime complexities of finding the equations and solving the linear system. We leave the adaptation of the individual logarithms computation as an exercise.

With this generalization, the complexity of computing discrete logarithms in arbitrary finite fields $\mathbb{F}_Q$, with $Q = p^n$ and $p = o(L_Q(1/3))$, is $L_Q(1/3, 2\theta_D) = L_Q\left(1/3, \sqrt[3]{32/9}\right)$. This generalization is a simplification of the **function field sieve** Algorithm [Adl94]. The important aspect of this simplification is that, as in Section 15.2, we are directly working in two different representations of the finite field: each element can be expressed either as a polynomial in x or as a polynomial in y. In general, the function field algorithm makes use of a more complicated mathematical structure than polynomial rings: function fields.

15.3.1 Overview of the regular function field sieve

Despite the fact that function fields can be avoided altogether for the computation of discrete logarithms in small characteristic, it is interesting to look at the function field sieve in order to introduce some of the technicalities that are needed in larger characteristic with the number field sieve. When comparing the function field sieve with our previous approach, the first difference is that we break the symmetry between the two unknowns that appear in the algorithm. To emphasize this breach of symmetry, we rename the unknowns t and X instead of x and y. The function field sieve is based on the following commutative diagram:

$$\mathbb{F}_p[t, X]$$
$$\swarrow \qquad\qquad \searrow$$
$$\frac{\mathbb{F}_p[t,X]}{(F_1(t,X))} \qquad\qquad\qquad \frac{\mathbb{F}_p[t,X]}{(F_2(t,X))} \tag{15.22}$$
$$\searrow \qquad\qquad \swarrow$$
$$\mathbb{F}_Q = \frac{\mathbb{F}_p[t]}{(g(t))}$$

Looking at the bottom line of this diagram, we see that t plays a special role because it is used to generate our representation of $\mathbb{F}_Q$. On the contrary, X is no longer used to generate $\mathbb{F}_Q$. The middle line of the diagram shows another crucial difference, instead of removing one of the unknowns on each side, we keep both and consider all polynomials modulo a bivariate polynomial either $F_1(t, X)$ or $F_2(t, X)$. From a mathematical point-of-view, $\frac{\mathbb{F}_p[t,X]}{(F_1(t,X))}$ is a ring.

If in addition, $F_1(t, X)$ is an irreducible polynomial, this quotient becomes an entire ring and we can form its field of fractions. This field of fractions is called a function field. We already introduced such a mathematical object in Chapter 14.

In the most general case, F_1 and F_2 can be arbitrary irreducible polynomials in t and X. In order to make the above diagram commutative, we need the two simultaneous conditions $F_1(t, X) = 0$ and $F_2(t, X) = 0$. Eliminating X from these two algebraic equations in two unknowns, we obtain an equation $G(t) = 0$. If G has an irreducible factor g of degree n, such that $p^n = Q$, we can keep F_1 and F_2 as an implicit definition of $\mathbb{F}_Q$. In practice, this general setting is rarely needed and one polynomial is usually chosen of a simpler form. A typical choice is the following:

$$F_1(t, X) = X - f_1(t) \quad \text{and} \quad F_2(t, X) = \sum_{i=0}^{D} f_2^{(i)}(t) X^i.$$

In other words, we view both F_1 and F_2 as polynomials in X whose coefficients are polynomials in t. The first polynomial f_1 is simpler since it is a linear polynomial. With this specific choice, several simplifications appear. First, G can now be computed without using resultants, more precisely, we have $G(t) = F_2(t, f_1(t))$. Second, the commutative diagram becomes simpler, because on the F_1 side, also called the linear side, with can remove X completely, replacing it by $f_1(t)$. Thus, on the linear side, we only need to factor polynomials and the technical difficulties related to the use of function fields only appear on the side of F_2.

With this setup, we are now ready to explain how the equations are generated with the function field sieve. In order to produce the equations, we start from linear polynomials in X of the form $a(t) + Xb(t)$. On the linear side, this becomes $a(t) + f_1(t)b(t)$ with can easily be factored. On the function field side, matters are more difficult. Indeed, unique factorization is not even guaranteed on that side. As a consequence, we need to decompose $a(t) + Xb(t)$ into a product of prime ideals in the function field[4].

Informally, in this context, an ideal I in the function field is a set of elements of the function field such that:

$$I + I \subset I \quad \text{and}$$
$$H(x, T)I \subset I, \quad \text{for all polynomial } H \in \mathbb{F}_p[t, X]. \qquad (15.23)$$

An ideal is prime if it cannot be decomposed into a non-trivial product of ideals. Thanks to the fact that $a(t) + Xb(t)$ is linear in X, we only need to

---

[4] In Chapter 14, we give a complete description of function fields of a special kind. More precisely, using the unknowns t and X, this chapter fully describes the function field of the curve associated with the irreducible polynomial $X^2 - t^3 - at - b$. In truth, decomposing an ideal as a product of prime ideals is equivalent to writing the divisor of a function.

consider prime ideals of a special form. These prime ideals correspond to a pair $(q(t), X - r(t))$, where $q(t)$ is an irreducible polynomial modulo p and $r(t)$ a root of F_2 modulo $q(t)$, i.e., $F_2(t, r(t))$ is a multiple of $q(t)$.

Testing whether $a(t) + Xb(t)$ can be decomposed into a product of prime ideals of this form is achieved by checking if the numerator of $F_2(t, -a(t)/b(t))$ is a product of polynomials of small enough degree. Thus, in practice, generating the equations with the function field sieve is not really more difficult. Interpreting the equations in the finite field is more delicate: it is not clear at all that an ideal in the function field can be sent to a element of the finite field. Indeed, in general, an ideal is not generated by a single element. However, with function fields, this difficulty can in fact be ignored. We simply write the equations in logarithmic form without worrying about the images of ideals in the finite field and everything works out. We will see that with the number field sieve, the situation is more complicated and that we need additional tools to obtain a working algorithm.

15.4 Introduction to the number field sieve

The basic number field sieve algorithm can be used for two main purposes: to factor composite numbers N or to compute discrete logarithm modulo a prime p. The overall outline of the algorithm is identical, however, at the detailed level there are some important differences. In all cases, the number field sieve algorithm relies on the following commutative diagram:

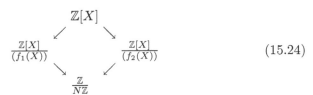

$$(15.24)$$

where N is either the composite to factor or a copy of the prime p. As in the case of the function field sieve, f_1 and f_2 are chosen to make sure that the diagram commute. In the context of the number field sieve, this means that f_1 and f_2 have a common root modulo N and thus that N divides the resultant of the two polynomials as explained in Chapter 11. The above commutative diagram is valid in the most general case with f_1 and f_2 of arbitrary degrees. However, very frequently, f_1 is a polynomial of degree 1, $f_1(X) = X - \alpha$ where α is a root of f_2 modulo N. In that special case, we say that f_1 defines a linear side and we can replace $\frac{\mathbb{Z}[X]}{(f_1(X))}$ by $\mathbb{Z}$ in the above commutative diagram. As with the function field sieve, equations are generated by looking at the images of elements from the top of the diagram, both on the left- and the right-hand sides. In all generality, we may consider polynomials of arbitrary degree in $\mathbb{Z}[X]$. However, in the most frequent cases, we restrict ourselves to linear polynomials of the form $a + bX$, where a and b are, of course, coprime integers. On the rational side, the image of $a + bX$ simply is the integer $a + b\alpha$. On the right-hand (algebraic) side, the image is the algebraic number $a + b\beta$, where β denotes a complex root of the polynomial f_2.

Any element $a + bX$ whose images $a + b\alpha$ and $a + b\beta$ can both be decomposed into small enough factors, yields an equation which can be interpreted in $\mathbb{Z}/N\mathbb{Z}$. Note that, as in the case of the function field sieve of Section 15.3.1, this interpretation is not always straightforward and requires a careful analysis of the underlying mathematical structure.

We give a sketch of a description for the number field sieve in Section 15.4.2.1. However, for a complete description, we refer the reader to [JLSV06] and [LL93].

15.4.1 Factoring with the quadratic sieve

Before looking at the number field sieve algorithm and its application to factoring and discrete logarithms computations, it is useful to review simpler special cases. The simpler situation that can be imagined involves two linear sides. It can be achieved by choosing $f_1(X) = X$ and $f_2(X) = X + N$. In that basic case, looking at the restricted set of values $a + bX$ with $b = 1$, we are simply looking at relations $a \equiv a + N \pmod{N}$. In that case, if both a and $a + N$ decompose into products of primes, all smaller than a smoothness bound B, we get a multiplicative relation between two products of primes. This approach is called the linear sieve. We leave its detailed analysis as an exercise for the reader.

Looking beyond the linear sieve, the next step is to choose a linear polynomial for f_1 and a quadratic polynomial for f_2. When N is a composite that we want to factor, let R be the nearest integer to $\sqrt{N}$ and let f_2 be the polynomial:

$$f_2(x) = x^2 + 2Rx + (R^2 - N). \tag{15.25}$$

We can see that coefficients of f_2 are not too large. More precisely, they are of the order of $O(\sqrt{N})$. Since $-R$ is a root of f_2 modulo N, it is natural to choose for f_1 the polynomial:

$$f_1(x) = x + R. \tag{15.26}$$

With this choice, we are now ready to generate multiplicative equations. At this point, the quadratic sieve slightly differs from our general framework. We may remark that for any integer a, we have:

$$f_1(a)^2 \equiv f_2(a) \pmod{N}. \tag{15.27}$$

Thanks to the square on the left-hand side of this equation, we do not need to factor $f_1(a)$ but only to look at the decomposition of $f_2(a)$. The quadratic sieve algorithm is parametrized by two values A and B. The parameter A

fixes the range $[-A, A]$ of values of a we consider. The parameter B is a bound on the size of small primes. When $f_2(a)$ (or its absolute value) factors into primes smaller than B, we write an equation:

$$f_1(a)^2 \equiv \prod_{i=1}^{n_a} \left(p_{c_i^{(a)}} \right)^{e_i^{(a)}} \pmod{N}, \tag{15.28}$$

where n_a is the number of primes in the equation, $c_i^{(a)}$ is the number of the i-th prime involved in the equation and $e_i^{(a)}$ is its multiplicity. Note that it is useful to add an additional element to our list of primes, $p_0 = -1$. This value of p_0 offers a convenient way a representing the sign of $f_2(a)$. We denote by N_B the total number of primes that may appear in the equations, including the additional p_0. Another interesting remark about the equations, is that since $f_2(a)$ is a number of the order of $2aR$, as long as a remains small compared to R, the number of primes involved in each instance of Equation (15.28) remains small.

Once N_B (or a few more) equations are found, we construct a Boolean matrix M. Each line of the matrix corresponds to an instance of Equation (15.28); each column to a prime from p_0 to p_{N_B-1}. For a line l, we denote by a_l the value of a that generates the corresponding equation. On this line, in the column corresponding to $p_{c_i^{(a_l)}}$ (for each possible value of i) we write the value $c_i^{(a_l)} \pmod 2$. Since the number of primes in each equation is small, the matrix M is a sparse matrix. Once M is constructed, we search for a sum of lines equal to zero in $\mathbb{F}_2$. This can be done by finding a non-trivial element of the kernel of $^\top M$. Given such an element of the kernel, we can compute the corresponding sum of lines over the integers and the product P (modulo N) of the values $f_1(a_l)$ for the lines l included in the sum. Clearly, the sum yields a vector in which each coordinate is even. Let V denote the vector where each coordinate is halved. Putting everything together, we see that:

$$P^2 \equiv \left(\prod_{i=0}^{N_B+1} p_i^{V_i} \right)^2 \pmod{N}. \tag{15.29}$$

We recall from Section 2.3.3.1 that two different square roots of the same number modulo N can be used to generate a square root of 1 and that random square roots of 1 modulo N can factor N with probability $1/2$ or more.

15.4.2 Discrete logarithms with the Gaussian integer method

The quadratic sieve as presented in Section 15.4.1 relies on the specific property that a congruence of square is sufficient for factoring. It is thus natural to wonder whether the special cases which make things simpler for factoring also apply for discrete logarithm computations. Clearly, the most special case of the linear sieve can be applied. Thus, we are left to consider whether the next step with a linear polynomial and a quadratic polynomial is applicable. In fact, it is and we proceed as follows: first, we choose a small positive integer $d > 0$ such that $-d$ is a quadratic residue in $\mathbb{F}_p$. Let R be a root of $-d$ modulo p. We know from Chapter 2 that R can be written as $R \equiv A/B \pmod{N}$ with A and B of the order of $\sqrt{p}$. With this choice, we can now let:

$$f_1(X) = A - BX \quad \text{and} \quad f_2(X) = X^2 + d. \tag{15.30}$$

We see that R is a root of both f_1 and f_2 modulo N. As usual, we take elements $a + bX$ and see how they can be sent into the sides defined by f_1 and f_2. Clearly, since f_1 is linear, this side should be simple. Indeed, replacing X by A/B in $a + bX$ we find that the left-hand image is $(aB + bA)/B$. Since B is a fixed value, we factor $aB + bA$ as a product of primes and when the factorization is acceptable, we consider B as additional prime which is included in all equations with exponent -1.

The side of f_2 is slightly more complicated. Since $f_2(X) = X^2 + d$, the quotient field of $\mathbb{Z}(X)/(f_2)$ is the number field $\mathbb{Q}[\sqrt{-d}]$ obtained by adjoining to $\mathbb{Q}$ (one of) the purely imaginary square root of $-d$. This specific kind of number field is called an imaginary quadratic field. In addition, when $d = 1, 2, 3, 7, 11, 19, 43, 67$ or 163, the ring of integers $\mathbb{Q}[\sqrt{-d}]$ is a unique factorization domain. As a consequence, in these nine special cases, the quadratic side essentially behaves as a linear side. The image of $a + bX$, i.e., $a + b\sqrt{-d}$ can be factored into a product of primes. For the sake of completeness, it is useful to describe the set of primes in $\mathbb{Q}[\sqrt{-d}]$; they are of two types. The first type is simply prime integers ℓ such that $-d$ is a non-quadratic residue modulo ℓ. Note that most of these primes cannot occur in the factorization of $a + b\sqrt{-d}$ when a and b are coprime, as a consequence, they can safely be excluded from the smoothness basis. The second type of primes is obtained by decomposing the prime integers ℓ such that $-d$ is a quadratic residue modulo ℓ as:

$$\ell = \frac{a_\ell + b_\ell \sqrt{-d}}{2} \cdot \frac{a_\ell - b_\ell \sqrt{-d}}{2}. \tag{15.31}$$

The two primes appearing in this equation are complex conjugates of each other. Note that a_ℓ/b_ℓ and $-a_\ell/b_\ell$ are the two square roots of $-d$ modulo ℓ.

Once we have the list of primes, assuming that d is one of the nine above cases and not equal to 1, $a + b\sqrt{-d}$ can be written up to sign as a product of primes. Thus, adding an additional "prime" $\ell_0 = -1$ we find ourselves exactly in the situation of the linear sieve. When $d = 1$, the situation is slightly different, $a + b\sqrt{-d}$ can be written as a product of primes multiplied by either 1, -1, i or $-i$ where i denotes $\sqrt{-1}$. Thus we need to add $\ell_0 = i$. Remark that ℓ_0 occurs with exponent 0, 1, 2 or 3 in the decomposition. From a mathematical point-of-view, we see that in both cases ℓ_0 generates the unit group in $\mathbb{Q}[\sqrt{-d}]$. Indeed, with imaginary quadratic fields, the unit group is either $\{1, -1\}$ when $d \neq 1$ or $\{1, i, -1, -i\}$ when $d = 1$.

The only remaining step to deal with $\mathbb{Q}[\sqrt{-d}]$ in the unique factorization domain case is to explicitly compute the factorization of $a + b\sqrt{-d}$. We first remark that when $(a_\ell + b_\ell\sqrt{-d})/2$ divides $a + b\sqrt{-d}$, then $(a_\ell - b_\ell\sqrt{-d})/2$ divides $a - b\sqrt{-d}$. Thus, ℓ divides $(a + b\sqrt{-d}) \cdot (a - b\sqrt{-d}) = a^2 + db^2$, the norm of $a + b\sqrt{-d}$. Conversely, if p divides the norm of $a + b\sqrt{-d}$ then either $(a_\ell + b_\ell\sqrt{-d})/2$ or $(a_\ell - b_\ell\sqrt{-d})/2$ divide $a + b\sqrt{-d}$. To distinguish between the two cases, it suffices to test whether a/b is equal to a_ℓ/b_ℓ or $-a_\ell/b_\ell$ modulo ℓ. Remarking that when a and b are coprime, two conjugate primes cannot both divide $a + b\sqrt{-d}$, we see that the multiplicity of a prime $a_\ell + b_\ell\sqrt{-d}$ in the decomposition is equal to the multiplicity of ℓ in the norm of $a + b\sqrt{-d}$.

Once we have written down exact multiplicative equalities modulo p, between $(aB + bA)/B$ and $a + b\sqrt{-d}$, we proceed to compute discrete logarithms as in Section 15.3. More precisely, taking logarithms, we obtain linear relations between the logarithms of all primes[5] modulo all prime factors Q of $p - 1$. Once we have enough relations, we can solve the linear system to obtain logarithms of primes. Note that as usual small prime factors of $p - 1$ do not need to be addressed by this method, but can be dealt with using a generic algorithm with square root complexity.

When none of $-1, -2, -3, -7, -11, -19, -43, -67$ and -163 is a quadratic residue in $\mathbb{F}_p$, it is no longer possible to directly work with a unique factorization domain. Of course, only a small proportion of primes (approximately 1 in 512 primes) are of this bad form. However, it is interesting to see how this difficulty can be overcome. In this case, we need to proceed with a value of d such that the ring of integers of $\mathbb{Q}[\sqrt{-d}]$ is not a unique factorization domain. For example, in $\mathbb{Q}[\sqrt{-5}]$ we may see that:

$$6 = (1 + \sqrt{-5}) \cdot (1 - \sqrt{-5}) = 2 \cdot 3, \tag{15.32}$$

and that the prime 2 does not have any not trivial divisor and divides neither $1 + \sqrt{-5}$ nor $1 - \sqrt{-5}$. As in the case of the function field sieve, we now need to factor the elements $a + b\sqrt{-d}$ into products of prime ideals. For example, the equality in Equation (15.32) corresponds to a decomposition of the ideal generated by 6 into a product $I_2^2 I_3^+$ and I_3^-, where each of the three ideals is generated by two elements, 2 and $1 + \sqrt{-5}$ for I_2, 3 and $1 + \sqrt{-5}$ for I_3^+, 3 and $1 - \sqrt{-5}$ for I_3^-. None of these ideals can be generated by a single element. Without going into the theory of ideals in number fields or even quadratic fields, let us say that the set of ideals forms an abelian group. Moreover, in this group we find a subgroup formed of ideals which can be generated by single elements, called principal ideals. The quotient of group of ideals by the subgroup of principal ideals is called the ideal class group of the number field. It is a finite group and its cardinality k is called the class number of the number field. A consequence of this is that the k-th power of any ideal is always a principal ideal. As a consequence, up to multiplication by a unit, i.e., up to sign, the k-th power of any $a + b\sqrt{-d}$ in a imaginary quadratic field

---

[5] Including the special "primes" corresponding to B and to the units in the quadratic imaginary field.

can be written as a product of k-th power of prime ideals, i.e., as a product of elements which generates these ideals. Going back to our example in $\mathbb{Q}[\sqrt{-5}]$, we have a class number $k = 2$ and we can find generators for the squares of the ideals appearing in the decomposition of 6. More precisely, I_2^2 is generated by 2, I_3^{+2} by $2 - \sqrt{-5}$ and I_3^{-2} by $2 + \sqrt{-5}$.

Thus, for all quadratic imaginary fields, we can write explicit multiplicative equations between the k-th powers of $(aB+bA)/B$ and $a+b\sqrt{-d}$. When taking the logarithm of such equations, we can clearly factor k out. Moreover, if the linear algebra is to be done modulo Q, we can in fact divide the equations by k as long Q and k are not coprime. Since Q is a large prime divisor of $p - 1$, it is clear that Q and k are coprime unless we get extremely unlucky. With this process we can associate with any prime ideal a value equal to the logarithm of a generator of the k-th power of this ideal divided by k. For convenience, we call this value the **virtual logarithm** of the ideal. A final important remark is that when computing modulo an odd prime Q, the units in the quadratic imaginary field can altogether be ignored. Indeed, since their order is either 2 or 4, their discrete logarithm modulo Q is necessarily equal to 0.

Note that to compute arbitrary discrete logarithm, we need to adapt the descent method for this case.

15.4.2.1   Obstructions to the number field sieve

Taking the quadratic sieve as a model, we may try to generalize it to arbitrary number fields. That is, given a polynomial f_2 irreducible over $\mathbb{Q}[X]$, we can consider the number field obtained as quotient field of $\mathbb{Z}[X]/(f_2)$ and denoted $\mathbb{Q}[\beta]$, where β is a complex (potential real) root of f_2. We are interested by the possible factorizations of elements $a + b\beta$. In general, as in the general quadratic case, we are not in a unique factorization domain and we can decompose $a + b\beta$ into a product of prime. However, we can write it as a product of prime ideals. Moreover, for a general number field, we also have a finite class number k and looking at k-th powers, we may write an ideal equality between two products, one on the linear side and the other in the number field. Thus, at first, it may seem that going to the number field sieve for discrete logarithm computations does not require any new tool. However, this first impression is not correct. The problem is that when two integral elements x and y in a number field generate the same ideal, we cannot say that $x = y$ but only that there exists a unit u such that $x = uy$. With imaginary quadratic fields, all units had finite order 2 or 4 and could easily be dealt with. With arbitrary number fields, there exists unit of non-finite order which cannot be removed when working modulo Q. This is the most important problem that needs to be overcome in order to use arbitrary number fields in index calculus.

When factoring with the number field sieve, there are in fact two potential obstructions. Clearly, the obstruction coming from units remains, but in addition, we cannot ignore the class group. Indeed, in the case of factoring, we want to compute the linear algebra modulo 2 and we cannot have any guarantee that the class number k is invertible modulo 2, i.e., odd. One way out of this problem would be to restart the computation with another number field should a problem occur. Since we expect an odd class number about half of the time, this would on average double the running time. However, this is not necessary. Instead, it suffices to add extra terms in the equations we are writing to get rid of the obstruction. These extra terms are computed using either characters [LL93] or Schirokauer's maps [Sch93].

15.4.3 Constructing number field sieve polynomials

The number field sieve is too complex to develop in details here. Instead, we refer the reader to [LL93] and [JLSV06] for complete descriptions in all accessible finite fields. However, the question of constructing polynomials f_1 and f_2 for number field sieve algorithms is easily accessible with the tools we have at hand. We first analyze theoretical bounds on the sizes of coefficients that can be achieved for these polynomials and then present the simple base m construction.

15.4.3.1 General setting

Throughout this section, we denote by d_1 the degree of f_1 and by d_2 the degree of f_2. We further write:

$$f_1(X) = \sum_{i=0}^{d_1} C_i^{(1)} X^i \text{ and } f_2(X) = \sum_{i=0}^{d_2} C_i^{(2)} X^i.$$

To optimize the efficiency of the sieving step with these polynomials, we need to minimize the size of the smooth numbers we need to find while sieving. These numbers are norms of linear number field elements obtained from $a+bX$. For simplicity, we assume that we are in the balanced case where a and b are chosen in a half-square, with $0 < a \leq S$ and $-S \leq b \leq S$. In this balanced case, since the norm on each side respectively is:

$$b^{d_1} f_1(-a/b) = \sum_{i=0}^{d_1} C_i^{(1)} (-a)^i b^{d_1-i} \text{ and} \tag{15.33}$$

$$b^{d_2} f_2(-a/b) = \sum_{i=0}^{d_2} C_i^{(2)} (-a)^i b^{d_2-i}, \tag{15.34}$$

it is natural to have all coefficients in f_1 of roughly the same size C_1 and all coefficients in f_2 of size C_2.

Since f_1 and f_2 have a common root modulo N or p but not over $\mathbb{C}$, the resultant of f_1 and f_2 is necessarily a multiple of N, resp. p. Since the resultant is the determinant of a matrix that contains d_2 copies of the coefficients of f_1 and d_1 copies of the coefficient of f_2, its order of magnitude is $C_1^{d_2} C_2^{d_1}$. Thus, $d_2 \log_2(C_1) + d_1 \log_2(C_2)$ is of the order of $\log_2(N)$ or larger.

Moreover, the above property shows that f_1 and f_2 form an encoding of N in the information theoretic sense: given f_1 and f_2 we can retrieve N or at least a small set of candidates containing N. Since N is essentially an arbitrary number, there should not exist any short encoding of N. As a consequence, the sizes of the coefficients C_1 and C_2 cannot be too small. More precisely, we expect that $(d_1 + 1) \log_2(C_1) + (d_2 + 1) \log_2(C_2)$ is near to $\log_2(N)$ or larger.

As a consequence, given d_1 and d_2, the minimum we can hope to achieve for $\gamma_1 = \log_2(C_1)$ and $\gamma_2 = \log_2(C_2)$ is a solution of the linear system of equations:

$$d_2 c_1 + d_1 c_2 = \log_2(N) \text{ and} \tag{15.35}$$

$$(d_1 + 1)c_1 + (d_2 + 1)c_2 = \log_2(N). \tag{15.36}$$

We can remark that for $d_1 = d_2$, this system of equation does not have a solution and the two linear constraints cannot be optimized simultaneously. Otherwise, we find a solution and in particular, we have:

$$(c_1 + c_2) = \frac{2 \log_2 N}{d_1 + d_2 + 1}. \tag{15.37}$$

Looking at Equations (15.33) and (15.34), we see that when a and b are upper bounded by a common bound B, the product of the norms is (ignoring small factors) $C_1 C_2 B^{d_1 + d_2} = 2^{c_1 + c_2} B^{d_1 + d_2}$. It is interesting to note that this expression only involves $d_1 + d_2$ and $c_1 + c_2$. Moreover Equation (15.37) above relates these two parameters. As a consequence, when presenting constructions of polynomials for the number field sieve, it is very useful to compare the value of $c_1 + c_2$ that can be achieved by a given construction to the limit given by Equation (15.37).

15.4.3.1.1   The special number field sieve   Note that the bound given by Equation (15.37) only holds for numbers N of a general form. Special numbers with a short description may correspond to better polynomials, i.e., polynomials with smaller coefficients. The number field sieve when applied to such numbers is called the special number field sieve or SNFS. In this special case, Equation (15.35) remains true, but Equation (15.36) no longer holds. Thus, to determine the best polynomials we can achieve, we need to minimize $c_1 + c_2$ under the constraints that $c_1 \geq 0$ and $c_2 \geq 0$, while respecting Equation (15.35). Assuming that $d_1 \leq d_2$, the best choice is $c_1 = \log_2(N)/d_2$ and $c_2 = 0$. Clearly, with such a choice, for a fixed value D of $d_1 + d_2$ the overall optimum is achieved by choosing $d_1 = 1$ and $d_2 = D - 1$. For example, see Exercise 6.

15.4.3.2 Base m construction

The simpler of all known constructions that can be used for the number field sieve is the base-m construction. With this construction, one of the two polynomials is linear, say $d_1 = 1$ and the other can be chosen with arbitrary degree d_2. The construction works as follows:

- Choose an integer m by rounding $\sqrt[d_2+1]{N}$ up.

- Write N in basis m as:

$$N = \sum_{i=0}^{d_2} M_i m^i, \tag{15.38}$$

  with $0 \le M_i < m$ for all i.

- Let $f_1(X) = X - m$.

- Let $f_2(X) = \sum_{i=0}^{d_2} M_i x^i$.

Clearly, m is a root of both f_1 and f_2 modulo N and the resultant of f_1 and f_2 is N. The sizes of the coefficients that are achieved are:

$$c_1 = c_2 = \frac{\log_2(N)}{d_2 + 1}. \tag{15.39}$$

Thus, $c_1 + c_2$ is equal to $\frac{2\log_2(N)}{d_1+d_2}$ and slightly above the theoretical minimum $\frac{2\log_2(N)}{d_1+d_2+1}$. For an asymptotic point-of-view, this small difference does not affect the overall complexity that can be achieved. However, in practice, its effect might be non-negligible.

15.5 Smoothness probabilities

15.5.1 Computing smoothness probabilities for polynomials

In order to precisely evaluate the behavior of index calculus algorithms based on the function field sieve, it is essential to give fine predictions for the probability for a polynomial of degree D to split into factors of maximum degree m. In practice, there exists a simple combinatorial algorithm that allows us to compute these probabilities exactly. In order to describe this algorithm, we introduce a few notations:

$$\mathcal{S}_D^m = \{f \in \mathbb{F}_p[x]| \deg f = D, \text{any irreducible } g|f \text{ has } \deg g \le m\},$$
$$\tilde{\mathcal{S}}_D^m = \{f \in \mathcal{S}_D^m| \deg f = D, \text{at least one } g \text{ of } f \text{ has } \deg g = m\},$$
$$N_D^m = \sharp \mathcal{S}_D^m \quad \text{and} \quad \tilde{N}_D^m = \sharp \tilde{\mathcal{S}}_D^m. \tag{15.40}$$

Remember that since $\mathbb{F}_p$ is a field, all irreducible polynomials can be chosen to be unitary polynomials. In addition to the above notations, it is convenient to extend the definitions and let $N_0^0 = \tilde{N}_0^0 = 1$ and $N_D^0 = \tilde{N}_D^0 = 0$ for $D > 0$. Indeed, no non-constant polynomial can be written as a product of constants.

The difference between S or N and $\tilde{S}$ or $\tilde{N}$ is that in the second case we insist on having at least one factor of exact degree m. From these definitions, a few properties can immediately be deduced:

$$N_D^m = p^D, \text{whenever } m \geq D, \text{ and } N_D^m = \sum_{j=1}^{m} \tilde{N}_D^j. \qquad (15.41)$$

Indeed, the first property follows from the remark that any polynomial of degree D can be decomposed in factors of degree at most D. The second property is that the factor of largest degree in this decomposition necessarily has a degree between 1 and D. Another noteworthy property is that $\tilde{N}_D^D$ is equal to the number of irreducible polynomials of degree D.

With these notations in mind, we can now proceed to compute the various values of N_D^m and $\tilde{N}_D^m$. Any polynomial f of degree N is either irreducible or reducible. When f is reducible and m is the degree of its largest irreducible factor, f can be decomposed as:

$$f = \prod_{i=1}^{t} F_i \times g, \qquad (15.42)$$

where each F_i has degree m and where g is a product of polynomials of degree at most $m - 1$. Moreover, this decomposition is unique up to a renumbering of the polynomials F_i. When m and t are fixed, the number of polynomials f which can be written as in Equation (15.42), is simply the product of the number of possible t-uples $(F_1, \ldots, F_t)$ by the number of possible polynomials g. The number of possible polynomials g is simply N_{D-tm}^{m-1}, since g is a polynomial of degree $D - tm$ with no irreducible factor of degree m or more. We now need to determine the number of unordered t-uples $(F_1, \ldots, F_t)$. Since each F_i is an irreducible polynomial of degree m, it suffices to choose t such polynomials. However, we should be careful that even though the t-uples are unordered, repetitions are allowed. A very classical combinatorial result is that an unordered choice with repetitions allowed of t elements in a set of $\tilde{N}_m^m$ elements is:

$$\binom{\tilde{N}_m^m + t - 1}{t} = \frac{(\tilde{N}_m^m + t - 1)!}{t!(\tilde{N}_m^m - 1)!}. \qquad (15.43)$$

One easy way to prove this result is to remark that if $(n_1, n_2, \ldots, n_t)$ is a unordered t-uples of t integers in $[1 \cdots M]$ with possible repetitions sorted in increasing order, then $(n_1 + 1, n_2 + 2, \ldots, n_t + t)$ is an unordered t-uples of t integers in $[2 \cdots M + t]$ without repetitions. Moreover the correspondence works both ways.

It is now very easy to sum up the contribution of the various values of t and obtain:

$$\tilde{N}_D^m = \sum_{t=1}^{\lfloor D/t \rfloor} \binom{\tilde{N}_m^m + t - 1}{t} N_{D-tm}^{m-1}. \tag{15.44}$$

Thanks to Equation (15.41), it suffices to add $\tilde{N}_D^m$ and N_D^{m-1} to obtain N_D^m. As a side bonus, the computation also yields the number of irreducible polynomials of degree D, thanks to the relation:

$$\tilde{N}_D^D = p^D - N_D^{D-1}. \tag{15.45}$$

Note that, by theoretical analysis we already know an exact formula for $\tilde{N}_D^D$. Indeed, by taking any element g of $\mathbb{F}_{p^D}$ that does not belong to any proper subfield, we can construct an irreducible polynomial by contructing the product:

$$\prod_{i=0}^{D-1} X - g^{p^i}.$$

Thus, the number of irreducible is equal to the number of such elements divided by D. For example, when D is prime, we find:

$$\tilde{N}_D^D = \frac{p^D - p}{D}. \tag{15.46}$$

When D is not prime, we need to remove each element which belongs to a subfield and we find:

$$\tilde{N}_D^D = \frac{1}{D} \left(\sum_{d|D} \mu(D/d) p^d \geq \frac{p^D - p^{D/2}}{D} \right), \tag{15.47}$$

where μ is the Moebius function defined on a number from its decomposition into primes as follows:

$$\mu(x) = \begin{cases} 0 \text{ if } x \text{ is not square free,} \\ 1 \text{ if } x \text{ has an even number of distinct prime factors,} \\ -1 \text{ if } x \text{ has an odd number of distinct prime factors.} \end{cases}$$

So, in all cases, $\tilde{N}_D^D$ is close to p^D/D.

The above equations can directly be translated into Algorithm 15.1. For simplicity, the variable V in this algorithm is assumed to be a rational. In order to work with integers only, one should slightly reorder the computations to avoid non-exact divisions. Note that in all cases, the numbers involved in this algorithm can become very large. As a consequence, one should preferably work with a multi-precision integer library. A much less preferable alternative is to perform the computations using floating point numbers. Indeed, once we exceed the accuracy limit, the results become completely meaningless.

Algorithm 15.1 Compute number of smooth polynomials

Require: Input: characteristic p, maximum degree D

  Create a two-dimensional $D \times D$ array N

  Create a vector of D elements I

  Let $I[1] \longleftarrow p$

  for i from 1 to D **do**

    for j from i to D **do**

      Let $N[i,j] \longleftarrow p^i$

    end for

  end for

  for i from 2 to D **do**

    Let $V \longleftarrow 1$

    Let $C \longleftarrow I[1] + i$

    for k from 1 to i **do**

      Let $V \longleftarrow (C - k)V/k$

    end for

    Let $N[i,1] \longleftarrow V$

    for m from 2 to $i-1$ **do**

      Let $N[i,m] \longleftarrow N[i,m-1]$

      Let $V \longleftarrow 1$

      for t from 1 to T **do**

        Let $C \longleftarrow I[m] + t$

        Let $V \longleftarrow N[i - tm, m-1]$

        for k from 1 to t **do**

          Let $V \longleftarrow (C - k)V/k$

        end for

        Let $N[i,m] \longleftarrow N[i,m] + V$

      end for

    end for

    Let $I[i] \longleftarrow N[i,i] - N[i,i-1]$

  end for

15.5.2 Asymptotic lower bound on the smoothness probability

Here, using the same notations as above, we prove the lower bound of the probability of smoothness of polynomials of degree D over the basis of monic irreducible polynomials of degree at most m. Thus, our goal is to give an asymptotic lower bound on N_D^m.

We first assume that D is a multiple of m, $D = \ell m$. In that case, the number of smooth polynomials is greater than the number of possible products of ℓ distinct irreducible polynomials of degree m. This number is:

$$\frac{1}{\ell!} \prod_{i=0}^{\ell-1} (\tilde{N}_m^m - i).$$

Remembering that $\tilde{N}_m^m$ is close to p^m/m, letting ℓ and m grow and dividing by p^D to get a probability, we obtain a lower bound of: $\frac{1}{\ell!(m+\epsilon)^\ell}$ for any value of $\epsilon > 0$. Taking the logarithm we find $\ell(\log \ell + m + \epsilon)$ which is asymptotically equivalent to $\ell \log \ell$ as expected.

In the general case, we write $D = \ell m + r$ with $r < m$ and proceed similarly with a product of a single irreducible of degree r and ℓ distinct irreducibles of degree m. The lower bounds immediately follows.

15.5.3 Smoothness probabilities for integers

To compute the probability of smoothness for integers, let us introduce the quantity $\Psi(x, y)$ to denote the number of positive integers up to x which are y-smooth. Then, we have the following theorem from [CEP83].

THEOREM 15.2 Canfield, Erdös, Pomerance
For every $\epsilon > 0$, there exists a constant C_ϵ such that for all $x \geq 1$ and $3 \leq u \leq (1 - \epsilon) \log(x)/\log\log(x)$ we have:

$$\Psi(x, x^{1/u}) \geq x \cdot e^{-u\left(\log(u) + \log\log(u) - 1 + \frac{\log\log(u)-1}{\log(u)} + E(x,u)\right)},$$

where

$$|E(x, u)| \leq C_\epsilon \frac{\log\log(u)^2}{\log(u)^2}.$$

As a consequence, to express $\Psi(x, y)$, we let $u = \log(x)/\log(y)$ and substitute u in the above theorem. After dividing by x to obtain a probability of smoothness and ignoring low order terms we find that:

$$\log(\Psi(x, y)/x) \approx -\frac{\log(x)}{\log(y)} \log\left(\frac{\log(x)}{\log(y)}\right).$$

To understand the relation between this smoothness probability and the functions $L_N(\alpha, c)$, see Exercise 9.

Exercises

1^h. Draw the graphs of $L_N(\alpha, c)$ as a function of N for several values of α and c. In particular, focus on the values of α and c that arise in the complexity analysis we have seen. In each case, what is the maximum value of N that can be reached if we want to ensure that $L_N(\alpha, c) \leq 2^{80}$? This gives an upper bound on the size of N that can be reached in the near future.

2. Repeat the complexity analysis of the function field sieve variation of Section 15.3 when D is a fixed integer. For $D = 1$, you should recover the results of Section 15.2. For each D, what is the range for p where this value of D is the best choice? Draw the corresponding asymptotic complexities on a graph.

3^h. For the case of $\mathbb{F}_{p^n}$ with $p = 65537$ and $n = 49$, construct two polynomials f_1 and f_2 allowing to apply the basic algorithm of Section 15.2. Write a program to generate equations from $xy + ax + by + c$. How long would it take to find enough equations? How long to solve the linear algebra problem and obtain discrete logarithms for the smoothness basis?

Turn to the descent problem and write code, possibly in a computer algebra system, to perform the descent step down to degree 2. Can you descend from degree 2 to degree 1?

4. Consider the nine fields $\mathbb{Q}[\sqrt{-d}]$ whose rings of integers are unique factorization domains. For each, list the primes of the form $(a_\ell + b_\ell \sqrt{-d})/2$. Remark that division by 2 is not always required. When is it needed? Write a program to explicitly factor any $a + b\sqrt{-d}$. Assuming that a and b are coprime, is there any integer prime appearing in these decompositions?

5^h. Consider the field $\mathbb{Q}[\sqrt{-5}]$. Characterize all the prime ideals in the ring of numbers of the form $a + b\sqrt{-5}$. Write a program to compute the decomposition into prime ideals of all numbers of this form when $|a|$ and $|b|$ are smaller than 1000.

6. Consider the p-th Fermat number $F_p = 2^{2^p} + 1$. Construct two polynomials with a common root modulo F_p which beat the bound of Equation (15.37). Explain why the special number field sieve outperforms the number field sieve in this case.

7^h. Let N be a fixed integer and let $f_1(X) = \alpha X - \beta$ for two integers α and β. How can we use lattice reduction to construct a second polynomial f_2 of degree d which also has $\beta/\alpha \pmod{N}$ as a root modulo N? Given N

and d, which size can be achieved for α and β. Compare with the base m construction. Assume that the sieving step considers elements of the form $a + bX$ with unbalanced bounds on $|a|$ and $|b|$. Show that in that case, using unbalanced (skewed) coefficients in f_1 and f_2 is important. Adapt the construction to the skewed case.

8. This exercise considers a contruction, due to Montgomery, of two quadratic polynomials usable for the number field sieve modulo N. Let p be a prime near $\sqrt{N}$ such that N is a quadratic residue modulo p. Let c be a square root of N modulo p. Show that the numbers p, c and $(c^2 - N)/p)$ form a geometry progression of ratio c/p (mod N). Consider the lattice formed of all vectors whose scalar product with $(p, c, (c^2 - N)/p)$ modulo N is 0. What is the rank of this lattice? What is the size of the short vectors it contains? Show that each short vector in this lattice can be transformed into a polynomial with root c/p modulo N. Conclude how we obtain two quadratic polynomials f_1 and f_2 for the number field sieve. Compare with the bound of Equation (15.37).

9[h]. This exercise studies the relation between functions $L_N(\alpha, c)$ and the smoothness probabilities given in Section 15.5.3. Assume that $x = L_N(\alpha, c)$ and $y = L_N(\beta, d)$ and compute $\Psi(x, y)/x$. For which values of α, β, c and d is this formula valid?

A natural implementation project for this topic is to implement one of the possible index calculus algorithms described in this chapter. Note that this is a long project, because it is necessary to implement at least three separate parts, sieving, linear algebra and final phase (individual logarithm or factor extraction). This can possibly be split into separate but well-coordinated subprojects.

References

Numbers within brackets after each reference indicate the citing pages.

[AB04]   A. O. L. Atkin and Daniel J. Bernstein. Prime sieves using binary quadratic forms. *Mathematics of Computation*, 73(246):1023–1030, 2004. [123, 133, 134, 135]

[ABW03]   Martín Abadi, Michael Burrows, and Ted Wobber. Moderately hard and memory-bound functions. In *NDSS 2003*, San Diego, California, USA, February 5–7, 2003. The Internet Society. [164]

[ADKF70] V. L. Arlazarov, E. A. Dinic, M. A. Kronod, and I. A. Faradzev. On economical construction of the transitive closure of an oriented graph. *Soviet Math. Dokl.*, 11:1209–1210, 1970. [85]

[Adl94]   Leonard M. Adleman. The function field sieve. In *First Algorithmic Number Theory Symposium (ANTS)*, volume 877 of *LNCS*, pages 108–121. Springer-Verlag, Berlin, Germany, 1994. [453]

[ADR02]   Jee Hea An, Yevgeniy Dodis, and Tal Rabin. On the security of joint signature and encryption. In Lars R. Knudsen, editor, *EUROCRYPT 2002*, volume 2332 of *LNCS*, pages 83–107, Amsterdam, The Netherlands, April 28–May 2, 2002. Springer-Verlag, Berlin, Germany. [20]

[AFK+07] Kazumaro Aoki, Jens Franke, Thorsten Kleinjung, Arjen K. Lenstra, and Dag Arne Osvik. A kilobit special number field sieve factorization. In Kaoru Kurosawa, editor, *ASIACRYPT 2007*, volume 4833 of *LNCS*, pages 1–12, Kuching, Malaysia, December 2–6, 2007. Springer-Verlag, Berlin, Germany. [113]

[AKS01]   Miklós Ajtai, Ravi Kumar, and D. Sivakumar. A sieve algorithm for the shortest lattice vector problem. In *33rd ACM STOC*, pages 601–610, Crete, Greece, July 6–8, 2001. ACM Press. [328]

[AKS02]   Manindra Agrawal, Neeraj Kayal, and Nitin Saxena. PRIMES is in P. *Ann. of Math*, 2:781–793, 2002. [38]

[AS08]   Onur Aciiçmez and Werner Schindler. A vulnerability in RSA implementations due to instruction cache analysis and its demonstration on OpenSSL. In Tal Malkin, editor, *CT-RSA 2008*, LNCS, pages 256–273, San Francisco, CA, USA, April 7–11, 2008. Springer-Verlag, Berlin, Germany. [92]

[AScKK07] Onur Aciiçmez, Werner Schindler, and Çetin Kaya Koç. Cache based remote timing attack on the AES. In Masayuki Abe, editor, *CT-RSA 2007*, volume 4377 of *LNCS*, pages 271–286, San

Francisco, CA, USA, February 5–9, 2007. Springer-Verlag, Berlin, Germany. [92]

[Bar04] Magali Turrel Bardet. *Étude des systèmes algébriques surdéterminés. Applications aux codes correcteurs et à la cryptographie*. PhD thesis, Université de Paris VI, 2004. [367]

[BBD07] Daniel J. Bernstein, Johannes Buchmann, and Erik Dahmen, editors. *Post Quantum Cryptography*. Springer-Verlag, Berlin, Germany, 2007. [363]

[BC04] Eli Biham and Rafi Chen. Near-collisions of SHA-0. In Matthew Franklin, editor, *CRYPTO 2004*, volume 3152 of *LNCS*, pages 290–305, Santa Barbara, CA, USA, August 15–19, 2004. Springer-Verlag, Berlin, Germany. [179]

[BCJ+05] Eli Biham, Rafi Chen, Antoine Joux, Patrick Carribault, Christophe Lemuet, and William Jalby. Collisions of SHA-0 and reduced SHA-1. In Ronald Cramer, editor, *EUROCRYPT 2005*, volume 3494 of *LNCS*, pages 36–57, Aarhus, Denmark, May 22–26, 2005. Springer-Verlag, Berlin, Germany. [179, 181]

[BCRL79] Dario Bini, Milvio Capovani, Francesco Romani, and Grazia Lotti. $o(n^{2.7799})$ complexity for $n \times n$ approximate matrix multiplication. *Information processing letters*, 8(5):234–235, 1979. [89]

[BD99] Dan Boneh and Glenn Durfee. Cryptanalysis of RSA with private key d less than $n^{0.292}$. In Jacques Stern, editor, *EUROCRYPT'99*, volume 1592 of *LNCS*, pages 1–11, Prague, Czech Republic, May 2–6, 1999. Springer-Verlag, Berlin, Germany. [414]

[BDJR97] Mihir Bellare, Anand Desai, Eric Jokipii, and Phillip Rogaway. A concrete security treatment of symmetric encryption. In *38th FOCS*, pages 394–403, Miami Beach, Florida, October 19–22, 1997. IEEE Computer Society Press. [15]

[Ber00] Daniel J. Bernstein. How to find small factors of integers. Available on cr.yp.to, 2000. [152]

[Ber07] Côme Berbain. *Analyse et conception d'algorithmes de chiffrement à flot*. PhD thesis, Université Paris Diderot, 2007. [289]

[BGP06] Côme Berbain, Henri Gilbert, and Jacques Patarin. QUAD: A practical stream cipher with provable security. In Serge Vaudenay, editor, *EUROCRYPT 2006*, volume 4004 of *LNCS*, pages 109–128, St. Petersburg, Russia, May 28–June 1, 2006. Springer-Verlag, Berlin, Germany. [289]

[Bih97] Eli Biham. A fast new DES implementation in software. In Eli Biham, editor, *FSE'97*, volume 1267 of *LNCS*, pages 260–

272, Haifa, Israel, January 20–22, 1997. Springer-Verlag, Berlin, Germany. [162]

[BJ07] Aurélie Bauer and Antoine Joux. Toward a rigorous variation of Coppersmith's algorithm on three variables. In Moni Naor, editor, *EUROCRYPT 2007*, volume 4515 of *LNCS*, pages 361–378, Barcelona, Spain, May 20–24, 2007. Springer-Verlag, Berlin, Germany. [413]

[BJN00] Dan Boneh, Antoine Joux, and Phong Q. Nguyen. Why textbook ElGamal and RSA encryption are insecure. In Tatsuaki Okamoto, editor, *ASIACRYPT 2000*, volume 1976 of *LNCS*, pages 30–43, Kyoto, Japan, December 3–7, 2000. Springer-Verlag, Berlin, Germany. [269]

[BK03] Mihir Bellare and Tadayoshi Kohno. A theoretical treatment of related-key attacks: RKA-PRPs, RKA-PRFs, and applications. In Eli Biham, editor, *EUROCRYPT 2003*, volume 2656 of *LNCS*, pages 491–506, Warsaw, Poland, May 4–8, 2003. Springer-Verlag, Berlin, Germany. [21]

[BK04] Mihir Bellare and Tadayoshi Kohno. Hash function balance and its impact on birthday attacks. In Christian Cachin and Jan Camenisch, editors, *EUROCRYPT 2004*, volume 3027 of *LNCS*, pages 401–418, Interlaken, Switzerland, May 2–6, 2004. Springer-Verlag, Berlin, Germany. [192]

[BKN04] Mihir Bellare, Tadayoshi Kohno, and Chanathip Namprempre. Authenticated encryption in SSH: provably fixing the SSH binary packet protocol. *ACM transactions on information and system security*, 7(2):206–241, May 2004. Full paper available at http://www.cse.ucsd.edu/users/mihir/papers/ssh.html. Earlier version appeared in *ACM CCS 02*. [238, 239]

[BKR00] Mihir Bellare, Joe Kilian, and Phillip Rogaway. The security of the cipher block chaining message authentication code. *Journal of Computer and System Sciences*, 61(3):362–399, 2000. [5]

[BM97] Mihir Bellare and Daniele Micciancio. A new paradigm for collision-free hashing: Incrementality at reduced cost. In Walter Fumy, editor, *EUROCRYPT'97*, volume 1233 of *LNCS*, pages 163–192, Konstanz, Germany, May 11–15, 1997. Springer-Verlag, Berlin, Germany. [266]

[BM05] Johannes Blömer and Alexander May. A tool kit for finding small roots of bivariate polynomials over the integers. In Ronald Cramer, editor, *EUROCRYPT 2005*, volume 3494 of *LNCS*, pages 251–267, Aarhus, Denmark, May 22–26, 2005. Springer-Verlag, Berlin, Germany. [412]

[BN00] Mihir Bellare and Chanathip Namprempre. Authenticated encryption: Relations among notions and analysis of the generic composition paradigm. In Tatsuaki Okamoto, editor, *ASIACRYPT 2000*, volume 1976 of *LNCS*, pages 531–545, Kyoto, Japan, December 3–7, 2000. Springer-Verlag, Berlin, Germany. [17, 18]

[BR94] Mihir Bellare and Phillip Rogaway. Optimal asymmetric encryption. In Alfredo De Santis, editor, *EUROCRYPT'94*, volume 950 of *LNCS*, pages 92–111, Perugia, Italy, May 9–12, 1994. Springer-Verlag, Berlin, Germany. [64]

[BR96] Mihir Bellare and Phillip Rogaway. The exact security of digital signatures: How to sign with RSA and Rabin. In Ueli M. Maurer, editor, *EUROCRYPT'96*, volume 1070 of *LNCS*, pages 399–416, Saragossa, Spain, May 12–16, 1996. Springer-Verlag, Berlin, Germany. [10, 64]

[BR06] Mihir Bellare and Phillip Rogaway. The security of triple encryption and a framework for code-based game-playing proofs. In Serge Vaudenay, editor, *EUROCRYPT 2006*, volume 4004 of *LNCS*, pages 409–426, St. Petersburg, Russia, May 28–June 1, 2006. Springer-Verlag, Berlin, Germany. [186]

[BS91a] Eli Biham and Adi Shamir. Differential cryptanalysis of DES-like cryptosystems. In Alfred J. Menezes and Scott A. Vanstone, editors, *CRYPTO'90*, volume 537 of *LNCS*, pages 2–21, Santa Barbara, CA, USA, August 11–15, 1991. Springer-Verlag, Berlin, Germany. [273]

[BS91b] Eli Biham and Adi Shamir. Differential cryptoanalysis of Feal and N-hash. In Donald W. Davies, editor, *EUROCRYPT'91*, volume 547 of *LNCS*, pages 1–16, Brighton, UK, April 8–11, 1991. Springer-Verlag, Berlin, Germany. [273]

[BS92] Eli Biham and Adi Shamir. Differential cryptanalysis of Snefru, Khafre, REDOC-II, LOKI and Lucifer. In Joan Feigenbaum, editor, *CRYPTO'91*, volume 576 of *LNCS*, pages 156–171, Santa Barbara, CA, USA, August 11–15, 1992. Springer-Verlag, Berlin, Germany. [273]

[BS93] Eli Biham and Adi Shamir. Differential cryptanalysis of the full 16-round DES. In Ernest F. Brickell, editor, *CRYPTO'92*, volume 740 of *LNCS*, pages 487–496, Santa Barbara, CA, USA, August 16–20, 1993. Springer-Verlag, Berlin, Germany. [273]

[BS00] Alex Biryukov and Adi Shamir. Cryptanalytic time/memory/data tradeoffs for stream ciphers. In Tatsuaki Okamoto, editor, *ASIACRYPT 2000*, volume 1976 of *LNCS*, pages 1–13, Kyoto, Japan, December 3–7, 2000. Springer-Verlag, Berlin, Germany. [394]

[BSS05] Ian F. Blake, Gadiel Seroussi, and Nigel P. Smart, editors. *Advances in Elliptic Curve Cryptography*, volume 317 of *London Mathematical Society Lecture Note Series*. Cambridge University Press, New York, 2005. [425]

[BSV07] Thomas Baignères, Jacques Stern, and Serge Vaudenay. Linear cryptanalysis of non binary ciphers. In Carlisle M. Adams, Ali Miri, and Michael J. Wiener, editors, *SAC 2007*, volume 4876 of *LNCS*, pages 184–211, Ottawa, Canada, August 16–17, 2007. Springer-Verlag, Berlin, Germany. [289]

[BT04] Alexandra Boldyreva and Nut Taesombut. Online encryption schemes: New security notions and constructions. In Tatsuaki Okamoto, editor, *CT-RSA 2004*, volume 2964 of *LNCS*, pages 1–14, San Francisco, CA, USA, February 23–27, 2004. Springer-Verlag, Berlin, Germany. [238]

[Buc04] Johannes Buchmann. *Introduction to Cryptography (Second edition)*. Undergraduate texts in Mathematics. Springer, New York, 2004. [3]

[BV07] Johannes Buchmann and Ulrich Vollmer. *Binary Quadratic Forms – An Algorithmic Approach*, volume 20 of *Algorithms and Computation in Mathematics*. Springer-Verlag, Berlin, Germany, 2007. [311]

[Cav00] Stefania Cavallar. Strategies in filtering in the number field sieve. In *Fourth Algorithmic Number Theory Symposium (ANTS)*, volume 1838 of *LNCS*, pages 209–232. Springer-Verlag, Berlin, Germany, 2000. [118]

[CEP83] E. R. Canfield, Paul Erdös, and Carl Pomerance. On a problem of Oppenheim concerning *factorisatio numerorum*. *Journal of Number Theory*, 17:1–28, 1983. [467]

[CF05] Henri Cohen and Gerhard Frey, editors. *Handbook of Elliptic and Hyperelliptic Curve Cryptography*, volume 34 of *Discrete Mathematics and its Applications*. Chapman & Hall, CRC, 2005. [417, 424, 432]

[CJ98] Florent Chabaud and Antoine Joux. Differential collisions in SHA-0. In Hugo Krawczyk, editor, *CRYPTO'98*, volume 1462 of *LNCS*, pages 56–71, Santa Barbara, CA, USA, August 23–27, 1998. Springer-Verlag, Berlin, Germany. [179]

[CJL⁺92] Matthijs J. Costerr, Antoine Joux, Brian A. LaMacchia, Andrew M. Odlyzko, Clauss-Peter Schnorr, and Jacques Stern. Improved low-density subset sum algorithms. *Computational Complexity*, 2:111–128, 1992. [402]

[CJM02] Philippe Chose, Antoine Joux, and Michel Mitton. Fast correlation attacks: An algorithmic point of view. In Lars R. Knudsen, editor, *EUROCRYPT 2002*, volume 2332 of *LNCS*, pages 209–221, Amsterdam, The Netherlands, April 28–May 2, 2002. Springer-Verlag, Berlin, Germany. [257, 385, 387]

[CKM97] Stéphane Collart, Michael Kalkbrener, and Daniel Mall. Converting bases with the Groöbner walk. *J. Symbolic Computation*, 24(3–4):465–469, 1997. [361]

[CKSU05] Henry Cohn, Robert Kleinberg, Balazs Szegedy, and Christopher Umans. Group-theoretic algorithms for matrix multiplication. In *Proceedings of the 46th Annual IEEE Symposium on Foundations of Computer Science*, pages 379–388, Washington, DC, USA, 2005. IEEE Computer Society. [93]

[CL09] Jean-Marc Couveignes and Reynald Lercier. Elliptic periods for finite fields. *Finite fields and their applications*, 15:1–22, 2009. [49]

[CLO07] David Cox, John Little, and Donal O'Shea. *Ideals, Varieties and Algorithms (Third edition)*. Undergraduate texts in Mathematics. Springer, New York, 2007. [345, 348, 350, 353, 354]

[Cop94] Don Coppersmith. Solving homogeneous linear equations over $gf(2)$ via block wiedemann algorithm. *Mathematics of Computation*, 62(205):333–350, 1994. [113]

[Cop96a] Don Coppersmith. Finding a small root of a bivariate integer equation; factoring with high bits known. In Ueli M. Maurer, editor, *EUROCRYPT'96*, volume 1070 of *LNCS*, pages 178–189, Saragossa, Spain, May 12–16, 1996. Springer-Verlag, Berlin, Germany. [412]

[Cop96b] Don Coppersmith. Finding a small root of a univariate modular equation. In Ueli M. Maurer, editor, *EUROCRYPT'96*, volume 1070 of *LNCS*, pages 155–165, Saragossa, Spain, May 12–16, 1996. Springer-Verlag, Berlin, Germany. [410]

[Cor04] Jean-Sébastien Coron. Finding small roots of bivariate integer polynomial equations revisited. In Christian Cachin and Jan Camenisch, editors, *EUROCRYPT 2004*, volume 3027 of *LNCS*, pages 492–505, Interlaken, Switzerland, May 2–6, 2004. Springer-Verlag, Berlin, Germany. [412]

[Cor07] Jean-Sébastien Coron. Finding small roots of bivariate integer polynomial equations: A direct approach. In Alfred Menezes, editor, *CRYPTO 2007*, volume 4622 of *LNCS*, pages 379–394, Santa Barbara, CA, USA, August 19–23, 2007. Springer-Verlag, Berlin, Germany. [412]

[CP91] Paul Camion and Jacques Patarin. The Knapsack hash function proposed at Crypto'89 can be broken. In Donald W. Davies, editor, *EUROCRYPT'91*, volume 547 of *LNCS*, pages 39–53, Brighton, UK, April 8–11, 1991. Springer-Verlag, Berlin, Germany. [264, 405]

[CPS08] Jean-Sébastien Coron, Jacques Patarin, and Yannick Seurin. The random oracle model and the ideal cipher model are equivalent. In David Wagner, editor, *CRYPTO 2008*, volume 5157 of *LNCS*, pages 1–20, Santa Barbara, CA, USA, August 17–21, 2008. Springer-Verlag, Berlin, Germany. [22]

[CT65] James W. Cooley and John W. Tukey. An algorithm for the machine calculation of complex Fourier series. *Mathematics of Computation*, 19:297–301, 1965. [296]

[CV94] Florent Chabaud and Serge Vaudenay. Links between differential and linear cryptoanalysis. In Alfredo De Santis, editor, *EURO-CRYPT'94*, volume 950 of *LNCS*, pages 356–365, Perugia, Italy, May 9–12, 1994. Springer-Verlag, Berlin, Germany. [279, 281]

[CW90] Don Coppersmith and Shmuel Winograd. Matrix multiplication via arithmetic progressions. *J. of Symbolic Computation*, 9(3):251–280, 1990. [93]

[Dam90] Ivan Damgård. A design principle for hash functions. In Gilles Brassard, editor, *CRYPTO'89*, volume 435 of *LNCS*, pages 416–427, Santa Barbara, CA, USA, August 20–24, 1990. Springer-Verlag, Berlin, Germany. [405, 406]

[DES77] Data encryption standard. National Bureau of Standards, NBS FIPS PUB 46, U.S. Department of Commerce, January 1977. [157]

[DFV97] Hervé Daudé, Philippe Flajolet, and Brigitte Vallée. An average-case analysis of the gaussian algorithm for lattice reduction. *Combinatorics, Probability & Computing*, 6(4):397–433, 1997. [318]

[DGV94] Joan Daemen, René Govaerts, and Joos Vandewalle. Correlation matrices. In Bart Preneel, editor, *FSE'94*, volume 1008 of *LNCS*, pages 275–285, Leuven, Belgium, December 14–16, 1994. Springer-Verlag, Berlin, Germany. [279]

[DH76] Whitfield Diffie and Martin E. Hellman. New directions in cryptography. *IEEE Transactions on Information Theory*, 22(6):644–654, 1976. [5, 9]

[DLC07] Frédéric Didier and Yann Laigle-Chapuy. Finding low-weight polynomial multiples using discrete logarithm. *Computing Research Repository (CoRR)*, abs/cs/0701069, 2007. [386]

[DN93] Cynthia Dwork and Moni Naor. Pricing via processing or combatting junk mail. In Ernest F. Brickell, editor, *CRYPTO'92*, volume 740 of *LNCS*, pages 139–147, Santa Barbara, CA, USA, August 16–20, 1993. Springer-Verlag, Berlin, Germany. [164]

[DS09] Itai Dinur and Adi Shamir. Cube attacks on tweakable black box polynomials. In Antoine Joux, editor, *EUROCRYPT 2009*, volume 5479 of *LNCS*, pages 278–299. Springer-Verlag, Berlin, Germany, 2009. [390, 391, 392, 396]

[Eis95] David Eisenbud. *Commutative Algebra with a View Towards Algebraic Geometry*, volume 150 of *Graduate Texts in Mathematics*. Springer, New York, 1995. [342]

[ElG85] Taher ElGamal. A public key cryptosystem and a signature scheme based on discrete logarithms. In G. R. Blakley and David Chaum, editors, *CRYPTO'84*, volume 196 of *LNCS*, pages 10–18, Santa Barbara, CA, USA, August 19–23, 1985. Springer-Verlag, Berlin, Germany. [66, 67]

[Fau99] Jean-Charles Faugère. A new efficient algorithm for computing Gröbner bases (F4). *J. Pure Appl. Algebra*, 139(1-3):61–88, 1999. Effective methods in algebraic geometry (Saint-Malo, 1998). [356, 357]

[Fau02] Jean-Charles Faugère. A new efficient algorithm for computing Gröbner bases without reduction to zero (F5). In T. Mora, editor, *ISSAC 2002*, pages 75–83, 2002. [356, 359]

[FGLM93] Jean-Charles Faugère, Patricia Gianni, Daniel Lazard, and Teo Mora. Efficient computation of zero-dimensional Groöbner bases by change of ordering. *J. Symbolic Computation*, 16(4):329–344, 1993. [361, 362]

[FJ03] Jean-Charles Faugère and Antoine Joux. Algebraic cryptanalysis of hidden field equation (HFE) cryptosystems using gröbner bases. In Dan Boneh, editor, *CRYPTO 2003*, volume 2729 of *LNCS*, pages 44–60, Santa Barbara, CA, USA, August 17–21, 2003. Springer-Verlag, Berlin, Germany. [365]

[FJMV04] Pierre-Alain Fouque, Antoine Joux, Gwenaëlle Martinet, and Frédéric Valette. Authenticated on-line encryption. In Mitsuru Matsui and Robert J. Zuccherato, editors, *SAC 2003*, volume 3006 of *LNCS*, pages 145–159, Ottawa, Ontario, Canada, August 14–15, 2004. Springer-Verlag, Berlin, Germany. [238]

[FJP04] Pierre-Alain Fouque, Antoine Joux, and Guillaume Poupard. Blockwise adversarial model for on-line ciphers and symmetric encryption schemes. In Helena Handschuh and Anwar Hasan, editors, *SAC 2004*, volume 3357 of *LNCS*, pages 212–226, Waterloo, Ontario, Canada, August 9–10, 2004. Springer-Verlag, Berlin, Germany. [238]

[FLPR99] Matteo Frigo, Charles E. Leiserson, Harald Prokop, and Sridhar Ramachandran. Cache-oblivious algorithms. In *40th FOCS*, pages 285–298, New York, New York, USA, October 17–19, 1999. IEEE Computer Society Press. [92]

[FM03] Joanne Fuller and William Millan. Linear redundancy in S-boxes. In Thomas Johansson, editor, *FSE 2003*, volume 2887 of *LNCS*, pages 74–86, Lund, Sweden, February 24–26, 2003. Springer-Verlag, Berlin, Germany. [282]

[FMP03] Pierre-Alain Fouque, Gwenaëlle Martinet, and Guillaume Poupard. Practical symmetric on-line encryption. In Thomas Johansson, editor, *FSE 2003*, volume 2887 of *LNCS*, pages 362–375, Lund, Sweden, February 24–26, 2003. Springer-Verlag, Berlin, Germany. [239]

[FO90] Philippe Flajolet and Andrew M. Odlyzko. Random mapping statistics. In Jean-Jacques Quisquater and Joos Vandewalle, editors, *EUROCRYPT'89*, volume 434 of *LNCS*, pages 329–354, Houthalen, Belgium, April 10–13, 1990. Springer-Verlag, Berlin, Germany. [231, 233, 234]

[FP85] U. Fincke and Michael E. Pohst. Improved methods for calculating vectors of short length in a lattice, including a complexity analysis. *Mathematics of Computation*, 44(170):463–471, 1985. [328]

[FS87] Amos Fiat and Adi Shamir. How to prove yourself: Practical solutions to identification and signature problems. In Andrew M. Odlyzko, editor, *CRYPTO'86*, volume 263 of *LNCS*, pages 186–194, Santa Barbara, CA, USA, August 1987. Springer-Verlag, Berlin, Germany. [10]

[Gal04] William F. Galway. *Ph. D. in mathematics*. PhD thesis, University of Illinois at Urbana-Champaign, 2004. [135]

[GC91] Henri Gilbert and Guy Chassé. A statistical attack of the FEAL-8 cryptosystem. In Alfred J. Menezes and Scott A. Vanstone, editors, *CRYPTO'90*, volume 537 of *LNCS*, pages 22–33, Santa Barbara, CA, USA, August 11–15, 1991. Springer-Verlag, Berlin, Germany. [273]

[GH05] Jovan Dj. Golic and Philip Hawkes. Vectorial approach to fast correlation attacks. *Des. Codes Cryptography*, 35(1):5–19, 2005. [380]

[GJS06] Louis Granboulan, Antoine Joux, and Jacques Stern. Inverting HFE is quasipolynomial. In Cynthia Dwork, editor, *CRYPTO 2006*, volume 4117 of *LNCS*, pages 345–356, Santa Barbara, CA, USA, August 20–24, 2006. Springer-Verlag, Berlin, Germany. [366]

[GL96] Gene H. Golub and Charles F. Van Loan. *Matrix Computations (third edition)*. The Johns Hopkins University Press, London, 1996. [71]

[GL03] Rosario Gennaro and Yehuda Lindell. A framework for password-based authenticated key exchange. In Eli Biham, editor, *EUROCRYPT 2003*, volume 2656 of *LNCS*, pages 524–543, Warsaw, Poland, May 4–8, 2003. Springer-Verlag, Berlin, Germany. http://eprint.iacr.org/2003/032.ps.gz. [156]

[GMR89] Shafi Goldwasser, Silvio Micali, and Charles Rackoff. The knowledge complexity of interactive proof systems. *SIAM Journal on Computing*, 18(1):186–208, 1989. [66]

[GN08] Nicolas Gama and Phong Q. Nguyen. Predicting lattice reduction. In Nigel P. Smart, editor, *EUROCRYPT 2008*, LNCS, pages 31–51, Istanbul, Turkey, April 13–17, 2008. Springer-Verlag, Berlin, Germany. [405]

[HG97] Nick Howgrave-Graham. Finding small roots of univariate modular equations revisited. In Michael Darnell, editor, *Cryptography and Coding, 6th IA International Conference*, volume 1355 of *LNCS*, pages 131–142, Cirencester, UK, December 1997. Springer-Verlag, Berlin, Germany. [407, 410]

[HG07] Nick Howgrave-Graham. A hybrid lattice-reduction and meet-in-the-middle attack against NTRU. In Alfred Menezes, editor, *CRYPTO 2007*, volume 4622 of *LNCS*, pages 150–169, Santa Barbara, CA, USA, August 19–23, 2007. Springer-Verlag, Berlin, Germany. [405]

[HILL99] Johan Håstad, Russell Impagliazzo, Leonid A. Levin, and Michael Luby. A Pseudorandom Generator from any One-way Function. *SIAM J. Comput.*, 28(4):1364–1396, 1999. [286]

[HS07] Guillaume Hanrot and Damien Stehlé. Improved analysis of kannan's shortest lattice vector algorithm. In Alfred Menezes, editor, *CRYPTO 2007*, volume 4622 of *LNCS*, pages 170–186, Santa Barbara, CA, USA, August 19–23, 2007. Springer-Verlag, Berlin, Germany. [331]

[JG94] Antoine Joux and Louis Granboulan. A practical attack against Knapsack based hash functions (extended abstract). In Alfredo De Santis, editor, *EUROCRYPT'94*, volume 950 of *LNCS*, pages 58–66, Perugia, Italy, May 9–12, 1994. Springer-Verlag, Berlin, Germany. [406]

[JL01] Antoine Joux and Reynald Lercier. "Chinese & Match," an alternative to Atkin's "Match and Sort" method used in the SEA algorithm. *Mathematics of Computation*, 70:827–836, 2001. [267, 268, 269]

[JLSV06] Antoine Joux, Reynald Lercier, Nigel Smart, and Frederik Vercauteren. The number field sieve in the medium prime case. In Cynthia Dwork, editor, *CRYPTO 2006*, volume 4117 of *LNCS*, pages 326–344, Santa Barbara, CA, USA, August 20–24, 2006. Springer-Verlag, Berlin, Germany. [452, 456, 461]

[JMV02] Antoine Joux, Gwenaëlle Martinet, and Frédéric Valette. Blockwise-adaptive attackers: Revisiting the (in)security of some provably secure encryption models: CBC, GEM, IACBC. In Moti Yung, editor, *CRYPTO 2002*, volume 2442 of *LNCS*, pages 17–30, Santa Barbara, CA, USA, August 18–22, 2002. Springer-Verlag, Berlin, Germany. [238, 239]

[JN08] Marc Joye and Gregory Neven, editors. *Identity-based Cryptography*, volume 2 of *Cryptology and Information Security Series*. IOS Press, Amsterdam, 2008. [417]

[JNT07] Antoine Joux, David Naccache, and Emmanuel Thomé. When e-th roots become easier than factoring. In Kaoru Kurosawa, editor, *ASIACRYPT 2007*, volume 4833 of *LNCS*, pages 13–28, Kuching, Malaysia, December 2–6, 2007. Springer-Verlag, Berlin, Germany. [439]

[JP07] Antoine Joux and Thomas Peyrin. Hash functions and the (amplified) boomerang attack. In Alfred Menezes, editor, *CRYPTO 2007*, volume 4622 of *LNCS*, pages 244–263, Santa Barbara, CA, USA, August 19–23, 2007. Springer-Verlag, Berlin, Germany. [182]

[Jut01] Charanjit S. Jutla. Encryption modes with almost free message integrity. In Birgit Pfitzmann, editor, *EUROCRYPT 2001*, volume 2045 of *LNCS*, pages 529–544, Innsbruck, Austria, May 6–10, 2001. Springer-Verlag, Berlin, Germany. [17]

[Kah67] David Kahn. *The Codebreakers: The Comprehensive History of Secret Communication from Ancient Times to the Internet.* Scribner, 1967. [11]

[Kan83] Ravi Kannan. Improved algorithms for integer programming and related lattice problems. In *Proc. 15th Symp. Theory of Comp.*, pages 193–206, 1983. [327, 328, 330]

[Ker83] Auguste Kerckhoffs. La cryptographie militaire. *Journal des sciences militaire*, IX, 1883. Article in two parts: Jan. and Feb. issues. [4]

[Knu94] Lars R. Knudsen. Truncated and higher order differentials. In Bart Preneel, editor, *FSE'94*, volume 1008 of *LNCS*, pages 196–211, Leuven, Belgium, December 14–16, 1994. Springer-Verlag, Berlin, Germany. [282, 392]

[KPT96] Jyrki Katajainen, Tomi Pasanen, and Jukka Teuhola. Practical in-place mergesort. *Nordic J. of Computing*, 3(1):27–40, 1996. [201]

[Kra01] Hugo Krawczyk. The order of encryption and authentication for protecting communications (or: How secure is SSL?). In Joe Kilian, editor, *CRYPTO 2001*, volume 2139 of *LNCS*, pages 310–331, Santa Barbara, CA, USA, August 19–23, 2001. Springer-Verlag, Berlin, Germany. [18]

[KS99] Aviad Kipnis and Adi Shamir. Cryptanalysis of the HFE public key cryptosystem by relinearization. In Michael J. Wiener, editor, *CRYPTO'99*, volume 1666 of *LNCS*, pages 19–30, Santa Barbara, CA, USA, August 15–19, 1999. Springer-Verlag, Berlin, Germany. [357]

[KVW04] Tadayoshi Kohno, John Viega, and Doug Whiting. CWC: A high-performance conventional authenticated encryption mode. In Bimal K. Roy and Willi Meier, editors, *FSE 2004*, volume 3017 of *LNCS*, pages 408–426, New Delhi, India, February 5–7, 2004. Springer-Verlag, Berlin, Germany. [17]

[Kwa00] Matthew Kwan. Reducing the gate count of bitslice DES. IACR eprint archive, 2000. Report 2000/051. [163, 183]

[Lai94] Xuejia Lai. Higher order derivatives and differential cryptanalysis. In *Communication and Cryptography – Two Sides of One Tapestry*, pages 227–233. Kluwer Academic Publisher, 1994. [392]

[Lan05] Serge Lang. *Algebra*, volume 211 of *Graduate Texts in Mathematics*. Springer, New York, 2005. Revised third edition. [37, 47, 48, 62, 110, 343]

[Laz83] Daniel Lazard. Gröbner bases, gaussian elimination and resolution of systems of algebraic equations. In *Computer algebra (London, 1983)*, volume 162 of *LNCS*, pages 146–156. Springer-Verlag, Berlin, Germany, 1983. [355]

[LG89] Leonid A. Levin and Oded Goldreich. A Hard-core Predicate for all One-way Functions. In D. S. Johnson, editor, *21th ACM Symposium on Theory of Computing - STOC '89*, pages 25–32. ACM Press, 1989. [286]

[LL93] Arjen K. Lenstra and Hendrick W. Lenstra, Jr., editors. *The development of the number field sieve*, volume 1554 of *Lecture Notes in Mathematics*. Springer-Verlag, Berlin, Germany, 1993. [456, 461]

[LLL82] Arjen K. Lenstra, Hendrick W. Lenstra, Jr., and László Lovász. Factoring polynomials with rational coefficients. *Math. Ann.*, 261:515–534, 1982. [319]

[LMV05] Yi Lu, Willi Meier, and Serge Vaudenay. The conditional correlation attack: A practical attack on bluetooth encryption. In Victor Shoup, editor, *CRYPTO 2005*, volume 3621 of *LNCS*, pages 97–117, Santa Barbara, CA, USA, August 14–18, 2005. Springer-Verlag, Berlin, Germany. [380]

[LO85] Jeffrey C. Lagarias and Andrew M. Odlyzko. Solving low-density subset sum problems. *Journal of the ACM*, 32(1):229–246, 1985. [402, 406]

[LO91] Brian A. LaMacchia and Andrew M. Odlyzko. Solving large sparse linear systems over finite fields. In Alfred J. Menezes and Scott A. Vanstone, editors, *CRYPTO'90*, volume 537 of *LNCS*, pages 109–133, Santa Barbara, CA, USA, August 11–15, 1991. Springer-Verlag, Berlin, Germany. [113, 115]

[Luc05] Stefan Lucks. Two-pass authenticated encryption faster than generic composition. In Henri Gilbert and Helena Handschuh, editors, *FSE 2005*, volume 3557 of *LNCS*, pages 284–298, Paris, France, February 21–23, 2005. Springer-Verlag, Berlin, Germany. [17]

[Mar57] Harry M. Markowitz. The elimination form of the inverse and its application to linear programming. *Management Science*, 3(3):255–269, 1957. [116]

[Mat93] Mitsuru Matsui. Linear cryptoanalysis method for DES cipher. In Tor Helleseth, editor, *EUROCRYPT'93*, volume 765 of *LNCS*, pages 386–397, Lofthus, Norway, May 23–27, 1993. Springer-Verlag, Berlin, Germany. [273]

[Mat94a] Mitsuru Matsui. The first experimental cryptanalysis of the data encryption standard. In Yvo Desmedt, editor, *CRYPTO'94*, volume 839 of *LNCS*, pages 1–11, Santa Barbara, CA, USA, August 21–25, 1994. Springer-Verlag, Berlin, Germany. [273]

[Mat94b] Mitsuru Matsui. On correlation between the order of S-boxes and the strength of DES. In Alfredo De Santis, editor, *EURO-CRYPT'94*, volume 950 of *LNCS*, pages 366–375, Perugia, Italy, May 9–12, 1994. Springer-Verlag, Berlin, Germany. [273]

[MG90] Miodrag J. Mihaljevic and Jovan Dj. Golic. A fast iterative algorithm for a shift register initial state reconstruction given the noisy output sequence. In Jennifer Seberry and Josef Pieprzyk, editors, *AUSCRYPT'90*, volume 453 of *LNCS*, pages 165–175, Sydney, Australia, January 8–11, 1990. Springer-Verlag, Berlin, Germany. [380]

[MG02] Daniele Micciancio and Shafi Goldwasser. *Complexity of Lattice Problems: A Cryptographic Perspective*, volume 671 of *The Kluwer International Series in Engineering and Computer Science*. Kluwer Academic Publishers, 2002. [311]

[Mil04] Victor S. Miller. The Weil pairing, and its efficient calculation. *Journal of Cryptology*, 17(4):235–261, September 2004. [431]

[Mon92] Peter L. Montgomery. *A FFT Extension of the Elliptic Curve Method of Factorization*. PhD thesis, University of California, Los Angeles, 1992. [236, 435]

[Mon95] Peter L. Montgomery. A block Lanczos algorithm for finding dependencies over GF(2). In Louis C. Guillou and Jean-Jacques Quisquater, editors, *EUROCRYPT'95*, volume 921 of *LNCS*, pages 106–120, Saint-Malo, France, May 21–25, 1995. Springer-Verlag, Berlin, Germany. [112]

[MP08] Stéphane Manuel and Thomas Peyrin. Collisions on SHA-0 in one hour. In Kaisa Nyberg, editor, *FSE 2008*, volume 5086 of *LNCS*, pages 16–35, Lausanne, Switzerland, February 10–13, 2008. Springer-Verlag, Berlin, Germany. [182]

[MS89] Willi Meier and Othmar Staffelbach. Fast correlation attacks on certain stream ciphers. *Journal of Cryptology*, 1(3):159–176, 1989. [380]

[MSK98] Shiho Moriai, Takeshi Shimoyama, and Toshinobu Kaneko. Higher order differential attak of CAST cipher. In Serge Vaudenay, editor, *FSE'98*, volume 1372 of *LNCS*, pages 17–31, Paris, France, March 23–25, 1998. Springer-Verlag, Berlin, Germany. [392]

[MT09] Ravi Montenegro and Prasad Tetali. How long does it take to catch a wild kangaroo? In Michael Mitzenmacher, editor, *41st ACM STOC*, pages 1–10, Bethesda, Maryland, USA, May 31–June 2 2009. ACM Press. [238]

[MvOV97] Aldred J. Menezes, Paul C. van Oorschot, and Scott A. Vanstone, editors. *Handbook of Applied Cryptography*. CRC Press LLC, Boca Raton, Florida, 1997. [3]

[MY92] Mitsuru Matsui and Atsuhiro Yamagishi. A new method for known plaintext attack of FEAL cipher. In Rainer A. Rueppel, editor, *EUROCRYPT'92*, volume 658 of *LNCS*, pages 81–91, Balatonfüred, Hungary, May 24–28, 1992. Springer-Verlag, Berlin, Germany. [273]

[Niv04] G. Nivasch. Cycle detection using a stack. *Information Processing Letter*, 90(3):135–140, 2004. [229, 242]

[NP99] Wim Nevelsteen and Bart Preneel. Software performance of universal hash functions. In Jacques Stern, editor, *EURO-CRYPT'99*, volume 1592 of *LNCS*, pages 24–41, Prague, Czech Republic, May 2–6, 1999. Springer-Verlag, Berlin, Germany. [8]

[NS05] Phong Q. Nguyen and Damien Stehlé. Floating-point LLL revisited. In Ronald Cramer, editor, *EUROCRYPT 2005*, volume 3494 of *LNCS*, pages 215–233, Aarhus, Denmark, May 22–26, 2005. Springer-Verlag, Berlin, Germany. [326]

[Odl85] Andrew M. Odlyzko. Discrete logarithms in finite fields and their cryptographic significance. In Thomas Beth, Norbert Cot, and Ingemar Ingemarsson, editors, *EUROCRYPT'84*, volume 209 of *LNCS*, pages 224–314, Paris, France, April 9–11, 1985. Springer-Verlag, Berlin, Germany. [113]

[OST06] Dag Arne Osvik, Adi Shamir, and Eran Tromer. Cache attacks and countermeasures: The case of AES. In David Pointcheval, editor, *CT-RSA 2006*, volume 3860 of *LNCS*, pages 1–20, San Jose, CA, USA, February 13–17, 2006. Springer-Verlag, Berlin, Germany. [92]

[Pai99] Pascal Paillier. Public-key cryptosystems based on composite degree residuosity classes. In Jacques Stern, editor, *EURO-CRYPT'99*, volume 1592 of *LNCS*, pages 223–238, Prague, Czech Republic, May 2–6, 1999. Springer-Verlag, Berlin, Germany. [64]

[Pan84] Victor Pan. *How to multiply matrix faster*, volume 179 of *LNCS*. Springer-Verlag, Berlin, Germany, 1984. [89]

[Pat96] Jacques Patarin. Hidden fields equations (HFE) and isomorphisms of polynomials (IP): Two new families of asymmetric algorithms. In Ueli M. Maurer, editor, *EUROCRYPT'96*, volume 1070 of *LNCS*, pages 33–48, Saragossa, Spain, May 12–16, 1996. Springer-Verlag, Berlin, Germany. [362, 363]

[PGF98] Daniel Panario, Xavier Gourdon, and Philippe Flajolet. An analytic approach to smooth polynomials over finite fields. In *Third Algorithmic Number Theory Symposium (ANTS)*, volume 1423 of *LNCS*, pages 226–236. Springer-Verlag, Berlin, Germany, 1998. [444]

[PK95] Walter T. Penzhorn and G. J. Kuhn. Computation of low-weight parity checks for correlation attacks on stream ciphers. In *Cryptography and Coding – 5th IMA Conference*, volume 1025 of *LNCS*, pages 74–83. Springer-Verlag, Berlin, Germany, 1995. [386]

[Pol75] John M. Pollard. A Monte Carlo method for factorization. *BIT Numerical Mathematics*, 15(3):331–334, 1975. [233]

[Pom82] Carl Pomerance. Analysis and comparison of some integer factoring methods. In Jr. Hendrik W. Lenstra and Robert Tijdeman, editors, *Computational methods in number theory – Part I*, volume 154 of *Mathematical centre tracts*, pages 8–139. Mathematisch Centrum, Amsterdam, 1982. [141]

[Pri81] Paul Pritchard. A sublinear additive sieve for finding prime numbers. *Communications of the ACM*, 24(1):18–23, 1981. [128, 133]

[Pri83] Paul Pritchard. Fast compact prime number sieves (among others). *Journal of algorithms*, 4:332–344, 1983. [133]

[QD90] Jean-Jacques Quisquater and Jean-Paul Delescaille. How easy is collision search. New results and applications to DES. In Gilles Brassard, editor, *CRYPTO'89*, volume 435 of *LNCS*, pages 408–413, Santa Barbara, CA, USA, August 20–24, 1990. Springer-Verlag, Berlin, Germany. [229, 244]

[RBBK01] Phillip Rogaway, Mihir Bellare, John Black, and Ted Krovetz. OCB: A block-cipher mode of operation for efficient authenticated encryption. In *ACM CCS 01*, pages 196–205, Philadelphia, PA, USA, November 5–8, 2001. ACM Press. [15, 17]

[RH07] Sondre Rønjom and Tor Helleseth. A new attack on the filter generator. *IEEE Transactions on Information Theory*, 53(5):1752–1758, 2007. [388, 389]

[Sch87] Claus-Peter Schnorr. A hierarchy of polynomial time lattice basis reduction algorithms. *Theoretical Computer Science*, 53:201–224, 1987. [331]

[Sch90] Claus-Peter Schnorr. Efficient identification and signatures for smart cards. In Gilles Brassard, editor, *CRYPTO'89*, volume 435 of *LNCS*, pages 239–252, Santa Barbara, CA, USA, August 20–24, 1990. Springer-Verlag, Berlin, Germany. [67]

[Sch91] Claus-Peter Schnorr. Efficient signature generation by smart cards. *Journal of Cryptology*, 4(3):161–174, 1991. [10]

[Sch93] Oliver Schirokauer. Discrete logarithms and local units. *Phil. Trans. R. Soc. Lond. A 345*, pages 409–423, 1993. [461]

[Sch96] Bruce Schneier. *Applied Cryptography (Second Edition)*. John Wiley & Sons, 1996. [3]

[SE94] Claus-Peter Schnorr and M. Euchner. Lattice basis reduction: Improved practical algorithms and solving subset sum problems. *Math. Program.*, 66:181–199, 1994. [326, 328]

[Sha49] Claude E. Shannon. Communication theory of secrecy systems. *Bell System Technical Journal*, 28:656–715, 1949. [4, 337]

[Sie84] T. Siegenthaler. Correlation-immunity of nonlinear combining functions for cryptographic applications. *IEEE Trans. on Information Theory*, IT-30:776–780, 1984. [378]

[Sie85] T. Siegenthaler. Decrypting a class of stream ciphers using ciphertext only. *IEEE Trans. Comput.*, C-34:81–85, 1985. [378]

[Sil86] Joseph H. Silverman. *The Arithmetic of Elliptic Curves*, volume 106 of *Graduate Texts in Mathematics*. Springer, New York, 1986. [417, 424, 431]

[Sim82] Gustavus J. Simmons. A system for point-of-sale or access user authentication and identification. In Allen Gersho, editor, *CRYPTO'81*, volume ECE Report 82-04, pages 31–37, Santa Barbara, CA, USA, 1982. U.C. Santa Barbara, Dept. of Elec. and Computer Eng. [8]

[Sim85] Gustavus J. Simmons. Authentication theory/coding theory. In G. R. Blakley and David Chaum, editors, *CRYPTO'84*, volume 196 of *LNCS*, pages 411–431, Santa Barbara, CA, USA, August 19–23, 1985. Springer-Verlag, Berlin, Germany. [8]

[Sim86] Gustavus J. Simmons. The practice of authentication. In Franz Pichler, editor, *EUROCRYPT'85*, volume 219 of *LNCS*, pages 261–272, Linz, Austria, April 1986. Springer-Verlag, Berlin, Germany. [8]

[Sor98] Jonathan P. Sorenson. Trading time for space in prime number sieves. In *Third Algorithmic Number Theory Symposium (ANTS)*, volume 1423 of *LNCS*, pages 179–195. Springer-Verlag, Berlin, Germany, 1998. [133]

[SS81] Richard Schroeppel and Adi Shamir. A $T = O(2^{n/2})$, $S = O(2^{n/4})$ algorithm for certain NP-complete problems. *SIAM Journal on Computing*, 10(3):456–464, 1981. [251]

[Sti02] Douglas Stinson. *Cryptography: Theory and Practice (Third Edition)*. CRC Press LLC, Boca Raton, Florida, 2002. [3]

[Str69] Volker Strassen. Gaussian elimination is not optimal. *Numer. Math.*, 13:354–356, 1969. [80]

[TCG92] Anne Tardy-Corfdir and Henri Gilbert. A known plaintext attack of FEAL-4 and FEAL-6. In Joan Feigenbaum, editor, *CRYPTO'91*, volume 576 of *LNCS*, pages 172–181, Santa Barbara, CA, USA, August 11–15, 1992. Springer-Verlag, Berlin, Germany. [273]

[TSM94] Toshio Tokita, Tohru Sorimachi, and Mitsuru Matsui. Linear cryptanalysis of LOKI and s2DES. In Josef Pieprzyk and Reihaneh Safavi-Naini, editors, *ASIACRYPT'94*, volume 917 of *LNCS*, pages 293–303, Wollongong, Australia, November 28 – December 1, 1994. Springer-Verlag, Berlin, Germany. [273]

[Val91] Brigitte Vallée. Gauss' algorithm revisited. *J. Algorithms*, 12(4), 1991. [318]

[vW96] Paul C. van Oorschot and Michael J. Wiener. Improving implementable meet-in-the-middle attacks by orders of magnitude. In Neal Koblitz, editor, *CRYPTO'96*, volume 1109 of *LNCS*, pages 229–236, Santa Barbara, CA, USA, August 18–22, 1996. Springer-Verlag, Berlin, Germany. [244]

[Wag99] David Wagner. The boomerang attack. In Lars R. Knudsen, editor, *FSE'99*, volume 1636 of *LNCS*, pages 156–170, Rome, Italy, March 24–26, 1999. Springer-Verlag, Berlin, Germany. [182]

[Wag02] David Wagner. A generalized birthday problem. In Moti Yung, editor, *CRYPTO 2002*, volume 2442 of *LNCS*, pages 288–303, Santa Barbara, CA, USA, August 18–22, 2002. Springer-Verlag, Berlin, Germany. [264, 265]

[Was03] Lawrence C. Washington. *Elliptic curves: number theory and cryptography*. CRC Press LLC, Boca Raton, Florida, 2003. [422]

[WC81] Mark N. Wegman and Larry Carter. New hash functions and their use in authentication and set equality. *Journal of Computer and System Sciences*, 22:265–279, 1981. [8]

[Wie90] Michael J. Wiener. Cryptanalysis of short RSA secret exponents (abstract). In Jean-Jacques Quisquater and Joos Vandewalle, editors, *EUROCRYPT'89*, volume 434 of *LNCS*, page 372, Houthalen, Belgium, April 10–13, 1990. Springer-Verlag, Berlin, Germany. [414]

[Wie04] Michael J. Wiener. The full cost of cryptanalytic attacks. *Journal of Cryptology*, 17(2):105–124, March 2004. [5]

[WYY05a] Xiaoyun Wang, Yiqun Lisa Yin, and Hongbo Yu. Efficient collision search attacks on SHA-0. In Victor Shoup, editor, *CRYPTO 2005*, volume 3621 of *LNCS*, pages 1–16, Santa Barbara, CA, USA, August 14–18, 2005. Springer-Verlag, Berlin, Germany. [179, 182]

[WYY05b] Xiaoyun Wang, Yiqun Lisa Yin, and Hongbo Yu. Finding collisions in the full SHA-1. In Victor Shoup, editor, *CRYPTO 2005*, volume 3621 of *LNCS*, pages 17–36, Santa Barbara, CA, USA, August 14–18, 2005. Springer-Verlag, Berlin, Germany. [179, 182]

[XM88] Guo-Zhen Xiao and James L. Massey. A spectral characterization of correlation-immune combining functions. *IEEE Transactions on Information Theory*, 34(3):569–571, 1988. [275]

[Yuv79] Gideon Yuval. How to swindle Rabin. *Cryptologia*, 3:187–189, 1979. [243]

[ZF06] Bin Zhang and Dengguo Feng. Multi-pass fast correlation attack on stream ciphers. In Eli Biham and Amr M. Youssef, editors, *SAC 2006*, volume 4356 of *LNCS*, pages 234–248, Montreal, Canada, August 17–18, 2006. Springer-Verlag, Berlin, Germany. [380]

[Zha05] Fuzhen Zhang, editor. *The Schur Complement and Its Applications (Numerical Methods and Algorithms)*. Springer, New York, 2005. [94]

[Zhe97] Yuliang Zheng. Digital signcryption or how to achieve cost(signature & encryption) < cost(signature) + cost(encryption). In Burton S. Kaliski Jr., editor, *CRYPTO'97*, volume 1294 of *LNCS*, pages 165–179, Santa Barbara, CA, USA, August 17–21, 1997. Springer-Verlag, Berlin, Germany. [20]

Lists

List of Algorithms

List of Figures

List of Programs

List of Tables

Index